# Springer Series in Materials Science

Volume 282

The Springer Series in Materials Science covers the complete spectrum of materials research and technology, including fundamental principles, physical properties, materials theory and design. Recognizing the increasing importance of materials science in future device technologies, the book titles in this series reflect the state-of-the-art in understanding and controlling the structure and properties of all important classes of materials.

More information about this series at http://www.springer.com/series/856

Yuri Estrin · Yves Bréchet ·
John Dunlop · Peter Fratzl
Editors

# Architectured Materials in Nature and Engineering

## Archimats

Springer

*Editors*
Yuri Estrin
Department of Materials Science
and Engineering
Monash University
Clayton, VIC, Australia

John Dunlop
Department of Chemistry and Physics
of Materials
University of Salzburg
Salzburg, Austria

Yves Bréchet
SIMAP
Institut Polytechnique de Grenoble
Saint-Martin d'Hères, France

Peter Fratzl
Department of Biomaterials
Max Planck Institute of Colloids
and Interfaces
Potsdam, Germany

ISSN 0933-033X ISSN 2196-2812 (electronic)
Springer Series in Materials Science
ISBN 978-3-030-11941-6 ISBN 978-3-030-11942-3 (eBook)
https://doi.org/10.1007/978-3-030-11942-3

Library of Congress Control Number: 2018968382



This Springer imprint is published by the registered company Springer Nature Switzerland AG
The registered company address is: Gewerbestrasse 11, 6330 Cham, Switzerland

# Preface

Over the last decade, the notion of 'architectured materials' has been steadily making its way into the mainstream materials science and engineering. This development came about due to the recognition that design of materials with new properties and functionalities can no longer rely only on traditional approaches based on manipulating a material's composition and microstructure. Following the pioneering views of M. F. Ashby, the inner architecture of the material at a length scale intermediate between the microstructural scale and the specimen dimensions is now considered as a new 'degree of freedom' in materials design. By using this degree of freedom, especially in the context of hybrid materials where the geometry and mutual arrangement of the constituents become the main elements of the inner architecture, it becomes possible to 'fill the holes' in material property charts, thus attaining new functionalities not possible with traditional materials design.

The workshop on *Designing Non-Traditional Materials Based on Geometrical Principles* organised by Yuri Estrin, Mike Ashby, Yves Bréchet and Arcady Dyskin in Hanover, Germany, in June 2005 has been inspired by the idea of architectured materials. Hardly noticed by the broad materials community at the time, the workshop nonetheless has defined the area and outlined possible directions of research on architectured materials. It also helped forming a fraternity of scientists and engineers who see the development of architectured materials ('archimats') as their mission. A workshop organised at Collège de France in 2012 as a closing event for a year of courses by Yves Bréchet as the chair in 'Innovation and Technology' funded by the Bettencourt Foundation provided an opportunity to witness the major progress in this new field. A number of forthcoming symposia at TMS, MRS, IUTAM, THERMEC and probably many other meetings are a testimony to the growing acceptance of the concept of architectured materials by a broader community. Its adoption as a solution to engineering design problems is indeed entering the mainstream research activities. This research naturally involves competences not only from materials science, but also from mechanics, chemistry, mathematics and biology.

Naturally grown materials, such as wood, bone, silk or seashells are generally based on a small selection of base materials, proteins, polysaccharides and minerals. From an engineering viewpoint, these base materials seem rather poor in terms of their intrinsic physical properties. Their power, however, resides in their self-assembly capacities, usually controlled and enhanced by living cells, into architectured materials with remarkable properties. Nature generates a multitude of functions not primarily by varying the chemical composition but rather by a diversity of mesoscopic structures. For this reason, biological materials are ideal sources of inspiration for the development of architectured engineering materials.

The area of architectured materials—both bioinspired and engineered ones—has now reached a degree of maturity and popularity, and we feel it is timely to review the progress made and provide an outlook to possible future developments. In this book that comprises 14 chapters, the various aspects of architectured materials are presented by authors who are active in the field. Important archetypes of archimats, such as microtruss composites and topological interlocking materials, are presented in the chapters written by Abu Samk and Hibbard (Chap. 1) and Dyskin, Estrin and Pasternak (Chap. 2), respectively. Chapter 3 (by Pasternak and Dyskin) and Chap. 4 (by Dirrenberger, Forest and Jeulin) deal with auxetic materials. Archimat design methodology is discussed by Kromm and Wargnier (Chap. 5), while Chap. 6 written by Vermaak et al. addresses topology optimisation, with an emphasis on the interfaces in architectured materials and the role they play in property improvement. A further group of chapters concern the methods by which archimats can be manufactured. These include friction stir processing (Simar and Avettand-Fènoël, Chap. 7), severe plastic deformation (Bygelzimer, Kulagin and Estrin, Chap. 8) and additive manufacturing (Molotnikov, Simon and Estrin, Chap. 9). Architectured materials and mechanisms occurring in living organisms, which may inspire engineering design, are discussed by Politi, Bar-On and Fabritius (Chap. 10), Seidel et al. (Chap. 11), and Krijnen, Steinmann, Jaganatharaja and Casas (Chap. 12).

While many chapters in this book touch upon possible applications of archimats, two of them discuss practical applications quite specifically. These are the contributions by Yrieix (Chap. 13), devoted to building materials, and Fallacara, Barberio and Colella (Chap. 14), which describes the use of topological interlocking in stereotomy-based architectural design and construction. Analytical and computer-based techniques enter several chapters, the one by Dirrenberger, Forest and Jeulin (Chap. 4) being specifically devoted to homogenisation techniques as a means for mathematical modelling of the mechanical behaviour of architectured materials.

Taken together, the contributions presented provide a rather full picture of the history, current state and future prospects of research on architectured materials. The book offers an instructive and entertaining reading both to a specialist in this field and to a person who just enters it. It provides a useful compendium of the results obtained to date owing to an intensive study conducted by several research groups. It is hoped that this book will promote the materials design principles underlying engineering of architectured materials and boost efforts to implement them in novel materials and structures, in a field which is now accepted as a mainstream strategy yet is still very open to further developments and new ideas.

Clayton, Australia — Yuri Estrin
Saint-Martin d'Hères, France — Yves Bréchet
Salzburg, Austria — John Dunlop
Potsdam, Germany — Peter Fratzl

# Contents

# Contributors

**Khaled Abu Samk** Department of Materials Science and Engineering, University of Toronto, Toronto, ON, Canada

**G. Allaire** CMAP, Ecole Polytechnique, Palaiseau, France

**Marie-Noëlle Avettand-Fènoël** Université de Lille, Unité Matériaux Et Transformations, UMR CNRS 8207, Villeneuve d'Ascq, France

**Maurizio Barberio** Dipartimento di Scienze dell'Ingegneria e dell'Architettura, Politecnico di Bari, Bari, Italy

**Benny Bar-On** Department of Mechanical Engineering, Ben-Gurion University of the Negev, Beer Sheva, Israel

**Yan Beygelzimer** Donetsk Institute for Physics and Engineering named after A. A. Galkin, National Academy of Sciences of Ukraine, Kiev, Ukraine

**Y. Bréchet** SIMaP, CNRS, Université Grenoble, Grenoble, France

**Jérôme Casas** Institute de Recherche en Biologie de l'Insecte, Université de Tours, Tours, France

**Micaela Colella** Dipartimento di Scienze dell'Ingegneria e dell'Architettura, Politecnico di Bari, Bari, Italy

**Mason N. Dean** Department of Biomaterials, Max Planck Institute of Colloids and Interfaces, Potsdam, Germany

**Justin Dirrenberger** Laboratoire PIMM, Arts et Métiers-ParisTech, CNRS, Cnam, Paris, France

**A. V. Dyskin** Department of Civil, Environment and Mining Engineering, The University of Western Australia, Perth, WA, Australia

**R. Estevez** SIMaP, CNRS, Université Grenoble, Grenoble, France

**Yuri Estrin** Department of Materials Science and Engineering, Monash University, Clayton, Australia

**Helge-Otto Fabritius** Department Microstructure Physics and Alloy Design, Max-Planck-Institut für Eisenforschung GmbH, Düsseldorf, Germany

**Giuseppe Fallacara** Dipartimento di Scienze dell'Ingegneria e dell'Architettura, Politecnico di Bari, Bari, Italy

**A. Faure** SIMaP, CNRS, Université Grenoble, Grenoble, France

**Samuel Forest** Centre des Matériaux, MINES-ParisTech, CNRS UMR 7633, Evry, France

**Glenn D. Hibbard** Department of Materials Science and Engineering, University of Toronto, Toronto, ON, Canada

**Ram K. Jaganatharaja** TechMed Centre, University of Twente, Enschede, The Netherlands

**Aravind K. Jayasankar** Department of Biomaterials, Max Planck Institute of Colloids and Interfaces, Potsdam, Germany

**Dominique Jeulin** Centre de Morphologie Mathématique, MINES-ParisTech, Fontainebleau, France

**F. Jouve** Laboratoire J. L. Lions (UMR 7598), University Paris Diderot, Paris, France

**Gijs J. M. Krijnen** TechMed Centre, University of Twente, Enschede, The Netherlands

**F. X. Kromm** Université de Bordeaux, Talence, France

**Roman Kulagin** Institute of Nanotechnology (INT), Karlsruhe Institute of Technology (KIT), Eggenstein-Leopoldshafen, Germany

**G. Michailidis** SIMaP, CNRS, Université Grenoble, Grenoble, France

**Andrey Molotnikov** Department of Materials Science and Engineering, Monash University, Clayton, Australia

**G. Parry** SIMaP, CNRS, Université Grenoble, Grenoble, France

**E. Pasternak** Department of Mechanical Engineering, The University of Western Australia, Perth, WA, Australia

**Yael Politi** Department of Biomaterials, Max Planck Institute of Colloids and Interfaces, Potsdam, Germany

**Ronald Seidel** Department of Biomaterials, Max Planck Institute of Colloids and Interfaces, Potsdam, Germany

**Ron Shahar** The Robert H. Smith Faculty of Agriculture, Food and Environmental Sciences, Koret School of Veterinary Medicine, The Hebrew University of Jerusalem, Rehovot, Israel

**Aude Simar** Institute of Mechanics, Materials and Civil Engineering (iMMC), Université catholique de Louvain, Louvain-la-Neuve, Belgium

**George P. Simon** Department of Materials Science and Engineering, Monash University, Clayton, Australia

**Thomas Steinmann** Institute de Recherche en Biologie de l'Insecte, Université de Tours, Tours, France

**N. Vermaak** Department of Mechanical Engineering and Mechanics, Lehigh University, Bethlehem, PA, USA

**H. Wargnier** Université de Bordeaux, Talence, France

**Bernard Yrieix** Materials and Mechanics of Components, EDF Lab Les Renardières, EDF R&D, Moret-Sur-Loing, France

# Chapter 1
# Microtruss Composites

**Khaled Abu Samk and Glenn D. Hibbard**

**Abstract** This chapter provides a framework for predicting buckling instabilities (global and local) in composite struts using a discretizing approach to divide the cross-section into finite regions or blocks (elements) with distinctive composition and material properties. This framework is applied to two compositing approaches: (1) an additive process, e.g. electrodeposition of nanocrystalline nickel (n-Ni) on steel or aluminum substrates creating a discrete interface, and (2) mass transport phenomena, e.g. gas carburization of steel microtrusses creating a continuous and gradual change in composition. The individual composite struts are prone to failure through inelastic/elastic global/local buckling instabilities due to their typically high slenderness ratio (for lightweight requirements). An optimal architecture is defined as the lightest architecture in the defined architectural space that can support a given load before failing; mathematically equivalent to satisfying the Kuhn-Tucker condition. For the electrodeposition example, the optimal trajectory was found to be dependent on the substrate material as well as the level of adhesion of the sleeve to the substrate. However, in the carburizing example, optimal architectures cannot be found using the same optimization approach due to the negligible weight penalty associated with the addition of carbon atoms (i.e. no trade-off between strength and mass). Other factors such as ductility and fracture toughness need to be addressed to find the optimal architectures.

## 1.1 Introduction

An underlying question in the development of hybrid materials is: to what extent can the strategies of microstructural and architectural design be combined to tailor

K. Abu Samk · G. D. Hibbard (✉)
Department of Materials Science and Engineering, University of Toronto,
184 College St, Toronto, ON M5S 3E4, Canada
e-mail: glenn.hibbard@utoronto.ca

K. Abu Samk
e-mail: khaled.abusamk@mail.utoronto.ca



Y. Estrin et al. (eds.), *Architectured Materials in Nature and Engineering*, Springer Series in Materials Science 282,
https://doi.org/10.1007/978-3-030-11942-3_1

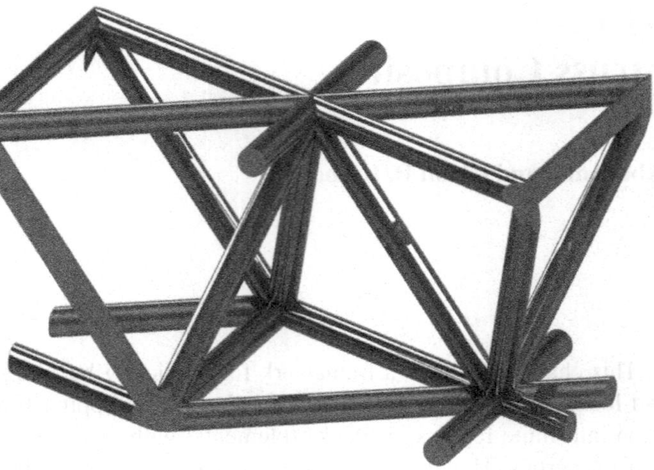

**Fig. 1.1** Schematic representation of a repeat unit cell in a tetrahedral microtruss panel. The strut cross-section can be any shape; solid cylinders are shown in the schematic

the intrinsic properties of a material? A hybrid composed of metal and open space would result in a lightweight cellular architecture potentially creating new applications in the transportation and aerospace industries. Foams, entangled materials, and microtrusses represent several types of cellular architectures. The deformation mechanism of most foams and entangled materials is bending-dominated, which means that applied loads are resolved transversely through the internal ligaments [1–3]. On the other hand, the deformation mechanism of microtrusses is stretch-dominated (because of the high degree of internal connectivity), which means that the applied loads are resolved axially through the struts [1, 2]. As a result, the stiffness and strength of metal microtrusses can be up to an order of magnitude higher than those of metal foams, at a given relative density [1, 2]. Figure 1.1 shows an example of the nodal connectivity in a tetrahedral microtruss. Deshpande et al. [1] showed that for a lattice to be stretch-dominated, its connectivity needs to satisfy the generalized Maxwell rule in three-dimensions [4], which is a modified version of the original Maxwell's criterion [5] to account for states of self-stress in the ligaments of the lattice:

$$b - 3j + 6 = s - m, \tag{1.1}$$

where $b$ and $j$ are the number of struts and frictionless joints in the structure, respectively, while $s$ and $m$ are the number of self-stress states and mechanisms, respectively. For an isolated tetrahedron, $b = 6$ and $j = 4$, thus the left side of (1.1) reduces to zero. This lattice framework provides the architectural backdrop and in the following we will be examining the composite design question.

The performance of a microtruss is dictated by the performance of the individual struts making up the lattice. Ashby categorized the failure of these struts into yielding, buckling, or fracture; whichever is reached first [2]. Since lightweight is a design necessity, microtrusses with low relative density tend to be favoured, which results in higher susceptibility to buckling failure mechanisms whether the applied load is compressive or bending load, e.g. [6, 7]. The simplest scenario occurs when one of the composite constituents is very much stronger and stiffer than the other, such that the mechanical contribution of the softer material can be ignored. An example is the electrodeposition of high strength nanocrystalline metal on a polymer template. Here the electrodeposited material is more than two orders of magnitude higher in stiffness and more than thirty times higher in tensile strength than the underlying polymeric template [8]. As a result, it is convenient to take the conservative approach and ignore the mechanical contribution from the polymer [8]. Optimization here would mean determining which architecture (of the possible defined configurations) is able to support a given load for the least possible mass. The trajectory of the optimal architecture through the architectural space would satisfy the Kuhn-Tucker condition (see [9] for more details about this classical optimization), $\nabla\widehat{M} = \lambda\nabla\widehat{F}$, which in plain words means that the gradient of the mass index ($\widehat{M}$) is parallel to the gradient of the load index ($\nabla\widehat{F}$), by some proportionality constant(s), the Lagrange multiplier ($\lambda$). Note that the polymer plays an interesting role in this type of composite. Its function is really only to provide the desired three-dimensional architectural template upon which the electrodeposition occurs. Once the first metal has been deposited, it is no longer necessary and contributes only its mass to the overall performance. This means that if one was to take an already optimally designed composite microtruss (i.e. that most efficient arrangement of metal and polymer for a given load) and then one changed the electrodeposition conditions to make the grain size smaller (and accordingly the metal stronger), one would be, yes, increasing the performance of the composite, but the hybrid would no longer be optimal. In other words, if one wanted to change the grain size for this type of composite microtruss, one would have needed to change the underlying architecture as well.

This chapter will focus on a different design scenario. We will start with a metallic microtruss core, and then look at creating a composite structure by reinforcing the struts within. Several strategies can be imagined, including: (1) additive processes such as the addition of a sleeve material on a substrate giving a discrete boundary between the substrate and the sleeve (e.g. electrodeposition of nanocrystalline nickel (n-Ni)), (2) mass transport phenomena such as the modification of the substrate composition via diffusion giving a continuous and gradual change in composition (e.g. gas carburization of steel microtrusses), or (3) conversion processes such as the conversion of the substrate surface resulting in a discrete boundary but at the expense of substrate consumption (e.g. anodic alumina formation through the anodization of aluminum microtrusses). Figure 1.2 shows a schematic cross-section illustrating the 'compositing' that would occur through each of the three routes. Due to the slenderness of the struts in the microtruss, we can assume negligible gradient or compositional variation along the length and around the longitudinal axis,

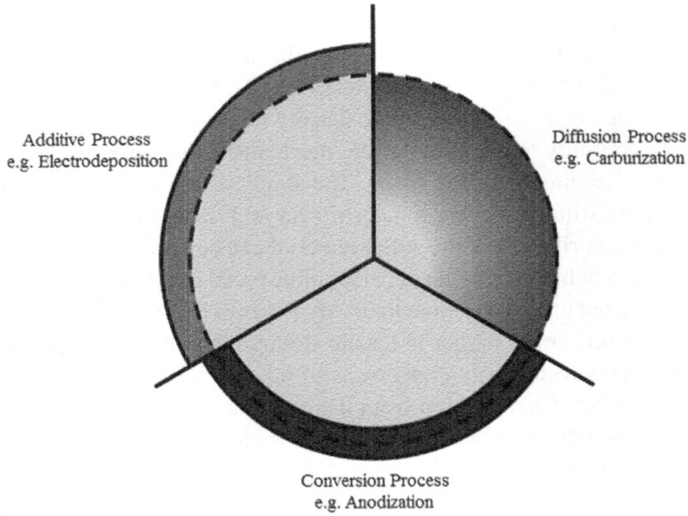


**Fig. 1.2** Schematic diagram illustrating the 'compositing' changes in composition and dimension for a cylindrical microtruss strut. The dashed line represents the starting strut surface and the compositional grading occurs only in the radial direction. Gradients and compositional variations along and around the cylinder axis are assumed to be negligible in sufficiently slender struts

with the compositional grading occurring only in the radial direction. In any of the above strategies, we must know the mechanical response of the constituents of the composite. For the first two design examples, the mechanical properties and failure mechanisms are reasonably well known and understood. However, in the third case, there is a completely different failure mechanism that results in the progressive fracture and rubblization of the oxide along the full length of the strut [10]; the weight specific mechanical performance of this composite type is quite promising (factor of two to three strength increase at minimal weight penalty), but we do not yet have reliable analytical models to predict the failure strength (more study is needed) and so it will not be discussed further in this chapter.

In all of the above processes, the composite performance will be determined by the properties of the phases or materials present. The first step is to determine the composite material response using an appropriate rule of mixtures; after this point is established, we can move towards translating the rule of mixtures onto the microtruss lattice. Especially relevant here are methods that can be used to analytically estimate the mechanical response of the composite through the transition from elastic to plastic deformation. There exist many approaches to predict the elastic and thermophysical properties of composites, including the Hashin-Shtrikman approach [11, 12] which assumes isotropic elastic behaviour. For plasticity, the approaches available are usually either empirical or numerical. The upper and lower bounds for the elastic modulus of the material are known as the Voigt [13] and Reuss [14] bounds, respectively, and they can also be applied to the elastic and plastic composite response [15, 16]. The upper (Voigt) bound is obtained by assuming that the substrate and reinforcement experience the same strain, i.e. isostrain, and consequently the stress is the volume-average of the local stresses. The lower (Reuss) bound is obtained by assuming that the substrate and reinforcement experience the same stress, i.e. isostress, and consequently the strain is the volume-average of the local strains. The maximum composite strength is achieved if both components reach failure at the same time (i.e. at equal strain); if one fails before the other, the strength is determined by the weaker component [15]. As a result, the upper limit for the strength of the composite is simply:

$$\sigma_c^{upper} = f_s\sigma_s + f_r\sigma_r, \tag{1.2}$$

where $f$ and $\sigma$ are the volume fraction and stress in the component, respectively, and the subscripts $c$, $s$ and $r$ correspond to the composite, substrate, and reinforcement, respectively. A lower limit for the strength is more difficult to obtain, and numerous studies and special cases are presented and summarized in literature [15]. An example is Kim et al. [16] who validated a lower limit rule of mixtures based on the Reuss method,

$$\sigma_c^{lower} = \left(\frac{f_s}{\sigma_s} + \frac{f_r}{\sigma_r}\right)^{-1}. \tag{1.3}$$

The isostrain Voigt rule of mixtures is the most widely used one due to its simplicity [17]. However, it becomes inaccurate when the constituents exhibit vastly different strengths and ductility [17–19]. As a result, many efforts were made to improve the accuracy of the isostrain rule of mixtures while maintaining its simplicity, such as [17–20]. One example is Bouaziz and Buessler [17] who presented an isowork rule of mixtures which assumes that for a disordered structure, the mechanical work increment is equal in each constituent, i.e. $\sigma_s d\varepsilon_s = \sigma_r d\varepsilon_r$. The isowork model is applicable to elasticity as well as monotonic plasticity.

For the purpose of this chapter, dividing the material into thin shells or blocks that experience the same amount of deflection upon compression is a convenient and simple approach that allows one to maintain compatibility at the interfaces between the said shells or blocks. As a result, the isostrain assumption, (1.2) is used for describing the composite material response in this chapter.

## 1.2 Example 1: Electrodeposition

The first composite microtruss design examples considered here are those where a metallic microtruss has been reinforced by nanocrystalline electrodeposition. In nanocrystalline electrodeposition, the bath chemistry and plating parameters are adjusted in order to control the migration and mobility of the metal cations to favor nucleation of new grains over the growth of already nucleated ones, thus allowing an ultra-fine and equiaxed nanostructure to be deposited [21]. In the case of Ni, grain size reduction from the μm-scale to the nm-scale can result in an order of magnitude yield strength increase (e.g. from 86 MPa [22] to 910 [23] when the grain size decreases from 10 μm to 10 nm). Furthermore, nanocrystalline electrodeposition places the high strength n-Ni sleeve away from the neutral bending axis of the individual strut, thus enhancing the global buckling resistance [24]. As we will see, however, the thin sleeve may be susceptible to local buckling depending on the quality of the adhesion between coating and substrate.

In this section two example substrate materials are considered: AISI1006 plain carbon steel and AA3003 aluminum alloy. The Ramberg-Osgood relation was selected for this part of the analysis because of its convenience in describing the elastic to plastic transition of these materials ($\varepsilon = \frac{\sigma}{E} + 0.002\left(\frac{\sigma}{K}\right)^n$, where $\varepsilon$, $\sigma$, and $E$ are the strain, stress, and Young's modulus, respectively, while $K$ and $n$ are the Ramberg-Osgood fitting parameters [25]). For the steel, $K_{\text{steel}} = 155$ MPa, $n_{\text{steel}} = 4.31$, and $E_{\text{steel}} = 200$ GPa, while for the aluminum, $K_{Al} = 49$ MPa, $n_{Al} = 4.88$, and $E_{Al} = 69$ GPa. The sleeve material being considered here is an electrodeposited nickel having grain size of 21 nm, whose tensile stress-strain curve was fit to the Voce constitutive relationship (it more accurately describes the knee of the stress-strain curve for this material) [26], $\sigma = B - (B - A)\exp(-C\varepsilon)$, giving $A_{\text{Ni}} = 61$ MPa, $B_{\text{Ni}} = 1435$ MPa, and $C_{Ni} = 120$ [8]. The two example substrates were selected because of their relative differences in strength, stiffness, and density when compared to the electrodeposited nickel, as well as the varying levels of adhesion that can be expected for the coating on each of these two substrates. In general, it is much easier to obtain good adhesion for Ni coatings on steel than on aluminum. The poorer adhesion when electrodepositing onto aluminum is related to the presence of the native aluminum oxide layer at the interface [27, 28].

Figure 1.3 shows two possible strut geometries, solid square columns and hollow circular tubes, with the labeled dimensions. For a square column, the dimensions of interest are $b$ (half of the square width), $n$ (the electrodeposited sleeve thickness), and $l$ (the column length). Similarly, for the hollow tube, the dimensions of interest are $r$ (the tube inner radius), $t$ (the tube wall thickness), $n$ (the electrodeposited sleeve thickness), and $l$ (the tube length). Note that at this point we are not considering how this metallic template would have been initially fabricated, we are simply dealing with the question of reinforcement by electrodeposition. It is convenient to transform the strut dimensions to non-dimensional terms in order to reduce the number of free variables, as well as to make the models and optimization results applicable to any range of architecture space. The non-dimensional variables define a design space that contains all possible strut geometries for a given set of materials. While it is possible to non-dimensionalize the strut dimensions with respect to any length, some choices may be favoured over others. For example, one may choose to non-dimensionalize the strut cross-sectional dimensions by the length. For a square column, this results in $\bar{b}_l = b/l$ and $\bar{n}_l = n/l$, while for a hollow tube, $\bar{r}_l = r/l$, $\bar{t}_l = t/l$, and $\bar{n}_l = n/l$. This definition does eliminate any interdependency between the dimensionless parameters, but a more meaningful description of the coating thickness $n$ would be

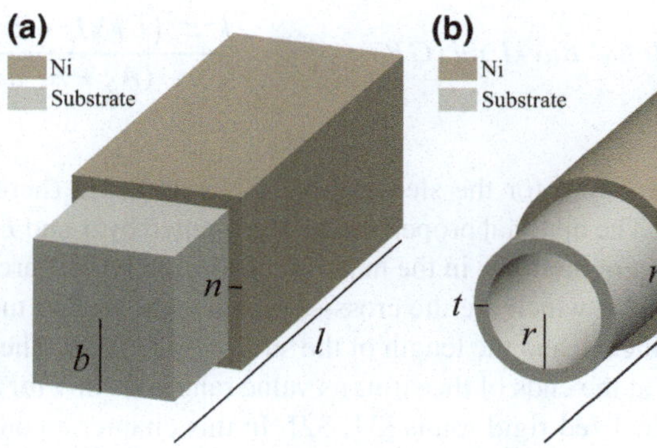


**Fig. 1.3** Nanocrystalline nickel coated **a** solid square column, and **b** round hollow tube. Labelled are the dimensions of interest

in terms of the substrate thickness dimension $b$ or $t$, i.e. $n_b = n/b$ or $n_t = n/t$, for a square column or a tube, respectively. The significance of the quantities $\bar{n}_b$ and $\bar{n}_t$ is that it is proportional to the area fraction of reinforcement. As a result, the non-dimensional variables for the case of the column (a) and the tube (b) become (with the subscripts eliminated to simplify the notation):

$$\bar{b} = \frac{b}{l}; \quad \bar{n} = \frac{n}{b}, \tag{1.4a}$$

$$\bar{r} = \frac{r}{l}; \quad \bar{t} = \frac{t}{r}; \quad \bar{n} = \frac{n}{t}. \tag{1.4b}$$

The $\bar{b}$ and $\bar{r}$ parameters are a representation of the slenderness of the column or the tube, while the $t$ term is a representation of the substrate material to open area ratio in the tube, and finally the $n$ is a representation of the reinforcement to substrate fraction.

A strut experiencing uniaxial compression will eventually yield as the applied load increases. However, depending on the strut geometry and material, global and/or local buckling instabilities could short-circuit yielding. The following section will focus on developing the relevant buckling equations and optimization procedure for the simpler solid square columns example. This geometry has two free variables instead of three (1.4a, b), making the visualizations of the design space much easier. At the same time, the failure of composite struts having square or rectangular cross-sections have been studied to date more completely, e.g. [24, 29, 30]. The critical stress for global buckling, $\sigma_{CR}^{GB}$, of a composite square column can be determined by solving the following system of equations, the isostrain rule of mixtures and the Shanley-Engesser buckling equation [31–33],

$$Rule\,of\,Mixture\,(RoM){:}\;\; \sigma_{CR}^{GB} = \frac{\sigma_S A_S + \sigma_C A_C}{A_S + A_C}, \tag{1.5a}$$

$$\textit{Shanley Engesser Global Buckling (GB)}: \ \sigma_{CR}^{GB} = \frac{k^2\pi^2(E_{T,S}I_S + E_{T,C}I_C)}{(A_S + A_C)l^2}. \tag{1.5b}$$

The subscripts $S$ and $C$, are for the sleeve (here n-Ni) and core (here steel or aluminum), respectively. The material properties are represented by $\sigma$ and $E_T$, which are the stress and the tangent modulus in the material, respectively. The architecture is represented by $A$, $I$, and $l$, which are the cross-sectional area, second moment of area of one of the constituents, and the length of the strut, respectively. The constant $k$ defines the constraints at the ends of the strut, its value ranges from 1 for idealized pin joints, up to 2 for idealized rigid joints [31, 32], In this chapter a conservative assumption is taken assuming $k = 1$. When the tangent modulus is equal to Young's modulus, (1.5b) becomes the Euler buckling relation for very slender struts. The development of (1.5a, b) is based on the following assumptions: (1) the deposited sleeve has uniform thickness and it shares the same neutral bending axis with the core, (2) the interfacial adhesion is sufficient to prevent delamination and achieve isostrain conditions, i.e. both the sleeve and the substrate will fail simultaneously at the same strain and by the same failure mechanism, and (3) the sleeve is ductile enough to reach the buckling instability point before any fracture occurs. In addition, there is also an implicit assumption that the material's mechanical behaviour is symmetric under tension and compression, which enables us to use a tensile stress-strain curve to predict the buckling strength of a material under uniaxial compression.

Equations (1.5a, b) can be rewritten as follows in terms of the non-dimensionalized parameters in (1.4a):

$$\textit{Column} - \textit{RoM}: \ \sigma_{CR}^{GB} = \frac{\sigma_S\left[(1+\bar{n})^2 - 1\right] + \sigma_C}{\left[(1+\bar{n})^2 - 1\right] + 1} \tag{1.6a}$$

$$\textit{Column} - \textit{Composite GB}: \ \sigma_{CR}^{GB} = \frac{k^2\pi^2}{3}\bar{b}^2 \frac{E_{T,S}\left[(1+\bar{n})^4 - 1\right] + E_{T,C}}{\left[(1+\bar{n})^2 - 1\right] + 1} \tag{1.6b}$$

Note that in the above, replacing $\bar{n}$ with zero reduces the equations to the case of a monolithic column of core material. The above equations provide an upper limiting case as it was assumed that the adhesion strength is sufficient to result in the sleeve and the substrate failing together, i.e. the isostrain assumption is valid up until the composite components fail simultaneously.

If the adhesion of the sleeve to the substrate is poor, debonding or delamination of the sleeve could short circuit the composite failure predicted by (1.6a, b). This provides a lower limiting case in which failure of the structure is decided by the weaker constituent. We predict the critical stress that would cause the sleeve to locally buckle, then we calculate the applied stress on the composite that is needed to reach this value. The major assumptions involved here are no adhesion and no interaction between the sleeve and the core (i.e. no shearing or frictional forces considered, which provides the lowest threshold regarding adhesion). The critical stress in the sleeve can be found by solving the following equation [34] along with the stress-strain curve of the sleeve:

$$\sigma_{CR,S} = \frac{k_p \pi^2 E_{t,S}}{12(1-\nu^2)} \left( \frac{n}{2b+2n} \right)^2 = \frac{k_p \pi^2 E_{t,S}}{48(1-\nu^2)} \left( \frac{\bar{n}}{1+\bar{n}} \right)^2, \tag{1.7}$$

where $k_p \approx 4$ for sleeve width to length ratios greater than two, and this value was used over the entire range of architectures considered. The stress in the composite at the point of wrinkling can be found by modifying (1.6a) to substitute the stress in the sleeve by the value obtained from (1.7), and the stress in the core is determined at the value of critical strain for the sleeve.

To summarize the above equations; we are looking at two cases for solid square columns: (1) the upper limiting case of perfect adhesion in which composite global buckling occurs and can be predicted by solving (1.6a, b) simultaneously, and (2) the lower limiting case of no adhesion with a buckling (1.6a, b) and sleeve local buckling or wrinkling (1.7).

Figure 1.4 shows an SEM image of an AISI 1006 steel microtruss reinforced with a 60 μm thick n-Ni coating (n-Ni/steel). The microtruss has been compressed to a point just past the inelastic buckling stability limit of the strut members as can be seen by the permanent deformation (curvature) of the struts. The coating remains intact, with no signs of delamination, debonding, or fracture. Previous experimental studies by Bouwhuis et al. [24] and Abu Samk et al. [29] showed that no delamination or debonding between the nickel and the steel was observed until well past the buckling instability and the good agreement between experimental, finite element, and analytical models suggests that an example of the upper limiting case is achieved in n-Ni/steel microtrusses the point of buckling instability An example of the upper limiting case is achieved in n-Ni/steel microtrusses [24, 29, 30]. On the other hand, an example of the lower limiting case is achieved in n-Ni/Al microtrusses. Previous experiments and finite element studies showed that the assumption of simultaneous nickel and aluminum failure breaks down, over predicting the buckling strength [30]. When loaded experimentally, the thinnest electrodeposited coatings of n-Ni on aluminum (10–30 μm) exhibit low residual stresses and remain intact with the substrate (i.e. relatively good adhesion), while thicker coatings were more likely to debond and delaminate (>40 μm) [30].

Figure 1.5 illustrates how the predictions are performed by simultaneously solving the appropriate pair of equations (i.e. 1.6a, b) for the case of n-Ni/steel. Figure 1.5a demonstrates the effect of changing $\bar{b}$ from 0.02 to 0.20 (equivalent to changing the substrate slenderness ratio *SR* from ~90 to ~9) while fixing $\bar{n}$ at 0.20 (equivalent to 0.31 nickel to steel area ratio). More slender struts are more susceptible to global buckling and therefore the critical stress, graphically illustrated by the intersection of (1.6a, b), occurs at a lower stress as $\bar{b}$ decreases (or equivalently as the slenderness ratio increases). Figure 1.5b demonstrates the effect of changing the fraction of reinforcement from $\bar{n}$ of 0.02–0.2 (equivalent to nickel to steel area ratios of 0.04 and 0.89, respectively) while fixing $\bar{b}$ at 0.05 (equivalent to substrate slenderness

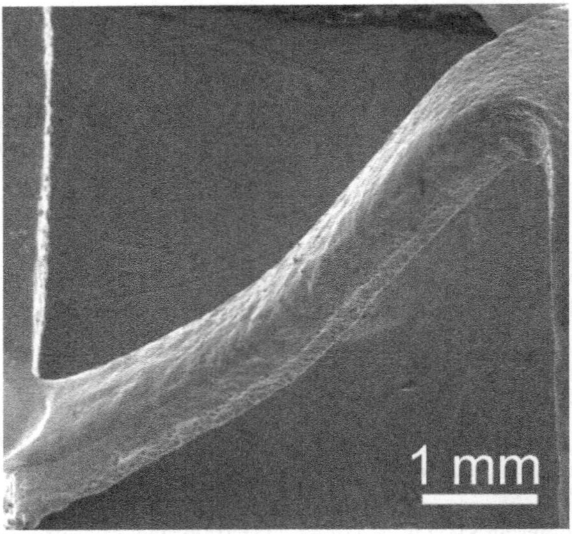


**Fig. 1.4** An SEM image showing n-Ni/steel composite microtruss that has been uniaxially compressed to a strain beyond the point of inelastic global buckling. No debonding, delamination, or cracks are observed in the coating. The sleeve and the core remain intact until beyond the buckling instability, supporting the isostrain assumption used in the composite design

ratio of ~35). As the nickel area fraction increases (proportional to increasing $\bar{n}$), the composite stress-strain curve becomes more dominated by the nickel properties and therefore the critical stress increases as $\bar{n}$ increases.

The predictions shown in Fig. 1.5 can be expanded over a defined architectural space, and failure maps can be plotted to predict the failure mechanism and strength for any architecture within that design space. Figure 1.6a, b show the strength prediction for a nickel plated square column over the $\bar{b} \times \bar{n}$ space for steel and aluminum, respectively. The highlighted yellow region indicates elastic global buckling of the composite, while the red surface is inelastic global buckling. The dark blue surface is inelastic local buckling (wrinkling) of the sleeve while the cyan blue region is the elastic local-buckling of the sleeve. The boundary between the elastic and inelastic regions is determined by the stress level. If the stress in the material(s) at the point of buckling is below the proportional limit (defined as the point at which the tangent modulus value is equal to 95% of the value of Young's modulus), then the buckling occurs elastically and the strut (or the sleeve) would fully recover if the load is released immediately after the instability occurred. Elastic global buckling of the composite occurs for very small values of $\bar{b}$ (very slender struts), while elastic local buckling of the sleeve occurs for very small values of $\bar{n}$ (very thin coatings). On the other hand, beyond the proportional limit, permanent deformation occurs upon buckling. Two additional white lines are shown on the surface. The dashed white line represents architectures where the substrate has just yielded according to the 0.2%-offset yield criterion. Similarly, the solid white line indicates that the sleeve

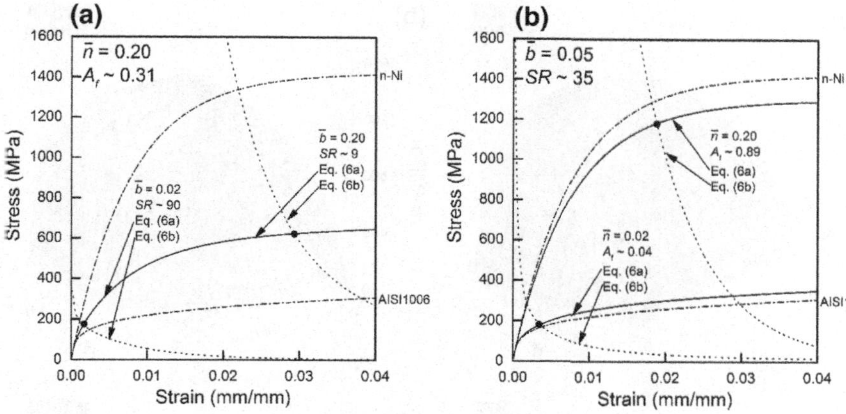


**Fig. 1.5** Global buckling of n-Ni/steel square columns. The dash-dot lines are the monolithic material properties for the n-Ni sleeve and the steel substrate. The solid lines are the composite stress-strain curves obtained by the rule of mixtures as shown in (1.6a). The dotted lines are the Shanley-Engesser inelastic buckling curves as shown in (1.6b). The black filled circles represent the intersection point and therefore the critical buckling stress. The effect of changing $b$ while $n$ is fixed (i.e. fixed area fraction of n-Ni, $A_f$) is shown in (**a**). Similarly, the effect of changing $n$ while fixing $\bar{b}$ (i.e. fixed substrate slenderness ratio, *SR*) is shown in (**b**)

has yielded. Any architecture with a buckling stress exceeding the white lines could also be considered to have failed by yielding (region above white line). Figure 1.6c, d are the top views of Fig. 1.6a, b, respectively, as this view is convenient to determine the failure mechanism zones over the design space.

In order to find the optimal trajectory, we seek the architecture with the lowest mass (objective) that can support a series of incrementally increasing loads (constraints). It is first convenient to define a dimensionless load index, $\widehat{F}$, and a dimensionless mass index, $\widehat{M}$, as follows:

$$\widehat{F} = \frac{\sigma_{CR} A_{\text{strut}}}{\sigma_{ref} l^2}, \tag{1.8a}$$

$$\widehat{M} = \frac{m_{strut}}{\rho_{ref} l^3}, \tag{1.8b}$$

where the reference stress, $\sigma_{\text{ref}}$, and reference density, $\rho_{\text{ref}}$, are the yield strength and density of the substrate parent material, respectively. The critical stress, $\sigma_{CR}$, is that of the active failure mechanism. The mass of the strut is determined by summing the mass of the substrate and the sleeve using the appropriate bulk density for each constituent, $\rho_{\text{steel}} = 7.87\ \text{Mg/m}^3$, $\rho_{\text{Al}} = 2.73\ \text{Mg/m}^3$, and $\rho_{\text{Ni}} = 8.90\ \text{Mg/m}^3$.

By applying the Kuhn-Tucker condition to the $\widehat{F}$ and $\widehat{M}$ surfaces as discussed in the introduction, the optimal trajectory can be found. Figure 1.7a, b show zoomed areas of interest in the architectural design space for n-Ni/steel and n-Ni/Al respectively. Plotted are the contours of constant dimensionless load and mass indices. The Kuhn-

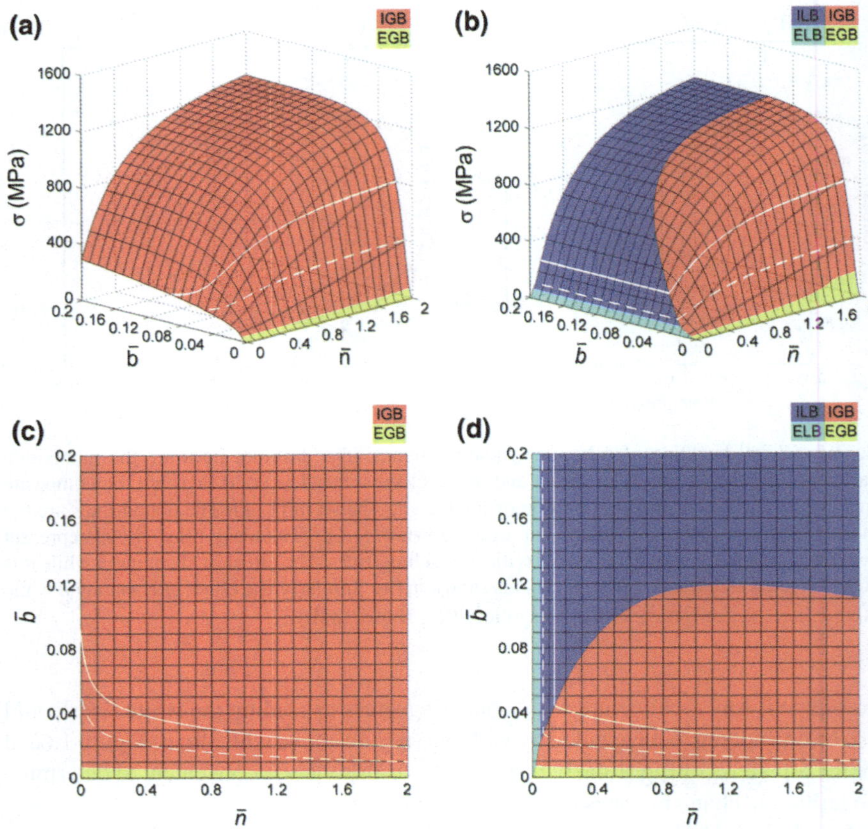


**Fig. 1.6** The failure surface for **a** n-Ni/steel (perfect adhesion) and **b** n-Ni/Al (no adhesion) square columns. The red region is inelastic global buckling (IGB) and the yellow region is elastic global buckling (EGB). The blue region is inelastic local buckling (ILB) of the sleeve (wrinkling) and the cyan region is the elastic local buckling (ELB). The white dashed and solid lines are the architectures at which the substrate and nickel have just yielded, respectively. The maps in **c** and **d** are the top-view of **a** and **b** surfaces, respectively

Tucker condition is satisfied and an optimal architecture is found at regions where the $\widehat{F}$ and $\widehat{M}$ contours are parallel. Optimal trajectories through the design space depend on the particular material combinations (Fig. 1.7). For example, in n-Ni/steel composites, Fig. 1.7a, for a given constraint, i.e. load index contour $\widehat{F} = 2 \times 10^{-4}$, the parallel mass index contour is $\widehat{M} = 5 \times 10^{-4}$ and this occurs at a point $(\bar{n}, \bar{b}) =$ (0.042, 0.011), see (i) on Fig. 1.7a. As the constraint increases, $\widehat{F} = 5 \times 10^{-4}$ is parallel to $\widehat{M} = 10 \times 10^{-4}$ at (1.141, 0.007), i.e. $\bar{b}$ decreased while $\bar{n}$ increased to move in the design space to match the new load constraint, see (ii) on Fig. 1.7a. On the other hand, in n-Ni/Al composites, for a given load index contour $\widehat{F} = 60 \times 10^{-4}$, the parallel mass index contour is $\widehat{M} = 35 \times 10^{-4}$ and this occurs at (0.214, 0.019) ((i) on Fig. 1.7b). As the constraint increases, $\widehat{F} = 2900 \times 10^{-4}$ is parallel to $\widehat{M} = 500 \times 10^{-4}$ at (0.823, 0.038), i.e. $\bar{b}$ increased while $\bar{n}$ increased ((ii) on Fig. 1.7b). To explain this behaviour we have to understand that there are many directions one can move through the design space and still end up on the new, increased constraint load index. However, only one of these directions will end up at a point that corresponds to an architecture with the least mass. In the n-Ni/steel case, it is beneficial to use a more slender starting steel strut and add more n-Ni, i.e. lower $\bar{b}$ and higher $\bar{n}$. This is because the weight penalty associated with adding n-Ni is actually close to that incurred when adding steel (or using a stockier column). The density of n-Ni is only 1.13 times that of steel, so it is more efficient to replace some of the steel in the cross section with n-Ni, since the weight penalty of adding n-Ni is at least associated with a significant increase in the buckling strength. In the n-Ni/Al case, the optimal trajectory is very different as shown on Fig. 1.7b. Since the density of aluminum is less than one third of that of nickel, there are scenarios where if the required load to be supported increases, it is worthwhile using a stockier aluminum column to distance the coating further away from the neutral bending axis combined with an increased coating thickness to carry the required load. This is very different than what was observed in the steel case in Fig. 1.7a where it was more efficient to reduce the stockiness of the steel column and replace it with nickel. In both of these scenarios however, if one was to simply add more coating in order to meet a greater load requirement, one would be creating a stronger structure, but it would no longer be optimal.

## 1.3 Example 2: Diffusion-Based Composites

Carburizing is a surface treatment process that induces the formation of ultra-fine length scale martensite laths at the outer surfaces of the specimen after quenching—in effect creating a composite structure having an ultra-high strength outer shell and ductile inner core that eliminates the strain compatibility and adhesion issues that arise when a different material is used as a coating. It could also be argued that a single material graded composite is more favourable from the recycling and disposal point of view at the end of the product's life cycle. While carburizing will be used as the example, the fundamentals discussed below could be equally well applied to other treatments such as boriding, nitriding, carbonitriding, and carboboriding.

In this scenario, a compositional gradient, such as that illustrated in Fig. 1.2 is created. The buckling composite is no longer a simple sleeve/core system with a discrete interface; instead, the composite can be imagined as a series of infinitesimal elements of varying composition (in compliance with the diffusion profile) to which the mechanical properties can be tied. If one can analytically describe the composition as a function of position in the strut cross-section, and the mechanical properties as a function of composition, one can develop an analytical prediction. Depending on the geometry of the strut, it may or may not be possible to conveniently develop

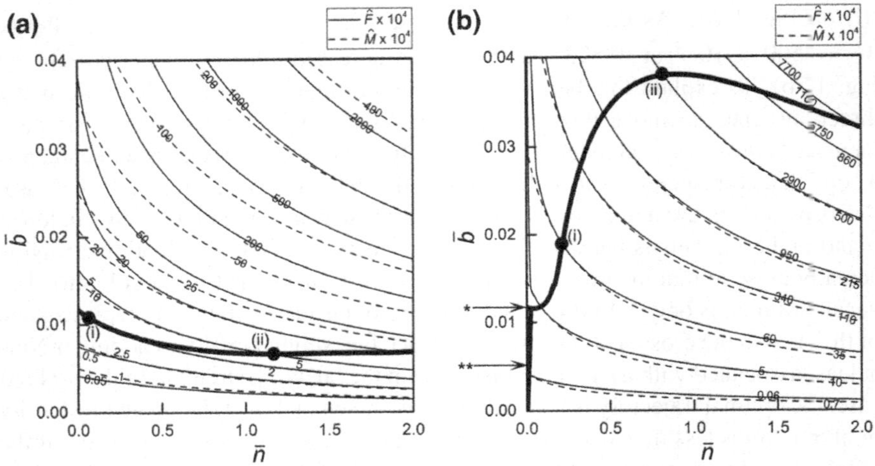


**Fig. 1.7** Contours of the dimensionless load (thin solid black lines) and mass (thin dashed black lines) indices for composite square columns of **a** n-Ni/steel (perfect adhesion) and **b** n-Ni/Al (no adhesion). The thick back lines are the optimal trajectories which depend on the material properties of both the substrate and the sleeve. The labels * and ** refer to the optimal trajectory under the assumption of perfect and no adhesion, respectively. Note that in case of perfect adhesion assumption in both composites, the trajectory starts at the same point on the axis. Black circles labelled (i) and (ii) are discussed in the text

an analytical relation that relates the mechanical properties to position. The following section introduces a manner to discretize the strut cross-section into a series of elements to predict the buckling strength after a diffusion-based reinforcement treatment. The optimization question for this type of composite will not be discussed in this chapter because: (1) the weight penalty associated with carburizing is negligible, making it always beneficial to fully carburize the strut cross-sections to the highest possible carbon content, and (2) the optimization has to take into account other properties that change as the carbon content increases, such as ductility and fracture toughness.

One of the key questions is: what would be the ideal distribution of added carbon within the microtruss? For example, the carbon could be distributed uniformly over the entire cross-section of the strut (useful perhaps if one had fabricated the microtruss by mechanical means introducing work-hardening into the design question, e.g. [35]), while another way is to create a high-carbon layer adjacent to the outer surface of the strut. One could also combine decarburizing and carburizing in order to introduce a soft layer on top of the carburized hard layer, which would be an interesting approach from the mechanical property perspective because it could inhibit the propagation of surface cracks and allow greater strains to be achieved before final fracture [36].

In the following, an analytical model is developed to predict the buckling resistance of compositionally and microstructurally graded carbon steel solid square columns. A finite system solution to Fick's second law is used to predict the carbon content profile after carburizing. The cross-section is then divided into finite regions and a constitutive model is assigned to each region based on its average carbon content. The buckling resistance of each region is then calculated using the appropriate buckling equations and the results are combined to give a microstructure-architecture-property map. If a carbon profile is known, and the material's response as a function of carbon content is available, this model can be used to predict the buckling strength for cases ranging in complexity from simple carburization (current example) to more complex combinations of carburization and decarburization. The dimensions and non-dimensional variables remain unchanged from the first design examples with the exception of eleminating $\bar{n}$ and introducing a dimensionless carburizing time as will be explained shortly.

It is first necessary to predict the carbon profile in the strut cross-section as this is what determines the mechanical properties as a function of position. A finite system solution to Fick's second law is used by superimposing the carbon content profiles in the $x$- and $y$-directions using a product solution, after [37]:

$$\frac{C(x,y,t)-C_s}{C_i-C_s}=X(x,t)\cdot Y(y,t) \tag{1.9a}$$

$$X(x,t)=\frac{4}{\pi}\sum_{n=0}^{\infty}\frac{(-1)^n}{2n+1}e^{\left[\frac{-(2n+1)^2\pi^2}{4}\frac{D_{C(\gamma-Fe)}t}{b_x^2}\right]}\cos\left[\frac{(2n+1)\pi}{2}\frac{x}{b_x}\right] \tag{1.9b}$$

$$Y(y,t)=\frac{4}{\pi}\sum_{n=0}^{\infty}\frac{(-1)^n}{2n+1}e^{\left[\frac{-(2n+1)^2\pi^2}{4}\frac{D_{C(\gamma-Fe)}t}{b_y^2}\right]}\cos\left[\frac{(2n+1)\pi}{2}\frac{y}{b_y}\right] \tag{1.9c}$$

where $C(x,y,t)$ is the carbon content at any point in the strut cross-section with the co-ordinates $(x,y)$ after a carburizing time $t$. $C_i$ is the initial carbon content in the strut before carburizing (0.06–0.08 wt%C for AISI1006 low carbon steel [38], minimum of 0.06 wt%C is assumed) and $C_s$ is the carbon content at the surface of the strut and is set by the composition of the carburizing atmosphere (0.40 wt%C is chosen for the current example). $X(x,t)$ is the 1-dimensional finite system solution to Fick's second law for a slab with finite width $2\times b_x$ and infinite length and thickness. Similarly, $Y(y,t)$ is the solution for a slab with finite thickness $2\times b_y$ and infinite length and width. For a square cross-section, $b_x=b_y=b$. $D_{C(\gamma-Fe)}$ is the diffusion coefficient (in m$^2$/s) of carbon in $\gamma$-Fe given by the following relation [39]:

$$D_{C(\gamma-Fe)}=4.53\times10^{-7}\left[1+Y_C(1-Y_C)\frac{8,339.9}{T}\right]\times e^{\left[-\left(\frac{1}{T}-2.221\times10^{-4}\right)\cdot(17,767-Y_C\times26,436)\right]} \tag{1.10}$$

where $T$ is the carburizing temperature in degrees Kelvin (assumed to be 1103 K in the current example) and $Y_C$ is equal to $X_C/(1-X_C)$, in which $X_C$ is the mole fraction of carbon. For the carburizing conditions assumed in the current example, $D_{C(\gamma-Fe)}$ ranges from $2.54\times10^{-12}$ to $3.78\times10^{-12}$ m$^2$/s; however, a constant average value of $3.16\times10^{-12}$ m$^2$/s was used to reduce the complexity of the model.

In order to generalize this model, it is convenient to non-dimensionalize the carburizing parameters following the method presented in [37]. A dimensionless carburizing time is defined as $\tau = \left(D_{C(\gamma-Fe)}t\right)/b^2$, and a dimensionless position in the strut, i.e. relative position, is defined as $\bar{x} = x/b$ and $\bar{y} = y/b$.

However, another level of complexity is introduced to the model since the diffusion problem under consideration is that for a multiphase binary system, rather than a simple single-phase case. At the selected carburizing temperature of 1103 K (830 °C), the boundary between the single-phase (γ-Fe) and the two-phase (α-Fe and γ-Fe) regions occurs at ~0.19 wt%C at the given temperature (1103 K), while the boundary between the single-phase (α-Fe) and the two-phase (α-Fe and γ-Fe) regions occurs at ~0.014 wt%C. Therefore, the AISI1006 exists as a two phase mixture consisting of ~26% austenite (γ-Fe) and ~74% ferrite (α-Fe), determined by applying the lever rule. The diffusion process will take place through the formation of austenite at the surface of the strut which will advance into the depth of the strut consisting of a ferrite and austenite mixture. In the single-phase part of the strut, (1.9a, b) and (1.10) can be applied to predict the carbon profile in the austenite. In the two-phase mixture, the compositions of the austenite and ferrite phases are constant and equal to the equilibrium compositions of 0.19 and 0.014 wt%C, respectively. As a result, the carbon activity is constant through the two-phase region and there is no driving force for diffusion to take place; the phase fractions and compositions remain constant, i.e. ~26% austenite (at 0.19 wt%C) and ~74% ferrite (at 0.014 wt%C) [40].

Upon quenching, fully austenitic regions are transformed into a fully martensitic structure in the single-phase region, while a dual-phase microstructure of ferrite and martensite remains in the mixed region. The strut becomes a compositionally and microstructurally graded structure. Figure 1.8 shows the predicted carbon profiles after carburizing for dimensionless times, $\tau$ of 0.025, 0.05, 0.1, and 0.2, which are equivalent to carburizing a 1 mm by 1 mm square strut for 33, 66, 132, and 264 min, respectively, at the given carburizing conditions. Note that only one quarter is necessary to be shown due to the four-fold symmetry of the problem.

In order to predict the inelastic buckling strength, the stress-strain response of the compositionally graded struts is required. The strut quarter is divided into $N \times N$ finite blocks of equal areas. Each block is then assigned its own continuum level mechanical properties depending on its composition and microstructure. The overall stress-strain behaviour of the composite strut can then be determined by using isostrain mixing to combine the properties of all the blocks, after [30]:

$$\sigma = \frac{1}{N^2}\sum_{j=1}^{N}\sum_{i=1}^{N}\sigma_{i,j}. \tag{1.11}$$

where $N^2$ is the number of blocks in the strut quarter, and $\sigma_{i,j}$ is the stress-strain behaviour for a block at a position $(i, j)$.

Blocks with carbon contents exceeding 0.19 wt%C are entirely martensitic and the stress-strain behaviour follows the as-quenched martensite behaviour law [41]:

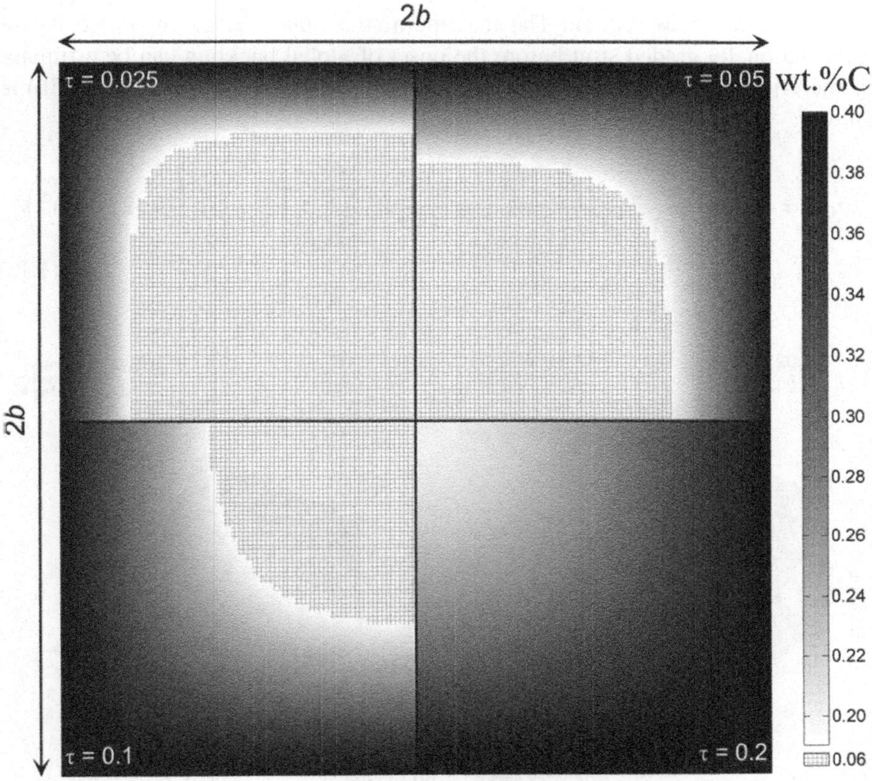


**Fig. 1.8** Carbon profiles in the strut cross-section after carburizing for $\tau$ (dimensionless time units) of 0.025, 0.05, 0.1, and 0.2. The grayscale gradient indicates the carbon content within the martensite phase, while the checkered pattern is the dual-phase core with the fixed carbon content 0.06 wt%C

$$\sigma_m = a + b(1 - \exp(-c \cdot \varepsilon)) \tag{1.12a}$$

$$\begin{aligned} a = 682 + 33(\%\text{Mn}) + 81(\%\text{Si}) + 58(\%\text{Cr} + \%\text{Mo})\text{MPa} \\ b = 2722(\%C)^{0.78}\,\text{MPa} \\ c = 179 + \frac{7.52 \times 10^{-4}}{(\%C)^{4.61}} \end{aligned} \tag{1.12b}$$

where the composition is in wt% and the law is valid for carbon contents in the range of 0.15–0.40 wt%C. For the AISI1006 alloy used, the manganese content is at a maximum of 0.45 wt%Mn, the carbon content before carburizing is at taken as 0.06 wt%C, while the silicon, chromium, and molybdenum contents are negligible [38].

Blocks that fall within the dual-phase region of the strut are assigned the core material properties prior to carburizing, i.e. the same material properties of the steel used in the previous section. The maximum stress that can be supported by the compositionally graded strut before the onset of global buckling can be estimated by the critical bifurcation stress ($\sigma_{CR}$), which is obtained by modifying (1.5b) as follows:

$$\sigma_{CR} = \frac{k^2\pi^2 \sum_{j=1}^{N} \sum_{i=1}^{N} E_{t,(i,j)} I_j}{A_{strut} l^2} = \frac{k^2\pi^2}{4N^4}\bar{b}^2 \sum_{j=1}^{N} \sum_{i=1}^{N} E_{t,(i,j)} \left(\frac{1}{3} + (2j-1)^2\right) \tag{1.13}$$

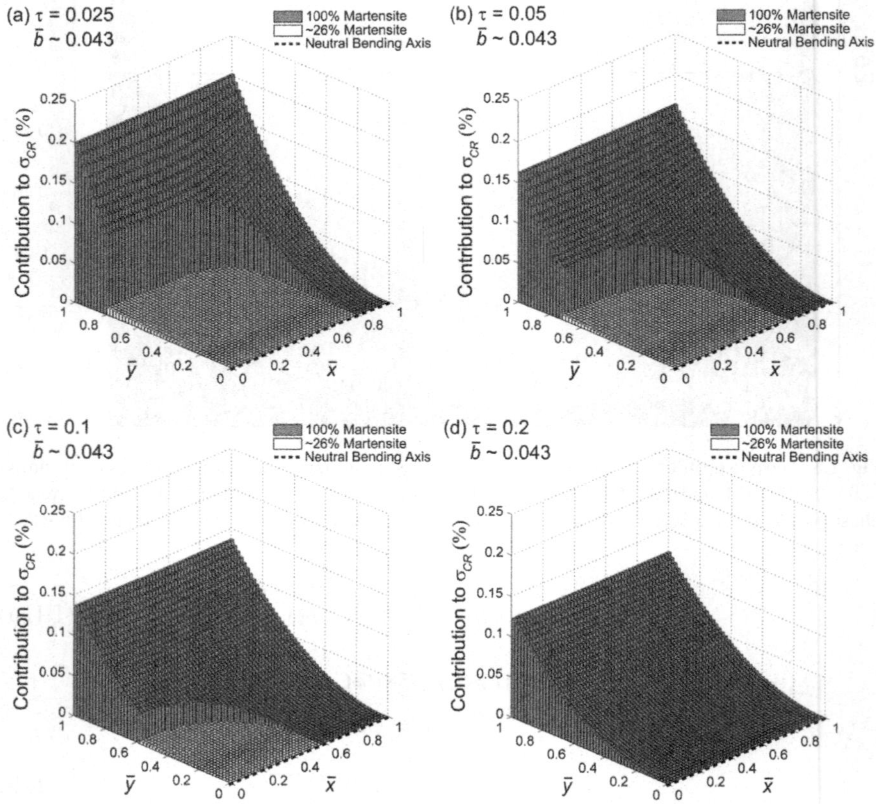


**Fig. 1.9** The percent contribution of each finite block in the strut cross-section to the overall critical buckling stress of the carburized strut, for $\bar{b} \sim 0.043$ (i.e. a slenderness ratio of 40) after carburizing for dimensionless time, $\tau$ of **a** 0.025, **b** 0.05, **c** 0.1, and **d** 0.2. Summing all the contributions from the 50 × 50 blocks gives a total of 100%. The origin (0, 0) is the center of the strut. $\bar{x}$ and $\bar{y}$ are relative positions in the strut, where $\bar{x}$ or $\bar{y}$ of 1 is equal to half the strut cross-sectional dimension, $b$. Note that given the square cross-section, the neutral bending axis could have equally well been placed along the $\bar{y}$ axis

To graphically illustrate the effect of carbon content and distribution, Fig. 1.9 shows the strength contribution at each point in the cross-section for a specific example in which a slenderness ratio of 40 or $\bar{b} \approx 0.043$ is assumed after dimensionless carburizing times and consequently composition profiles equal to those in Fig. 1.8. A position of (0, 0) is located at the center of the strut. The highest strength contribution is always associated with the outermost regions at any carburizing time or slenderness ratio. This is due to the higher second moment of area (further from the neutral bending axis) associated with outer regions compared to inner ones, as well as the higher carbon content, resulting in higher buckling resistance. Upon fixing the slenderness ratio and increasing the carburizing time, the transition from 100% martensite to dual-phase takes place further inside the strut beneath the surface, as represented by the transition from dark to light grey shading on the bars in Fig. 1.9. The carbon is allowed to diffuse further into the strut during the carburizing treatment, thus resulting in the growth of the austenite (and consequently martensite) region into the dual-phase region. Note that a decrease in the contributing percentage of a given outer block as $\tau$ increases does not indicate that it carries less stress; the drop is because other blocks start to contribute more to the overall strength as their carbon content increases. Discretizing the strut cross-section by composition allows us to determine the contribution each block provides towards the buckling strength of the compositionally graded strut, and the total buckling strength can be determined by summing the contributions from all the blocks. This approach enables us to predict the critical buckling stress for compositionally graded strut cross-sections.

The overall buckling strength of the compositionally graded struts is plotted as contours of equal strengths against $\tau$ and $\bar{b}$ in Fig. 1.10. Carburizing for times

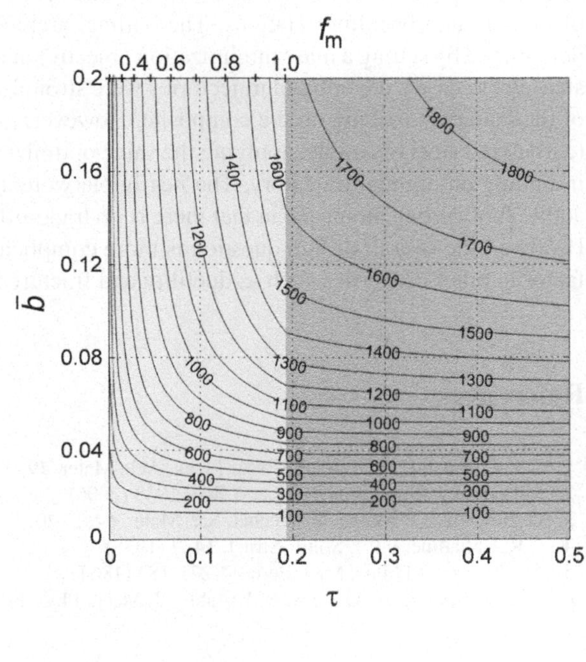


**Fig. 1.10** Contour plots of strut buckling strength (in MPa) plotted against dimensionless carburizing time $\tau$ and strut dimension $\bar{b}$. The axis on the top shows the martensite fraction, $f_m$ in the strut corresponding to the carburizing times below. The dark grey shaded region on the right represents fully martensitic struts, while the light grey shaded region on the left represents struts with outer single-phase martensitic case and inner dual-phase mixed core

longer than $\tau \sim 0.19$ result in an entirely martensitic structure (as shown by the top axis and the darker shaded region on the graph). Figure 1.10 is in effect a type of microstructure-architecture-property map in which the dimensionless carburizing time $\tau$ determines the martensite fraction and composition, while $\bar{b}$ (inversely proportional to slenderness ratio) represents the internal structure of the lattice.

## 1.4 Summary

Composite microtrusses can be fabricated using a combination of different strategies. Regardless of which strategy is selected, the overall performance of the microtruss is dictated by the performance of the strut, and the question of strut design needs to be addressed in terms of the architectural and material parameters which will affect the resistance of the struts to the various potential failure mechanisms acting on the lattice assembly, with the active failure mechanism being the one at the lowest applied load. This chapter provided a framework for predicting buckling instabilities in composite struts by discretizing the cross-section into finite regions or blocks with assigned composition and material properties. This method was applied to examples of n-Ni/steel, n-Ni/Al, and carburized steel microtrusses with square cross-section. For n-Ni/steel and n-Ni/Al, the presence of the discrete interface between the two constituents of the composite allowed us to divide the material into two regions, the core (steel or Al substrate), and the shell (n-Ni), then the rule of mixture and buckling relations are applied. On the other hand, in the carburized steel case, there was a gradient in carbon content and no distinct interface, so the material was first discretized into a series of finite compositionally homogeneous blocks before applying the rule of mixture and buckling relations. The optimal architectures and trajectories were determined by setting a mass minimization objective at a given load constraint. In the sleeve/core cases, the optimal trajectories were strongly dependent on the properties of the materials making up the composite. However, in the compositionally graded (carburized steel) example, applying the same optimization procedure doesn't result in finding an optimal trajectory. The negligible weight penalty associated with the addition of carbon atoms mean that there is no trade-off between strength and mass. However, the overall design question is more complicated since one would need to factor in other properties such as ductility and fracture toughness.

## References

1. V.S. Deshpande, M.F. Ashby, N.A. Fleck, Acta Mater. **49**, 6 (2001)
2. M.F. Ashby, Phil. Trans. R. Soc. A **364**, 1838 (2006)
3. O. Bouaziz, J.P. Masse, Y. Brechet, Scr. Mater. **64**, 2 (2011)
4. C.R. Calladine, Int. J. Solids Struct. **14**, 2 (1978)
5. J.C. Maxwell, Philos. Mag. Series **4**(27), 182 (1864)
6. V.S. Deshpande, N.A. Fleck, M.F. Ashby, J. Mech. Phys. Solids **49**, 8 (2001)

7. V.S. Deshpande, N.A. Fleck, Int. J. Solids Struct. **38**, 36–37 (2001)
8. A.T. Lausic, C.A. Steeves, G.D. Hibbard, J. Sandw. Struct. Mater. **16**, 3 (2014)
9. P.M. Weaver, M.F. Ashby, Prog. Mater Sci. **41**, 1 (1997)
10. E. Bele, B.A. Bouwhuis, C. Codd, G.D. Hibbard, Acta Mater. **59**, 15 (2011)
11. Z. Hashin, S. Shtrikman, J. Appl. Phys. **33**, 10 (1962)
12. Z. Hashin, S. Shtrikman, J. Mech. Phys. Solids **11**, 2 (1963)
13. W. Voigt, *Lehrbuch der Krystallphysik* (B. G. Teuber, Leipzig, 1928), p. 962
14. A. Reuss, Z. Angew, Math. Mech. **9**, 1 (1929)
15. M.F. Ashby, Acta Metall. Mater. **41**, 5 (1993)
16. H.S. Kim, S.I. Hong, an S.J. Kim, J. Mater. Process. Technol. **112**, 1 (2001)
17. O. Bouaziz, P. Buessler, Adv. Eng. Mater. **6**, 1–2 (2004)
18. T. Gladman, I.D. McIvor, F.B. Pickering, J. Iron Steel Inst. **210**, 12 (1972)
19. B. Karlsson, G. Linden, Mater. Sci. Eng. **17**, 2 (1975)
20. I. Tamura, Y. Tomota, A. Akao, Y. Yamaoka, M. Ozawa, S. Kanatani, Trans. ISIJ **13**, 4 (1973)
21. R.T.C. Choo, J.M. Toguri, A.M. Elsherik, U. Erb, J. Appl. Electrochem. **25**, 4 (1995)
22. A.W. Thompson, Acta Metall. **25**, 1 (1977)
23. N. Wang, Z.R. Wang, K.T. Aust, U. Erb, Mater. Sci. Eng., A **237**, 2 (1997)
24. B.A. Bouwhuis, T. Ronis, J.L. McCrea, G. Palumbo, G.D. Hibbard, Compos. Sci. Technol. **69**, 3–4 (2009)
25. W. Ramberg, W.R. Osgood, National Advisory Committee for Aeronautics Technical Note No. 902 (1943)
26. E. Voce, Metall. **51** (1955)
27. J.W. Dini, *Electrodeposition: the materials science of coatings and substrates* (Noyes Publications, Park Ridge, NJ, 1993)
28. *Modern electroplating*, 4th ed. edited by M. Schlesinger, M. Paunovic (Wiley Interscience, Toronto, ON, 2000)
29. K. Abu Samk, B. Yu, G.D. Hibbard, Composites Part A **43**, 6 (2012)
30. E. Bele, B.A. Bouwhuis, G.D. Hibbard, Acta Mater. **57**, 19 (2009)
31. F.R. Shanley, J. Aeronaut. Sci. **14**, 5 (1947)
32. F.R. Shanley, *Strength of Materials* (McGraw-Hill, New York, 1957)
33. J.M. Gere, S.P. Timoshenko, *Mechanics of Materials*, 4th edn. (PWS, Boston, MA, 1997)
34. S.P. Timoshenko, J.M. Gere, *Theory of Elastic Stability*, 2nd edn. (Dover Publications, Mineola, NY, 2009)
35. B. Yu, K. Abu Samk, G.D. Hibbard, Metall. Mater. Trans. A **46**, 5 (2015)
36. B. Chehab, H. Zurob, D. Embury, O. Bouaziz, Y. Brechet, Adv. Eng. Mater. **11**, 12 (2009)
37. D.R. Poirier, G.H. Geiger, *Transport Phenomena in Materials Processing* (TMS, Warrendale, 1994)
38. ASM Handbook—*Properties and Selection: Irons, Steels, and High-Performance Alloys* (Materials Park, Ohio, 1990), Vol. 1
39. J. Agren, Scr. Metall. **20**, 11 (1986)
40. D.A. Porter, K.E. Easterling, *Phase Transformations in Metals and Alloys*, 2nd edn. (Chapman & Hall, London, UK, 1992)
41. S. Allain, O. Bouaziz, M. Takahashi, ISIJ Int. **52**, 4 (2012)

# Chapter 2
# Topological Interlocking Materials

A. V. Dyskin, Yuri Estrin and E. Pasternak

**Abstract** In this chapter, we present a materials design principle which is based on the use of segmented, rather than monolithic, structures consisting of identical blocks locked within the assembly by virtue of their special geometry and mutual arrangement. First, a brief history of the concept of topological interlocking materials and structures is presented and current trends in research are outlined. Recent work of the authors and colleagues directed at the variation of the shape of interlockable building blocks and the mechanical performance of structures—either assembled from them or 3D printed—are overviewed. Special emphasis is put on materials responsive to external stimuli. Finally, an outlook to possible new designs of topological interlocking materials and their engineering applications is given.

## 2.1 Introduction

In a book dedicated to architectured materials, a chapter on topological interlocking should not be missing. Indeed, the design principle, dubbed *topological interlocking,* that we introduced in 2001 [1] and developed over the years [2] has taken a prominent place in the arsenal of material design tools [3, 4]. The underlying mathematical concept is relatively simple: a material or structure is assembled from identical blocks whose geometrical shape and mutual arrangement provide kinematic constraints that make it impossible to remove any block from the assembly (apart

A. V. Dyskin
Department of Civil, Environment and Mining Engineering, The University of Western Australia, 35 Stirling Highway, Perth, WA 6009, Australia

Y. Estrin (✉)
Department of Materials Science and Engineering, Monash University, Wellington Road, Clayton, VIC 3800, Australia
e-mail: Yuri.estrin@monash.edu

Y. Estrin · E. Pasternak
Department of Mechanical Engineering, The University of Western Australia, 35 Stirling Highway, Perth, WA 6009, Australia



Y. Estrin et al. (eds.), *Architectured Materials in Nature and Engineering*, Springer Series in Materials Science 282,
https://doi.org/10.1007/978-3-030-11942-3_2

from those at its periphery, which need to be handled separately). No connectors or binders are required, as interlocking of the blocks arresting them within the structure is provided by sheer geometry and relative positions of the blocks. Starting with an assembly of interlocked tetrahedron-shaped blocks assembled to a monolayer [1] we later demonstrated that blocks with the outer geometry of any of the five platonic bodies (tetrahedron, cube, octahedron, dodecahedron, and icosahedron) are interlockable to a monolayer structure [5]. Furthermore, truncated platonic bodies are also interlockable, including backyballs, which permit three different types of topological interlocking [6]. Examples of such structures are shown in Fig. 2.1.

A whole range of block geometries can be derived from these basic ones by transforming them, as demonstrated in [7] for the case of interlockable tetrahedra. As interlocking is provided by the parts of the blocks close to the middle plane of the assembly, truncation of the blocks by planes parallel to the middle plane does not change their capacity of being interlockable, thus generating a continuum of 'derivate' interlocked assemblies. The pattern of cross-sections of the blocks in the middle plane is a defining feature of the entire assembly. By considering possible tessellations of the middle plane in a regular pattern and establishing rules for trans-

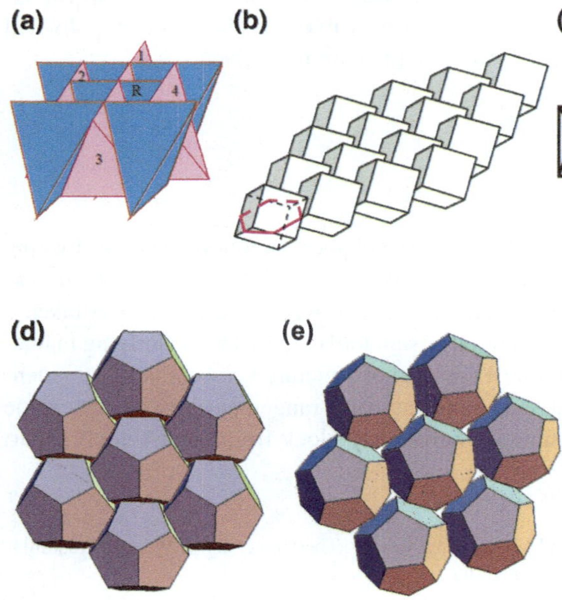


**Fig. 2.1** Schematics of monolayers of topologically interlocked blocks with the shape of one of the five platonic solids: **a** interlocking of tetrahedra; the reference block R movement upwards is constrained by blocks 1 and 3; its movement downwards is constrained by blocks 2 and 4; **b** assembly of interlocked cubes; **c** assembly of interlocked octahedra; **d** assembly of interlocked dodecahedra with the interlocking plane normal to the 3-rd order symmetry axis; **e** assembly of interlocked dodecahedra with the interlocking plane normal to the 5-th order symmetry axis; **f** assembly of interlocked isocahedra. (After [5])

forming them in the planes parallel to the middle one ('method of moving sections' [8]), a recipe for generating an assembly of interlocked convex polyhedra was proposed [9].

Another family of topologically interlockable blocks is represented by osteomorphic blocks (called so because of their semblance with a bone) and their further modification—Tetraloc bricks—whose ability to get interlocked within a structure owes to their matching concavo-convex contact surfaces [10], Fig. 2.2.

The topological interlocking structures employing the two distinct types of an elementary building block—shaped as a platonic body or an osteomorphic brick—enjoy considerable attention due to the remarkable properties they possess and the potential to create new materials and structures they offer [11–15]. The benefits of topological interlocking materials include their high resistance to crack propagation and tolerance to local failures, tuneable bending rigidity, ease of assembly and disassembly, and thus nearly perfect recyclability. In addition, topological interlocking can be used to combine different—even entirely dissimilar—materials within an integral multi-material structure, as blocks of the same shape and size made from different constituent materials can be blended together in virtually any proportion and pattern.

Below we provide an overview of research on topological interlocking materials from the conception of this notion to the most recent developments.

## 2.2 A Brief History of the Concept of Topological Interlocking Materials

The idea of topological interlocking as a means to capitalise on improved mechanical properties of a material enabled by its segmented structure was proposed in our first publications [1, 7]. At that time, little did we know that the concept had been introduced in the context of vault design in architecture more than three centuries ago! In the year 1699 flat vaults formed by topological interlocking of both truncated tetrahedra and blocks with concavo-convex surfaces similar to our osteomorphic design were proposed independently by Abeille and Truchet, respectively [16–18], cf. Fig. 2.3. Closer to the modern-day engineering, pavements with interlocked osteomorphic-type blocks [19] and truncated tetrahedra [20] were proposed. However, these concepts were not followed and did not find entry in the engineering practice.

The rich palette of possible geometries of topologically interlockable blocks developed through our systematic work on the subject, cf. [2], gradually garnered recognition by the architectural design community. Various groups of architects and civil engineers, including Fallacara [16, 17], Tessmann [21], Grobman et al. [22], Miodragovic and Kotnik [23], Alothman and Chavan [24] and others are now using the topological interlocking principle in their creative work. Aspects of topological interlocking design were also employed in or proposed for a variety of other

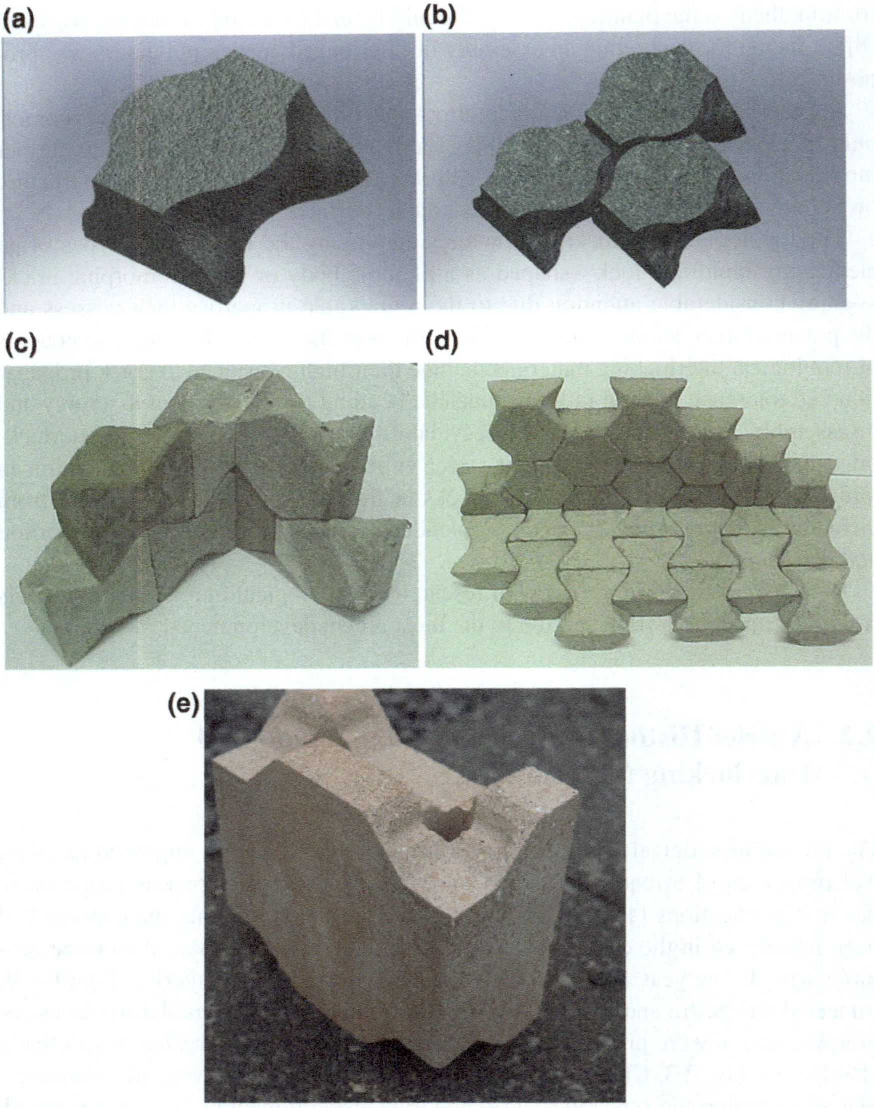


**Fig. 2.2** Osteomorphic blocks: **a** an individual block; **b** planar assembly of osteomorphic blocks enabling interlocking; **c** corner assembly; **d** wall-floor assembly; **e** Tetraloc block, https://tetraloc.com

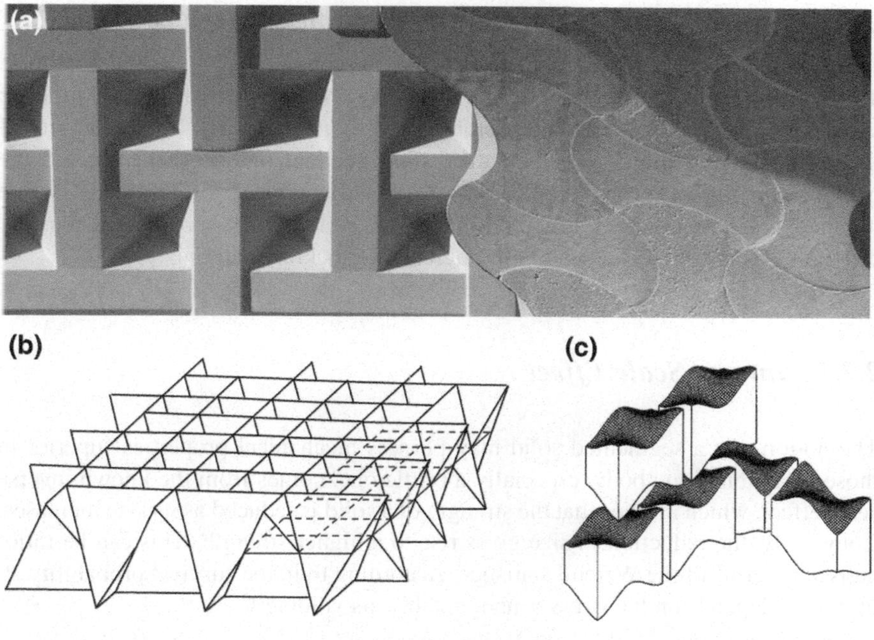


Fig. 2.3 Similarities between the medieval vault designs and the topological interlocking structures based on tetrahedra and osteomorphic blocks: **a** Abeille's vault formed by truncated tetrahedra; **b** interlocking tetrahedra; **c** interlockable osteomorphic blocks

engineering applications [25–30], particularly protective tiles [31], extraterrestrial construction [32, 33], coastal protection [34, 35], design of glass structures [36], and toy manufacturing [37, 38].

In the realm of materials engineering, an important contribution to the mechanics of topological interlocking was made by the research group of Thomas Siegmund [39–43] who also discussed recipes for manufacturing of such materials [38, 39]. A recent paper published by that group [42] provides a comprehensive overview of the area. Attempts at representing topological interlocking materials as an effective continuum, which is a formidable task due to intrinsic anisotropy of such structures, are emerging [44, 45].

In what follows, our aim is to highlight the potential of the topological interlocking concept as applied to engineering materials. The extraordinary properties of such materials will be discussed based on the outcomes of our own research and some results reported in literature.

## 2.3 Mechanics of Topological Interlocking Materials

The main attraction of the topological interlocking principle is its capacity to produce segmented materials without considerable stress concentrations. It is the ability of the segments to exhibit a small relatively independent motion that underlies the useful properties of the material, such as elevated energy absorption and fracture toughness, and provides its structural integrity. The following subsections consider different aspects of mechanics of topological interlocking materials.

### 2.3.1 *Inverse Scale Effect*

The notion that a segmented solid may possess mechanical properties superior to those of a monolithic body, especially a brittle one, comes from the known inverse scale effect, which implies that the strength of a solid is reduced as its size increases. Conversely, the reduction in size gives rise to a higher strength. This can be rationalised in terms of the Weibull statistics. According to it, the survival probability $P_s$ of a solid depends on the stress $\sigma$ and, notably, its volume $V$:

$$P_s = \exp\left\{-\frac{V}{V_0}\left(\frac{\sigma}{\sigma_0}\right)^m\right\} \tag{2.1}$$

Here $V_0$ and $\sigma_o$ are normalisation parameters, which are constant for a given material. The Weibull exponent $m$ characterises the brittleness of the material: the lower the magnitude of $m$, the more brittle the material is. Based on (2.1), Ashby and Bréchet [3] evaluated the benefits of segmenting a monolithic solid into interlocked blocks quantitatively and showed that the gain in strength increases with $1/m$ exponentially. In other words, the more brittle the material, the greater is the effect of its segmentation.

Subdividing a monolithic plate in topologically interlocked blocks is arguably the best way of segmentation. Indeed, it does not require the use of any glue or binder. Neither are connectors, e.g. of the pin-and-hole (LEGO) type, required, so that the attendant stress concentrations are avoided. The blocks are held in place by kinematic constraints imposed by their neighbours due to their shape and mutual arrangement, and only small relative displacements are allowed. These displacements are important, as they provide the segmented plate with a degree of flexibility—a desirable property that effectively makes rigid and brittle materials to compliant and ductile ones.

As mentioned above, the blocks at the periphery of an assembly need to be dealt with separately, as they do not have the neighbourhood necessary to arrest them. Therefore, a lateral constraint is required. The force $P$ to be exerted by the lateral constraint to ensure that the removal of a block from the assembly by a normal force $F$ can be calculated using the following simple model, Fig. 2.4 [46, 47].

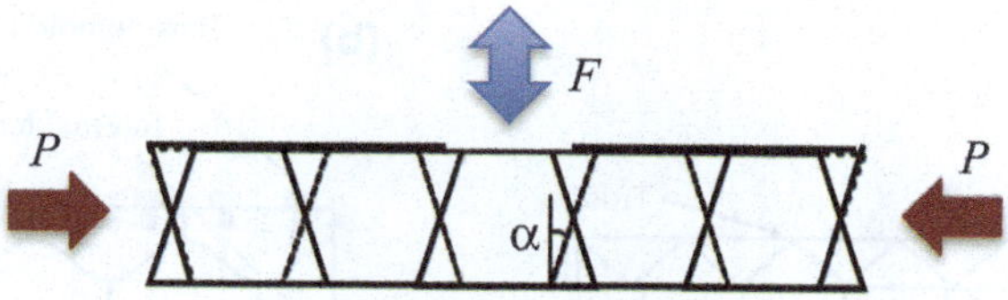


**Fig. 2.4** A simplified model of the effect of lateral constraining force of an assembly of topologically interlocked blocks [46, 47]. Here $P$ is the constraining force, $F$ is the (upward or downward) normal force required to remove the block it is applied to, and $\alpha$ is the 'angle of interlocking'

The force necessary to remove a block from the assembly is given by

$$F = nP\tan(\alpha + \varphi) \tag{2.2}$$

Here $\alpha$ is either the 'angle of interlocking' if it is constant (as for example in platonic or truncated platonic solids) or the maximum angle of interlocking as, for instance, in the case of osteomorphic blocks, $\varphi$ is the friction angle (for the sake of simplicity we neglect adhesion) and $n$ is the number of surfaces of a block in contact with its neighbours (for instance, in Fig. 2.4, $n = 2$).

It is seen that the effect of topological interlocking is similar to conventional friction, only that the friction angle is now equal to $\alpha + \varphi$ and the coefficient of friction $\tan(\varphi)$ is replaced with $\tan(\alpha + \varphi)$. Therefore even if the actual friction between the contacting surfaces is zero, topological interlocking still provides a friction-type resistance to sliding of a block with respect to its neighbours. One can thus refer to the effect of topological interlocking as large-scale friction. In essence, the interlocking assemblies can be treated in multiscale fashion (see [48] for the application of multiscale modelling to interlocking structures): at "small" scale it is an assembly of discrete elements, while at "large" scale it can be regarded as a continuum with pressure-dependent yield stress obtained from (2.2) and, possibly, with non-linear stiffness, as discussed in Sect. 3.3.

While the resistance to the block removal increases with the angle of interlocking, there are limitations on the magnitude of $\alpha$ posed by the build-up of stress concentrations created by partial sliding over the contacting surfaces [48], as well as technical difficulties of manufacturing and assembling of blocks with large interlocking angles.

To provide the necessary lateral constraint, a frame can be placed around a planar array of topologically interlocked blocks (Fig. 2.5a). A benefit of such a solution is that the bending stiffness of the structure can be tuned by applying a controlled lateral load through the frame. A further advantage is that a bigger structure can be assembled from 'cassettes' consisting of such framed groups of interlocked blocks by joining them together. The resulting cellular architecture with a rigid skeleton (frames) encasing flexible cell interiors can provide a range of tuneable mechanical properties. However, as a heavy element of a structure, a rigid frame may be cumbersome and undesirable. A more elegant solution in such a case can be provided by a set of tensioning cables threaded through an assembly of topologically interlocked

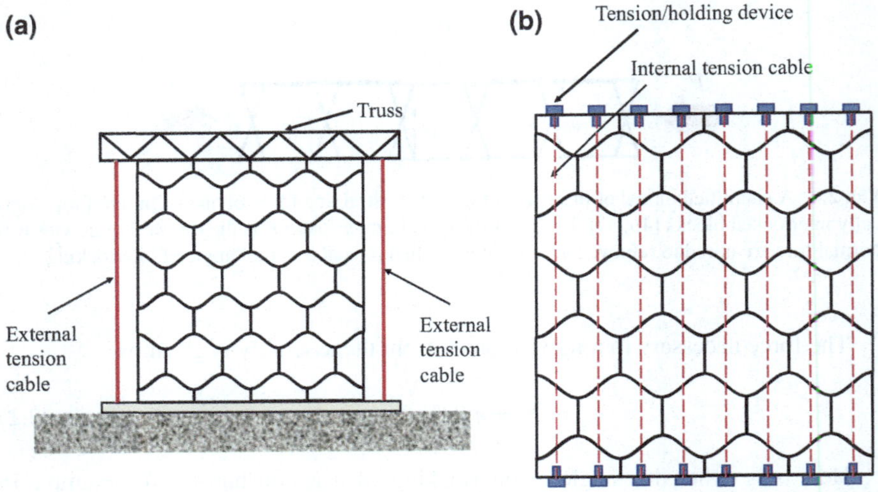


**Fig. 2.5** Types of peripheral constraint: **a** lateral constraint imposed by a specially built frame; **b** constraint produced by internal tensioned cables [30, 36, 50]

blocks (Fig. 2.5b). By varying the magnitude of the load put on the structure through the wires its bending rigidity can be controlled in a broad range, from that of a stiff plate to that of a rollable compliant mat.

The holes made in the osteomorphic blocks to accommodate the cables can in principle induce stress concentrations. It was shown that the smallest stress concentration is achieved with the hole radius of about 16% of the block thickness [49].

### 2.3.2 *Enhanced Fracture Toughness*

Macroscopic fracture propagation in topologically interlocking materials involves cracks growing over distances exceeding the block size and hence covering the neighbouring blocks. As the blocks are not bonded to each other, boundaries between them are able to hinder crack propagation thus increasing effective (macroscopic) fracture toughness [7, 10]. Elevated fracture toughness owes to the fact that a Mode I crack running into an interface also produces a concentration of tensile stress on the areas normal to the crack, thus opening a secondary crack at the interface. This prevents the stress component driving the primary crack from being transferred into the neighbouring block across the interface [51, 52]. Indeed, consider a block with a Mode I crack propagating along the $x$-axis normal to an interface. At a distance $r$ from the crack tip, it creates normal tensile stresses: $\sigma_x = \sigma_y = K_I(2\pi r)^{-1/2}$, where the stress component $\sigma_y$ is the one that drives the crack propagation. As the crack tip approaches the interface, the magnitude of $\sigma_x$ increases and eventually becomes sufficient to separate the blocks along part of the interface (see [53, 54] for details). This mechanism of crack arrest was experimentally demonstrated in [55] by indentation and cracking of a concrete osteomorphic block when the interfaces isolated it from the neighbouring blocks and prevented crack propagation. Therefore, it is the lack of material continuity or bonding that isolates the crack-containing block and preserves the macroscopic integrity of the assembly as a whole.

### *2.3.3 Tolerance to Missing Blocks*

Direct experiments [56] showed that some assemblies, such as assemblies of interlocking tetrahedral and osteomorphic blocks, are tolerant to missing blocks. That is to say, the assemblies would maintain their integrity even when some vacancies were left deliberately or some of the blocks failed under load. (Thus, in the tests reported in [56] up to 8 blocks were left out from an assembly of 400 tetrahedron-shaped blocks without causing its collapse.) When more blocks are missing, the possibility must be considered that, in the terminology of the percolation theory, an infinite or bone cluster will be formed, ultimately causing a collapse of the structure. Numerical simulations [57] showed that under random removal of elements of a planar assembly of osteomorphic blocks such a percolation limit, giving rise to failure of the assembly, is reached when nearly 25% of the blocks are removed.

### *2.3.4 Out-of-Plane Deformation of Topological Interlocking Assemblies*

Out-of-plane deformation of plate-like interlocking assemblies was investigated by applying indentation to the centre of a plate held and constrained at the periphery [10, 46–48, 50–55]. The loading configuration is presented in Fig. 2.6. Figure 2.7 shows the force-deflection curve under indentation of an interlocking assembly of cubes [50]. Two main features can be highlighted. The first one is that the force-deflection dependence is highly non-linear from the very beginning of loading. This indicates that bending rigidity of the assembly, which is typically very low [1, 10, 44, 50, 54, 55], drops off as the loading proceeds. The second prominent feature is the occurrence of a peak in the curve followed by post-peak softening, as reflected in a negative slope of the curve beyond the peak. Post-peak softening is a typical feature of loading of brittle materials (rocks, concrete) using a sufficiently stiff or servo-controlled loading frame. It is commonly related to damage accumulation or fracturing. By contrast, in an assembly of interlocking blocks no damage is observed. It is firstly evident from the unloading curve part of which also exhibits a negative slope. Secondly, the negative slopes in both loading and unloading are also seen in finite element modelling [44, 58] of an interlocking assembly where the cubes are assumed to deform purely elastically, and no damage is included. Therefore, the

**Fig. 2.6** Experimental set-up for concentrated loading (indentation) of a planar assembly of topologically interlocked cube-shaped blocks

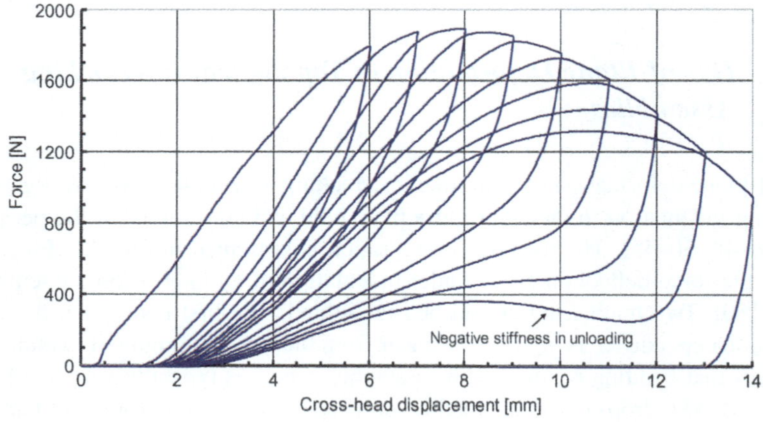


**Fig. 2.7** Force-deflection dependence observed in indentation of a planar assembly of topologically interlocked cubes

occurrence of a negative slope is to be considered as a purely structural phenomenon, which can be referred to as a negative stiffness in strong bending of assemblies of topologically interlocking blocks.

Low bending rigidity is associated with the fact that the blocks are not bonded to each other and can rotate under bending, which continually diminishes the contact

**Fig. 2.8** Delamination between the blocks of a 1D segmented structure caused by the applied bending moment $M$. The constraining force, $P$, is the only factor resisting delamination. (In this sketch, bending and delamination are exaggerated.)

**Fig. 2.9** Partially delaminated rectangular section of a segmented beam of thickness $t$: $f$ is the delamination length; $\sigma_y(y)$ is the stress acting on the contact surface

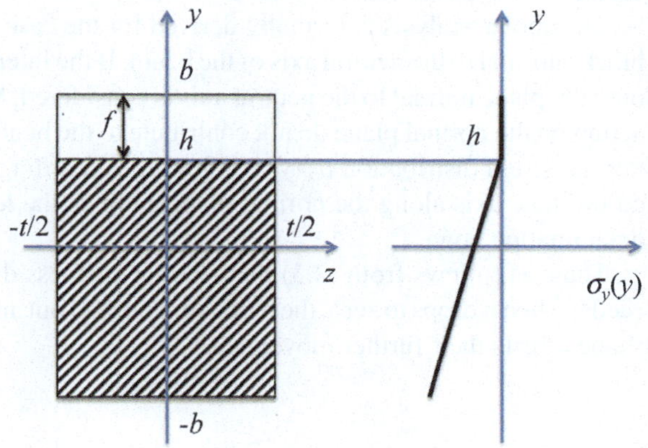


area in the course of loading. This can loosely be referred to as 'delamination' at block interfaces. The only moment that acts against delamination is that created by the constraint force. This leads to a dependence of the bending rigidity upon the bending moment, which can be modelled by representing the interlocking assembly as a fragmented plate, where the role of topological interlocking is simply to provide structural integrity of the plate. Consequently, the simplest model of bending rigidity can be based on considering mutual rotations of blocks against the moment created by externally applied constraint force $P$. A further simplification was to consider a one-dimensional model, i.e. that of a segmented beam [47, 59, 60]. Figure 2.8 shows a 1D model configuration.

Another simplification made in [47] is the assumption that the size of the contact area is given by that of the beam cross-section, which makes the delamination zone rectangular, Fig. 2.9. As follows from the elementary beam theory, the distribution of contact stress is approximated by a piece-wise linear function whose linear part ends at the beginning of the delamination zone, $y = h$: $\sigma(y) = s^{-1}(y - h)$ when $-b \leq y \leq h$. Here $s^{-1}$ is the stress gradient. For the rest of the contact zone, $y > h$, $\sigma(y) = 0$.

Both the stress gradient and the length of the delamination zone, $f = b\text{-}h$, are determined from the equations of force and moment equilibrium, which gives [41] $s = t(2b - f)^2/2P$, $f = 3|M/P| - b$, where $P$ is the constraining force and $M$

is the applied bending moment. We note that condition $f = 2b$ corresponds to full delamination.

From here it follows for the bending stiffness, see Dyskin et al. [52]:

$$EI = \begin{cases} \frac{2}{3}Etb^3, & \text{if } f = 0 \\ \frac{9}{4}Et\left(b - \left|\frac{M}{P}\right|\right)^3, & \text{if } 0 \le f < 2b \\ 0, & \text{if } f \ge 2b \end{cases}, \tag{2.3}$$

where $t$ is the beam thickness, cf. Fig. 2.9.

The above analysis is formally derived for the case of the interfaces between the blocks normal to the neutral axis of the beam. If the interface is inclined, its projection onto the plane normal to the neutral axis is considered. We note that the shear stresses acting on the normal plane do not contribute to the bending moment. Therefore, only normal stress distribution $\sigma_y(y)$ is relevant. Accordingly, (2.3) holds where $y$ is the co-ordinate axis along the normal plane and $f$ is the length of the projection of the delamination zone.

Thus, as follows from (2.3) the bending stiffness diminishes as the loading proceeds; when it drops to zero, the blocks are pushed out and move, until the interlocking shape arrests their further movement.

### 2.3.5 *Modelling of Vibrations in Topological Interlocking Assemblies*

The absence of bonding between the building blocks leads to a rather complex pattern of possible oscillations in topological interlocking assemblies. For instance, plate-like interlocking assemblies can undergo in-plane vibrations enforcing in-plane displacements of the blocks, bending vibration, and vibrations associated with relative rotational oscillations of the blocks. (The bending vibrations essentially correspond to macroscopic rotational vibrations.) These types of vibrations are characterised by strong non-linearity, which presents considerable challenge to modelling. So far, only one type of non-linearity has been considered, viz. the bilinear character of in-plane oscillations [61–66].

Figure 2.10a illustrates the nature of non-linearity of in-plane vibrations. As a result of the ability of blocks to part by delamination, the stiffness in tension is controlled by the effective stiffness of the constraint, $k_F$. It is much smaller than the stiffness in compression determined by both the constraint stiffness $k_F$ and the high stiffness $k_B$ of the material of the blocks. This suggests that the simplest model of in-plane vibrations can be constructed by considering a bilinear oscillator, Fig. 2.10b, or a system of bilinear oscillators, Fig. 2.10c.

A single (undamped) bilinear oscillator is described by the equation of motion

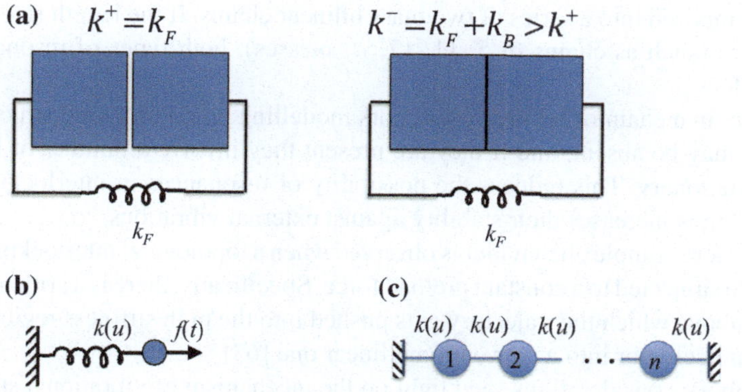


**Fig. 2.10** Modelling of in-plane oscillations of topological interlocking assemblies as bilinear oscillators: **a** a pair of neighbouring blocks with low stiffness in tension (which represents the constraint) and high stiffness in compression (which is controlled by the resistance of both the constraint and the block material); **b** bilinear oscillator representing a pair of blocks with bilinear stiffness given by (2.4); **c** a chain of bilinear oscillators (*n* masses) with fixed ends

$$M\ddot{u} + k(u)u = f(t), \quad k(u) = \begin{cases} k^+ & u > 0 \\ k^- & u < 0 \end{cases}, \tag{2.4}$$

where $u$ is the displacement of the mass $M$ of the oscillator. The discontinuity in stiffness $k(u)$ at $u = 0$ is eliminated since the stiffness enters (2.4) only through the product $uk(u)$. This equation is linear everywhere but at point $u = 0$.

By introducing angular frequencies for positive and negative displacements, $(\omega^\pm)^2 = k^\pm/M$, one obtains the main resonance frequency of the bilinear oscillator (cf., e.g. [65])

$$\omega = \frac{2\omega^+\omega^-}{\omega^+ + \omega^-} \tag{2.5}$$

When the bilinear oscillator is driven by a harmonic force, e.g. proportional to $\sin \omega t$, it will exhibit resonance with linear envelopes, which, is, however, different for positive and negative displacements. It is remarkable that a bilinear oscillator can be put in resonance by all multiple frequencies, $2\omega, 3\omega, \ldots$, although the rate of amplitude increase will be reduced for higher harmonics. Also, there is a half frequency resonance, that is a resonance at the frequency of $\omega/2$. The envelope is non-linear in this case but rather increases as square root of time [60]. This suggests that topological interlocking assemblies can exhibit resonances at multiple frequencies.

When bilinear oscillators form a finite chain with fixed ends, Fig. 2.10c, the structure of resonances becomes even more interesting. Firstly, not all lengths support resonances. They only occur if (1) the chain has an odd number of masses ($n = 2m + 1$); then it gets decomposed into a series of single mass bilinear chains with stationary masses in between; (2) the chain has a number of masses given by $n = 3m + 2$, then it is decomposed into a series of two mass bilinear chains. If the length satisfies both conditions (such as chains of 5, 11, 17, ... masses), both types of resonances are accessible.

Hence, in a chain of bilinear oscillators modelling an interlocking structure, resonances may be absent, and if they are present they involve a number of blocks to remain stationary. This reduces the possibility of resonances in interlocking structures and thus increases their stability against external vibrations.

Even more complex behaviour is observed when a topological interlocking assembly is also subjected to a constant preload force. Specifically, there is a critical value of preload above which the trajectory gets pushed into the high stiffness region turning a bilinear oscillator into a conventional linear one [63].

The above considerations shed light on the mechanism of vibrational stability of topological interlocking structures and, more generally, segmented structures (e.g., classical columns under harmonic and earthquake excitations [67]). Vibrational stability is further assisted by possible energy losses upon low restitution impacts of contacting surfaces. (Initial modelling of this damping mechanism is presented in [68].) Further, these mechanisms can shed light on the increased impact resistance of topological interlocking structures observed in [69].

## 2.4 Responsive Materials Based on Topological Interlocking

The sensitivity of the mechanical response of assemblies of topologically interlocked blocks under normal load to lateral constraint conditions opens up an interesting possibility of controlling their load bearing capability and bending stiffness through varying the magnitude of the lateral constraint forces. Owing to this sensitivity, a responsiveness of topological interlocking assemblies to external stimuli can be achieved by embedding in them some actuating elements. For instance, inclusion of shape memory or piezoelectric materials, which would change their volume as a response to temperature changes or application of an electric field [70], can effectively alter the lateral constraint forces, thus affecting mechanical response in the transverse direction. In particular, tensioning wires from a shape memory alloy (SMA) threaded through a planar assembly of topologically interlocked blocks [30] can be used to actuate such response.

This idea was realised in [71] where a lateral constraint on a planar assembly of ceramic GC Fujirock osteomorphic blocks was provided by tensioning wires made from a NiTi-based SMA. A total of 67 full blocks and 26 half-blocks of 10 mm thickness were assembled to a 160 mm $\times$ 150 mm rectangular plate armoured with 16 pre-tensioned NiTi wires and loaded vertically using a three-point bending support.

In the tests, the shape memory effect of the wires activated by heating was utilised. Heating was provided by an electric current passed through the SMA wires. The loading response was compared with that of an assembly in which the SMA wires were

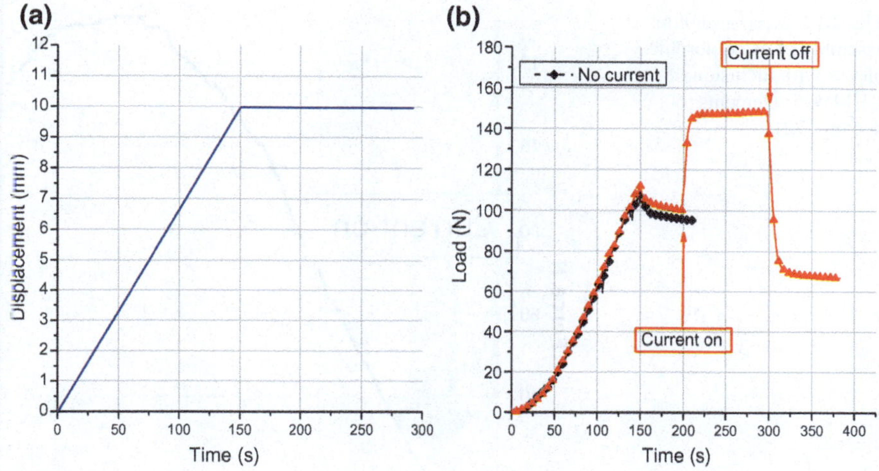


**Fig. 2.11** Demonstration of responsiveness of an assembly of osteomorphic GC Fujirock blocks armoured with SMA wires to heating by an electric current. **a** Indenter displacement as a function of time; **b** Response of the assembly to activation of the shape memory effect by the electric current [71]

used just as armour and a tensioning tool, without inducing the shape memory effect by heating. As the wires were extended during bending of the assembly at ambient temperature until the cross-head of the machine was stopped at a pre-set displacement of the loading device. After a certain hold time during which stress relaxation took place, the wires were heated above the martensite-to-austenite transformation temperature by the electric current. In an unconstrained state this would force the wires to shrink to their initial length, but in the experiment in question they developed a reaction stress, which acted as an increment of the constraint force on the assembly. The resulting response in terms of the load versus displacement curve is seen on Fig. 2.11.

Similar results, but with a much more pronounced effect, were also obtained for an assembly of osteomorphic GC Fujirock blocks cladded on its periphery with platelets made from ultra-high-molecular-weight polyethylene (UHMWPE) that exhibits a shape memory effect, cf. Fig. 2.12 [72].

## 2.5 Assemblies of Modified Topologically Interlocked Blocks: Shape Variations, Soft Interlayers, Secondary Surface Profiles

Some ways in which the mechanical performance of ensembles of interlocked blocks can be influenced will be considered here. An obvious one is variation of the geometry

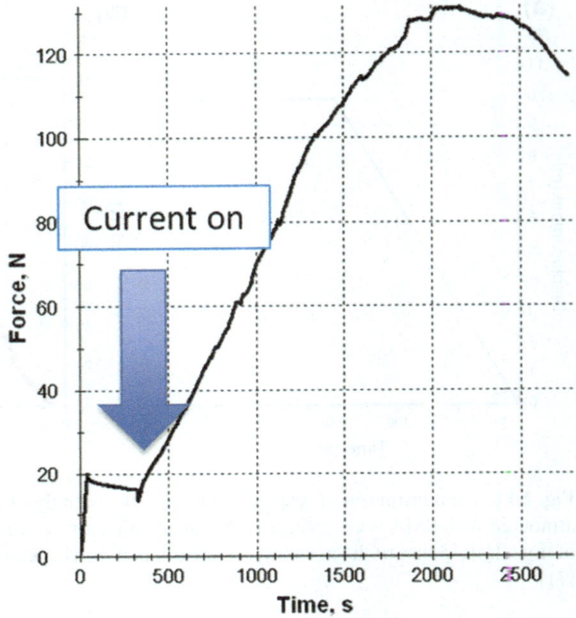


**Fig. 2.12** Response of an assembly of osteomorphic blocks with inclusion of UHMWPE elements to heating [72]

of the building blocks [73, 74]. To elucidate the effect of block geometry, 3D-printed planar assemblies of hard polymer blocks with four different geometries, cf. Fig. 2.13, were tested under concentrated load in the same way as in the experiments described above. The tests have demonstrated the superior performance of an assembly of osteomorphic blocks under concentrated load over all other geometries tested in terms of the maximum load and deflection they sustained, as represented by the peak stress and the strain at failure, Fig. 2.14. Both geometries involving topological interlocking, viz. the 'hourglass' and the osteomorphic ones, are seen to exhibit the highest peak stress. Further block geometries [74] shown in Fig. 2.15 were also considered. The flexural performance of an assembly of modified, osteomorphic-like concrete blocks, yet with interlocking at four, rather than two contact surfaces, was studied in [75]. A substantial improvement over the performance of a monolithic plate in terms of bending compliance and impact energy absorption was demonstrated.

A very attractive pathway to further improving the already outstanding mechanical properties of topological interlocking materials is provided by a biomimetic approach [76, 77]. It suggests that by combining hard interlocking blocks as a majority phase with a small fraction of a soft phase in interlayers between the hard blocks improved ductility and enhanced fracture toughness of a block assembly may be attained. Interleaving hard blocks with a soft polymer was shown to provide the assemblies with extra compliance [72]. A spectacular example of this effect is seen in Fig. 2.16. It shows how segmenting of a monolithic ceramic plate into an assembly of osteomorphic blocks provides the brittle plate with substantial bending ductility under a concentrated load. What is particularly remarkable is that by interleaving the

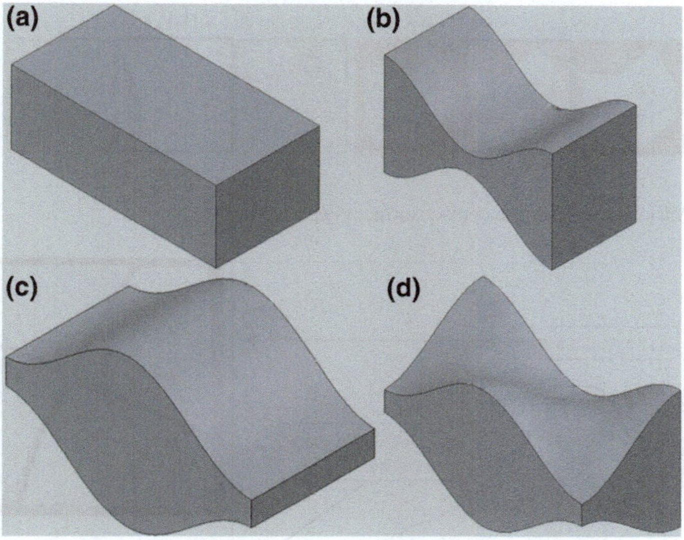


**Fig. 2.13** Prototype geometries of hard blocks as designated in [74]. **a** Reference rectangular block; **b** 'hourglass' block, **c** 'smooth honeycomb' block; **d** modified osteomorphic block

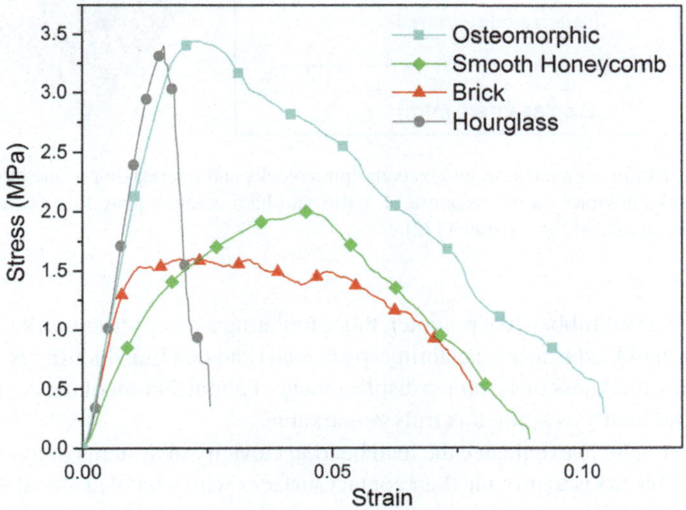


**Fig. 2.14** Stress-strain curves for planar assemblies of blocks with different geometry [74]

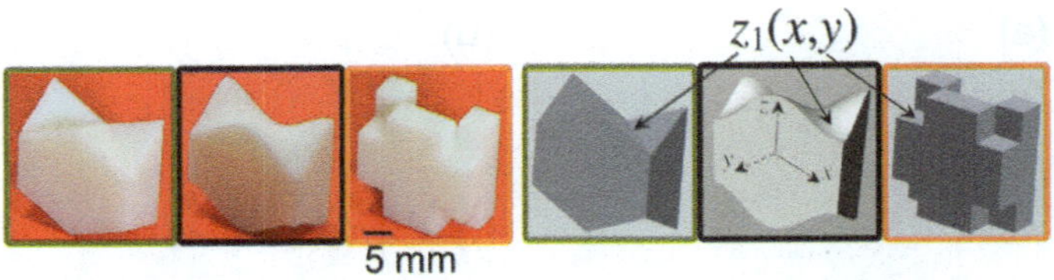


**Fig. 2.15** Various shapes of the contact surfaces explored in [74]

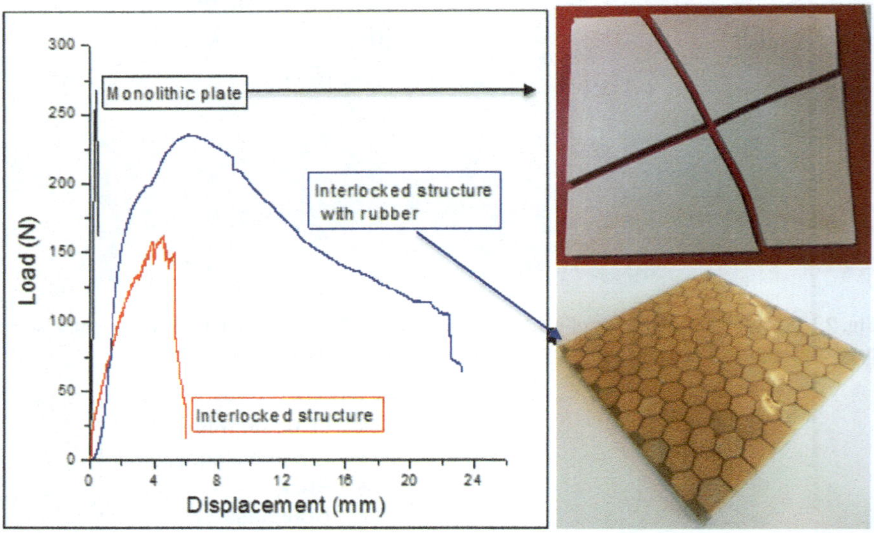


**Fig. 2.16** Effect of segmentation into osteomorphic blocks and interleaving of the blocks with a soft rubber-like polymer on the response of a stiff and hard ceramic plate to concentrated load applied to the middle of the specimen [72]

blocks with a soft rubber-like polymer, thus emulating a nacre structure, a favourable combination of load bearing capability (peak load) and bending ductility is achieved. With a block thickness of 12 mm, a displacement of about 24 mm of the central block to which the load was applied is truly remarkable.

A further vehicle to enhance the load bearing capacity of systems of topologically interlocked blocks is to provide their contact surfaces with a set of artificial asperities, which effectively increases friction between neighbouring blocks. An example of the surface morphology of this kind designed and 3D-printed in a recent work [74] is shown in Fig. 2.17. The asperity height of 2 mm was an order of magnitude smaller than the block dimensions. An increase in the peak load sustained by an assembly of osteomorphic blocks whose contact surfaces were patterned in this way was found. When only planar side surfaces were patterned, the effect was the smallest. Patterning of the concavo-convex contact surfaces gave rise to a larger effect. The greatest increment of the load bearing capacity (about 40%) occurred when all four contact surfaces had the patterned morphology. Pronounced load drops near the peak load were associated with precipitous slip of the block to which the load was applied [74].

The results presented in this section demonstrate that there are various possible ways in which the mechanical performance of assemblies of topological interlocking blocks can be improved, and one can draw from this wealth of designs to achieve the desired mechanical properties of the assemblies in terms of stiffness, load bearing capability, energy absorption, etc.

## 2.6 Sound Absorption

One of the most important practical advantages of sets of topologically interlocked blocks is their high sound absorption capability. The great potential of topological interlocking in this regard was demonstrated by testing sound absorption of plates assembled from ceramic osteomorphic blocks produced by freeze gelation technique [78, 79]. Both monomodal porosity blocks (with pores stemming from the freeze gelation process) and bimodal porosity ones were produced. In the latter case sacrificial polystyrene spheres were added to the slurry used in the process shown schematically in Fig. 2.18. The sound absorption coefficient was measured using a classical Kundt tube set-up. The effect of segmenting of a monolithic plate into a planar assembly of interlocked osteomorphic blocks is seen in Fig. 2.19a. Segmentation alone brought about a rise of the absorption coefficient to a level of about 0.6 in a frequency range of audible sound, just below 200 Hz. Combined with an additional intrinsic effect caused by bimodal porosity, this produced an impressive level of sound absorption above 95%. A shift of the absorption peak to higher frequencies and peak broadening associated with bimodal porosity (Fig. 2.19b) should be noted. These experiments give reasons to believe that by using topological interlocking design in conjunction with block material design, sound absorption can be tuned to desirable levels in the frequency ranges of interest for practical applications in sound insulations in buildings and engineering structures. An example of such applications in a sound absorbing ceramic lining of a cylindrical combustion chamber was reported in [80].

## 2.7 Manufacturing of Topological Interlocking Materials

Successful application of the material and structure design principle presented above hinges on the feasibility of manufacturing processes by which the individual blocks and assemblies thereof can be produced. In a recent publication [42], Siegmund et al. gave a detailed overview of the possible techniques for manufacturing of topological interlocking materials. In a bottom-up approach, individual blocks whose geometry permits topological interlocking are assembled to a final structure, which may be both planar and non-planar. Assembly of such structures requires the use of templates or scaffolds which keep the blocks in place until the necessary confine-

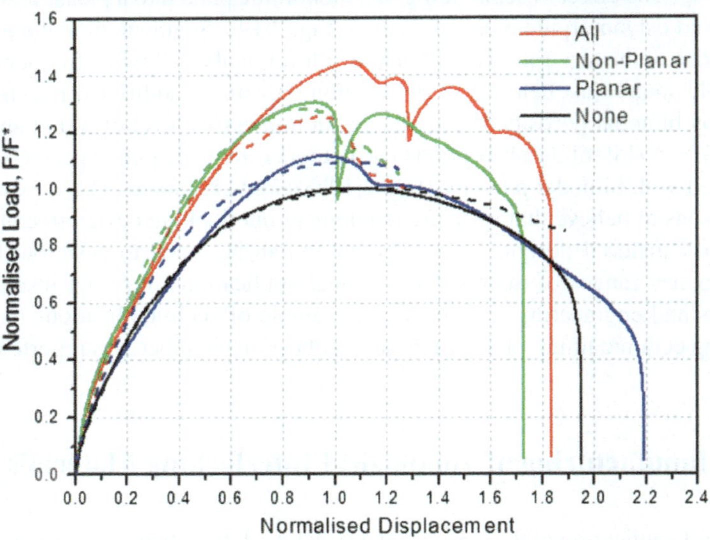


**Fig. 2.17** Variation of surface morphology (left picture) and its effect on the mechanical response of an assembly of blocks to concentrated load (right picture). Block size: ~20 mm; asperity height: ~2 mm. The load $F$ was normalised with respect to the peak load $F^*$; the displacement of the indenter was normalised with respect to the block thickness [73]

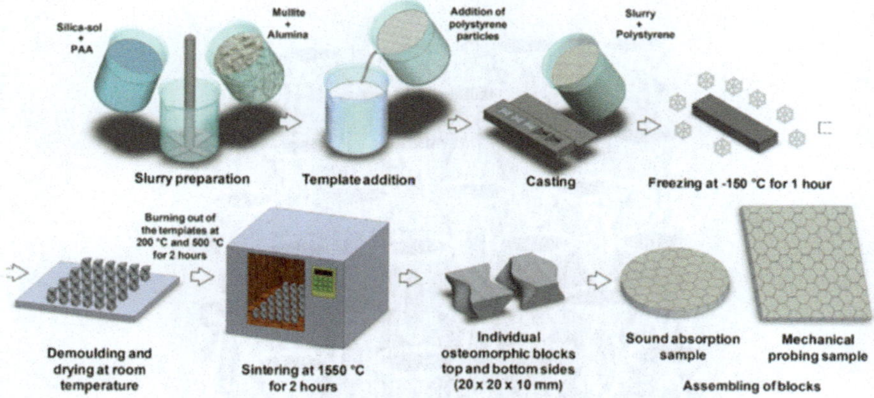


Fig. 2.18 Manufacturing of osteomorphic ceramic blocks by freeze gelation for use in planar assemblies. The processing chains includes sacrificial templating with polystyrene spheres aimed at producing bimodal porosity [79]

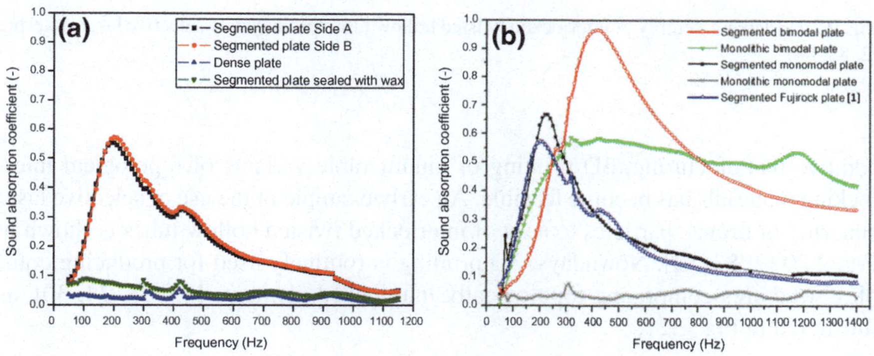


Fig. 2.19 Enhancement of sound absorption by **a** segmentation into interlocked osteomorphic blocks and **b** segmentation combined with introduction of bimodal porosity [78, 79]

ment is established by an external frame or a set of tensioning cables [30, 42, 47, 81]. The feasibility of robotic assembly of topological interlocking materials and structures in a pick-and-place process discussed in [42] was demonstrated in [30]. Obviously, directionality of a building block and the prescribed mutual orientation of the blocks within a layer provides substantial challenges to robotic assembly. A top-down approach, in which the contours of the blocks are carved within a monolithic material, e.g. by a laser beam [82], provides an interesting alternative to the 'bottom-up' piece-by-piece assembly. First attempts of self-assembly of tetrahedron-shaped particles to a planar panel [42] are seen as an interesting avenue for future work.

Individual interlockable blocks can be produced by a multitude of techniques, including CNC-machining [1] and casting [79], cf. Fig. 2.18. With the advent of

**Fig. 2.20** Planar assembly of interlocked twisted hollow tubes produced by selective laser sintering [2, 83, 84]

additive manufacturing, 3D printing of innumerable variants of topological interlocking materials has become feasible. An early example of the use of selective laser sintering of bronze particles to a set of interlocked twisted hollow tubes is shown in Fig. 2.20 [2, 83, 84]. Nowadays, 3D printing is routinely used for producing complex structures comprising topologically interlocked elements [42, 73, 74, 30], as illustrated in Fig. 2.21.

## 2.8 Conclusion

Topological interlocking has firmly established itself as a powerful concept of design of architectured materials. The advantageous properties of topological interlocking materials and structures include high tolerance to local failure, tuneable bending stiffness and load-bearing capability, high impact energy absorption, and exceptional sound insulation properties. Some of these properties have been discussed in this chapter. Due to ease of assembly and disassembly, topological interlocking materials offer themselves for construction in earthquake-prone areas, extra-terrestrial construction, rapidly deployable anti-riot structures, protective wear for first respondents, and many other engineering applications. A particular area of opportunity for use of topological interlocking materials is in manufacturing of hybrid materials combining constituents that may be entirely dissimilar and not otherwise compati-

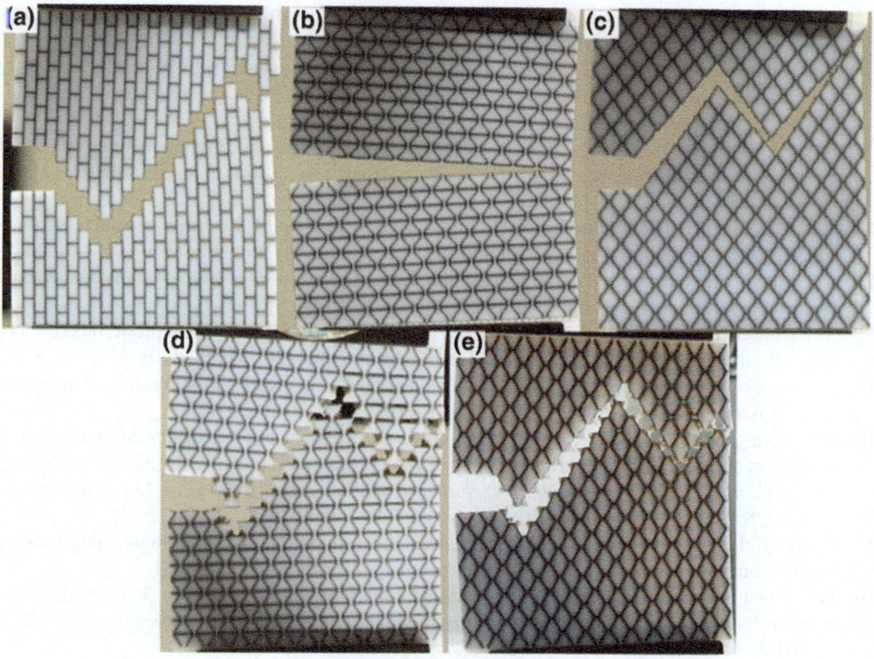


**Fig. 2.21** Polymeric panels consisting of hard blocks interleaved with soft layers after testing in Mode I fracture. Each panel was 3D printed in a single build [74]

ble. By blending blocks identical in shape but made from different materials, which is possible in virtually any proportion, the mechanical and physical properties of such a hybrid material can be varied in a broad range. Multifunctionality of the hybrid can also be achieved in this way. Research in this area is ongoing and further breakthroughs are to be expected.

## References

1. A.V. Dyskin, Y. Estrin, A.J. Kanel-Belov, E. Pasternak, A new concept in design of materials and structures: assemblies of interlocked tetrahedron-shaped elements. Scripta Mater. **44**, 2689–2694 (2001)
2. Y. Estrin, A.V. Dyskin, E. Pasternak, Topological interlocking as a material design concept. Mater. Sci. Eng. C, **31**(6), 1189–1194 (2011)
3. M.F. Ashby, Y.J.M. Bréchet, Designing hybrid materials. Acta Mater. **51**, 5801–5821 (2003)
4. M.F. Ashby, *Materials Selection for Mechanical Design*, 4th edn. (Elsevier Ltd., 2011)
5. A.V. Dyskin, Y. Estrin, A.J. Kanel-Belov, E. Pasternak, Topological interlocking of platonic solids: a way to new materials and structures. Philos. Mag. Lett. **83**(3), 197–203 (2003)
6. A.V. Dyskin, Y. Estrin, A.J. Kanel-Belov, E. Pasternak, Interlocking properties of buckyballs. Phys. Lett. A **319**, 373–378 (2003)

7. A.V. Dyskin, Y. Estrin, A.J. Kanel-Belov, E. Pasternak. Toughening by fragmentation—how topology helps. Adv. Eng. Mater. **3**(11), 885–888 (2001)
8. A. Dyskin, E. Pasternak, Y. Estrin, Topological interlocking as a design principle for hybrid materials in *Proceedings of the 8th Pacific Rim International Congress on Advanced Materials and Processing*, ed. by F. Marquis (Springer, Cham, 2013), pp. 1525–1534
9. A.J. Kanel-Belov, A. Dyskin, Y. Estrin, E. Pasternak, I.A. Ivanov-Pogodaev, Interlocking of convex polyhedra: towards a geometric theory of fragmented solids. Moscow Math. J. **10**(2), 337–342 (2010)
10. A.V. Dyskin, Y. Estrin, E. Pasternak, H.C. Khor, A.J. Kanel-Belov, Fracture resistant structures based on topological interlocking with non-planar contacts. Adv. Eng. Mater. **5**, 116–119 (2003)
11. Y. Feng, T. Siegmund, E. Habtour, J. Riddick, Impact mechanics of topologically interlocked material assemblies. Int. J. Impact Eng **75**, 140–149 (2015)
12. S. Khandelwal, T. Siegmund, R.J. Cipra, J.S. Bolton, Adaptive mechanical properties of topologically interlocking material systems. Smart Mater. Struct. **24**, 045037 (2015)
13. S. Leo, C. Tallon, N. Stone, G.V. Franks, Near-net-shaping methods for ceramic elements of (Body) Armor systems. J. Am. Ceram. Soc. **97**(10), 3013–3033 (2014)
14. H.-C. Ries, M.V. Carlesso, C. Eigenbrod, C. Kroll, K. Rezwan (2013). On the performance of porous sound absorbent ceramic lining in a combustion chamber test rig, in *Proceedings of ASME turbo expo 2013: Turbine Technical Conference and Exposition* (At San Antonio, Texas, USA, 2013) GT2013–95492, Volume: GT2013–95492
15. W. Feng-chun, Y. Zhi-hong, Z. Na, Z. Heng, Design and experiment for topological interlocking fragile composites structure. Acta Armamentarii **29**(12), 1454–1457 (2008)
16. G. Fallacara, Digital stereotomy and topological transformations: reasoning about shape building, in *Proceedings of the second international congress on construction history*, vol 1. pp. 1075–1092 (2006)
17. G. Fallacara, Toward a stereotomic design: experimental constructions and didactic experiences, in *Proceedings of the Third International Congress on Construction History* (Cottbus, May 2009), pp. 553–559
18. Y. Estrin, Topological interlocking and osteomorphic blocks, in *Preface to Stereotomic design—Catalogo della mostra omonima per Inside Marmomacc and Abitare il Tempo Veronafiere*, ed. by G. Fallacara, V. Minenna (Verona, Edizioni Giofredda 2014)
19. D.A. Robson, German patent DE 2554516; British patent GB1533980 (1978)
20. M. Glickman, The G-block system of vertically interlocking paving, in *2nd International Conference on Concrete Block Paving, Delft University of Technology, Apr. 10–12, American Society for Testing and Materials* (Delft, The Netherlands, 1984), pp. 345–348.
21. O. Tessmann, M. Becker, Extremely heavy and incredibly light: performative assemblies in dynamic environments, in *Proceedings of the 18th International Conference on Computer-Aided Architectural Design Research in Asia (CAADRIA 2013)* (Department of Architecture, National University of Singapore, Singapore, May 15–18, 2013), pp. 469–478
22. M. Weizmann, O. Amir, Y.J. Grobman, Topological interlocking in buildings: a case for the design and construction of floors. Autom. Constr. **72**(Part 1), 18–25 (2016)
23. I. Miodragovic Vella, T. Kotnik, Geometric Versatility of abeille vault. A stereotomic, topological interlocking assembly, in *34th Annual eCAADe Conference, Oulu School of Architecture, Finland, Shape, Form and Geometry*. Applications, Volume 2, eCAADe 34, 391–397 (2016).
24. S. Alothman, C. Chavan, Topological interlocking systems for the construction of seismic-proof shell structures, in *Proceedings of the IASS Symposium 2018: Creativity in Structural Design, July 16–20, 2018* (MIT, Boston, USA, 2018)
25. F. Oikonomopoulou, T. Bristogianni, L. Barou, E. Jacobs, G. Frigo, F.A. Veer, R. Nijsse, A novel, demountable structural glass system out of dry-assembly, in *Interlocking Cast Glass Components, Challenging Glass 6—Conference on Architectural and Structural Applications of Glass Louter*, ed. by B., Bos, N., Veer (Delft University of Technology, May 2018). https://doi.org/10.7480/cgc.6.2118
26. A.R. Javan, H. Seifi, S. Xu, Y.M. Xie. Design of a new type of interlocking brick and evaluation of its dynamic performance, in *Proceedings of the IASS Annual Symposium 2016 'Spatial*

*Structures in the 21st Century*, ed. by J.K. Kawaguchi, M. Ohsaki, T. Takeuchi (Tokyo, 26–30 September, 2016)
27. S. Weir, D. Moult, S. Fernando, Stereotomy of wave jointed blocks—toward a wave-jointed stone construction using wire cutter toolpath generation robotic fabrication in architecture, *Art and Design* 284–293 (2016)
28. Y. Totoev, A. Al Harthy, Semi interlocking masonry as infill wall system for earthquake resistant buildings: a review. J. Eng. Res. (TJER) **13**(1), 33–41 (2016)
29. A.M. Sá, K..R. Echavarria, M. Griffin, D. Covill, J. Kaminski, D. Arnold (2012) Parametric 3D-fitted frames for packaging heritage artefacts in *The 13th International Symposium on Virtual Reality, Archaeology and Cultural Heritage (VAST)*, ed. by D. Arnold, J. Kaminski, F. Niccolucci, A. Stork (2012)
30. B. Zareiyan, B. Khoshnevis, Effects of interlocking on interlayer adhesion and strength of structures in 3D printing of concrete. Autom. Constr. **83**, 212–221 (2017)
31. Y. Estrin, A.V. Dyskin, E. Pasternak, H.C. Khor, A.J. Kanel-Belov, Topological interlocking of protective tiles for space shuttle. Phil. Mag. Lett. **83**, 351–355 (2003)
32. A. Dyskin, Y. Estrin, E. Pasternak, H.C. Khor, A.J. Kanel-Belov, The principle of topological interlocking in extraterrestrial construction. Acta Astronaut. **57**(1), 1–64 (2005)
33. A.V. Dyskin, H.C. Khor, D. Yong, E. Pasternak, Y. Estrin, A.J. Kanel-Belov, Deployable interlocking structures for Martian bases, in *Proceedings of 7th Australian Mars Exploration Conference, July 13–15, 2007*, (Trinity College Perth, Western Australia (CD) 2007)
34. E. Pasternak, A.V. Dyskin, C. Pattiaratchi,, E. Pelinovsky, Coastal protection using topological interlocking blocks. EGU General Assembly 2013, held 7–12 April, 2013 in Vienna, Austria, Paper ID: EGU2013-8048, (2013)
35. V.Y. Piirainen, Y. Estrin, Topological interlocking as a principle of engineering design in construction of marine and coastal structures. J. Min. Inst. **226**, 480–486 (2017)
36. H.C. Dyskin, D. Khor, E. Yong, Y.E Pasternak, A.J. Kanel-Belov, Deployable interlocking structures for Martian bases *in Proeedings of 7th Australian Mars Exploration Conference*, (Trinity College Perth, Western Australia (CD-ROM), July 13–15, 2007)
37. P. Houlis, A.V. Dyskin, A. Kanel-Belov, E. Pasternak, Y. Estrin, Puzzle DESIGN Competition, http://www.kastellorizo.org/puzzleuniversity/html/slickred/Puzzle_OneFourAll.html (2010)
38. P. Houlis, A.V. Dyskin, E. Pasternak, Y. Estrin, A. Kanel-Belov. Topological interlocking puzzle. Australian Patent application 20109–2951 (2010)
39. S. Khandelwal, T. Siegmund, R.J. Cipra, J.S. Bolton, Transverse Loading of Cellular Topologically Interlocked Materials. Int. J. Solids Struct. **49**(18), 2394–2403 (2012)
40. S. Khandelwal, R.J. Cipra, J.S. Bolton, T. Siegmund, Adaptive mechanical properties of topologically interlocking material systems. Smart Mater. Struct. **24**(4), 045037 (2015)
41. Y. Feng, T. Siegmund, E.E. Habtour, J, Riddick, (2015) Impact mechanics of topologically interlocked material assemblies. Intl. J. Impact Eng. **75**, 140–149
42. T. Siegmund, F. Barthelat, R.J. Cipra, E. Habtour, J. Riddick, Manufacture and mechanics of topologically interlocked material assemblies. Appl. Mech. Rev. **68**(4), 041401–12016 (2016)
43. A. Mather, R.J. Cipra, T. Siegmund, Structural integrity during remanufacture of a topologically interlocked material. Int. J. Struct. Integr. **3**(1), 61–78 (2012)
44. A.V. Dyskin, Y. Estrin, A.J. Kanel-Belov, E. Pasternak, A new principle in design of composite materials: reinforcement by interlocked elements. Compos. Sci. Technol. **63**(3–4), 483–491 (2003)
45. M. Brocato, A continuum model of interlocking structural systems. Rend. Lincei Mat. Appl. **29**, 63–83 (2018). https://doi.org/10.4171/RLM/793
46. A.V. Dyskin, E. Pasternak, H.C. Khor, Y. Estrin, A.J. Kanel-Belov, Mortar-free construction based on topological interlocking, in *Developments in Mechanics of Structures and Materials*, ed. by A. Deeks, H. Hao (Taylor and Francis Group, London, 2005), pp. 665–670
47. A.V. Dyskin, E. Pasternak, Y. Estrin, Mortarless structures based on topological interlocking. Front. Struct. Civil Eng. **6**(2), 188–197 (2012)
48. A.V. Dyskin, D. Yong, E. Pasternak, Y. Estrin, Stresses in topologically interlocking structures: two scale approach, in *ICTAM 2008, XXII International Congress of Theoretical and Applied*

*Mechanics,* ed. by Denier, J., Finn, M.D., Mattner T., Adelaide, August 24–29, 2008, CD-ROM Proceedings ISBN 978-0-9805142-1-6, paper 10134 (2008)
49. M. Khudyakov, A.V. Dyskin, E. Pasternak, B. Lehane, Optimal through holes in osteomorphic elements in *Proceeding of the 10th International Conference on Structural Integrity and Failure (SIF2016)*, ed. by A. Kotousov, J Ma. Adelaide, Australia, 2016, Paper #36 (2016)
50. Y. Estrin, A.V. Dyskin, E. Pasternak, Topological interlocking in design of structures and materials, in *Architectured Multifunctional Materials*, ed. by Y. Brechet, J.D. Embury, P.R. Onck (Mater. Res. Soc. Symp. Proc. Volume 1188, Warrendale, PA, 2009, 1188-LL05–06) Paper T40.014
51. Y. Estrin, A.V. Dyskin, A.J. Kanel-Belov, E. Pasternak, Materials with novel architectonics: Assemblies of interlocked elements, in *IUTAM Symposium on Analytical and Computational Fracture Mechanics of Non-homogeneous Materials*, ed. by B. Karihaloo, Cardiff, UK, June 18–22, 2001 (Kluwer Academic Press, 2002), pp. 51–56
52. A.V. Dyskin, Y. Estrin, A.J. Kanel-Belov, E. Pasternak, A new class of composite materials based on topological interlocking, in *Applied mechanics—Progress and application. ACAM 2002 The Third Australasian Congress on Applied Mechanics Sydney, February 20–22, 2002,* ed. by L. Zhang, L. Tong, J. Gal (World Scientific, Singapore, New Jersey, London, Hong Kong, 2002), pp.485–490
53. A.V. Dyskin, A. Caballero, Orthogonal crack approaching an interface. Eng. Fract. Mech. **76**(16), 2476–2485 (2009)
54. H.C. Khor, A.V. Dyskin, Y. Estrin, E. Pasternak, Mechanisms of fracturing in structures built from topologically interlocked blocks, in *Structural Integrity and Fracture, SIF 2004* ed by A. Atrens, J.N. Boland, R. Clegg, J.R. Griffiths, pp. 189–194
55. H.C. Khor, A.V. Dyskin, E. Pasternak, Y. Estrin, A.J. Kanel-Belov, Integrity and fracture of plate-like assemblies of topologically interlocked elements, in *Structural Integrity and Fracture*, ed. by A.V. Dyskin, X.Z. Hu, E. Sahouryeh (Swets & Zeitlinger, Lisse, 2002), pp. 449–456
56. A. Molotnikov, Y. Estrin, A.V. Dyskin, E. Pasternak, A.J. Kanel-Belov, Percolation mechanism of failure of a planar assembly of interlocked osteomorphic elements. Eng. Fract. Mech. **74**, 1222–1232 (2007)
57. Y. Estrin, A.V. Dyskin, E. Pasternak, S. Schaare, S. Stanchits, A.J. Kanel-Belov, Negative stiffness of a layer with topologically interlocked elements. Scripta Mater. **50**, 291–294 (2004)
58. S. Schaare, A.V. Dyskin, Y. Estrin, S. Arndt, E. Pasternak, A.J. Kanel-Belov, Point loading of assemblies of interlocked cube-shaped elements. Int. J. Eng. Sciences **46**, 1228–1238 (2008)
59. E. Pasternak, A.V. Dyskin, I. Shufrin, Homogenisation methods in mechanics of fragmented solids and hybrid materials in 7th Australasian Congress on Applied Mechanics, ACAM 7, 9–12 December 2012, Adelaide, Australia. Paper 138, pp. 563–573 (2012)
60. I. Shufrin, E. Pasternak, A.V. Dyskin, Bending and stability of fragmented beams, in 8th *Australasian Congress of Applied Mechanics ACAM 8 Melbourne 23–26 November 2014* ed by R. Das, S. John, 8 pp (2014)
61. A.V. Dyskin, E. Pasternak, E. Pelinovsky. Modelling resonances in topological interlocking structures. *ACAM2007*, in *Proceeding 5th Australasian Congress on Applied Mechanics, 10–12 December 2007, Brisbane, Australia,* vol. 2, ed by F. Albermani, B. Daniel, J. Griffiths, D. Hargreaves, P. Meehan, A. Tan, M. Veidt, (2007), pp. 408–413
62. A.V. Dyskin, E. Pasternak, E. Pelinovsky, Coupled bilinear oscillators, their resonances and controlling parameters, in *Proceedings. 6th Australasian Congress on Applied Mechanics, ACAM 6*, ed by K. Teh, I. Davies I. Howard 12–15 December 2010, Perth, Paper 1170, 9 pp. ISBN/ISSN 978-0-85825-941-6, 12/12/2010
63. A.V. Dyskin, E. Pasternak, E. Pelinovsky, Periodic motions and resonances of impact oscillators. J. Sound Vibr. **331**(12) 2856–2873 (2012)
64. I. Shufrin, A.V. Dyskin, E. Pasternak, Stationary points created by resonances in a chain of bilinear oscillators, in 7th *Australasian Congress on Applied Mechanics, ACAM 7,* 9–12 December 2012, Adelaide, Australia. Paper 131 (2012)
65. A.V. Dyskin, E. Pasternak, I. Shufrin, Structure of resonances and formation of stationary points in symmetrical chains of bilinear oscillators. J. Sound Vib. **333**, 6590–6606 (2014)

66. A. Guzek, A.V. Dyskin, E. Pasternak, I. Shufrin, Asymptotic analysis of bilinear oscillators with preload. Int. J. Eng. Sci. **106**, 125–141 (2016)
67. I.N. Psycharis, D.Y. Papastamatiou, A.P. Alexandris, Parametric investigation of the stability of classical columns under harmonic and earthquake excitations. Earthquake Engng. Struct. Dyn. **29**, 1093–1109 (2000)
68. M. Khudyakov, A.V. Dyskin, E. Pasternak, Continuum model of wave propagation in fragmented media: linear damping approximation. Nonlinear Proc. Geophys. (NPG) **24**, 461–466 (2017)
69. H.C. Khor, A.V. Dyskin, N. Nofal, E. Pasternak, Y. Estrin, Topological Interlocking—A New Principle in Design of Concrete Structures, *Futures in Mechanics of Structures and Materials,* in *Proceedings 20th Australasian Conference on the Mechanics of Structures and Materials (ACMSM20)*, Toowoomba, Queensland, Australia, 2–5 December 2008, 5 pp (2008)
70. W. Quan, W. Nan, A review on structural enhancement and repair using piezoelectric materials and shape memory alloys. Smart Mater. Struct. **21**, 013001 (2012)
71. A. Molotnikov, R. Gerbrand, Y. Qi, G.P. Simon, Y. Estrin, Design of responsive materials using topologically interlocked elements. Smart Mater. Struct. **24**, 25034 (2015)
72. Y. Estrin, A. Molotnikov, G.P. Simon, S. Kaloshkin, F. Senatov, A. Maksimkin, Flexible ceramics with self-stiffening capability, Presented at: European Symposium on Intelligent Materials 2015, 10–12 June 2015, Kiel, Germany (2015)
73. Djumas, L., Molotnikov, A., Simon, G.P., Estrin, Y., Enhanced mechanical performance of bio-inspired hybrid structures utilising topological interlocking geometry, Scientific Reports 6, Article number: 26706 (2016)
74. L. Djumas, G.P. Simon, Y. Estrin, A. Molotnikov, Deformation mechanics of non-planar topologically interlocked assemblies with structural hierarchy and varying geometry, Scientific Reports, 7, Article number: 11844 (2017)
75. A. Rezaee Javan, H. Seifi, S. Xu, D. Ruan, Y.M. Xie, The impact behaviour of plate-like assemblies made of newinterlocking bricks: an experimental study. Mater. Des. **134**, 361–373 (2017)
76. H.D. Espinosa, J.E. Rim, F. Barthelat, M.J. Buehler, Merger of structure and material in nacre and bone—Perspectives on de novo biomimetic materials. Prog. Mater Sci. **54**, 1059–1100 (2009)
77. Y. Beygelzimer, Y. Estrin, R. Kulagin, Synthesis of hybrid materials by severe plastic deformation: a new paradigm of SPD processing. Adv. Eng. Mater. **17**(12), 1852–1861 (2015)
78. M. Carlesso, A. Molotnikov, T. Krause, K. Tushtev, S. Kroll, K. Rezwan, Y. Estrin, Enhancement of sound absorption properties using topologically interlocked elements. Scripta Mater. **66**, 483–486 (2012)
79. M. Carlesso, R. Giacomelli, T. Krause, A. Molotnikov, D. Koch, K. Tushtev, S. Kroll, Y. Estrin, K. Rezwan, Improvement of sound absorption and flexural compliance of porous alumina-mullite ceramics by engineering the microstructure and segmentation into topologically interlocked blocks. J. Eur. Ceram. Soc. **33**, 2549–2558 (2013)
80. H.-C. Ries, M. Carlesso, C. Eigenbrod, S. Kroll, K. Rezwan, On the performance of porous sound absorbent ceramic lining in a combustion chamber test rig, in *Proceedings of ASME Turbo Expo 2013 GT2013* (San Antonio, Texas, USA, June 3–7, 2013)
81. Y. Estrin, A.V. Dyskin, E. Pasternak, S. Schaare, Topological interlocking in design of structures and materials, in *Architectured Multifunctional Materials*, ed by Y. Brechet, J.D. Embury, P.R. Onck Mater. Res. Soc. Symp. Proc. Volume 1188, Warrendale, PA, 2009, 1188-LL05–06. Paper T40.014, 10 pp (2009)
82. M. Mirkhalaf, J. Tanguay, F. Barthelat, Carving 3D architectures within glass: exploring new strategies to transform the mechanics and performance of materials, extreme mech. Lett. **7**, 104–113 (2016)
83. D. Trenke, J. Estrin, Selbsttragende Strukturen und deren Fertigung durch Rapid Tooling IMW, TU Clausthal—Institutsmitteilung Nr. 28 (2003)
84. Y. Estrin, N. Müller, D. Trenke, A. Dyskin, E. Pasternak, Structure composed of elements and method of its production, US Patent #6884486 (26.4.2005) (2005)

# Chapter 3
# Architectured Materials with Inclusions Having Negative Poisson's Ratio or Negative Stiffness

**E. Pasternak and A. V. Dyskin**

**Abstract** Architectured materials with negative Poisson's ratio (auxetic materials) have been subject of interest for quite some time. The effect of negative Poisson's ratio is achieved macroscopically through various types of microstructure made of conventional materials. There also exist (unstable) microstructures that, under certain boundary conditions, exhibit negative stiffness. In this chapter we review the microstructures that which generate macroscopic negative Poisson's ratio and negative stiffness, determine their effective moduli and discuss general properties of the materials with such microstructures. We then consider hybrid materials consisting of conventional (positive Poisson's ratio and positive moduli) matrix and randomly positioned inclusions having either negative Poisson's ratio or a negative stiffness (one of the moduli being negative). We use the differential scheme of the self-consisting method to derive the effective moduli of such hybrids keeping in the framework of linear time-independent theory. We demonstrate that the inclusions of both types can, depending on their properties, either increase or decrease the effective moduli.

## 3.1 Introduction

A prospective direction in designing of architectured materials is developing special microstructures of such materials to achieve desired properties unachievable otherwise. Such an approach created a class of metamaterials with special mechanical and acoustic characteristics. Another way of designing materials with the desired mechanical behaviour is to fill the matrix made of a conventional material with inclusions possessing specific microstructure that turns them into microscopic mech-

E. Pasternak
Department of Mechanical Engineering, The University of Western Australia,
35 Stirling Hwy, Perth, WA 6009, Australia
e-mail: elena.pasternak@uwa.edu.au

A. V. Dyskin (✉)
Department of Civil, Environment and Mining Engineering, The
University of Western Australia, 35 Stirling Hwy, Perth, WA 6009, Australia
e-mail: arcady.dyskin@uwa.edu.au


Y. Estrin et al. (eds.), *Architectured Materials in Nature and Engineering*, Springer Series in Materials Science 282,
https://doi.org/10.1007/978-3-030-11942-3_3

anisms or machines giving rise to so-called machine augmented composites (MACs). This concept was first introduced by Hawkins et al. [114], McCutcheon [167, 233] and then developed by Hawkes et al. [113] and Bafekrpour et al. [22]. In this paper two special types of micro-mechanisms (or micromachines) are considered whose mechanical behaviour exhibits either negative Poisson's ratio or apparent negative stiffness. In the second case, the matrix plays the role of stabiliser thus making the total energy of the composite with negative stiffness inclusions positive definite. We concentrate on the elastic or incremental elastic behaviour without dwelling into non-elastic or rate effects.

The approach discussed will be viewed in the frame of the paradigm of hybrid materials as outlined by Ashby and Bréchet [15]—instead of an expensive and uncertain traditional process of developing new alloys, new polymer chemistries and new compositions of glass and ceramics, a hybrid material as 'a combination of two or more materials in a predetermined geometry and scale, optimally serving a specific engineering purpose' is created. In what follows we firstly consider materials and structures exhibiting negative Poisson's ratio (auxetic materials) and the properties of composites with auxetic inclusions. We will then proceed with more exotic structures that exhibit negative stiffness or in continuum case the negative shear or Cosserat shear modulus and the composites with negative stiffness inclusions. It should be noted that so-called double-negative metamaterials have been proposed which possess both negative Poisson's ratio and negative stiffness, e.g. Hewage et al. [115].

## 3.2 Negative Poisson's Ratio

### *3.2.1 Structures Exhibiting the Effect of Negative Poisson's Ratio*

Negative Poisson's ratio (NPR), while thermodynamically permitted (in isotropic materials the permitted range is $-1 < \nu < 0.5$) has only been observed in some natural materials. For instance some crystals exhibit NPR in certain directions (historically, negative principal Poisson's ratio of $-1/7$ was observed in a cubic crystal of Pyrites $FeS_2$, [162]; other cases are described by Gunton and Saunders [108], Milstein and Huang [173], Evans et al. [80], Keskar and Chelikowsky [132], Yeganeh-Haeri et al. [5], Yeganeh-Haeri [273], Baughman et al. [27, 28], Alderson and Evans [5]; see also review by Greaves et al. [91]). Another class of auxetic materials is based on polymers. Alderson and Evans [1] reported technology of producing small samples of ultra high molecular weight polyethylene anisotropic auxetic material with high values of NPR (see [2] for modelling the behaviour of this material). The method was further developed by Alderson et al. [11] and Webber et al. [256]. Simkins et al. [225] proposed a method of producing polymer auxetic fibres. Hirotsu [118] found that the poly (N-isopropylacrylamide)/water gel (NIP A gel) exhibits NPR close to $-1$ at phase transition. NPR polyester fibres were produced and tested by Ravirala et al. [205], although values of NPR were small (up to $-0.002$). Small values of the NPR were predicted in carbon nitride by using ab initio calculations, Guo and Goddard III [109].

Alderson et al. [10] experimentally found that the auxetic microporous polyethylene was both more difficult to indent than the other materials at low loads (from 10–100 N) and was the least plastic with the most rapid viscoelastic creep recovery of any residual deformation. Alderson et al. [12] designed a method for production of thin flat and curved auxetic foam sheets. Bezazi and Scarpa [31] found that auxetic foams have enhanced tensile fatigue properties. Coenen and Alderson [53] considered auxetic carbon fibre laminates and the mechanisms of their failure under indentation. It needs to be noted that in this case of non-elastic behavior, the NPR is a structure identifier rather than the reason for the reported advantageous non-elastic properties. Scarpa et al. [214] using Monte-Carlo simulation found the extreme auxetic behaviour of non-reconstructed defective single-walled carbon nanotubes.

Grima et al. [94] used molecular modelling to find NPR in zeolites (aluminosilicate structures). Grima et al. [104] used computer simulations to investigate the auxetic character of NAT-type silicates and the effect of pressure. Using numerical simulations with Monte-Carlo technique the NPR was detected in a number of 2D and 3D auxetic crystal structures with different inter-molecular potentials (e.g., [35, 36, 119, 149, 178, 179, 201, 234, 238–241, 262–266]). Scarpa's et al. [216] numerical simulations suggested a possibility of NPR in single wall carbon nanotube (SWCNT) heterojunctions. Grima et al. [101] showed by numerical simulations a possibility of turning graphene auxetic by introducing vacancy defects thus making the graphene sheets similar to crumpled paper.

Recently, some indication emerged of the existence of negative Poisson's ratio in certain soils, Essien et al. [74] and rocks, Zaitsev et al. [274, 275]. All other known cases of negative Poisson's ratio are the man-made structures exhibiting this property macroscopically. The majority of such structures are anisotropic, however the isotropic matrix filled with chaotically oriented anisotropic inclusions is isotropic and can exhibit negative Poisson's ratio. Therefore, hereafter the analysis of behaviour of negative Poisson's ratio materials and structures will mostly be based on the isotropic case.

Baughman et al. [26] suggested that 69% of the cubic elemental metals have a negative Poisson's ratio when stretched along the [110] direction. Further Zhang et al. [276] noted that if anisotropic (transverse isotropic in their case) materials are axially loaded at an angle to the plane of isotropy, the values of elastic parameters exist that make the ratio of lateral to axial strain positive. It needs to be noted here that in isotropic materials the positive ratio of lateral to axial strain does indeed mean NPR, however, in anisotropic materials this ratio depends upon the direction of loading. For this reason the anisotropic materials showing positive ratios of lateral to axial strain only in some directions are referred to as partially auxetics (e.g. [36]) since the Poisson's ratios (the parameters of the Hooke's law) of anisotropic material do not depend upon the loading direction. Same remarks can be made with regard to Yao's et al. [270] modelling predicting NPR in single wall nanotubes, Chen's et al. [47] experiments with the CNT paper (buckypaper) where the ratio of lateral to axial strain varied from positive to negative depending upon whether the nano-fibres were aligned to the load direction or were directed at 45° to them, as well as Chen's et al. [48] auxetic skin. Dong et al. [60] reported the NPR in tetragonal $BaTiO_3$ after ageing under bending. Jayanty et al. [128] produced anisotropic auxetic stainless steel fibre mats and obtained very high values of NPR (up to −18.6) in the out-of-plane direction. Given that the mats are essentially thin plates one needs to consider a possibility of buckling affecting the measurements. The composites made from these mats and polymer also showed large NPR (up to −9.0).

An interesting method of creating a 2D structure with the out-of-plane negative stiffness was proposed by Grima et al. [100]. The structure consists of a conventional 2D hexagonal mesh and a very sift material filling the cells. A uniaxial tensile in-plane stress applied to the mesh makes the cells to contact in the lateral direction squeezing out the soft filler thus creating tensile average strain in the out-of-plane direction. Verma et al. [244] developed a heat-compression method of treating commercial nonwoven fabrics that produces out-of-plane NPR.

Alderson et al. [7] produced composites made of matrix (Araldite LY 5052 resin with HY 5084 hardener) filled with auxetic fibres and reported a two-fold increase in peak load as compared to the same matrix with non-auxetic fibres. We note however that the reported NPR of the fibres was only −0.156, which suggests that other factors contributed to the reported strength increase. Alderson and Coenen [8] reported increased energy absorption of auxetic composite carbon fibre laminates on low velocity as compared to similar composited with positive Poisson's ratio.

Babaee et al. [21] designed and manufactured prototype auxetic structure called Bucklicrystal based on buckling of the structural elements. The NPR up to −0.5 were achieved. It should be noted that buckling is one of the mechanisms of negative stiffness, Bažant and Cedolin [29], so one can expect that the Bucklicrystal would show negative Young's modulus as well.

Sigmund [224] and Larsen et al. [151] proposed a method of designing high compliance structures, in particular auxetic structures by employing a special optimization problem formulated as minimization of least-squares errors to obtain the prescribed elastic behaviour; the optimisation problem is solved using a sequential linear programming. Shokrieh and Assadi [218] proposed algorithm of designing composites with maximum value of NPR. Further, the optimal design for achieving NPR was proposed by Czarnecki and Wawruch [57]. Theocaris et al. [235] proposed a method of designing structures with desired macroscopic values of NPRs using reverse analysis based on numerical homogenisation.

We concentrate on mechanical nature of auxetic structures and propose a new classification of the structures. Some previous classifications of auxetics are based on the major classes of materials (polymers, composites, metals and ceramics) existing in auxetic form (e.g., [230]). Blumenfeld [33], Blumenfeld and Edwards [34] proposed to single out so-called iso-auxetic structures, essentially mesh structures where the stress and strain are defined by a sort of homogenisation (see e.g. Pasternak and Mühlhaus [190] for a review of the methods of homogenisation) as opposite to elasto-auxetic structures, where notionally the stress and strain are determinable. Liu and Hu [161] reviewed geometric auxetic structures developed by the time. Further review by Prawoto [202] considered mechanisms of NPR and possible applications.

Novak et al. [181] provided a classification of internal structures of cellular solids leading to NPR. Here we classify the auxetic materials in general (beyond the cellular structures) by the mechanisms of producing the NPR. The basic methods of designing the NPR are: (1) structures based on high ratio of shear to Young's moduli; (2) structures shrinking/expanding in lateral directions under axial compressive/tensile load; (3) structures based on rotating rigid units and; (4) materials with specially arranged interacting defects.

#### 3.2.1.1 Structures Based on High Ratios of Shear to Young's Moduli

From a basic relation between moduli of isotropic material

$$E/G = 2(1 + \upsilon) \tag{3.1}$$

one can see that negative Poisson's ratio corresponds to the relation $E/G < 2$. This suggests a general method of designing the NPR structures based on achieving high $G/E$ ratio; as $G/E \rightarrow \infty, \nu \rightarrow -1$. The simplest method of achieving this is to construct an NPR structure of particles connected by links with high ratios of shear to normal stiffnesses as implemented in a number of works [185–187, 192]. Another way of using this method is to construct the NPR material as an assembly of rigidly connected thin hollow spheres [221] whereby the effect of normal stiffness is achieved by low bending rigidity of thin wall spheres. Similar effect can be reached by using thin arches [220, 223] or various folded units (e.g., [192]) as links.

Another obvious but underappreciated consequence of relation (3.1) is that as $\nu \rightarrow -1$ only ratio $E/G$ tends to zero. That means that reaching the extreme negative Poisson's ratio, $\nu \rightarrow -1$, is equivalent to either very small $E$ or very large $G$. (For some reasons, Spadoni and Ruzzene [229] considered only one way of realising this limiting transition: $G \rightarrow \infty$). Subsequently, the claims that materials with NPR have high fracture toughness (e.g., [51, 141]), high energy absorption on impact [144] or high hardness (infinite when $\nu \rightarrow -1$, e.g., [3, 6, 9, 78]; see also the review by Yang et al. [268], Prawoto [202]) can only be substantiated when the limiting transition $\nu \rightarrow -1$ corresponds to the situation when Young's modulus $E$ is either constant or tends to zero slower than $(1 - \nu^2)$. Therefore, any increase in fracture resistance, hardness or similar properties observed experimentally and in numerical simulations, in particular the elevated fracture toughness experimentally observed in auxetic materials (e.g., [61]) or numerically predicted increase buckling load [71] should be regarded as a property of a particular microstructure that happens to produce the effect of negative Poisson's ratio as well; see Pasternak and Dyskin [187] for details. For instance, if a material with a disc-like crack of radius $R$ is subjected to uniform tensile load, the critical stress that causes the crack growth can be represented as $\sigma_c^2 = \pi E\gamma[2(1 - \nu^2)R]^{-1}$. Then indeed, as $\nu \rightarrow -1$, the critical stress *increases* tending to infinity. On the other hand, if the critical stress is expressed through the shear modulus $\sigma_c^2 = \pi G\gamma[(1+\nu)R]^{-1}$, the critical stress *decreases* as $\nu \to -1$.

Separately stands the dependence of thermal stresses on the Poisson's ratio. Pasternak and Dyskin [187] used a well known fact that the thermal stress is proportional to $\alpha E(1+\nu)^{-1} = 2\alpha G(1-\nu)(1+\nu)^{-1}$, where $\alpha$ is coefficient of linear thermal expansion. Then in both cases of constant $E$ or constant $G$, tending the Poisson's ratio to $-1$ leads to reduction of thermal stresses. In this case the strongest effect is achieved when $G = const$: the thermal stress tends to zero as $\nu \to -1$. A similar conclusion was reached by Lim [156] for auxetic thin plates. (It should be noted that some structures such as foams can show strongly non-linear deformation and exhibit for instance high damping and simultaneously negative Poisson's ratio, e.g., [146, 164], however the NPR does not form a mechanism of such a behaviour.)

The above example shows the importance of separate values of $E$ and $G$ whose combination gives the value of negative Poisson's ratio. The relationship between Young's modulus $E$ and shear modulus $G$ of isotropic auxetics can, following Ashby and Bréchet [15] be illustrated by a parametric space shown in Fig. 3.1. The combinations of $E$ and $G$ that correspond to the given values of $\nu$ are represented by radial lines. The auxetic materials that correspond to fixed $G$ could be called the *G-auxetics*, they are represented by the horizontal lines. Likewise, the vertical lines represent materials with fixed $E$; they should be called the *E-auxetics*.

The direct implementation of the idea of high ratios of shear to Young's moduli in constructing auxetic structures leads to the development of granular materials in which the grains are connected with the links with high ratios of shear to normal stiffnesses. This is achieved by considering a set of grains/particles and applying a homogenisation procedure based on differential expansions (see details in [185–187, 189, 190]). In this case

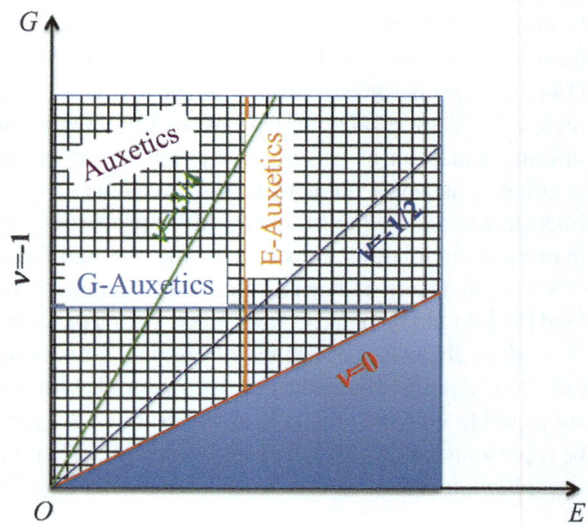


**Fig. 3.1** Parametric space for isotropic auxetics. The radial lines correspond to combinations of $E$ and $G$ associated with the given values of $\nu$; the horizontal lines correspond to the *G-auxetics* (class of materials with fixed $G$), while the vertical lines correspond to the *E-auxetics* (class of materials with fixed $E$)

$$\nu = \frac{k_n - k_s}{4k_n + k_s} \tag{3.2}$$

Therefore designing structures whose NPR reaches $-1$ becomes possible as a result of the limiting transition $k_n/k_s \to 0$.

Some known auxetic structures are based on different implementations of low normal to shear stiffness ratios. Bathurst and Rothenburg [24] proposed auxetic materials with interacting particles with low normal to shear stiffness ratios. Rothenburg et al. [207] proposed an auxetic structure based on indirect implementation of low normal to shear stiffness ratios. Evans and Alderson [77] proposed another version of this design principle resulting in an auxetic isotropic microstructure consisting of nodules and fibrils.

Another implementation of this idea is to connect the grains with a stiff rod that can freely move into the grains; the movement being resisted by a soft spring, Fig. 3.2a [186, 187, 192]; see also [115]. In this case the shear stiffness is controlled by the (high) stiffness of the rod, while the normal stiffness corresponds to the spring stiffness that can be made as low as necessary. The same principle underpins 2D auxetic system based on interlocking hexagons, Ravirala et al. [204], where the male-female interlock ensures large shear stiffness alongside with zero normal stiffness. The idea of high shear to normal stiffness ratio explains the mechanics of NPR of sheets with high concentrations of multiwall carbon nanotubes, Hall et al. [111] where low normal stiffness is provided by sliding of walls in the multiwall nanotube. Another example is the double-helix yarn double layer composite Miller et al. [171], Wright et al. [267] where the elastomeric core ensures high shear stiffness, while the stiffer wrap produces low normal stiffness.

An extreme case a ball-rod structure of multipods was proposed by Pikhitsa et al. [198]. This consists of stiff balls freely sliding on long stiff rods leading to $k_n = 0$ and, hence $\nu = -1$. Pasternak and Dyskin [186, 187] produced a 3D prototype implementing this design. Another variation of this principle consists of a matrix filled with special cracks whose sliding is prevented by for instance small rods piercing the crack in the normal direction, Fig. 3.2c, Pasternak and Dyskin [187]. Isotropic materials with multiscale distribution of randomly situated and oriented disk-like cracks of this type remain isotropic and permit very large crack concentrations. At large concentration the Poisson's ratio tends to $-1/3$. It is interesting that the crack sliding can also be prevented if the crack surfaces are rough, which is typical for geomaterials, Fig. 3.2d. This can form a mechanism of NPRs found in rocks by Zaitsev et al. [274, 275].

Another design realising low normal to shear stiffness ratios leads to structures consisting of thin wall spheres in direct contact with each other, Pasternak and Dyskin [186]. In this case the ratio of stiffnesses $k_n/k_s \sim h/R$, where $R$ is the sphere radius and $h$ it the thickness of its shell. Therefore thin enough spheres could potentially produce NPRs close to $-1$. This conclusion was confirmed by Shufrin et al. [221] by final element modelling (using ABAQUS). They obtained the NPRs close to $-1$. Yet, a method of producing hollow sphere structures with random packing of the spheres was earlier proposed by Andersen et al. [14]. In 2D probably the simplest auxetic

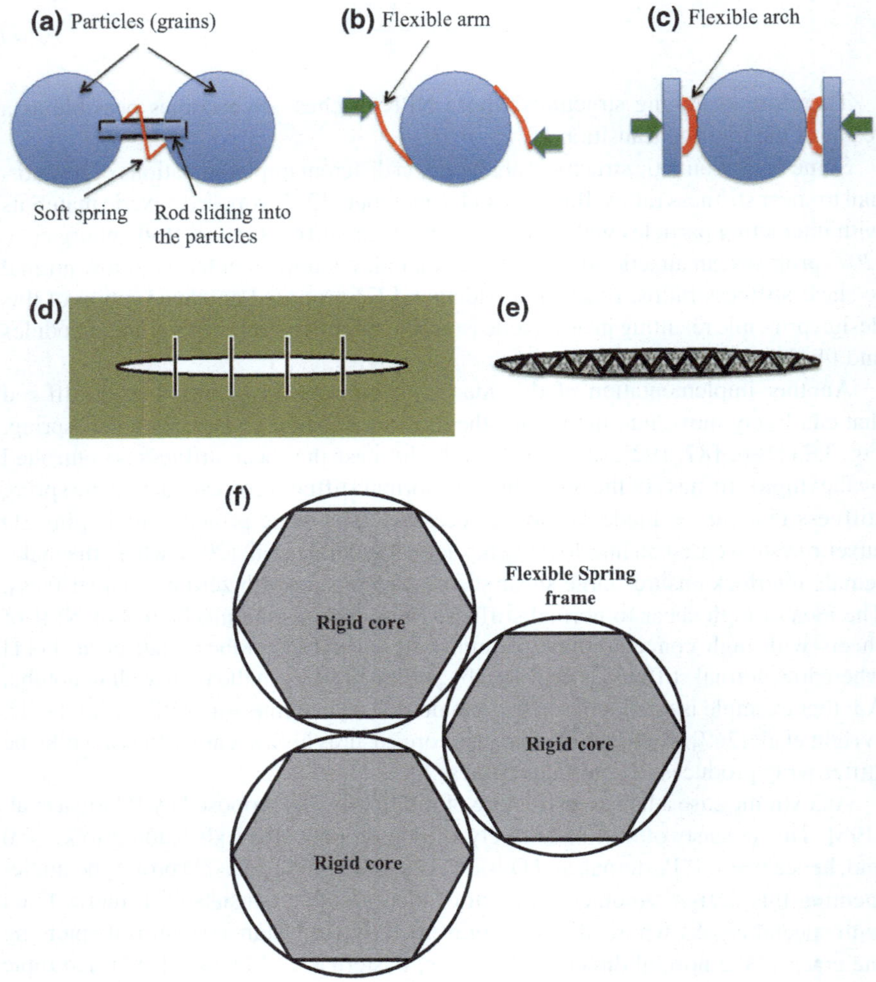


**Fig. 3.2** Basic auxetic structures ensuring low normal to shear stiffness ratios by different designs of links connecting the particles (grains): **a** stiff rods sliding into the particles provide high shear stiffness, while the soft spring ensures low shear stiffness; **b** flexible arms provide high shear stiffness through their axial stiffness, while low normal stiffness is ensured by low bending stiffness; **c** the same as (**b**) but with arches replacing the arms; **d** crack with pierced faces to exclude sliding; **e** schematics of a possible realisation of a crack with sliding blocked by asperities in rocks; **f** fragment of 2D hexagonal reinforced core structure; rigid cores in a shape of hexagon are connected by flexible spring frames

structure of this kind was proposed by Shufrin et al. [223] and named the reinforced core structures. In this structure hexagonal rigid cores were placed in circular flexible spring frames, Fig. 3.2f. The frames provide high shear to normal stiffness ratio. The structure was modelled by the finite element method and calibrated against laboratory experiments (the structure was made from polymers using selective laser sintering and stereolithography; the displacements were measured using the close range photogrammetry). The structures tested reached the NPR of about −0.5.

Hughes et al. [122] considered a cubic structures of nodes and introduced beams of 3 types: type 1 connecting the immediate neighbours, type 2 connecting the second neighbours and type 3 connecting the third neighbours. Then by playing with beam stiffnesses the NPR of near −1 was obtained in finite element simulations. Gaspar [85] suggested making the inter-grain links in the shape of asymmetric elastic arms and produced a 2D prototype. In this design, the required low normal stiffness is achieved by the arm bending; the shear stiffness is close to the axial stiffness of the arms. If the arm length is $L$ and its thickness $h$ then the stiffness ratio can be estimated as $k_n/k_s \sim h^2/L^2$. Again, the NPR close to −1 can be obtained using thin enough arms. Here reaching the limiting value of NPR is faster than in the case of the thin wall spheres, however the DEM simulations by Gaspar [85] could only reach the NPR of −0.1.

Similar but symmetric design, Fig. 3.2c, was proposed by Shufrin et al. [219, 220]. Numerical simulations showed that for different shapes of grains the NPR of −0.66 could be achieved. Further numerical simulation showed that replacing curved arch with a couple of pinned beams allows approaching the value of NPR = −1.

#### 3.2.1.2 Structures Based on the 'Anchor' Effect

The Poisson's ratio can be expressed in isotropic material (and in a properly chosen co-ordinate frame in anisotropic materials) as $\nu = -\varepsilon_x/\varepsilon_z$, where $\varepsilon_z$ is the normal strain component in the direction of loading (we will call it the axial strain), while $\varepsilon_x$ is a normal strain component in a perpendicular direction (we will call it the lateral strain). Negative Poisson's ratio corresponds to the same signs of both strains, which is a movement characteristic of an anchor, Fig. 3.3a. Based on this, Phan-Thien and Karihaloo [197] proposed an auxetic microstructure consisting of anchor-shaped inclusions in a matrix. Baughman et al. [26] by considering the possibility of NPR in the cubic elemental metals essentially refers to the crystals where some atoms are arranged in the anchor-type configurations presuming that the atomic connections can play the role of ridged links as in the anchor-type and re-entrant (see below) structures.

Similar to the anchor structure is the inverted/re-entrant hexagonal/honeycomb mesh, Fig. 3.3b, initially suggested by Kolpakov [137]. Almgren [13] used the idea of re-entrant structure for 2D and 3D structures of hinges, springs and sliding collars with negative Poisson's ratio of −1. Two years later it was re-invented by Lakes [139]. He manufactured an open cell polymer and metallic foams with negative Poisson's ratio of −0.7. Further methods for manufacturing 2D inverted honeycomb structures

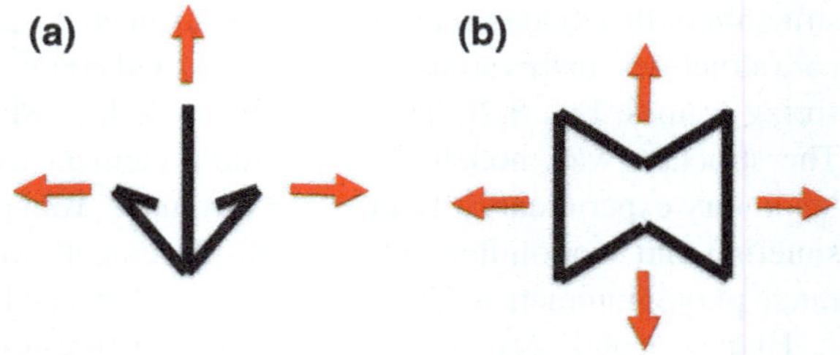


**Fig. 3.3** Basic elements of auxetic structures: **a** anchor—the movement up leads to lateral expansion of the anchor; **b** inverted (re-entrant) hexagonal element

were proposed by Miller et al. [172]. Choi and Lakes [49, 50] and Chan and Evans [46] proposed technology of producing auxetic foams with values of NPR reaching −0.82 in some cases. Scarpa et al. [215] produced auxetic foam with NPR reaching −0.3. A further production method was considered by Grima et al. [103]. Re-entrant foams were also discussed by Bianchi et al. [32].

Some modifications of re-entrant structures were proposed by Scarpa and Tomlin [213], Attard and Grima [17], Mitschke et al. [176]. Re-entrant honeycombs on molecular level produce auxetic microporous polymers and polymer laminates, Caddock and Evans [42], Evans et al. [80], Baughman and Galvao [25]. Alderson and Evans [2] used what is essentially an anchor-type structure to model the manufactured ultra high molecular weight auxetic polyethylene. Another type of inverted 2D honeycomb structure was proposed by Lira et al. [159, 160]. The auxetic triangular lattice core structure, Michelis and Spitas [169] and the triangular lattice structure, Mitschke et al. [175], Cabras and Brun [40, 41] are other examples of structures of this kind. Hou et al. [120, 121] proposed an auxetic material consisting of a matrix filled with triangular thin wall inclusions. It should be noted that while 2D hexagonal structures are elastically isotropic (see, e.g. [150]) the periodic re-entrant hexagonal structure loses its hexagonal symmetry and therefore is no longer isotropic, even in 2D.

#### 3.2.1.3 Structures Based on Rotating Rigid Units

Another class of auxetic structures is formed by the structures whose units can rotate under normal load. In this way, effectively, low shear stiffness is achieved. What distinguish these structures are their asymmetry and the involvement of internal rotations of either parts of the network (the chiral lattice) or rigid inclusions. Evans and Caddock [79] proposed tensile networks, some based on rotating units. Lakes [140] proposed a 2D chiral honeycomb structure that exhibits a negative Poisson's ratio by imposing (rigid) rotations on their units. This is somewhat similar to the 2D structure proposed by Gaspar [85], Fig. 3.2b. Further structures based on rotating unites were proposed by Baughman et al. [27]. Alderson and Evans [4, 5] reported auxetic structures based on rotating tetrahedral framework that is analogous to the molecular structure of $\alpha$-cristobalite. Nazare and Alderson [180] and Gatt et al. [87] proposed models for analysing its Poisson's ratio. Smith et al. [227] considered 2D networks with asymmetric units whose rotation creates NPR. Gaspar et al. [83]

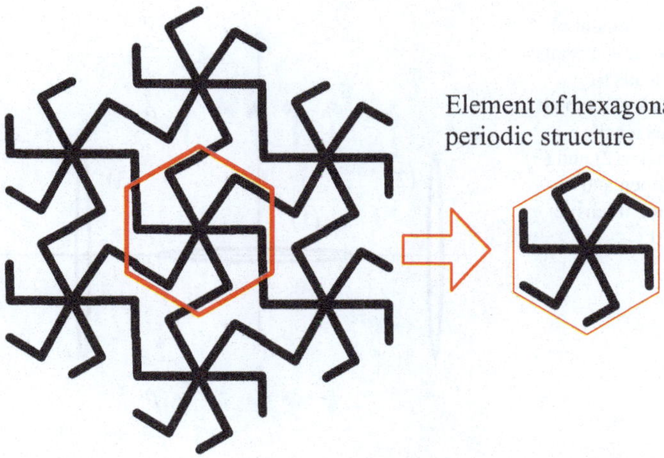


**Fig. 3.4** Auxetic hexagonal periodic structure based on rotating elements of the mesh [192]

proposed 2D auxetic networks based on regular (periodic) meshes with partially removed links, which leads to rotations of fragments of the mesh effectively as rigid inclusions, and manufactured some prototypes. Grima et al. [95, 106] considered NPR produced by rotation of non-regular polyhedral inclusions. Grima and Evans [92] considered auxetic behaviour caused by rotating rigid triangles. Similar 2D structures were analysed by Williams et al. [260]. The rotational mechanism of NPR was further considered by Pasternak et al. [192] leading to design of a 2D hexagonal periodic auxetic structure, Fig. 3.4. Attard et al. [18, 19] and Grima et al. [107] considered structures made of connected and rotating rectangular stiff units. Rotating mesh elements are also realised in Triangle-Square-Wheels structures, Mitschke et al. [175, 176]. Mizzi et al. [177] presented finite element modelling of structures with rotating units. The rotating units can be equipped with magnets; this opens up interesting avenue of controlling the material behavior with external magnetic fields, Grima et al. [98].

Gaspar et al. [86] presented theoretical foundations for the NPR created by rotating units. Spadoni and Ruzzene [229] proposed a micropolar (Cosserat) theory for modelling the NPR-produced rotations. Further, Milton [174] presented a general theory of (non-linear) macroscopic deformations of periodic affine unimode metamaterials constructed from rigid bars and pivots. Kochmann and Venturini [135] proposed a homogenisation scheme applicable for finite deformations.

Grima et al. [96] used the concept of rotating rigid units to explain the NPR in the auxetic foams, which are manufactured from conventional foams by hydrostatic compression and subsequent thermal treatment [139]. Grima and Gatt [93] attributed considered auxetic behaviour of plates with diamond or star-shaped perforations to rotation of solid parts of the plate between the perforations. Azzopardi et al. [20] used this mechanism to explain the deformation in stishovite The following subsection will offer another explanation of auxetic behaviour of perforated plates.

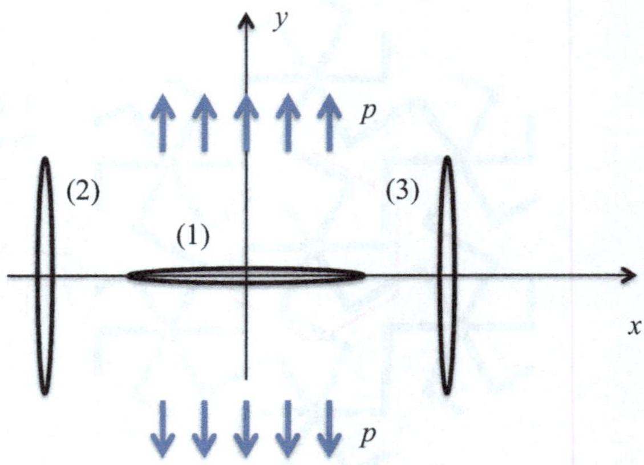


**Fig. 3.5** Basic element of perforated auxetic materials: tensile load $p$ leads to opening of crack (1), which, in turn, induces tensile stresses on cracks (2) and (3) making them open and produce additional tensile strain $\Delta\epsilon_x > 0$

#### 3.2.1.4 NPR Produced by Crack/Void Interaction

An interesting class of auxetic materials is based on special distributions of cracks (slits) and holes made in a conventional (continuous) material rather than on purposely designed structures. A basic element in such an auxetic is formed by cracks oriented perpendicularly to each other, Fig. 3.5. In this picture crack (1) opens under tensile stress $p$ and produces tensile stress $\Delta\sigma_x > 0$, which is proportional to $p$. This additional stress opens cracks (2) and (3) resulting in additional tensile strain $\Delta\epsilon_x > 0$ also proportional to $p$. Grima et al. [102] verified the auxeticity of such perforated sheets with 2D numerical simulations and experiments and demonstrated that NPRs close to $-1$ can be achieved. Slan et al. [226] confirmed this experimentally. Grima et al. [102] also showed that the NPR survives randomisation of locations and orientation of the cracks, although the value of NPR decreases with the increase of the randomness. The latter conclusion is not surprising given that the plates with randomly oriented and randomly situated cracks will have the Poisson's ratio tending to zero as the crack concentration increases (e.g. [210]). Bertoldi et al. [30] used similar mechanism and observed NPR in a 2D perforated plate, but attributed this effect to elastic instability owing to low stiffness of and hence large deformation in the produced system.

Gaspar et al. [84] used Kröner's homogenization solution and showed that it is sufficient to add random anisotropic fluctuations to originally isotropic material to make the Poisson's ratio negative. It is not clear though what was the influence of the fact that the Kröner's method is approximate.

## 3.2.2 *Properties of Composites with Negative Poisson's Ratio Inclusions*

This section deals with composite materials consisting of a conventional (positive Poisson's ratio) matrix filled with auxetic inclusions. The importance of such composites is that macroscopically matrix filled with negative Poisson's fibre networks [128] or auxetic inclusions [257–259] can become auxetic as well. Furthermore, it was demonstrated that placing the auxetic inclusions of various shapes in a positive Poisson's ratio elastic isotropic matrix increases the effective Young's modulus even when the Young's moduli of the matrix and inclusions are the same [185, 186, 192, 257].

### 3.2.2.1 Effective Moduli

Central for analysing such composites is the determination of their effective moduli, that is the coefficients of the Hooke's law relating stress and strain averaged over the volume elements large enough to encompass a significant number of inclusions, elements of fibre network or layers (so-called representative volume elements, RVEs). According to Hill's lemma [117], the average stress within the RVE is solely determined by the tractions applied to the boundary and hence in order to find the effective moduli it is sufficient to load the RVE by uniform tractions and determine the average strain. (Some papers use the term effective moduli in a different sense; for instance [165] used this term for frequency-dependent moduli in forced vibration of auxetic thermoelastic plate.) In general, such a problem does not have an analytical solution except for a simple case of layered materials [90, 153] and thus requires numerical modelling.

A number of approximate methods were used to compute effective moduli of composites with auxetic inclusions. The simplest one is the rule of mixtures. It is however well known that the rule of mixtures is not applicable for determining effective moduli of highly heterogeneous materials. Lim [154], Lim and Acharya [158] used the rule of mixtures for considering effective moduli of layered materials with alternating auxetic and non-auxetic layers. (We note in passing that the problem of effective moduli of layered material with general anisotropy has been solved in closed form by [153] and later by [90].) Wei and Edwards [257] used the mean field theory based on the assumption that each inclusion can be considered as being immersed in a certain effective stress or strain field. Shufrin et al. [222] compared the mean field theory with the results of finite element simulations and found that in the case of large contrast between the inclusions and matrix the mean field theory can suffer from significant errors; see discussion below.)

An important method of estimating the effective characteristics is based on the determination of lower and upper bounds. The most popular are the Hashin-Strikman bounds, Hashin and Shtrikman [112]. The bounds are widely used for composites with inclusions. For instance Scarpa et al. [214] used the Hashin-Strikman bounds to estimate effective moduli of composites whose matrix contains unidirectional non-reconstructed defective single-walled carbon nanotubes and found the in-plane NPR reaching −0.2. However, the bounds may not work in some cases such as the materials with cracks or composites with negative stiffness inclusions (considered later in this chapter).

Often the finite element analysis is used to determine the effective moduli, e.g., Wang and Pan [246], Kanouté et al. [131], Strek and Jopek [231]. Shufrin et al. [222] used the finite element method to find the effective moduli of a composite consisting of isotropic matrix and cubic inclusions. The isotropic matrix with cubic inclusions become orthotropic and its effective shear modulus becomes different from that of isotropic material, however the difference becomes appreciable only when the NPR of inclusions becomes close to −1. Furthermore, Shufrin et al. [222] compared their computations with the Hashin-Strikman bounds and found that the bounds are quite narrow even in the case of auxetic inclusions.

We concentrate here on analytical (and often approximate) methods of determining the effective moduli, which make it simpler to perform parametric analysis and consider the extreme cases. (Limitations of analytical models were discussed in [246].) In order to explain difference between various analytical methods of calculating effective moduli of a matrix filled with inclusions we consider a simple model example of an isotropic matrix containing parallel isotropic cylindrical inclusions in the anti-plane strain approximation. This example will also be used later to analyse the matrix with negative stiffness inclusions.

In anti-plane strain the only modulus that controls the elastic behaviour of an isotropic material is the shear modulus $G$. For a matrix of shear modulus $G_0$ with low concentration of cylindrical inclusions of shear modulus $G_1$ the expression for the effective shear modulus accurate to the linear term with respect to the concentration (area fraction of inclusions) $c$ reads (e.g., [88, 89]):

$$G = G_0\left[1 + 2\frac{G_1 - G_0}{G_1 + G_0}c\right] \tag{3.3}$$

Equation (3.3) is based on a solution of a problem for the inclusion in an infinite elastic space (plane in this case) under uniform tractions applied at infinity. It provides the constant and linear terms of what is known as virial series. Here the linear term is obtained by considering each inclusion as being under the original uniform stress field equal to the stress filed that would be produced in pure matrix by uniform traction applied to the surface of the RVE. The next, quadratic in concentration term is responsible for pairwise interaction between the inclusions, that is each inclusion is considered as being under the additional stress field induced by all other inclusions which are themselves acted upon by the original stress field. The cubic term represents interactions between triplets of inclusions and so on. Germanovich and Dyskin [89] calculated the quadratic terms for two extreme cases: equal cylinders and the multiscale distribution of cylinders. In the latter case it is assumed that the cylinders of close diameters do not interact directly, that is their concentration is infinitesimal, while the interaction is important only between cylinders of markedly different sizes. Such a distribution was termed 'wide distribution of sizes', Salganik [210]. In a way this case is a double asymptotics of large ratios of inclusions' radii and small centration of inclusions of equal radii. The results are:

For equal inclusions

$$G = G_0\left[1 + 2\frac{G_1 - G_0}{G_1 + G_0}c + \frac{4}{3}\frac{(G_1 - G_0)^2(2G_1 + G_0)}{(G_1 + G_0)^3}c^2\right] \tag{3.4}$$

For wide distribution of radii

$$G = G_0\left[1 + 2\frac{G_1 - G_0}{G_1 + G_0}c + \frac{(G_1 - G_0)^2(3G_1 + G_0)}{(G_1 + G_0)^3}c^2\right] \tag{3.5}$$

The determination of higher terms usually meets with considerable technical difficulties. That is why for practical usage simple approximate methods are used based on the solution of single inclusion in uniform stress field, which is the approximation of non-interacting inclusions, given for the example of parallel cylinders in anti-plane strain by (3.3). Interaction between the inclusions is accounted for by either a uniform effective stress different from the applied one (the mean field theories) or by changing the moduli of the medium in which the inclusion is considered (effective medium theories). In the latter case the most popular are the so-called self-consistent methods.

There are two main self-consistent methods: the symmetric (algebraic) scheme and asymmetric (differential) scheme. The symmetric scheme (e.g., [38]) considers the inclusion as being in an equivalent medium with the yet unknown effective moduli. Essentially, this scheme treats each inclusion equally and is seem to be suitable for every size distribution, in particular equal inclusions. This approach leads to a system of algebraic equations (hence the name for the scheme), which for the example considered reduces to the following equation:

$$G = G_0\left[1 + 2\frac{G_1 - G}{G_1 + G}c\right]. \tag{3.6}$$

Solution of this equations keeping only quadratic term reads

$$G = G_0\left[1 + 2\frac{G_1 - G_0}{G_1 + G_0}c + 4\frac{(G_1 - G_0)^2 G_1}{(G_1 + G_0)^3}c^2\right] \tag{3.7}$$

One can see that the difference from the independently computed quadratic term in (3.4) can reach as much as 100%. Furthermore, the implementation of this scheme for equal spherical particles (e.g., [38]) and randomly oriented equal disc-like cracks (e.g., [39, 183]) for large inclusion/crack concentrations gives negative values for the effective moduli; see also Bruner [37]. In other words for large concentrations of inclusions or cracks the symmetric scheme predicts the NPR where it should not be present. A reason for this is that the scheme overestimates the interaction between the inclusions/cracks.

The differential scheme is best understood for the case of multiscale or wide distribution of inclusion sizes [210, 242, 243]. For this distribution the effective moduli are calculated step-by-step. Firstly, the smallest inclusions are inserted in the original material. Since, by assumption the concentration of such inclusions is infinitesimal, $dc$, and hence the effect of interaction between them is negligible, the effective moduli are computed using the approximation of non-interacting inclusions. Then the group of inclusions of the next size also of concentration $dc$ is immersed in an equivalent medium with the effective moduli determined at the previous step. This step is repeated by induction until all inclusions are immersed.

The introduction of new inclusions into the equivalent material corresponds to the removal of a portion $dc$ of the equivalent material and replacing it with the material of inclusions. Together with the equivalent material some previously introduced inclusions get removed as well. Therefore, at each step the concentration previously immersed inclusions gets reduced by a factor $(1 - dc)$. Thus for the example at hand this procedure leads to the following differential equation (hence the name of the scheme):

$$\frac{dG}{dc} = \frac{2G}{1-c}\frac{G_1 - G}{G_1 + G} \tag{3.8}$$

with initial conditions $G = G_0$ at $c = 0$.

Solution of this equation reads

$$\sqrt{\frac{G}{G_0}\frac{G_1 - G}{G_1 + G_0}} = 1 - c \tag{3.9}$$

Expanding $G$ in power series in $c$ and keeping only the quadratic term one obtains expression (3.5). Thus the differential scheme gives exact quadratic term obtained for the wide distribution of sizes. It is not surprising that the quadratic term obtained by the differential scheme is different from the one computed for equal inclusions. What is interesting though is that the difference is at most 25% that is 4 times smaller than the symmetric scheme gives. Furthermore, other methods, like the mean field methods still give errors higher than the differential scheme [89].

The differential scheme has another interesting feature: a solution like solution (3.9) can be used all the way through up to extreme concentration of inclusions $c = 1$. (In the paradigm of wide distributions of sizes such concentration is theoretically possible since small inclusions get fitted between the large ones.) For this concentration (3.9) gives $G = G_1$, which corresponds to the case when the material of the whole matrix is replaced with the material of inclusions.

Given the above reasons we only use the differential scheme in the analysis. Using this scheme Pasternak and Dyskin [187] determined the effective moduli of isotropic material of shear modulus $G_0$ and bulk modulus $\kappa_0$ with multiscale distribution of

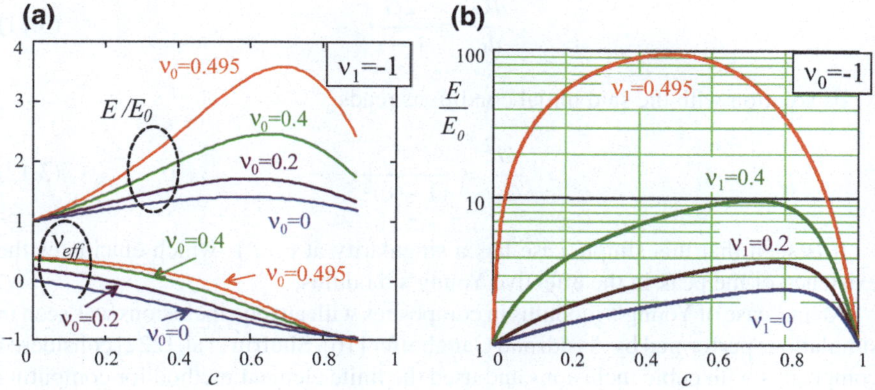


**Fig. 3.6** Dependence of effective Young's modulus of isotropic composites with spherical inclusions made of the same material with Young's modulus equal that of the matrix, Pasternak and Dyskin [187]: **a** extreme NPR $= -1$ inclusions in a matrix with different positive Poisson's ratios; **b** extreme NPR $= -1$ matrix with different positive Poisson's ratio inclusions

spherical auxetic inclusions of concentration (volumetric fraction) $c$ whose shear and bulk moduli are denoted $G_1$ and $\kappa_1$. The analysis was based on differential equations derived by McLaughlin [168] for spherical inclusions using the differential scheme; in notations adopted here they read:

$$\begin{cases} \dfrac{d\kappa}{dc} = \dfrac{\kappa+\kappa^*}{1-c}\dfrac{\kappa_1-\kappa}{\kappa_1+\kappa^*}, & \kappa^* = \frac{4}{3}G \\ \dfrac{dG}{dc} = \dfrac{G+G^*}{1-c}\dfrac{G_1-G}{G_1+G^*}, & G^* = \dfrac{G}{6}\dfrac{9\kappa+8G}{\kappa+2G} \\ \kappa|_{c=0} = \kappa_0, & G|_{c=0} = G_0 \end{cases} \tag{3.10}$$

Numerical solution of this system is shown in Fig. 3.6, Pasternak et al. [192], Pasternak and Dyskin [187]. Figure 3.6a shows the extreme case of inclusions with Poisson's ratio of $-1$ for different positive Poisson's ratios of the matrix. The Young's moduli of matrix and inclusions are assumed to be equal. It is seen that that extreme NPR inclusions reduce the effective Poisson's ratio of the composite that becomes $-1$ at the maximum concentration of inclusions. It is interesting that the effective Young's modulus has a peak at certain concentration of inclusions; when the matrix is nearly incompressible the effective Young's modulus exceeds that of the matrix more than 3 times.

The observed increase of effective Young's modulus is not surprising as the Young's modulus of inclusions and the matrix are assumed to coincide. Therefore, as the Poisson's ratio of inclusions tends to $-1$, the ratio between the shear and Young's moduli of inclusions tends to infinity. This situation can be further clarified by returning to the above example of anti-plane strain problem of composite with cylindrical inclusions. Extreme NPR corresponds to infinite shear modulus; performing limiting transition $G_1 \to \infty$ in (3.8) one obtains

$$\frac{dG}{dc} = \frac{2G}{1-c} \tag{3.11}$$

Its solution with the said initial conditions reads

$$\frac{G}{G_0} = \frac{1}{(1-c)^2}. \tag{3.12}$$

It is seen that this simple case has a singularity at $c = 1$, which elucidates the existence of the peak in the effective Young's modulus.

An increase of Young's modulus in composites with auxetic inclusions was seen in simulations performed by Assidi and Ganghoffer [16]. Shufrin et al. [222] considered composites with cubic inclusions and used the finite element method for computing the effective moduli. They obtained results similar to shown in Fig. 3.6a and showed that cubic inclusions produce significantly stronger increase in the effective Young's modulus than the spherical ones, while the difference in effective Poisson's ratios is minor. This was further confirmed by Pasternak et al. [194].

Interesting is also the opposite case when the extreme auxetic matrix is filled with positive Poisson's ratio inclusions, Fig. 3.6b. In this case a strong increase in the effective Young's modulus is observed as the inclusions tends to incompressible. Pasternak and Dyskin [187] demonstrated the existence of singularity as $\nu_1 \rightarrow 0.5$.

#### 3.2.2.2 Effective Coefficient of Thermal Expansion

Pasternak and Dyskin [187] put forward general arguments as to why the auxetic materials have reduced thermal stress as compared with simular positive Poisson's ratio material. Pasternak et al. [194] conducted finite element numerical simulations of an isotropic positive Poisson's ratio matrix with randomly distributed auxetic inclusions of cubic shape with positive or negative coefficient of thermal expansion (CTE) and determined the effective CTE and the thermal stress (see e.g., [23, 76, 97, 99, 105, 138, 157, 166, 170, 232] for structures producing negative CTE). It was assumed that the Young's moduli of inclusions and the matrix coincide. The model was calibrated against experimental results found in the literature. Also the results were compared with the Rosen-Hashin bounds, Rosen and Hashin [206] and found to be well within the bounds. It was found that the auxetic inclusions with positive CTE in the conventional matrix with positive Poisson's ratio are capable of reducing the thermal stress. Lim [155] investigated layered materials with alternating auxetic and positive Poisson's ratio isotropic layers with positive and negative CTE and showed that the combination of auxetic, positive CTE layers and positive Poisson's ratio, negative CTE layers lead to stronger negative effective CTE than the combination of auxetic, negative CTE layers and positive Poisson's ratio, positive CTE layers. Ha et al. [110] proposed to use chiral structures to ensure large thermal expansion.

### 3.2.3 Discussion

The above review shows that it is important to distinguish between the effects of microstructure producing the NPR, including non-elastic effects; energy absorption and various strength properties form the elastic effect of NPR per se. The latter is not profound and in isotropic case is tight to relationship (3.1). When the NPR is specified, the corresponding pairs of Young's and shear moduli can be found from the two-parametric space, Fig. 3.1. When the extreme NPR $= -1$ is considered, the effect strongly depends upon how limiting transition $G/E \to \infty$ is performed. If for instance $G$ is fixed and $E$ tends to zero the claimed improvements in e.g., fracture toughness and indentation resistance disappear. Only the reduction in thermal stresses survives, but the extent of the reduction does depend on the way the above limiting transition is performed. These considerations somewhat limit the applicability of auxetic elastic materials.

From this point of view it is advantageous to consider auxetic-inspired structures in their own right and the ensuing applications. In particular, in space engineering and construction engineering such structures belong to a class of deployable structures, as contraction in lateral directions induced by axial compression can be used to fold the structure into a compact configuration (e.g., [82, 196, 212, 228, 261]). Figure 3.7 shows an example of a deployable structure copied from Gantes et al. [82].

Another perspective application is in making composites (or hybrid materials) with auxetic phase (or phases). The presence of auxetic phase can modify various mechanical properties of the material, subsequently it is important to devise methods ensuring the optimal choice of the concentrations and properties of the auxetic phases needed to achieve the desired effective material characteristics in the spirit of Ashby and Bréchet [15]. In this respect, vital are the efficient methods of computing the effective characteristics (e.g. moduli, CTE). The above analysis shows that the differential scheme can provide a simple and sufficiently accurate method, which makes it a method of choice for in initial investigations with subsequent refinement using numerical simulations.

## 3.3 Negative Stiffness

### 3.3.1 Structures Exhibiting the Effect of Negative Stiffness

Term "negative stiffness" is used here to indicate the elastic materials with one or several moduli negative such that formally the elastic energy is no longer positive definite. The absence of positive definiteness means that a material with such a property cannot be in equilibrium on its own and is rather a mechanism that requires the presence of a special external system capable of providing stabilisation. In other words, the external system plays a role of energy reservoir, Dyskin and Pasternak [65], which makes the energy of the whole system (the material + the energy reservoir)

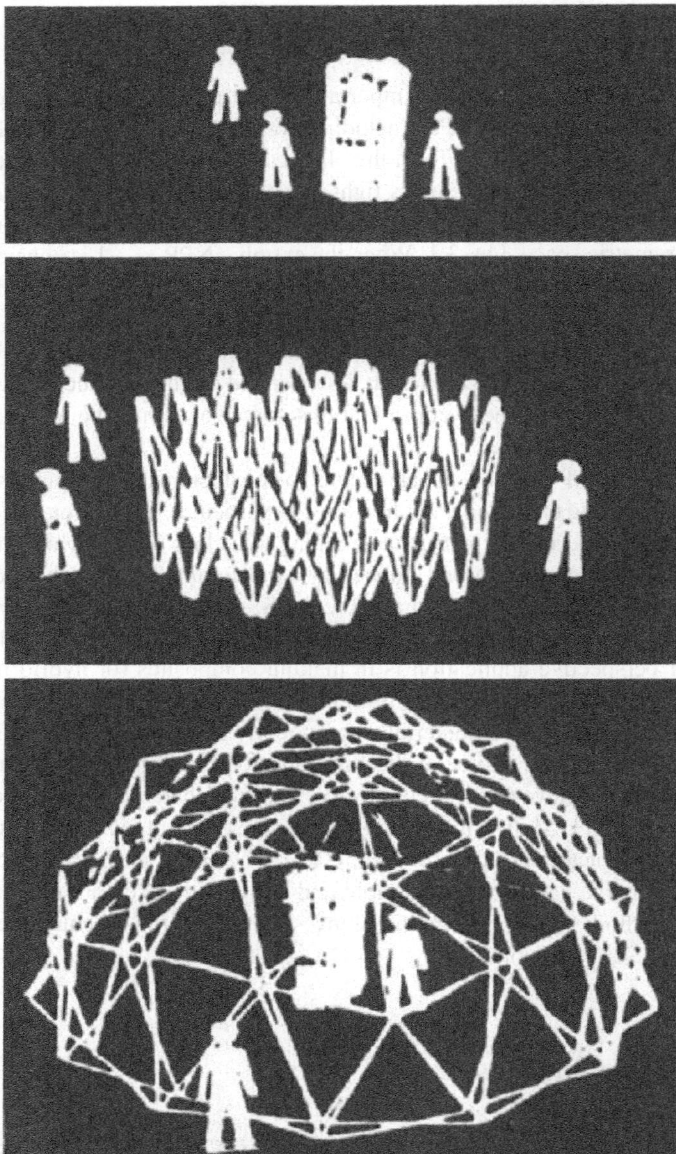

Fig. 3.7 A deployable dome, Gantes et al. [82]

positive definite. The first obvious example of such a combination is the post-peak stress-strain curve observed in uniaxial and some types of triaxial compression of heterogeneous materials such as rocks, soils and concretes. The descending character of the stress-strain curve, that is decrease of the stress with increase in the strain provides an example of incremental negative stiffness. It is well known (e.g., [54, 209]) that obtaining the post-peak softening branch requires a stiff loading frame otherwise the tested sample will be broken at the peak (in soils this failure is usually not observed owing to the considerable softness of the soil).

Further examples of structures exhibiting negative stiffness are given by various mechanisms made of levers and arches and contain preloaded elastic springs (e.g., [44, 45, 123, 145, 184, 211, 236, 248]) or arches [115, 200] needed to store the energy that gets released in loading thus exhibiting apparent negative stiffness. Indeed, if an equivalent spring modelling the negative stiffness system has stiffness $-k < 0$, then if the loading device increases force from $F$ to $F + \Delta F$ it will cause displacement increment $\Delta u = -\Delta F/k < 0$. Then work performed by the loading device on the displacement increment $\Delta u$ is $A = F\Delta u = -F\Delta F/k < 0$. This means that the energy pumped to the loading device should come from somewhere; that is why preloaded springs are necessary. (In the case of post-peak softening of a heterogeneous material in compression the energy gets released by damage accumulation, i.e. the fracture growth.)

A different type of negative stiffness structures is based on pre-buckled tubes and columns (e.g., [29, 81, 143]) and nanotubes [237, 271, 272], see also [58, 59, 129, 130]. In this case reaching the buckled state is accompanied by energy accumulation, whose release manifests itself as apparent negative stiffness. Similarly, elastomeric cellular solids, Lakes et al. [147], exhibit the negative stiffness due to buckling. It can be said that the appearance of negative stiffness is an indicator of the state of instability, buckling being one of them. For the same reason, the unstable regions in deformation of non-linear springs with non-convex energy (e.g. [43, 203]) show negative stiffness.

Another type of structures exhibiting negative stiffness is plate-like interlocking structures of cubic elements constrained by a frame under indentation as demonstrated experimentally and in numerical simulations [75, 217]. The role of energy storage here is played by the elasticity of stiff frame.

Dyskin and Pasternak [64–66, 69], Pasternak and Dyskin [188] suggested that apparent negative stiffness could be produced in rotating non-spherical grains in rocks (or blocks in rock mass [67, 68]) in the presence of compression. This mechanism is schematically explained in Fig. 3.8. It consists of a lever subjected to a horizontal displacement in the presence of a vertical compressive force. (The lever can be thought of as a suitable grain diameter drawn from a point of contact, point $O$ in Fig. 3.8a.) The force-displacement curve is shown in Fig. 3.8b. The descending branch clearly indicates the presence of incremental negative stiffness (the slope of the tangent line is proportional to magnitude of compressive force $P$). Furthermore the dependence is reversible and, in this set-up, time-independent, which makes it elastic (even nearly linear elastic). The energy balance is ensured by the work of the second loading device that provides compressive force $P$.

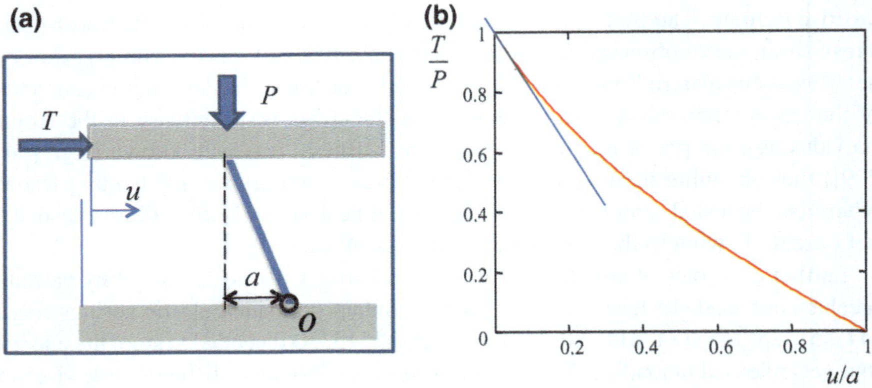


**Fig. 3.8** A mechanism exhibiting negative stiffness relating the shear force and displacement: **a** lever rotating about point $O$, subjected to normal (compressive) force $P$ and sear displacement $u$; **b** shear force-displacement diagram and a tangent line at $u = 0$

In living nature negative stiffness is hypothesised to characterise the deformation of hair-bundle in the ear providing an amplification mechanism in hearing [163, 255] and, in some cases, of joints [52, 152]. At microlevel, apparent negative stiffness at certain parts of the stress-strain curve can be exhibited during phase transformations accompanied by transitional deformation (e.g., [126, 208]). Here the energy is balanced by the energy of phase transformation. Lakes et al. [148] and Jaglinski et al. [125] suggested that ferromagnetic crystals that change volume due to phase transformation at certain temperatures could also exhibit negative stiffness.

## 3.4 Matrix with Negative Stiffness Inclusions

Using a stiff loading frame is not the only way to stabilise negative stiffness elements. Another method is to create a composite consisting of conventional positive stiffness matrix filled with negative stiffness elements (e.g., [44, 62, 65, 66, 133, 142, 143, 145, 148, 248, 254]). The stabilisation effect of the matrix obviously depends upon the concentration (volumetric fraction) of the negative stiffness elements. On top of that Kochmann and Drugan [136] proposed to increase the range of stability by rotating the whole composite. Jaglinski and Lakes [124] associated instability with the composite reaching negative stiffness. The implementation of this principle was based on Hashin-Strikman bounds.

The notion that matrix filled with negative stiffness inclusions can possess unusual mechanical and physical properties was widely discussed in the literature involving theoretical reasoning, numerical simulations and experimentation (e.g., [60, 125, 127, 148, 184, 251, 269]). Such materials were anticipated to have elevated stiffness, thermal expansion and damping [55, 56, 125, 142, 145, 245, 247, 249, 250, 252, 253,

272] and can provide sound attenuation (e.g., [269] and vibration isolation (e.g., [184, 199]).

Lakes [142, 143] and Lakes et al. [148] using lower and upper bounds to estimate the effective Young's modulus came to a conclusion that the negative stiffness inclusions could increase the effective stiffness of the composite as compared to the initial stiffness of the matrix. It should be noted however that the derivation of the bounds relies on positive definiteness of elastic energy (e.g., [116]), which can be violated within the material of negative stiffness inclusions. Kochmann and Milton [134] showed that the bounds are valid as long as the composite with negative stiffness inclusions is stable). Lakes and Drugan [145] obtained the effect of increase in effective stiffness in a matrix with spheres of negative stiffness using the approximation of non-interacting inclusions and found a critical concentration when the effective Young's modulus becomes infinite and then abruptly becomes negative. A similar behaviour (without involving the infinite modulus) was predicted by Wang and Lakes [248] by numerically simulating a system consisting of one positive and one negative stiffness element. Wang [245] considered a double hexagonal structure with positive stiffness external hexagon and negative stiffness internal hexagon and showed that depending on the value of the negative stiffness there exist zones of increased effective stiffness followed by abrupt transition to negative effective stiffness.

Dyskin and Pasternak [63, 65] analysed the effect of negative stiffness inclusions on the effective elastic properties of such composites using a simple model of anti-plane strain of the composite consisting of a conventional isotropic matrix of shear modulus $G_0$ with cylindrical inclusions of negative shear modulus. It was shown that despite possible non-uniqueness of the corresponding elastic solution, the average stress and strain and hence the effective modulus can be determined uniquely. The computation of the effective shear modulus was based on the differential scheme, (3.8) with $G_1 = -mG_0 < 0$. Figure 3.9 reproduces the results. It is seen that there exists critical concentration representing the transition from positive to negative effective shear modulus, however opposite to the previously believed the effective shear modulus can either increase, when the inclusions have high value of negative shear modulus or decrease, when the value of negative shear modulus of inclusions is low compared to the modulus of the matrix. Also, no unphysical infinite modulus is observed; the singularity is probably an artefact of approximation of non-interacting inclusions used by Lakes and Drugan [145]. It can be concluded that the interaction between the inclusions is always important and should not be neglected. After passing the critical concentration the effective shear modulus abruptly becomes negative; its value depends on parameter $B$ controlled by the loading device that ensures the stability of the negative stiffness composite (see [65] for details). Also it was found that when $m = 1$ the effective shear modulus is negative for all concentrations of inclusions. Dyskin and Pasternak [65, 66], following Jaglinski and Lakes [124] associated this state with instability.

Dyskin and Pasternak [66] considered conventional isotropic matrix with spherical inclusions of negative shear modulus whose bulk modulus equal to that of the matrix and found qualitatively similar behaviour. They also found the same type of behaviour

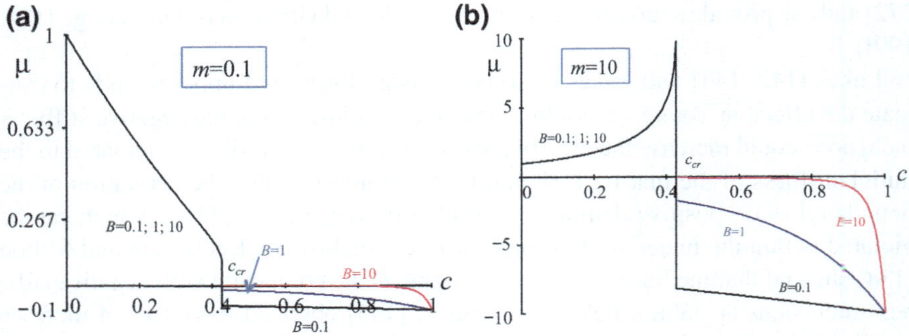


**Fig. 3.9** Normalised effective shear modulus of an isotropic matrix of shear modulus $G_0$ with different negative shear modulus cylindrical inclusions versus volumetric fraction (concentration) $c$ of inclusions: **a** low negative stiffness inclusions, $m = -G_1/G_0 = 0.1$; **b** high negative stiffness inclusions, $m = -G_1/G_0 = 10$

of matrix with disc-like shear cracks with negative factor relating the applied shear load and average displacement of the crack faces.

The stability of composites with negative stiffness inclusions was further investigated by Pasternak et al. [193], Dyskin and Pasternak [70] and Pasternak and Dyskin [191]. This was accomplished by using a simple model of a chain of springs with fixed ends. It was shown that in a stable chain no more than one spring could be negative with the value of negative stiffness not exceeding a certain critical value determined by the stiffnesses of all other springs. Furthermore, when the negative stiffness reaches its critical value one of the eigenfrequencies becomes zero. In systems with light viscous damping vanishing of an eigenfrequency does not yet lead to instability. Further increase in the value of negative stiffness leads to the appearance of aperiodic eigenmodes with light and high damping. At the critical negative stiffness the low dissipative mode becomes non-dissipative, while in the high dissipative mode the damping coefficient becomes as twice as high as the damping coefficient of the system. Therefore, negative stiffness inclusions can both increase and decrease damping, opposite to some claims in the literature that stiffness inclusions just increase damping and energy absorption (e.g., [125, 142, 145, 245, 247, 249, 250, 252, 272]). Esin et al. [72, 73] extended this analysis to some 1D systems with longitudinal, shear and rotational springs and 2D systems.

Pasternak et al. [195] by analysing the model of rotating non-spherical particles (see Fig. 3.8a, where the lever corresponds to a diameter of the particle) noticed that the dependence on Fig. 3.8b could be interpreted as negative factor between the shear force and rotation angle instead of displacement. When the material particles are able to rotate then its continuum model, is the Cosserat or micropolar continuum (e.g., [182]) in the case of isotropic matrix an randomly situated particles the Cosserat continuum is isotropic. However when the particles are not spherical, their rotation in the presence of compression will lead to a Cosserat continuum with negative Cosserat shear modulus that relates the non-symmetrical parts of Cosserat strain and shear stress. Based on that notion Pasternak et al. [195] investigated the wave propagation in such a continuum and found that the material is stable and the waves can propagate as long as the value of the negative shear modulus does not exceed the shear modulus. Furthermore, all 4 types of Cosserat waves (p-wave, two s-waves and a twist wave) can propagate at any frequencies (in the conventional positive modulus isotropic Cosserat continuum the second s-wave and the twist wave only exist at frequencies exceeding a certain threshold).

## 3.5 Discussion

The structures exhibiting negative stiffness (apparent negative stiffness) are unstable but can be stabilised by an external system stiff enough to make the total energy positive definite. Yet, within the negative stiffness material or structure the energy is not positive definite which can induce some form of internal instability like strain localisation observed in stiff compressive loading of rocks, soils or concrete. The role of the stabilising system can also be played by the matrix, if its stiffness is high enough and the concentration of inclusions is not very high.

The negative stiffness structures can be classified into two types: (1) structures that have pre-loaded energy elements (e.g. structures with pre-loaded springs or pre-buckled tubes, materials with phase transformation) and (2) structures that use an externally applied (compressive) load such that the energy is balanced by the work of an additional loading device; e.g. the lever system, Fig. 3.8 and rotation of non-spherical particles. In type (2) systems the value of negative stiffness depends upon the compressive force applied. Therefore the compressive force can be used to control negative stiffness. This opens up an avenue of designing hybrid materials with controllable properties based on type (2) negative stiffness systems, Pasternak et al. [195]. In particular, controllable are the effective stiffness, eigenmodes and damping; especially interesting effects can be expected when the value of negative stiffness is such that an eigenmodes is near zero. Furthermore this control can be flexible and operate in real time.

## 3.6 Conclusions

Auxetic materials – materials with negative Poisson's ratio(s) are elastic materials with moduli within admissible ranges. The effect of negative Poisson's ratio(s) on the mechanical behaviour strongly depends upon the combination of the other elastic moduli (e.g. Young's and shear moduli in isotropic case). For instance the fracture toughness or indentation resistance can both increase or stay the same depending upon the path in the parametric space, Fig. 3.1, that leads to the specified negative Poisson's ratio. The same is valid for composites containing negative Poisson's ratio inclusions. The experimental and numerical results suggesting extreme effects mostly concern non-elastic regimes and relate to the behaviour of particular structures that within the elastic regime exhibit negative Poisson's ratio(s). Such structures, including deployable structures have a great importance per se and numerous interesting applications, which however cannot be reduced to negative Poisson's ratio(s).

Negative stiffness on the other hand is a convenient measure characterising various structures (rather mechanisms or being brought to the stage when they become mechanisms). These structures being immersed in a matrix (made of a conventional material) can provide markedly different stiffnesses, eigenmodes and energy absorption characteristics. Most important is that some of the structures rely on additional external load, which can be used to control the properties of the composite (or rather hybrid) even in real time. Such structures can of course be modelled in full without resorting to the negative stiffness concept, but the notion of negative stiffness may simplify the analysis.

An efficient approximate method for the determination of effective moduli of matrix with negative Poisson's ratio or negative stiffness inclusions is the differential scheme based on consecutive immersing small portions of inclusions in the equivalent medium with effective moduli determined on the previous step.

Further development in mathematical modelling of composites and hybrids with negative stiffness elements can provide tools for designing materials with unusual and controllable properties.

## References


1. K.L. Alderson, K.E. Evans, The fabrication of microporous polyethylene having a negative Poisson's ratio. Polymer **33**(20), 4435–4438 (1992)
2. K.L. Alderson, K.E. Evans, Modelling concurrent deformation mechanism in auxetic microporous polymers. J. Mater. Sci. **32**, 2797–2809 (1997)
3. K.L. Alderson, K.E. Evans, Auxetic materials: the positive side of being negative. Eng. Sci. Educ. J. **9**(4), 148–154 (2000)
4. A. Alderson, K.E. Evans, Rotation and dilation deformation mechanism for auxetic behaviour in the $\alpha$-cristolobite tetrahedral framework structure. Phys. Chem. Miner. **28**(10), 711–718 (2001)
5. A. Alderson, K.E. Evans, Molecular origin of auxetic behavior in tetrahedral framework silicates. Phys. Rev. Lett. **89**, 225503 (2002)
6. A. Alderson, K.L. Alderson, Auxetic materials. Proc. Inst. Mech. Eng. Part G: J. Aerosp. Eng. **221**(4), 565–575 (2007)
7. K.L. Alderson, V.R. Simkins, V.L. Coenen, P.J. Davies, A. Alderson, K.E. Evans, How to make auxetic fibre reinforced composites. Phys. Status Solidi B **242**(3), 509–518 (2005)
8. K.L. Alderson, V.L. Coenen, The low velocity impact response of auxetic carbon fibre laminates. Phys. Status Solidi B **245**(3), 489–496 (2008)
9. K.L. Alderson, A.P. Pickles, P.J. Neale, K.E. Evans, Auxetic polyethylene: the effect of a negative Poisson's ratio on hardness. Acta Metall. Mater **42**(7), 2261–2266 (1994)
10. K.L. Alderson, A. Fitzgerald, K.E. Evans, The strain dependent indentation resilience of auxetic microporous polyethylene. J. Mater. Sci. **35**, 4039–4047 (2000)
11. K.L. Alderson, R.S. Webber, A.P. Kettle, K.E. Evans, Novel fabrication route for auxetic polyethylene. Part 1. Process. Microstruct. Polym. Eng. Sci. **45**(4), 568–578 (2005)

12. K. Alderson, A. Alderson, N. Ravirala, V. Simkins, P. Davies, Manufacture and characterisation of thin flat and curved auxetic foam sheets. Phys. Status Solidi (B) **249**(7), 1315–1321 (2012). https://doi.org/10.1002/pssb.201084215
13. R.F. Almgren, An isotropic three-dimensional structure with Poisson's ratio $= -1$. J. Elast. **15**, 427–430 (1985)
14. O. Andersen, U. Waag, L. Schneider, G. Stephani, B. Kieback, Novel metallic hollow sphere structures. Adv. Eng. Mater. **2**(4), 192–195 (2000)
15. M.F. Ashby, Y.J.M. Bréchet, Designing hybrid materials. Acta Mater. **51**, 5801–5821 (2003)
16. M. Assidi, J.-F. Ganghoffer, Composites with auxetic inclusions showing both an auxetic behaviour and enhancement of their mechanical properties. Compos. Struct. **94**, 2373–2382 (2012)
17. D. Attard, J.N. Grima, Modelling of hexagonal honeycombs exhibiting zero Poisson's ratio. Phys. Status Solidi B **248**(1), 52–59 (2011)
18. D. Attard, E. Manicaro, J.N. Grima, On rotating rigid parallelograms and their potential for exhibiting auxetic behaviour. Phys. Status Solidi B **246**(9), 2033–2044 (2009)
19. D. Attard, E. Manicaro, R. Gatt, J.N. Grima, On the properties of auxetic rotating stretching squares. Phys. Status Solidi B **246**(9), 2045–2054 (2009)
20. K.M. Azzopardi, G.-P. Brincat, J.N. Grima, R. Gatt, Advances in the study of deformation mechanism of stishovite. Phys. Status Solidi B **252**, 1486–1491 (2015)
21. S. Babaee, J. Shim, J.C. Weaver, E.R. Chen, N. Patel, K. Bertoldi, 3D soft metamaterials with negative Poisson's ratio. Adv. Mater. **25**, 5044–5049 (2013)
22. E. Bafekrpour, A.V. Dyskin, E. Pasternak, A. Molotnikov, Y. Estrin, Internally architectured materials with directionally asymmetric friction. Sci. Rep. **5**, 10732 (2015)
23. G.D. Barrera, J.A.O. Bruno, T.H.K. Barron, N.L. Allan, Negative thermal expansion: a topical review. J. Phys.: Condens. Matter **17**, R217–R252 (2005)
24. R. Bathurst, L. Rothenburg, Note on a random isotropic granular material with negative Poisson's ratio. Int. J. Eng. Sci. **26**, 373–383 (1988)
25. R.H. Baughman, D.S. Galvao, Crystalline networks with unusual predicted mechanical and thermal properties. Nature **365**, 735–737 (1993)
26. R.H. Baughman, J.M. Shacklette, A.A. Zakhidov, S. Stafström, Negative Poisson's ratios as a common feature of cubic metals. Nature **392**, 362–365 (1998)
27. R.H. Baughman, S. Stafström, C. Cui, S.O. Dantas, Materials with negative compressibilities in one or more dimensions. Science **279**, 1522–1524 (1998)
28. R.H. Baughman, S.O. Socrates, S. Stafström, A.A. Zakhidov, T.B. Mitchell, D.H.E. Dubin, Negative Poisson's ratios for extreme states of matter. Science **288**, 2018–2022 (2000)
29. Z.P. Bažant, L. Cedolin, *Stability of Structures: Elastic, Inelastic, Fracture, and Damage Theories* (Oxford University Press, Oxford, 1991)
30. K. Bertoldi, P.M. Reis, S. Willshaw, T. Mullin, Negative Poisson's ratio behavior induced by an elastic instability. Adv. Mater. **22**, 361–366 (2010)
31. A. Bezazi, F. Scarpa, Tensile fatigue of conventional and negative Poisson's ratio open cell PU foams. Int. J. Fatigue **31**, 488–494 (2009)
32. M. Bianchi, F. Scarpa, M. Banse, C.W. Smith, Novel generation of auxetic open cell foams for curved and arbitrary shapes. Acta Mater. **59**, 686–691 (2011)
33. R. Blumenfeld, Auxetic strains—insight from iso-auxetic materials. Mol. Simul. **31**(13), 867–871 (2005)
34. R. Blumenfeld, S.F. Edwards, Theory of strains in auxetic materials. J. Supercond. Novel Magn. **25**(3), 565–571 (2012)
35. A.C. Brańka, K.W. Wojciechowski, Auxeticity of cubic materials: the role of repulsive core interaction. J. Non-Cryst. Solids **354**, 4143–4145 (2008)
36. A.C. Brańka, D.M. Heyes, K.W. Wojciechowski, Auxeticity of cubic materials under pressure. Phys. Status Solidi B **248**, 96–104 (2011)
37. W.M. Bruner, Comment on 'Seismic velocities in dry and saturated cracked solids' by Richard J. O'Connell and Bernard Budiansky. J. Geophys. Res. **81**(14), 2573–2576 (1976)

38. B. Budiansky, On the elastic moduli of some heterogeneous materials. J. Mech. Phys. Solids **13**(4), 223–227 (1965)
39. B. Budiansky, R.J. O'Connell, Elastic moduli of a cracked solid. Int. J. Solid Struct. **12**(2), 81–97 (1976)
40. L. Cabras, M. Brun, Auxetic two-dimensional lattices with Poisson's ratio arbitrarily close to −1. Proc. R. Soc. London. Sect. B **470**, 20140538 (2014)
41. L. Cabras, M. Brun, Effective properties of a new auxetic triangular lattice: an analytical approach. Frattura ed Integrità Strutturale **29**, 9–18 (2014)
42. B.D. Caddock, K.E. Evans, Microporous materials with negative Poisson's ratios. I. Microstructure and mechanical properties. J. Phys. D Appl. Phys. **22**, 1877–1882 (1989)
43. F. Cardin, M. Favretti, Dynamics of a chain of springs with non-convex potential energy. Math. Mech. Solids **8**, 651–668 (2003)
44. A. Carrella, T.P. Brennan, T.P. Waters, Demonstrator to show the effects of negative stiffness on the natural frequency of a simple oscillator. Proc. IMechE Part C J. Mech. Eng. Sci. **222**, 1189–1192 (2008)
45. A.R. Champneys, G.W. Hunt, J.M.T. Thompson, *Localisation and Solitary Waves in Solid Mechanics* (World Scientific, Singapore, 1984), pp. 1–28
46. N. Chan, K.E. Evans, Fabrication methods for auxetic foams. J. Mater. Sci. **32**, 5945–5953 (1997)
47. L. Chen, C. Liu, J. Wang, W. Zhang, C. Hu, S. Fan, Auxetic materials with large negative Poisson's ratios based on highly oriented carbon nanotube structures. Appl. Phys. Lett. **94**, 253111 (2009)
48. Y.J. Chen, F. Scarpa, I.R. Farrow, Y.J. Liu, J.S. Leng, Composite flexible skin with large negative Poisson's ratio range: numerical and experimental analysis. Smart Mater. Struct. **22**, 045005 (2013)
49. J.B. Choi, R.S. Lakes, Nonlinear properties of metallic cellular materials with a negative Poisson's ratio. J. Mater. Sci. **27**(19), 5375–5381 (1992)
50. J.B. Choi, R.S. Lakes, Nonlinear properties of polymer cellular materials with a negative Poisson's ratio. J. Mater. Sci. **27**(19), 4678–4684 (1992)
51. J.B. Choi, R.S. Lakes, Fracture toughness of re-entrant foam materials with a negative Poisson's ratio: experiment and analysis. Int. J. Fract. **80**, 73–83 (1996)
52. A. Choi, T. Sim, J.H. Mun, Quasi-stiffness of the knee joint in flexion and extension during the golf swing. J. Sports Sci. **33**(16), 1682–1691 (2015)
53. V.L. Coenen, K.L. Alderson, Mechanisms of failure in the static indentation resistance of auxetic carbon fibre laminates. Phys. Status Solidi B **248**(1), 66–72 (2011)
54. N.G.W. Cook, The failure of rock. Int. J. Rock Mech. Min. Sci. **2**, 389–403 (1965)
55. D.M. Correa, T.D. Klatt, S. Cortes, M. Haberman, D. Kovar, C. Seepersad, Negative stiffness honeycombs for recoverable shock isolation. Rapid Prototyping J. **21**(2), 193–200 (2015)
56. D.M. Correa, C. Seepersad, M. Haberman, Mechanical design of negative stiffness honeycomb materials. Integrating Mater. Manuf. Innov. **4**(1), 1–11 (2015)
57. S. Czarnecki, P. Wawruch, The emergence of auxetic material as a result of optimal isotropic design. Phys. Status Solidi B **252**, 1620–1630 (2015)
58. L. Dong, R.S. Lakes, Advanced damper with negative structural stiffness elements. Smart Mater. Struct. **21**(7), 075026 (2012)
59. L. Dong, R.S. Lakes, Advanced damper with high stiffness and high hysteresis damping based on negative structural stiffness. Int. J. Solids Struct. **50**(14–15), 2416–2423 (2013)
60. L. Dong, D.S. Stone, R.S. Lakes, Anelastic anomalies and negative Poisson's ratio in tetragonal $BaTiO_3$ ceramics. Appl. Phys. Lett. **96**, 141904 (2010)
61. J.P. Donoghue, K.L. Alderson, K.E. Evans, The fracture toughness of composite laminates with a negative Poisson's ratio. Phys. Status Solidi B **246**, 2011–2017 (2009)
62. W.J. Drugan, Elastic composite materials having a negative stiffness phase can be stable. Phys. Rev. Lett. **98**(5), 055502 (2007)

63. A.V. Dyskin, E. Pasternak, Effective anti-plane shear modulus of a material with negative stiffness inclusions, in *9th HSTAM10, Limassol, Cyprus 12–14 July, 2010, Vardoulakis mini-symposia—Wave Propagation*, paper 116, ed. by P. Papanastasiou, E. Papamichos, A. Zervos, M. Stavropoulou (2010), pp. 129–136
64. A.V. Dyskin, E. Pasternak, Friction and localisation associated with non-spherical particles, in *Advances in Bifurcation and Degradation in Geomaterials. Proceedings of the 9th International Workshop on Bifurcation and Degradation in Geomaterials*, ed. by S. Bonelli, C. Dascalu, F. Nicot (Springer, 2011), pp. 53–58. ISBN/ISSN 978-94-007-1420-5
65. A.V. Dyskin, E. Pasternak, Elastic composite with negative stiffness inclusions in antiplane strain. Int. J. Eng. Sci. **58**, 45–56 (2012)
66. A.V. Dyskin, E. Pasternak, Mechanical effect of rotating non-spherical particles on failure in compression. Phil. Mag. **92**(28–30), 3451–3473 (2012)
67. A.V. Dyskin, E. Pasternak, Rock mass instability caused by incipient block rotation, in *Harmonising Rock Engineering and the Environment, Proceedings of 12th International Congress on Rock Mechanics*, ed. by Q. Qian, Y. Zhou (CRC Press, Balkema, 2012c), pp. 201–204
68. A.V. Dyskin, E. Pasternak, Rock and rock mass instability caused by rotation of non-spherical grains or blocks, in *Rock Engineering & Technology for Sustainable Underground Construction. Proceedings of Eurock 2012*, paper 102P (2012d)
69. A.V. Dyskin, E. Pasternak, Bifurcation in rolling of non-spherical grains and fluctuations in macroscopic friction. Acta Mech. **225**(8), 2217–2226 (2014)
70. A.V. Dyskin, E. Pasternak, Negative stiffness: Is thermodynamics defeated? in *Proceedings of 8th Australasian Congress on Applied Mechanics: ACAM 8* (Melbourne, Australia, 2014)
71. B. Ellul, M. Muscat, J.N. Grima, On the effect of the Poisson's ratio (positive and negative) on the stability of pressure vessel heads. Phys. Status Solidi (B) **246**(9), 2025–2032 (2009). https://doi.org/10.1002/pssb.200982033
72. M. Esin, E. Pasternak, A. Dyskin, Stability of chains of oscillators with negative stiffness normal, shear and rotational springs. Int. J. Eng. Sci. **108**, 16–33 (2016)
73. M. Esin, E. Pasternak, A.V. Dyskin, Stability of 2D discrete mass-spring systems with negative stiffness springs. Phys. Status Solidi (B) Basic Res. **253**(7), 1395–1409 (2016b)
74. U.E. Essien, A.O. Akankpo, M.U. Igboekwe, Poisson's ratio of surface soils and shallow sediments determined from seismic compressional and shear wave velocities. Int. J. Geosci. **5**(12), 1540–1546 (2014)
75. Y. Estrin, A.V. Dyskin, E. Pasternak, S. Schaare, S. Stanchits, A.J. Kanel-Belov, Negative stiffness of a layer with topologically interlocked elements. Scripta Mater. **50**(2), 291–294 (2003)
76. J.S.O. Evans, T.A. Mary, A.W. Sleight, Negative thermal expansion materials. Phys. B **241–243**, 311–316 (1998)
77. K.E. Evans, K.L. Alderson, The static and dynamic moduli of auxetic microporous polyethylene. J. Mater. Sci. **11**, 1721–1724 (1992)
78. K.E. Evans, A. Alderson, Auxetic materials: Functional materials and structures from lateral thinking! Adv. Mater. **12**(9), 617–628 (2000)
79. K.E. Evans, B.D. Caddock, Microporous materials with negative Poisson's ratios: II. Mechanisms and interpretation. J. Phys. D: Appl. Phys. **22**(12), 1883–1887 (1989)
80. K.E. Evans, M.A. Nkansah, I.J. Hutchinson, S.C. Rogers, Molecular network design. Nature **353**, 124 (1991)
81. B.A. Fulcher, D.W. Shahan, M.R. Haberman, C.C. Seepersad, P.S. Wilson, Analytical and experimental investigation of buckled beams as negative stiffness elements for passive vibration and shock isolation systems. J. Vibr. Acoust.-Trans. ASME **136**(3), 31009 (2014)
82. C.J. Gantes, J.J. Connor, L.D. Logcher, Y. Rosenfeld, Structural analysis and design of deployable structures. Comput. Struct. **32**(3–4), 661–669 (1989)
83. N. Gaspar, X.J. Ren, C.W. Smith, J.N. Grima, K.E. Evans, Novel honeycombs with auxetic behaviour. Acta Mater. **53**, 2439–2445 (2005)
84. N. Gaspar, C.W. Smith, K.E. Evans, Auxetic behaviour and anisotropic heterogeneity. Acta Mater. **57**, 875–880 (2009)

85. N. Gaspar, A granular material with a negative Poisson's ratio. Mech. Mater. **42**, 673–677 (2010)
86. N. Gaspar, C.W. Smith, A. Alderson, J.N. Grima, K.E. Evans, A generalised three-dimensional tethered-nodule model for auxetic materials. J. Mater. Sci. **46**, 372–384 (2011)
87. R. Gatt, L. Mizzi, K.M. Azzopardi, J.N. Grima, A force-field based analysis of the deformation mechanism in $\alpha$-cristobalite. Phys. Status Solidi B **252**, 1479–1485 (2015)
88. L.N. Germanovich, A.V. Dyskin, Virial expansions in problems of effective characteristics. Part I. General concepts. J. Mech. Compos. Mater. **30**(2), 222–237 (1994)
89. L.N. Germanovich, A.V. Dyskin, Virial expansions in problems of effective characteristics. Part II. Anti-plane deformation of fibre composite. Analysis of self-consistent methods. J. Mech. Compos. Mater. **30**(2), 234–243 (1994)
90. C.M. Gerrard, Elastic models of rock masses having one, two and three sets of joints. Int. J. Rock Mech. Mining Sci. Geomech. Abstr. **19**(1), 15–23 (1982). https://doi.org/10.1016/0148-9062(82)90706-9
91. G.N. Greaves, A.L. Greer, R.S. Lakes, T. Rouxel, Poisson's ratio and modern materials. Nat. Mater. **10**, 823–837 (2011)
92. J.N. Grima, K.E. Evans, Auxetic behavior from rotating triangles. J. Mater. Sci. **41**, 3193–3196 (2006)
93. J.N. Grima, R. Gatt, Perforated sheets exhibiting negative Poisson's ratios. Adv. Eng. Mater. **12**(6), 460–464 (2010). https://doi.org/10.1002/adem.201000005
94. J.N. Grima, R. Jackson, A. Alderson, K.E. Evans, Do zeolites have negative Poisson's ratios? Adv. Mater. **12**(24), 1912–1918 (2000)
95. J.N. Grima, A. Alderson, K.E. Evans, Auxetic behaviour from rotating rigid units. Phys. Status Solidi B **242**(3), 561–575 (2005)
96. J.N. Grima, R. Gatt, N. Ravirala, A. Alderson, K.E. Evans, Negative Poisson's ratios in cellular foam materials. Mater. Sci. Eng., A **423**(1–2), 214–218 (2006)
97. J.N. Grima, P.S. Farrugia, R. Gatt, V. Zammit, A system with adjustable positive or negative thermal expansion. Proc. R. Soc. Lond. A: Math. Phys. Eng. Sci. **463**, 1585–1596 (2007)
98. J.N. Grima, R. Caruana-Gauci, M.R. Dudek, K.W. Wojciechowski, R. Gatt, Smart metamaterials with tunable auxetic and other properties. Smart Mater. Struct. **22**, 084016 (2013)
99. J.N. Grima, B. Ellul, R. Gatt, D. Attard, Negative thermal expansion from disc, cylindrical, and needle shaped inclusions. Phys. Status Solidi B **250**, 2051–2056 (2013)
100. J.N. Grima, R. Cauchi, R. Gatt, D. Attard, Honeycomb composites with auxetic out-of-plane characteristics. Compos. Struct. **106**, 150–159 (2013)
101. J.N. Grima, S. Winczewski, L. Mizzi, M.C. Grech, R. Cauchi, R. Gatt, Tailoring graphene to achieve negative Poisson's ratio properties. Adv. Mater. **27**, 1455–1459 (2015)
102. J.N. Grima, L. Mizzi, K.M. Azzopardi, R. Gatt, Auxetic perforated mechanical metamaterials with randomly oriented cuts. Adv. Mater. **28**(2), 385–389 (2015)
103. J.N. Grima, D. Attard, R. Gatt, R.N. Cassar, A novel process for the manufacture of auxetic foams and for their re-conversion to conventional form. Adv. Eng. Mater. **11**(7), 533–535 (2009)
104. J.N. Grima, R.N. Cassar, R. Gatt, On the effect of hydrostatic pressure on the auxetic character of NAT-type silicates. J. Non-Cryst. Solids **355**, 1307–1312 (2009)
105. J.N. Grima, B. Ellul, R. Gatt, D. Attard, Negative thermal expansion from disc cylindrical, and needle shaped inclusions. Phys. Status Solidi B **250**(10), 2051–2056 (2012)
106. J.N. Grima, E. Chetcuti, E. Manicaro, D. Attard, M. Camilleri, R. Gatt, K.E. Evans, On the auxetic properties of generic rotating rigid triangles. Proc. R. Soc. A **468**, 810–830 (2012)
107. J.N. Grima, E. Manicaro, D. Attard, Auxetic behaviour from connected different-sized squares and rectangles. Proc. R. Soc. A **467**, 439–458 (2010)
108. D.J. Gunton, G.A. Saunders, The Young's modulus and Poisson's ratio of arsenic, antimony and bismuth. J. Mater. Sci. **7**, 1061–1068 (1972)
109. Y. Guo, W.A. Goddard III, Is carbon nitride harder than diamond? No, but its girth increases when stretched (negative Poisson ratio). Chem. Phys. Let. **237**, 72–76 (1995)

110. C.S. Ha, E. Hestekin, J. Li, M.E. Plesha, R.S. Lakes, Controllable thermal expansion of large magnitude in chiral negative Poisson's ratio lattices. Phys. Status Solidi B **252**, 1431–1434 (2015)
111. L.J. Hall, V.R. Coluci, D.S. Galvão, M.E. Kozlov, M. Zhang, S.O. Dantas, R.H. Baughman, Sign change of Poisson's ratio for carbon nanotube sheets. Science **320**, 504–507 (2008)
112. Z. Hashin, S. Shtrikman, A variational approach to the theory of the elastic behaviour of multiphase materials. J. Mech. Phys. Solids **11**(2), 127–140 (1963)
113. E.W. Hawkes, E.V. Eason, A.T. Asbeck, M.R. Cutkosky, IEEE/ASME Trans. Mechatron. **18**, 518 (2013)
114. G.F. Hawkins, M.J. O'Brien, C.Y. Tang, Proceedings of SPIE. Smart Mater. III **5648**, 37 (2004)
115. T.A.M. Hewage, K.L. Alderson, A. Alderson, F. Scarpa, Double-negative mechanical metamaterials displaying simultaneous negative stiffness and negative Poisson's ratio properties. Adv. Mater. **28**, 10323–10332 (2016)
116. R. Hill, The elastic behaviour of a crystalline aggregate. Proc. Phys. Soc. (Lond.) **A65**, 349–354 (1952)
117. R. Hill, Elastic properties of reinforced solids: some theoretical principles. J. Mech. Phys. Solids **11**, 357–372 (1963)
118. S. Hirotsu, Softening of bulk modulus and negative Poisson's ratio near the volume phase transition of polymer gels. J. Chem. Phys. **94**(5), 3949–3957 (1991). https://doi.org/10.1063/1.460672
119. D.T. Ho, H. Kim, S.-Y. Kwon, S.Y. Kim, Auxeticity of face-centered cubic metal (001) nanoplates. Phys. Status Solidi B **252**, 1492–1501 (2015)
120. X. Hou, H. Hu, V. Silberschmidt, A novel concept to develop composite structures with isotropic negative Poisson's ratio: effects of random inclusions. Compos. Sci. Technol. **72**, 1848–1854 (2012)
121. X. Hou, H. Hu, V. Silberschmidt, Numerical analysis of composite structure with in-plane isotropic negative Poisson's ratio: effects of materials properties and geometry features of inclusions. Compos.: Part B **58**, 152–159 (2014)
122. T.P. Hughes, A. Marmier, K.E. Evans, Auxetic frameworks inspired by cubic crystals. Int. J. Solids Struct. **47**, 1469–1476 (2010)
123. G.W. Hunt, H.-B. Mühlhaus, A.I.M. Whiting, Folding processes and solitary waves in structural geology, in *Localisation and Solitary Waves in Solid Mechanics*, ed. by A.R. Champneys, G.W. Hunt, J.M.T. Thompson (World Scientific, Singapore, 1984), pp. 332–348
124. T. Jaglinski, R.S. Lakes, Anelastic instability in composites with negative stiffness inclusions. Philos. Mag. Lett. **84**(12), 803–810 (2004)
125. T. Jaglinski, D. Kochmann, D. Stone, R.S. Lakes, Composite materials with viscoelastic stiffness greater than diamond. Science **315**, 620–622 (2007)
126. T. Jaglinski, P. Frascone, B. Moore, D.S. Stone, R.S. Lakes, Internal friction due to negative stiffness in the indium–thallium martensitic phase transformation. Phil. Mag. **86**(27), 4285–4303 (2006)
127. T. Jaglinski, D. Stone, R.S. Lakes, Internal friction study of a composite with a negative stiffness constituent. J. Mater. Res. **20**(09), 2523–2533 (2005)
128. S. Jayanty, J. Crowe, L. Berhan, Auxetic fibre networks and their composites. Phys. Status Solidi B **248**(1), 73–81 (2011)
129. H. Kalathur, R.S. Lakes, Column dampers with negative stiffness: high damping at small amplitude. Smart Mater. Struct. **22**(8), 084013 (2013)
130. H. Kalathur, T.M. Hoang, R.S. Lakes, W.J. Drugan, Buckling mode jump at very close load values in unattached flat-end columns: theory and experiment. J. Appl. Mech.-Trans. ASME **81**(4), 41010 (2014)
131. P. Kanouté, D.P. Boso, J.L. Chaboche, B.A. Schrefler, Multiscale methods for composites: a review. Arch. Comput. Methods Eng. **16**, 31–75 (2009)
132. N. Keskar, J.R. Chelikowsky, Negative Poisson's ratio in crystalline $SiO_2$ from first-principles calculations. Nature **358**, 222–224 (1992)

133. D.M. Kochmann, W.J. Drugan, Dynamic stability analysis of an elastic composite material having a negative-stiffness phase. J. Mech. Phys. Solids **57**(7), 1122–1138 (2009)
134. D.M. Kochmann, G.W. Milton, Rigorous bounds on the effective moduli of composites and inhomogeneous bodies with negative-stiffness phases. J. Mech. Phys. Solids **71**, 46–63 (2014)
135. D.M. Kochmann, G.N. Venturini, Homogenized mechanical properties of auxetic composite materials in finite-strain elasticity. Smart Mater. Struct. **22**, 084004–084011 (2013)
136. D.M. Kochmann, W.J. Drugan, Infinitely stiff composite via a rotation-stabilized negative-stiffness phase. Appl. Phys. Lett. **99**, 011909 (2011)
137. A.G. Kolpakov, Determination of the average characteristics of elastic frameworks. Appl. Math. Mech. **49**, 739–745 (1985)
138. R.S. Lakes, Cellular solids with tunable positive or negative thermal expansion of unbounded magnitude. Appl. Phys. Lett. **90**, 221905 (2007)
139. R.S. Lakes, Foam structures with a negative Poisson's ratio. Science **235**, 1038–1040 (1987)
140. R.S. Lakes, Deformation mechanisms of negative Poisson's ratio materials: structural aspects. J. Mater. Sci. **26**, 2287–2292 (1991)
141. R.S. Lakes, Design considerations for materials with negative Poisson's ratios. J. Mech. Des. **115**(4), 696–700 (1993). https://doi.org/10.1115/1.2919256
142. R.S. Lakes, Extreme damping in composite materials with a negative stiffness phase. Phys. Rev. Let. **86**(13), 2897–2900 (2001)
143. R.S. Lakes, Extreme damping in compliant composites with a negative-stiffness phase. Phil. Mag. Lett. **81**(2), 95–100 (2001)
144. R.S. Lakes, K. Elms, Indentability of conventional and negative Poisson's ratio foams. J. Compos. Mater. **27**, 1193–1202 (1993)
145. R.S. Lakes, W.J. Drugan, Dramatically stiffer elastic composite materials due to a negative stiffness phase? J. Mech. Phys. Solids **50**, 979–1009 (2002)
146. R.S. Lakes, A. Lowe, Negative Poisson's ratio foam as seat cushion material. Cell. Polym. **19**, 157–167 (2000)
147. R.S. Lakes, P. Rosakis, A. Ruina, Microbuckling instability in elastomeric cellular solids. J. Mat. Sci **28**, 4667–4672 (1993)
148. R.S. Lakes, T. Lee, A. Bersie, Y.C. Wang, Extreme damping in composite materials with negative stiffness inclusions. Nature **410**, 565–567 (2001)
149. R.S. Lakes, K.W. Wojciechowski, Negative compressibility, negative Poisson's ratio, and stability. Phys. Status Solidi **245**, 545–551 (2008)
150. L.D. Landau, E.M. Lifshitz, *Theory of Elasticity* (Oxford, London, Edinburgh, New York, Toronto, Sydney, Paris, Braunschweig, 1959)
151. U.D. Larsen, O. Sigmund, S. Bouwstra, Design and fabrication of compliant micromechanisms and structures with negative Poisson's ratio. J. Microelectromech. Systems **6**, 99–106 (1997)
152. M.L. Latash, V.M. Zatsiorsky, Joint stiffness: myth or reality? Hum. Mov. Sci. **12**(6), 653–692 (1993)
153. I.M. Lifshits, L.N. Rosentsveig, Zur Theorie der elastischen Eigenschaften yon Polykristallen. Zh. Eksp. Teor. Fiz. **16**, 967–975 (1946)
154. T.-C. Lim, Out-of-plane modulus of semi-auxetic laminates. Eur. J. Mech. A/Solids **28**, 752–756 (2009)
155. T.-C. Lim, Coefficient of thermal expansion of stacked auxetic and negative thermal expansion laminates. Phys. Status Solidi B **248**(1), 140–147 (2011)
156. T.-C. Lim, Thermal stresses in thin auxetic plates. J. Therm. Stresses **36**(11), 1131–1140 (2012)
157. T.-C. Lim, Negative thermal expansion structures constructed from positive thermal expansion trusses. J. Mater. Sci. **47**, 368–373 (2012)
158. T.-C. Lim, U. Rajendra Acharya, Counterintuitive modulus from semi-auxetic laminates. Phys. Status Solidi B **248**(1), 60–65 (2011)
159. C. Lira, P. Innocenti, F. Scarpa, Transverse elastic shear of auxetic multi re-entrant honeycombs. Compos. Struct. **90**, 314–322 (2010)

160. C. Lira, F. Scarpa, M. Olszewska, M. Celuch, The SILICOMB cellular structure: mechanical and dielectric properties. Phys. Status Solidi B **246**, 2055–2062 (2009)
161. Y. Liu, H. Hu, A review on auxetic structures and polymeric materials. Sci. Res. Essays **5**(10), 1052–1063 (2010)
162. A.E.H. Love, *A Treatise on Mathematical Theory of Elasticity* (Dover, New York, 1927)
163. P. Martin, A.D. Mehta, A.J. Hudspeth, Negative hair-bundle stiffness betrays a mechanism for mechanical amplification by the hair cell. Proc. Natl. Acad. Sci. U.S.A. **97**(22), 12026–12031 (2000)
164. E.O. Martz, R.S. Lakes, J.B. Park, Hysteresis behaviour and specific damping capacity of negative Poisson's ratio foams. Cell. Polym. **15**, 349–364 (1996)
165. B.T. Maruszewski, A. Drzewiecki, R. Starosta, On effective Young's modulus and Poisson's ratio of the auxetic thermoelastic material. Comput. Methods Sci. Technol. **22**, 233–237 (2016)
166. T.A. Mary, J.S.O. Evans, T. Vogt, A.W. Sleight, Negative thermal expansion from 0.3 to 1050 K in $ZrW_2O_8$. Science **272**, 90–92 (1996)
167. D.M. McCutcheon, J.N. Reddy, M.J. O'Brien, T.S. Creasy, G.F. Hawkins, Damping composite materials by machine augmentation. J. Sound Vib. **294**(4–5), 828–840 (2006). https://doi.org/10.1016/j.jsv.2005.12.029
168. R. McLaughlin, A study of the differential scheme in composite materials. Int. J. Eng. Sci. **15**, 237–244 (1977)
169. P. Michelis, V. Spitas, Numerical and experimental analysis of a triangular auxetic core made of CFR-PEEK using the directionally reinforced integrated single-yarn (DIRIS) architecture. Compos. Sci. Technol. **70**, 1064–1071 (2010)
170. W. Miller, C.W. Smith, D.S. Mackenzie, K.E. Evans, Negative thermal expansion: a review. J. Mater. Sci. **44**(20), 5441–5451 (2009)
171. W. Miller, P.B. Hook, C.W. Smith, X. Wanga, K.E. Evans, The manufacture and characterisation of a novel, low modulus, negative Poisson's ratio composite. Compos. Sci. Technol. **69**, 651–655 (2009)
172. W. Miller, C.W. Smith, K.E. Evans, Honeycomb cores with enhanced buckling strength. Compos. Struct. **93**, 1072–1077 (2011)
173. F. Milstein, K. Huang, Existence of a negative Poisson's ratio in fcc crystals. Phys. Rev. B **19**, 2030–2033 (1979)
174. G.W. Milton, Complete characterization of the macroscopic deformations of periodic unimode metamaterials of rigid bars and pivots. J. Mech. Phys. Solids **61**(7), 1543–1560 (2013)
175. H. Mitschke, V. Robins, K. Mecke, G.E. Schröder-Turk, Finite auxetic deformations of plane tessellations. Proc. R. Soc. Lond A: Math. Phys. Eng. Sci. **469**(2149), 20120465
176. H. Mitschke, J. Schwerdtfeger, F. Schury, M. Stingl, C. Körner, R.F. Singer et al., Finding auxetic frameworks in periodic tessellations. Adv. Mater. **23**(22–23), 2669–2674 (2011)
177. L. Mizzi, R. Gatt, J.N. Grima, Non-porous grooved single-material auxetics. Phys. Status Solidi B **252**, 1559–1564 (2015)
178. J.W. Narojczyk, K.W. Wojciechowski, Elastic properties of degenerate f.c.c. crystal of polydisperse soft dimers at zero temperature. J. Non-Cryst. Solids **356**, 2026–2032 (2010)
179. J.W. Narojczyk, A. Alderson, A.R. Imre, F. Scarpa, K.W. Wojciechowski, Negative Poisson's ratio behavior in the planar model of asymmetric trimers at zero temperature. J. Non-Cryst. Solids **354**, 4242–4248 (2008)
180. F. Nazare, A. Alderson, Models for the prediction of Poisson's ratio in the '$\alpha$-cristobalite' tetrahedral network. Phys. Status Solidi B **252**, 1465–1478 (2015)
181. N. Novak, M. Vesenjak, Z. Ren, Auxetic cellular materials—a review. Strojniski Vestnik-J. Mech. Eng. **62**(9), 485–493 (2016)
182. W. Nowacki, *The Linear Theory of Micropolar Elasticity* (Springer, Wien, New York), pp. 1–43
183. R.J. O'Connell, B. Budiansky, Seismic velocities in dry and saturated cracked solids. J. Geophys. Res. **79**, 5412–5426 (1974)
184. S.-T. Park, T-T. Luu, Techniques for optimizing parameters of negative stiffness. Proc. IMechE Part C: J. Mech. Eng. Sci. **221**, 505–511 (2007)

185. E. Pasternak, A.V. Dyskin, Multiscale hybrid materials with negative Poisson's ratio, in *IUTAM Symposium on Scaling in Solid Mechanics*, ed. by F. Borodich (Springer, 2008a), pp. 49–58
186. E. Pasternak, A.V. Dyskin, Materials with Poisson's ratio near −1: properties and possible realisations, in *ICTAM 2008, XXII International Congress of Theoretical and Applied Mechanics*, paper 11982, CD-ROM Proceedings, August 24–29, 2008, ed. by J. Denier, M.D. Finn, T. Mattner (Adelaide, 2008b), 2 p. ISBN 978-0-9805142-1-6
187. E. Pasternak, A.V. Dyskin, Materials and structures with macroscopic negative Poisson's ratio. Int. J. Eng. Sci. **52**, 103–114 (2012)
188. E. Pasternak, E., A.V. Dyskin, Instability and failure of particulate materials caused by rolling of non-spherical particles, in *Proceedings of the 13th International Conference on Fracture* (Beijing, China, 2013)
189. E. Pasternak, H.-B. Mühlhaus, Cosserat continuum modelling of granulate materials, in *Computational Mechanics—New Frontiers for New Millennium*, ed. by S. Valliappan, N. Khalili (Elsevier, 2001), pp. 1189–1194
190. E. Pasternak, H.-B. Mühlhaus, Generalised homogenisation procedures for granular materials. Eng. Math. **52**, 199–229 (2005)
191. E. Pasternak, A.V. Dyskin, Dynamic instability in geomaterials associated with the presence of negative stiffness elements, in *Bifurcation and Degradation of Geomaterials in the New Millennium*, ed. by K.-T. Chau, J. Zhao (Springer, 2015), pp. 155–160
192. E. Pasternak, A.V. Dyskin, I. Shufrin, Negative Poisson's ratio materials' design principles and possible applications, in *Proceedings of the 6th Australasian Congress on Applied Mechanics, ACAM 6*, 12–15 December 2010, Perth, Paper 1266, ed. by K. Teh, I. Davies, I. Howard (2010), 10 pp
193. E. Pasternak, A.V. Dyskin, G. Sevel, Chains of oscillators with negative stiffness elements. J. Sound Vib. **333**(24), 6676–6687 (2014)
194. E. Pasternak, I. Shufrin, A.V. Dyskin, Thermal stresses in hybrid materials with auxetic inclusions. Composite Structures **138**, 313–321 (2016)
195. E. Pasternak, A.V. Dyskin, M. Esin, Wave propagation in materials with negative Cosserat shear modulus. Int. J. Eng. Sci. **100**, 152–161 (2016)
196. S. Pellegrino, Deployable structures in engineering, in *Deployable structures*, ed. by S. Pellegrino (Springer, Wien GmbH, 2014)
197. N. Phan-Thien, B.L. Karihaloo, Materials with negative Poisson's ratio: a qualitative microstructural model. J. Appl. Mech. Trans. ASME **61**, 1001–1004 (1994)
198. P.V. Pikhitsa, M. Choi, H.-J. Kim, S.-H. Ahn, Auxetic lattice of multipods. Phys. Status Solidi B **246**, 2098–2101 (2009)
199. D.L. Platus, Negative-stiffness-mechanism vibration isolation systems, in *SPIE Conference on Current Developments in Vibration Control 98 for Optomechanical Systems, Denver, Colorado, July 1999, SPIE*, vol. 3786 (1999), pp. 98–105
200. M.E. Pontecorvo, S. Barbarino, G.J. Murray, F.S. Gandhi, Bistable arches for morphing applications. J. Intell. Mater. Syst. Struct. **24**(3), 274–286 (2013)
201. A.A. Pozniak, J. Smardzewski, K.W. Wojciechowski, Computer simulations of auxetic foams in two dimensions. Smart Mater. Struct. **22**, 084009 (2013)
202. Y. Prawoto, Seeing auxetic materials from the mechanics point of view: a structural review on the negative Poisson's ratio. Comput. Mater. Sci. **58**, 140–153 (2012)
203. G. Puglisi, L. Truskinovsky, Mechanics of a discrete chain with bi-stable elements. J. Mech. Phys. Solids **48**, 1–27 (2000)
204. N. Ravirala, A. Alderson, K.L. Alderson, Interlocking hexagons model for auxetic behaviour. J. Mater. Sci. **42**, 7433–7445 (2007)
205. N. Ravirala, K.L. Alderson, P.J. Davies, V.R. Simkins, A. Alderson, Negative Poisson's ratio polyester fibers. Text. Res. J. **76**, 540–546 (2006)
206. B.W. Rosen, Z. Hashin, Effective thermal expansion coefficients and specific heats of composite materials. Int. J. Eng. Sci. **8**, 157–173 (1970)

207. L. Rothenburg, A.L. Berlin, R. Bathurst, Microstructure of isotropic materials with negative Poisson's ratio. Nature **325**, 470–472 (1991)
208. A.L. Roytburd, Deformation through transformations. J. Phys. IV **C1**, 11–25 (1996). https://doi.org/10.1051/jp4:1996102
209. M.D.G. Salamon, Stability, instability and design of pillar workings. Int. J. Rock Mech. Min. Sci. **7**, 613–631 (1970)
210. R.L. Salganik, Mechanics of bodies with many cracks. Mech. Solids **8**, 135–143 (1973)
211. A.A. Sarlis, D.T.R. Pasala, M.C. Constantinou, A.M. Reinhorn, S. Nagarajaiah, D.P. Taylor, Negative stiffness device for seismic protection of structures. J. Struct. Eng. **139**(7), 1124–1133 (2013)
212. F. Scarpa, G. Tomlinson, Sandwich structures with negative Poisson's ratio for deployable structures, in *IUTAM-IASS Symposium on Deployable Structures: Theory and Applications* (2000), pp. 335–343
213. F. Scarpa, P.J. Tomlin, On the transverse shear modulus of negative Poisson's ratio honeycomb structures. Fatigue Fract. Eng. Mater. Struct. **23**(8), 717–720 (2000)
214. F. Scarpa, S. Adhikari, C.Y. Wang, Nanocomposites with auxetic nanotubes. Int J. Smart Nano Mater. **1**(2), 83–94 (2010)
215. F. Scarpa, P. Pastorino, A. Garelli, S. Patsias, M. Ruzzene, Auxetic compliant flexible PU foams: static and dynamic properties. Phys. Status Solidi B **242**(3), 681–694 (2005)
216. F. Scarpa, J.W. Narojczyk, K.W. Wojciechowski, Unusual deformation mechanisms in carbon nanotube heterojunctions (5,5)–(10,10) under tensile loading. Phys. Status Solidi B **248**(1), 82–87 (2011)
217. S. Schaare, A.V. Dyskin, Y. Estrin, E. Pasternak, A. Kanel-Belov, Point loading of assemblies of interlocked cube-shaped elements. Int. J. Eng. Sci. **46**, 1228–1238 (2008). https://doi.org/10.1016/j.ijengsci.2008.06.012
218. M.M. Shokrieh, A. Assadi, Determination of maximum negative Poisson's ratio for laminated fiber composites. Phys. Status Solidi B **248**(5), 1237–1241 (2011)
219. I. Shufrin, E. Pasternak, A.V. Dyskin, Symmetric structures with negative Poisson's ratio, in *Australian and New Zealand Industrial and Applied Mathematics Conference, ANZIAM—2010, Queenstown, New Zealand* (2010)
220. I. Shufrin, E. Pasternak, A.V. Dyskin, Planar isotropic structures with negative Poisson's ratio. Int. J. Solids Struct. **49**(17), 2239–2253 (2012)
221. I. Shufrin, E. Pasternak, A.V. Dyskin, Negative Poisson's ratio in hollow sphere materials. Int. J. Solids Struct. **54**, 192–214 (2015)
222. I. Shufrin, E. Pasternak, A.V. Dyskin, Hybrid materials with negative Poisson's ratio inclusions. Int. J. Eng. Sci. **89**, 100–120 (2015)
223. I. Shufrin, E. Pasternak, A.V. Dyskin, Deformation analysis of reinforced-core auxetic assemblies by close-range photogrammetry. Phys. Status Solidi (B) Basic Res. **253**(7), 1342–1358 (2016)
224. O. Sigmund, Tailoring materials with prescribed elastic properties. Mech. Mater. **20**, 351–368 (1995)
225. V.R. Simkins, A. Alderson, P.J. Davies, K.L. Alderson, Single fibre pullout tests on auxetic polymeric fibres. J. Mater. Sci. **40**, 4355–4364 (2005)
226. A. Slan, W. White, F. Scarpa, K. Boba, I. Farrow, Cellular plates with auxetic rectangular perforations. Phys. Status Solidi B **252**, 1533–1539 (2015)
227. C.W. Smith, J.N. Grima, K.E. Evans, A novel mechanism for generating auxetic behaviour in reticulated foams: missing rib foam model. Acta Mater. **48**, 4349–4356 (2000)
228. A. Sogame, H. Furuya, Conceptual study on cylindrical deployable space structures, in *IUTAM-IASS Symposium on Deployable Structures: Theory and Applications* (2000), pp 383–392
229. A. Spadoni, M. Ruzzene, Elasto-static micropolar behaviour of a chiral auxetic lattice. J. Mech. Phys. Solids. **60**, 156–171 (2012)
230. P.J. Stott, R. Mitchell, K. Alderson, A. Alderson, A growing industry. Mater. World **8**, 12–14 (2000)

231. T. Strek, H. Jopek, Effective mechanical properties of concentric cylindrical composites with auxetic phase. Phys. Status Solidi B **249**(7), 1359–1365 (2012)
232. K. Takenaka, Negative thermal expansion materials: technological key for control of thermal expansion'. Sci. Technol. Adv. Mater. **13**, 013001–013012 (2012)
233. C.Y. Tang, M.J. O'Brien, G.F. Hawkins, Embedding simple machines to add novel dynamic functions to composites. JOM **57**, 32 (2005)
234. M. Tatlier, L. Berhan, Modelling the negative Poisson's ratio of compressed fused fibre networks. Phys. Status Solidi B **246**, 2018–2024 (2009)
235. P.S. Theocaris, G.E. Stavroulakis, P.D. Panagiotopoulos, Negative Poisson's ratios in composites with star-shaped inclusions: a numerical homogenization approach. Arch. Appl. Mech. **67**, 274–286 (1997)
236. J.M.T. Thompson, G.W. Hunt, *A General Theory of Elastic Stability* (Wiley, London, 1973)
237. K. Toru, M. Yoshitaka, Nanoscale mechanics of carbon nanotube evaluated by nanoprobe manipulation in transmission electron microscope. *Japanese Journal of Applied* (2006)
238. K.V. Tretiakov, Negative Poisson's ratio of two-dimensional hard cyclic tetramers. J. Non-Cryst. Solids **355**, 1435–1438 (2009)
239. K.V. Tretiakov, K.W. Wojciechowski, Poisson's ratio of simple planar 'isotropic' solids in two dimensions. Phys. Status Solidi B **244**(3), 1038–1046 (2007)
240. K.V. Tretiakov, K.W. Wojciechowski, Elastic properties of soft sphere crystal from Monte Carlo simulations. J. Phys. Chem. B **112**, 1699–1705 (2009)
241. K.V. Tretiakov, K.W. Wojciechowski, Elastic properties of fcc crystals of polydisperse soft spheres. Phys. Status Solidi B, 1–10 (2013)
242. A.S. Vavakin, R.L. Salganik, Effective characteristics of nonhomogeneous media with isolated nonhomogeneities. Mech. Solids **10**(3), 58–66 (1975)
243. A.S. Vavakin, R.L. Salganik, Effective elastic characteristics of bodies with isolated cracks, cavities, and rigid nonhomogeneities. Mech. Solids **13**, 87–97 (1978)
244. P. Verma, M.L. Shofner, A. Lin, K.B. Wagner, A.C. Griffin, Inducing out-of-plane auxetic behaviour in needle-punched nonwovens. Phys. Status Solidi B **252**, 1455–1464 (2015)
245. Y.-C. Wang, Influences of negative stiffness on a two-dimensional hexagonal lattice cell. Philos. Mag. **87**(24), 3671–3688 (2007)
246. M. Wang, N. Pan, Predictions of effective physical properties of complex multiphase materials. Mat. Sci. Eng. R **63**, 1–30 (2008)
247. Y.C. Wang, R.S. Lakes, Extreme thermal expansion, piezoelectricity, and other coupled field properties in composites with a negative stiffness phase. J. Appl. Phys. **90**, 6458–6465 (2001)
248. Y.C. Wang, R.S. Lakes, Extreme stiffness systems due to negative stiffness elements. Am. J. Phys. **72**, 40–50 (2004)
249. Y.-C. Wang, R.S. Lakes, Stable extremely-high-damping discrete viscoelastic systems due to negative stiffness elements. Appl. Phys. Lett. **84**(22), 4451–4453 (2004)
250. Y.-C. Wang, R.S. Lakes, Composites with inclusions of negative bulk modulus: extreme damping and negative Poisson's ratio. J. Compos. Mater. **39**(18), 1645–1657 (2005)
251. Y.-C. Wang, C.-C. Ko, K.-W. Chang, Anomalous effective viscoelastic, thermoelastic, dielectric and piezoelectric properties of negative-stiffness composites and their stability. Phys. Status Solidi B **252**, 1640–1655 (2015)
252. Y.-C. Wang, M. Ludwigson, R.S. Lakes, Deformation of extreme viscoelastic metals and composites. Mater. Sci. Eng. A **370**, 41–49 (2004)
253. Y.C. Wang, J.G. Swadener, R.S. Lakes, Two-dimensional viscoelastic discrete triangular system with negative-stiffness components. Philos. Mag. Lett. **86**(2), 99–112 (2006)
254. Y.-C. Wang, J.G. Swadener, R.S. Lakes, Anomalies in stiffness and damping of a 2D discrete viscoelastic system due to negative stiffness components. Thin Solid Films **515**(6), 3171–3178 (2007)
255. Z.G. Wang, C.K. Kim, P. Martin, A.D. Mehta, A.J. Hudspeth, Negative hair-bundle stiffness betrays a mechanism for mechanical amplification by the hair cell. Proc. Natl. Acad. Sci. U.S.A **97**, 12026–12031 (2000)

256. R.S. Webber, K.L. Alderson, K.E. Evans, A novel fabrication route for auxetic polyethylene, part 2: mechanical properties. Polym. Eng. Sci. **48**(7), 1351–1358 (2008). https://doi.org/10.1002/pen.21110
257. G. Wei, S.F. Edwards, Poisson ratio in composites of auxetics. Phys. Rev. E **58**(5), 6173–6181 (1998)
258. G. Wei, S.F. Edwards, Effective elastic properties of composites of ellipsoids (I). Nearly spherical inclusions. Phys. A **264**, 388–403 (1999)
259. G. Wei, S.F. Edwards, Effective elastic properties of composites of ellipsoids (II). Nearly disk and needle-like inclusions. Phys. A **264**, 404–423 (1999)
260. J.J. Williams, C.W. Smith, K.E. Evans, Z.A.D. Lethbridge, R.I. Walton, An analytical model for producing negative Poisson's ratios and its application in explaining off-axis elastic properties of the NAT-type zeolites. Acta Mater. **55**, 5697–5707 (2007)
261. K. Wohlhart, Double-chain mechanisms, in *IUTAM-IASS Symposium on Deployable Structures: Theory and Applications* (2000), pp. 457–466
262. K.W. Wojciechowski, Two-dimensional isotropic system with a negative Poisson's ratio. Phys. Lett. A **137**, 60–64 (1989)
263. K.W. Wojciechowski, Non-chiral, molecular model of negative Poisson ratio in two dimensions. J. Phys. A: Math. Gen. **36**, 11765–11778 (2003)
264. K.W. Wojciechowski, A.C. Brańka, Elastic moduli of a perfect hard disc crystal in two dimensions. Phys. Lett. A **134**, 314–318 (1988)
265. K.W. Wojciechowski, A.C. Brańka, Negative Poisson's ratio in a two-dimensional "isotropic" solid. Phys. Rev. A **40**, 7222–7225 (1989)
266. K.W. Wojciechowski, K.V. Tretiakov, M. Kowalik, Elastic properties of dense solid phases of hard cyclic pentamers and heptamers in two dimensions. Phys. Rev. E **67**, 036121 (2003)
267. J.R. Wright, M.R. Sloan, K.E. Evans, Tensile properties of helical auxetic structures: a numerical study. J. Appl. Phys. **108**, 044905 (2010)
268. W. Yang, Z.-M. Li, W. Shi, B.-H. Xie, M.-B. Yang, Review on auxetic materials. J. Mater. Sci. **39**(10), 3269–3279 (2004)
269. Z. Yang, H.M. Dai, N.H. Chan, G.C. Ma, P. Sheng, Acoustic metamaterial panels for sound attenuation in the 50–1000 Hz regime. Appl. Phys. Lett. **96**, 041906 (2010)
270. Y.T. Yao, A. Alderson, K.L. Alderson, Can nanotubes display auxetic behaviour? Phys. Status Solidi B **245**, 2373–2382 (2008)
271. H.W. Yap, R.S. Lakes, R.W. Carpick, Mechanical instabilities of individual multiwalled carbon nanotubes under cyclic axial compression. Nano Lett. **7**(5), 1149–1154 (2007)
272. H.W. Yap, R.S. Lakes, R.W. Carpick, Negative stiffness and enhanced damping of individual multiwalled carbon nanotubes. Phys. Rev. B **77**, 045423 (2008)
273. A. Yeganeh-Haeri, D.J. Weidner, J.B. Parise, Elasticity of $\alpha$-cristobalite: a silicon dioxide with a negative Poisson's ratio. Science **257**, 650–652 (1992)
274. V.Y. Zaitsev, A.V. Radostin, E. Pasternak, A.V. Dyskin, Extracting real-crack properties from nonlinear elastic behavior of rocks: abundance of cracks with dominating normal compliance and rocks with negative Poisson's ratio. Nonlinear Process. Geophys. (NPG) **24**, 543–551 (2017)
275. V.Y. Zaitsev, A.V. Radostin, E. Pasternak, A.V. Dyskin, Extracting properties of crack-like defects from pressure dependences of elastic-wave velocities using an effective-medium model with decoupled shear and normal compliances of cracks. Int. J. Rock Mech. Min. Sci. **97**, 122–133 (2017)
276. R. Zhang, H.-L. Yeh, H.-Y. Yeh, A discussion of negative Poisson's ratio design for composites. J. Reinf. Plast. Compos. **18**, 1546–1556 (1999)

# Chapter 4
# Computational Homogenization of Architectured Materials

**Justin Dirrenberger, Samuel Forest and Dominique Jeulin**

**Abstract** Architectured materials involve geometrically engineered distributions of microstructural phases at a scale comparable to the scale of the component, thus calling for new models in order to determine the effective properties of materials. The present chapter aims at providing such models, in the case of mechanical properties. As a matter of fact, one engineering challenge is to predict the effective properties of such materials; computational homogenization using finite element analysis is a powerful tool to do so. Homogenized behavior of architectured materials can thus be used in large structural computations, hence enabling the dissemination of architectured materials in the industry. Furthermore, computational homogenization is the basis for computational topology optimization which will give rise to the next generation of architectured materials. This chapter covers the computational homogenization of periodic architectured materials in elasticity and plasticity, as well as the homogenization and representativity of random architectured materials.

## 4.1 Introduction

Architectured materials are a rising class of materials that bring new possibilities in terms of functional properties, filling the gaps and pushing the limits of Ashby's materials performance maps [11]. The term architectured materials encompasses

J. Dirrenberger (✉)
Laboratoire PIMM, Arts et Métiers-ParisTech, CNRS, Cnam,
151 boulevard de l'Hôpital, 75013 Paris, France
e-mail: justin.dirrenberger@ensam.eu

S. Forest
Centre des Matériaux, MINES-ParisTech, CNRS UMR 7633, BP 87,
91003 Evry cedex, France
e-mail: samuel.forest@mines-paristech.fr

D. Jeulin
Centre de Morphologie Mathématique, MINES-ParisTech, 35 rue St-Honoré,
77300 Fontainebleau, France
e-mail: dominique.jeulin@mines-paristech.fr



Y. Estrin et al. (eds.), *Architectured Materials in Nature and Engineering*, Springer Series in Materials Science 282,
https://doi.org/10.1007/978-3-030-11942-3_4

any microstructure designed in a thoughtful fashion, such that some of its materials properties, e.g. yield strength/density, have been improved in comparison to those of its constituents, due to both structure and composite effects, which depend on the multiphase morphology, i.e. the relative topological arrangement between each phase [11, 33, 35].

There are many examples: particulate and fibrous composites, foams, sandwich structures, woven materials, lattice structures, etc. with different objectives. For instance, developing architectured porous materials for structural, acoustic and insulation properties [40, 70], entangled monofilament of pearlitic steel [47, 173], sandwich composite structures [118, 163–165], segmented interlocking structures [59, 60, 62, 65, 66, 72, 117, 123, 136, 145–147, 189], asymmetric frictional materials [18, 19], woven and non-woven textile composites [57, 130, 141], porous metallic glasses [187], hierarchical composites [92], crumpled metallic foils [34]. Much more examples can be found in the present book.

One can play on many parameters in order to obtain architectured materials, but all of them are related either to the microstructure or the geometry. Parameters related to the microstructure can be optimised for specific needs using a materials-by-design approach, which has been thoroughly developed by chemists, materials scientists and metallurgists. Properties improvements related to microstructural design are intrinsically linked to the synthesis and processing of materials and are therefore due to micro and nanoscale phenomena, taking place at a scale ranging from 1 nm to 10 μm. This scale is below the scope of the present chapter, but has been extensively studied in the literature [63, 78, 148].

From a macroscopic viewpoint, parameters related to the geometry have mainly been the responsibility of structural and civil engineers for centuries: to efficiently distribute materials within structures. An obvious example would be the many different strategies available for building bridges. At the millimetre scale, materials can be considered as structures, i.e. one can enhance the bending stiffness of a component by modifying its geometry while keeping the lineic mass (for beams) or surfacic mass (for plates) unchanged [209]. On the other hand, one might need a lower flexural strength for specific applications, with the same lineic and/or surfacic masses. This can be achieved with strand structures, i.e. by creating topological interfaces in the material. Processing remains the key technological issue for further development of architectured materials as the microstructure, the shape, and the scale of the material depend on it. Progress is made every day in terms of material processing at the lab scale, as it was done in [191] by using a bottom-up approach of sequential processing techniques in order to fabricate ultralight metallic microlattice materials [190]. There is still a long way to go for the industry to actually apply architectured materials in product manufacturing.

Architectured materials lie between the microscale and the macroscale. This class of materials involves geometrically engineered distributions of microstructural phases at a scale comparable to the scale of the component [11, 33, 35], thus calling for enriched models of continuum mechanics in order to determine the effective properties of materials [81, 140], e.g. generalised continua theories, in order to describe the behaviour of architectured materials, such as strain-gradient elasticity [13], and strain-gradient plasticity. This topic has been especially fruitful these last few years in the mechanics of materials community [9, 13, 45, 51, 129, 166, 167, 174, 203]; this results in the availability of versatile models able to describe the various situations encountered with architectured materials. Given mature processing techniques, architectured materials are promised to a bright future in industrial applications due to their enticing customisable specific properties and the opportunity for multifunctionality.

When considering actual applications, one engineering challenge is to predict the effective properties of such materials; computational homogenisation using finite element analysis is a powerful tool to do so. Homogenised behaviour of architectured materials can thus be used in large structural computations, hence enabling the dissemination of architectured materials in the industry. Furthermore, computational homogenisation is the basis for computational topology optimisation [3, 10, 24, 41, 86, 116, 184, 205, 208, 210, 211] which will give rise to the next generation of architectured materials as it can already be seen in the works of [8, 48, 71, 83, 96, 119, 120, 128, 132, 149, 171, 204, 207].

Therefore, the development of architectured materials is related to the availability of appropriate computational tools for both design and modelling, but also for computerised manufacturing as for the various additive manufacturing techniques considered for producing architectured materials. In order to foster the development of architectured materials within an industrial framework, the availability of modelling bricks describing the underlying behaviour of such materials, is a necessity in order for these to be used in structural simulation codes. We are aiming at developing product design methodologies taking into account the specificities of architectured materials. To do so, we focused on computational approaches to modelling and optimising architectured materials, as well as novel processing techniques.

Materials science comes from the following fact: microstructural heterogeneities play a critical role in the macroscopic behaviour of a material [29, 32, 77, 112, 151, 202]. Constitutive modelling, thanks to an interaction between experiments and simulation, is usually able to describe the response of most materials in use. Such phenomenological models, including little to no information about the microstructure, cannot necessarily account for local fluctuation of properties. In this case, the material is considered as a homogeneous medium. Studying the behaviour of heterogeneous materials involves developing enriched models including morphological information about the microstructure [22, 49, 64, 105, 108, 115, 159, 195, 197, 201, 213]. These models should be robust enough to predict effective properties depending on statistical data (volume fraction, $n$-point correlation function, etc.) and the physical nature of each phase or constituent. As a matter of fact, advanced models are often restricted to a limited variety of materials. Although isotropic and anisotropic polycrystalline metals, for instance, have been extensively studied by the means of both analytical and computational tools [25, 27, 38, 79, 97, 114, 121, 133, 157], some material configurations (architectured materials, materials with infinite contrast of properties, nanocomposites, materials exhibiting nonlinear behaviour, etc.) call for further development of models and tools for describing their effective behaviour. The purpose of this chapter is to present existing homogenization methods applicable to architectured materials.

The chapter was conceived as a basis for the enquiring reader to find out the main concepts related to computational homogenization of architectured materials, as well as the bibliographical references necessary to further his/her understanding of the present topic. It is based on the Ph.D. dissertation of the first author [54]. The chapter is organized as follows: firstly, classical homogenization for linear elasticity is presented in Sect. 4.2 and applied to the case of periodic architectured materials with negative Poisson's ratio, or auxetics. For a similar case of application, computational homogenization for elastoplasticity is introduced in Sect. 4.3. Since architectured materials are not always periodic, considerations regarding the of random media based on computational homogenization are presented in Sect. 4.4. Conclusions and perspectives are postponed to Sect. 4.5.

In this chapter, zeroth, first, and second order tensors are denoted by $a$, $\underline{a}$, $\underset{\sim}{a}$ respectively. The simple and double contractions are written . and : respectively. In index form with respect to an orthonormal Cartesian basis, these notations correspond to

$$\underline{a}.\underline{b} = a_i b_j, \quad \underset{\sim}{a} : \underset{\sim}{b} = a_{ij} b_{ij} \tag{4.1}$$

where repeated indices are summed up. The tensor product is denoted by $\otimes$. For example, the component $(\underset{\sim}{a} \otimes \underset{\sim}{b})_{ijkl}$ is $a_{ij} b_{kl}$. The small strain assumption is made in this chapter. The nabla operator is denoted by $\nabla_X$ (resp. $\nabla_x$) when partial derivation is computed with respect to macroscopic (resp. microscopic) coordinates. For example $\underset{\sim}{\sigma}.\nabla$ is the divergence of the second order tensor $\underset{\sim}{\sigma}$. The index form of $\underset{\sim}{\sigma}.\nabla$ is $\sigma_{ij,j}$. Similarly, $\underline{u} \otimes \nabla$ means $u_{i,j}$. The sign := defines the quantity on the left-hand side.

## 4.2 Computational Homogenization for Linear Elasticity

### *4.2.1 Constitutive Equations*

When heterogeneous materials are assumed to respond linearly to mechanical loading, constitutive relations are expressed locally for each phase in a linear elasticity framework using the generalized Hooke law:

$$\underset{\sim}{\sigma} = \underset{\approx}{c} : \underset{\sim}{\varepsilon} \tag{4.2}$$

with $\underset{\sim}{\sigma}$ second-order symmetric Cauchy stress tensor, $\underset{\sim}{\varepsilon}$ second-order symmetric engineering strain tensor and $\underset{\approx}{c}$, fourth-order positive definite tensor of elastic moduli, also known as the elastic stiffness tensor. It is possible to express strain as a function of stress using the Compliance tensor $\underset{\approx}{s}$, which is defined as the inverse of tensor $\underset{\approx}{c}$:

$$\underset{\sim}{\varepsilon} = \underset{\approx}{s} : \underset{\sim}{\sigma} \quad \text{with,} \quad \underset{\approx}{s} := \underset{\approx}{c}^{-1} \tag{4.3}$$

such that,

$$\underset{\approx}{s} \cdot \underset{\approx}{c} = \underset{\approx}{I} \tag{4.4}$$

with $\underset{\approx}{I}$, fourth-order identity tensor operating on symmetric second-order tensors such that:

$$\underset{\approx}{I} = \frac{1}{2}\left(\delta_{ik}\delta_{jl} + \delta_{il}\delta_{jk}\right)\ \underline{e}_i \otimes \underline{e}_j \otimes \underline{e}_k \otimes \underline{e}_l \tag{4.5}$$

The 81 components of $c_{ijkl}$ can be thinned-down to 21 for the most anisotropic case (triclinic elasticity) due to symmetries of $\underset{\sim}{\sigma}$ and $\underset{\sim}{\varepsilon}$. By isomorphism, these 21 components can be written as a symmetric second-order tensor (matrix) $c_{IJ}$ with 21 independent components using Voigt's notation:

$$\begin{bmatrix} \sigma_{11} \\ \sigma_{22} \\ \sigma_{33} \\ \sigma_{23} \\ \sigma_{31} \\ \sigma_{12} \end{bmatrix} = \begin{bmatrix} c_{11} & c_{12} & c_{13} & c_{14} & c_{15} & c_{16} \\ \bullet & c_{22} & c_{23} & c_{24} & c_{25} & c_{26} \\ \bullet & \bullet & c_{33} & c_{34} & c_{35} & c_{36} \\ \bullet & \bullet & \bullet & c_{44} & c_{45} & c_{46} \\ \bullet & \bullet & \bullet & \bullet & c_{55} & c_{56} \\ \bullet & \bullet & \bullet & \bullet & \bullet & c_{66} \end{bmatrix} \begin{bmatrix} \varepsilon_{11} \\ \varepsilon_{22} \\ \varepsilon_{33} \\ \gamma_{23} \\ \gamma_{31} \\ \gamma_{12} \end{bmatrix} \tag{4.6}$$

Engineering shear strain is used in the strain column-vector:

$$\begin{aligned} \gamma_{23} &= 2\varepsilon_{23} \\ \gamma_{31} &= 2\varepsilon_{31} \\ \gamma_{12} &= 2\varepsilon_{12} \end{aligned}$$

The matrix form of the compliance tensor is obtained by inverting (4.6):

$$\begin{bmatrix} \varepsilon_{11} \\ \varepsilon_{22} \\ \varepsilon_{33} \\ \gamma_{23} \\ \gamma_{31} \\ \gamma_{12} \end{bmatrix} = \begin{bmatrix} s_{11} & s_{12} & s_{13} & s_{14} & s_{15} & s_{16} \\ \bullet & s_{22} & s_{23} & s_{24} & s_{25} & s_{26} \\ \bullet & \bullet & s_{33} & s_{34} & s_{35} & s_{36} \\ \bullet & \bullet & \bullet & s_{44} & s_{45} & s_{46} \\ \bullet & \bullet & \bullet & \bullet & s_{55} & s_{56} \\ \bullet & \bullet & \bullet & \bullet & \bullet & s_{66} \end{bmatrix} \begin{bmatrix} \sigma_{11} \\ \sigma_{22} \\ \sigma_{33} \\ \sigma_{23} \\ \sigma_{31} \\ \sigma_{12} \end{bmatrix} \tag{4.7}$$

The finite element code used in the applications is actually making use of the Mandel notation presented in (4.8) and (4.9):

$$\begin{bmatrix} \sigma_{11} \\ \sigma_{22} \\ \sigma_{33} \\ \sqrt{2}\sigma_{23} \\ \sqrt{2}\sigma_{31} \\ \sqrt{2}\sigma_{12} \end{bmatrix} = \begin{bmatrix} c_{11} & c_{12} & c_{13} & \sqrt{2}c_{14} & \sqrt{2}c_{15} & \sqrt{2}c_{16} \\ \bullet & c_{22} & c_{23} & \sqrt{2}c_{24} & \sqrt{2}c_{25} & \sqrt{2}c_{26} \\ \bullet & \bullet & c_{33} & \sqrt{2}c_{34} & \sqrt{2}c_{35} & \sqrt{2}c_{36} \\ \bullet & \bullet & \bullet & 2c_{44} & 2c_{45} & 2c_{46} \\ \bullet & \bullet & \bullet & \bullet & 2c_{55} & 2c_{56} \\ \bullet & \bullet & \bullet & \bullet & \bullet & 2c_{66} \end{bmatrix} \begin{bmatrix} \varepsilon_{11} \\ \varepsilon_{22} \\ \varepsilon_{33} \\ \sqrt{2}\varepsilon_{23} \\ \sqrt{2}\varepsilon_{31} \\ \sqrt{2}\varepsilon_{12} \end{bmatrix} \tag{4.8}$$

The matrix form of the compliance tensor is obtained by inverting (4.8):

$$\begin{bmatrix} \varepsilon_{11} \\ \varepsilon_{22} \\ \varepsilon_{33} \\ \sqrt{2}\varepsilon_{23} \\ \sqrt{2}\varepsilon_{31} \\ \sqrt{2}\varepsilon_{12} \end{bmatrix} = \begin{bmatrix} s_{11} & s_{12} & s_{13} & \sqrt{2}s_{14} & \sqrt{2}s_{15} & \sqrt{2}s_{16} \\ \bullet & s_{22} & s_{23} & \sqrt{2}s_{24} & \sqrt{2}s_{25} & \sqrt{2}s_{26} \\ \bullet & \bullet & s_{33} & \sqrt{2}s_{34} & \sqrt{2}s_{35} & \sqrt{2}s_{36} \\ \bullet & \bullet & \bullet & 2s_{44} & 2s_{45} & 2s_{46} \\ \bullet & \bullet & \bullet & \bullet & 2s_{55} & 2s_{56} \\ \bullet & \bullet & \bullet & \bullet & \bullet & 2s_{66} \end{bmatrix} \begin{bmatrix} \sigma_{11} \\ \sigma_{22} \\ \sigma_{33} \\ \sqrt{2}\sigma_{23} \\ \sqrt{2}\sigma_{31} \\ \sqrt{2}\sigma_{12} \end{bmatrix} \tag{4.9}$$

In the isotropic case, $\underset{\approx}{\boldsymbol{c}}$ can be rewritten as follows:

$$\underset{\approx}{\boldsymbol{c}} = 3k\underset{\approx}{\boldsymbol{J}} + 2\mu\underset{\approx}{\boldsymbol{K}} \tag{4.10}$$

with $k$ bulk modulus, $\mu$ shear modulus, $\underset{\approx}{\boldsymbol{J}}$ and $\underset{\approx}{\boldsymbol{K}}$ respectively spherical and deviatoric fourth-order tensorial projectors such that,

$$\underset{\approx}{\boldsymbol{J}} = \frac{1}{3}\delta_{ij}\delta_{kl}\, \underline{\boldsymbol{e}}_i \otimes \underline{\boldsymbol{e}}_j \otimes \underline{\boldsymbol{e}}_k \otimes \underline{\boldsymbol{e}}_l \tag{4.11}$$

and

$$\underset{\approx}{\boldsymbol{K}} = \underset{\approx}{\boldsymbol{I}} - \underset{\approx}{\boldsymbol{J}} \tag{4.12}$$

### 4.2.2 The Representative Volume Element

The question of representativity has been a topic of interest in scientific communities for half a century, especially in the field of materials science, micromechanics and microscopy. Indeed, microstructural heterogeneities play a critical role on the macroscopic physical properties of materials. One common way to account for this underlying complexity is resorting to homogenisation techniques. Most homogenisation approaches, including analytical and computational, require the existence of a representative volume element (RVE). Several definitions have been given for the

RVE over the past 50 years. A review of this topic can be found in [84]. The classical definition of RVE is attributed to [94], who stated that for a given material the RVE is a sample that is structurally typical of the whole microstructure, i.e. containing a sufficient number of heterogeneities for the macroscopic properties to be independent of the boundary values of traction and displacement. Later, [26] emphasised the role of statistical homogeneity, especially in a volume-averaged sense. This also means that the RVE size considered should be larger than a certain microstructural length for which moduli fluctuate. Hashin [88] made a review on analysis of composite materials in which he referred to statistical homogeneity as a practical necessity. Sab [179] considered that the classical RVE definition for a heterogeneous medium holds only if the homogenised properties tend towards those of a similar periodic medium. This entails that the response over an RVE should be independent of boundary conditions (BC). From numerical simulations on VEs of various sizes, [200] concluded that from a practical viewpoint RVE should be as large as possible. Ostoja-Starzewski [150] considers the RVE to be only defined over a periodic unit-cell or a non-periodic cell containing an infinite number of heterogeneities. Drugan and Willis [61] introduced explicitly the idea of minimising the RVE size, meaning that the RVE would be the smallest material volume for which the apparent and effective properties coincide. Besides, it is worth noticing that for a given material the RVE size for physical property A, e.g. thermal conductivity, is a priori different from the RVE size for physical property B, e.g. elastic moduli. Thus, one has to consider an RVE that depends on the specific investigated property.

Many definitions refer to the separation of scales as a necessary condition for the existence of a RVE. This condition is not always met, i.e. with percolating media or materials with microstructural gradient of properties. This separation of scale involves a comparison between different characteristic lengths:

- $d$, size of microstructural heterogeneities;
- $l$, size of the RVE considered;
- $L$, characteristic length of the applied load.

Previous considerations regarding characteristic lengths can be summarized as follow:

$$d \ll l \ll L \tag{4.13}$$

Nevertheless, Inequality (4.13) is a necessary but not sufficient condition for the applicability of homogenisation. As a matter of fact, quasi-uniform loading, i.e. $l \ll L$, has to be enforced. Let us consider a measurable property, such as a mechanical strain field. The spatial average of its measured value over a finite volume $V$ converges towards the mathematical expectation of its measured value over a series of samples smaller than $V$ (ensemble average). It is the ergodicity hypothesis. Moreover, ergodicity implies that one sample (or realisation) of volume $V$ contains the statistical information necessary for the description of its microstructure. Also, this entails that heterogeneities are small enough in comparison to the RVE size, i.e. $d \ll l$. If and only if these two conditions are met ($d \ll l$ and $l \ll L$), the existence and uniqueness of an equivalent homogeneous medium for both cases of random and periodic materials can be rigorously proven [179]. Homogenisation is therefore possible.

Taking into account these definitions, and assuming ergodicity for the heterogeneous media considered, [114] proposed a method based on a statistical analysis for computing the minimal RVE size for a given physical property $Z(x), \forall x \in V$ and precision in the estimate of effective properties. The computed RVE size was found to be proportional to the integral range [137], which corresponds to a volume of statistical correlation. This statistical approach is presented in Sect. 4.4.

### 4.2.3 Averaging Relations

Let us consider a given volume element (VE) of volume $V$ without voids or rigid inclusions, for the sake of simplicity. For the spatial average over $V$ of a kinematically compatible strain field $\underset{\sim}{\varepsilon}'$ which is defined as the symmetric part of the gradient of a kinematically admissible displacement field $\underline{u}'$:

$$\begin{aligned}
\langle \underset{\sim}{\varepsilon}' \rangle &= \frac{1}{V}\int_V \underset{\sim}{\varepsilon}' dV = \frac{1}{V}\int_V u'_{(i,j)} dV \, \underline{e}_i \otimes \underline{e}_j \\
&= \frac{1}{V}\int_{\partial V} u'_{(i}n_{j)} dS \, \underline{e}_i \otimes \underline{e}_j \\
&= \frac{1}{V}\int_{\partial V} \underline{u}' \overset{s}{\otimes} \underline{n} \, dS
\end{aligned} \tag{4.14}$$

with $\underline{u}' \overset{s}{\otimes} \underline{n}$ and $u'_{(i,j)}$ denoting the symmetric part of the resulting tensor. If one considers now the spatial average of a statically admissible stress field $\underset{\sim}{\sigma}^*$, i.e. $\underset{\sim}{\sigma}^*.\nabla = \underline{0}$ in $V$, it yields:

$$\begin{aligned}
\langle \underset{\sim}{\sigma}^* \rangle &= \frac{1}{V}\int_V \underset{\sim}{\sigma}^* dV = \frac{1}{V}\int_V \sigma^*_{ij} dV \, \underline{e}_i \otimes \underline{e}_j \\
&= \frac{1}{V}\int_V \sigma^*_{(ik}\delta_{j)k} dV \, \underline{e}_i \otimes \underline{e}_j \\
&= \frac{1}{V}\int_V \sigma^*_{(ik}x_{j),k} dV \, \underline{e}_i \otimes \underline{e}_j \\
&= \frac{1}{V}\int_{\partial V} \sigma^*_{(ik}n_k x_{j)} dS \, \underline{e}_i \otimes \underline{e}_j \\
&= \frac{1}{V}\int_{\partial V} \left(\underset{\sim}{\sigma}^* \cdot \underline{n}\right) \overset{s}{\otimes} \underline{x} \, dS
\end{aligned} \tag{4.15}$$

From these averaging relations, we can define the elastic strain energy density $\mathcal{E}^{el}$ such that,

$$
\begin{aligned}
2\mathcal{E}^{el} &= \langle \underset{\sim}{\sigma}^* : \underset{\sim}{\varepsilon}' \rangle \\
&= \frac{1}{V}\int_V \underset{\sim}{\sigma}^* : \underset{\sim}{\varepsilon}' dV \\
&= \frac{1}{V}\int_V \sigma^*_{ij} u'_{(i,j)} dV \\
&= \frac{1}{V}\int_V \left(\sigma^*{}_{ij} u'_i\right)_{,j} dV \\
&= \frac{1}{V}\int_{\partial V} \sigma^*{}_{ij} n_j u'_i dS \\
&= \frac{1}{V}\int_{\partial V} \left(\underset{\sim}{\sigma}^* \cdot \underline{n}\right) \cdot \underline{u}' dS
\end{aligned} \tag{4.16}
$$

### 4.2.4 *Boundary Conditions*

It is necessary to set boundary conditions to the volume $V$ considered in order to solve the constitutive equations in the case of statics. Let us consider three types of boundary conditions.

#### 4.2.4.1 Kinematic Uniform Boundary Conditions—KUBC

Displacement $\underline{\boldsymbol{u}}$ is prescribed for any material point $\underline{\boldsymbol{x}}$ on the boundary $\partial V$ such that,

$$
\underline{\boldsymbol{u}} = \underset{\sim}{\boldsymbol{E}} \cdot \underline{\boldsymbol{x}} \quad \forall \underline{\boldsymbol{x}} \in \partial V \tag{4.17}
$$

with $\underset{\sim}{\boldsymbol{E}}$ second-order macroscopic strain tensor, which is symmetric and independent of $\underline{\boldsymbol{x}}$. It follows from (4.17) and (4.14):

$$
\langle \underset{\sim}{\varepsilon} \rangle = \frac{1}{V}\int_V \underset{\sim}{\varepsilon} dV = \underset{\sim}{\boldsymbol{E}} \tag{4.18}
$$

The macroscopic stress tensor is then defined as the spatial average of the local stress field:

$$
\underset{\sim}{\boldsymbol{\Sigma}} := \langle \underset{\sim}{\sigma} \rangle = \frac{1}{V}\int_V \underset{\sim}{\sigma} dV \tag{4.19}
$$

#### 4.2.4.2 Static Uniform Boundary Conditions—SUBC

Traction $\underline{t}$ is prescribed for any material point $\underline{x}$ on $\partial V$ such that,

$$\underline{t} = \underset{\sim}{\Sigma} \cdot \underline{n} \quad \forall \underline{x} \in \partial V \tag{4.20}$$

with $\underset{\sim}{\Sigma}$ second-order macroscopic stress tensor, which is symmetric and independent of $\underline{x}$. It follows from (4.20) and (4.15):

$$\langle \underset{\sim}{\sigma} \rangle = \frac{1}{V} \int_V \underset{\sim}{\sigma} dV = \underset{\sim}{\Sigma} \tag{4.21}$$

The macroscopic strain tensor is then defined as the spatial average of the local strain field:

$$\underset{\sim}{E} := \langle \underset{\sim}{\varepsilon} \rangle = \frac{1}{V} \int_V \underset{\sim}{\varepsilon} dV \tag{4.22}$$

#### 4.2.4.3 Periodic Boundary Conditions—PBC

For PBC, the displacement field $\underline{u}$ can be dissociated into a part given by the macroscopic strain tensor $\underset{\sim}{E}$ and a periodic fluctuation field for any material point $\underline{x}$ of $V$, such that:

$$\underline{u} = \underset{\sim}{E} \cdot \underline{x} + \underline{v} \quad \forall \underline{x} \in V \tag{4.23}$$

with $\underline{v}$ the periodic fluctuations vector, i.e. taking the same value on two homologous points $\underline{x}^+$ and $\underline{x}^-$ of $\partial V$. Furthermore, the traction vector $\underline{t} = \underset{\sim}{\sigma} \cdot \underline{n}$ fulfills anti-periodic conditions such that,

$$\underset{\sim}{\sigma}^+ \cdot \underline{n}^+ + \underset{\sim}{\sigma}^- \cdot \underline{n}^- = \underline{0} \tag{4.24}$$

$$\underline{v}^+ - \underline{v}^- = \underline{0} \tag{4.25}$$

Periodicity is denoted by # while anti-periodicity is denoted by −#. A dual approach exists; it consists in prescribing a macroscopic stress to the cell. However we do not develop this approach here, cf. [142] for details.

### 4.2.5 Hill–Mandel Condition

Let us consider a volume $V$ with two independent local fields $\underset{\sim}{\varepsilon}'$ and $\underset{\sim}{\sigma}^*$ such that $\underset{\sim}{\varepsilon}'$ is kinematically compatible and $\underset{\sim}{\sigma}^*$ is statically admissible. If $\underset{\sim}{\sigma}^*$ verifies SUBC, or $\underset{\sim}{\varepsilon}'$ verifies KUBC, or if $\underset{\sim}{\sigma}^*$ and $\underset{\sim}{\varepsilon}'$ verify simultaneously the periodic boundary conditions, then:

$$\langle \underset{\sim}{\sigma}^* : \underset{\sim}{\varepsilon}' \rangle = \langle \underset{\sim}{\sigma}^* \rangle : \langle \underset{\sim}{\varepsilon}' \rangle \tag{4.26}$$

Thus, one obtains the following equivalence for the three types of boundary conditions:

$$\langle \underset{\sim}{\sigma} : \underset{\sim}{\varepsilon} \rangle = \langle \underset{\sim}{\sigma} \rangle : \langle \underset{\sim}{\varepsilon} \rangle \tag{4.27}$$

which corresponds to the Hill macrohomogeneity condition [95]. This ensures that the mechanical work density at the microscale is preserved while scaling up to the macroscopic level.

### 4.2.6 *Effective Properties Versus Apparent Properties*

When determining the properties of the volume $V$ smaller than the RVE, apparent properties are considered. The apparent properties converge towards the effective properties once $V \geq V_{\mathrm{RVE}}$.

The micromechanical linear elastic problem admits a unique solution, up to a rigid body displacement for SUBC and a periodic translation for PBC. Let us consider two fourth-order tensors $\underset{\approx}{A}$ and $\underset{\approx}{B}$ accounting respectively for strain localization and stress concentration:

$$\underset{\sim}{\varepsilon}(\underline{x}) = \underset{\approx}{A}(\underline{x}) : \underset{\sim}{E} \qquad \forall \underline{x} \in V \text{ and } \forall \underset{\sim}{E} \tag{4.28}$$

and

$$\underset{\sim}{\sigma}(\underline{x}) = \underset{\approx}{B}(\underline{x}) : \underset{\sim}{\Sigma} \qquad \forall \underline{x} \in V \text{ and } \forall \underset{\sim}{\Sigma} \tag{4.29}$$

such that,

$$\langle \underset{\approx}{A} \rangle = \langle \underset{\approx}{B} \rangle = \underset{\approx}{I} \tag{4.30}$$

Let us consider the elastic moduli $\underset{\approx}{c}(\underline{x})$ and the compliances $\underset{\approx}{s}(\underline{x})$, then:

$$\underset{\sim}{\sigma}(\underline{x}) = \underset{\approx}{c}(\underline{x}) : \underset{\sim}{\varepsilon}(\underline{x}) \qquad \forall \underline{x} \in V \tag{4.31}$$

and

$$\underset{\sim}{\varepsilon}(\underline{x}) = \underset{\approx}{s}(\underline{x}) : \underset{\sim}{\sigma}(\underline{x}) \qquad \forall \underline{x} \in V \tag{4.32}$$

Thus,

$$\underset{\sim}{\Sigma} = \langle \underset{\sim}{\sigma} \rangle = \langle \underset{\approx}{C} : \underset{\sim}{\varepsilon} \rangle = \langle \underset{\approx}{c} : \underset{\approx}{A} : \underset{\sim}{E} \rangle = \langle \underset{\approx}{c} : \underset{\approx}{A} \rangle : \underset{\sim}{E} \tag{4.33}$$

and

$$\underset{\sim}{E} = \langle \underset{\sim}{\varepsilon} \rangle = \langle \underset{\approx}{S} : \underset{\sim}{\sigma} \rangle = \langle \underset{\approx}{s} : \underset{\approx}{B} : \underset{\sim}{\Sigma} \rangle = \langle \underset{\approx}{s} : \underset{\approx}{B} \rangle : \underset{\sim}{\Sigma} \tag{4.34}$$

We can define $\underset{\approx}{C}^{\text{app}}_{E}$ and $\underset{\approx}{S}^{\text{app}}_{\Sigma}$, fourth-order symmetric tensors, accounting respectively for the apparent elastic moduli and compliances of the volume $V$ considered such that,

$$\underset{\approx}{C}^{\text{app}}_{E} = \langle \underset{\approx}{c} : \underset{\approx}{A} \rangle \tag{4.35}$$

and

$$\underset{\approx}{S}^{\text{app}}_{\Sigma} = \langle \underset{\approx}{s} : \underset{\approx}{B} \rangle \tag{4.36}$$

These equations show that homogenized properties are not usually obtained by a simple rule of mixtures.

Also, one can define the apparent properties from the elastic strain energy density $\mathcal{E}^{\text{el}}$:

$$\mathcal{E}^{\text{el}} = \frac{1}{2} \langle \underset{\sim}{\sigma} : \underset{\sim}{\varepsilon} \rangle = \frac{1}{2} \langle \underset{\sim}{\varepsilon} : \underset{\approx}{c} : \underset{\sim}{\varepsilon} \rangle = \frac{1}{2} \underset{\sim}{E} : \langle \underset{\approx}{A}^T : \underset{\approx}{c} : \underset{\approx}{A} \rangle : \underset{\sim}{E} \tag{4.37}$$

and

$$\mathcal{E}^{\text{el}} = \frac{1}{2} \langle \underset{\sim}{\sigma} : \underset{\sim}{\varepsilon} \rangle = \frac{1}{2} \langle \underset{\sim}{\sigma} : \underset{\approx}{s} : \underset{\sim}{\sigma} \rangle = \frac{1}{2} \underset{\sim}{\Sigma} : \langle \underset{\approx}{B}^T : \underset{\approx}{s} : \underset{\approx}{B} \rangle : \underset{\sim}{\Sigma} \tag{4.38}$$

This way, we obtain a new definition of the apparent elastic moduli and compliances:

$$\underset{\approx}{C}^{\text{app}}_{E} = \langle \underset{\approx}{A}^T : \underset{\approx}{c} : \underset{\approx}{A} \rangle \tag{4.39}$$

and

$$\underset{\approx}{S}^{\text{app}}_{\Sigma} = \langle \underset{\approx}{B}^T : \underset{\approx}{s} : \underset{\approx}{B} \rangle \tag{4.40}$$

This new definition justifies the symmetric nature of the apparent elastic moduli and compliance tensors. By applying the Hill–Mandel lemma (cf. Sect. 4.2.5) one can prove the equivalence between direct and energetic definitions [185].

According to [179], for an elementary volume $V$ large enough ($V > V_{\text{RVE}}$), the apparent properties do not depend on the boundary conditions and match with the effective properties of the considered material, then:

$$\underset{\approx}{C}^{\text{app}}_{\Sigma} = \underset{\approx}{S}^{\text{app}\,-1}_{\Sigma} = \underset{\approx}{C}^{\text{app}}_{E} = \underset{\approx}{S}^{\text{app}\,-1}_{E} = \underset{\approx}{C}^{\text{eff}} = \underset{\approx}{S}^{\text{eff}\,-1} \tag{4.41}$$

For volumes ($V \geq V_{\text{RVE}}$), based on energetic considerations and the subadditivity property of the effective elastic moduli tensor, [98] proposed the so-called partition theorem. The effective properties can be bounded by the following inequalities:

$$\underset{\approx}{C}^{\mathrm{app}}_{\Sigma} \le \underset{\approx}{C}^{\mathrm{eff}} \le \underset{\approx}{C}^{\mathrm{app}}_{E} \tag{4.42}$$

$$\underset{\approx}{S}^{\mathrm{app}}_{E} \le \underset{\approx}{S}^{\mathrm{eff}} \le \underset{\approx}{S}^{\mathrm{app}}_{\Sigma} \tag{4.43}$$

These inequalities have to be considered in the sense of quadratic forms. For elementary volumes smaller than the RVE, using the same arguments but for partitions of different sizes, [98] derived hierarchical inequalities regarding apparent and effective properties. Coarse and fine partitions are considered and their respective statistical apparent properties are denoted by indices $_c$ and $_f$:

$$\underset{\approx}{C}^{\mathrm{Reuss}} \le \underset{\approx}{C}^{\mathrm{app}}_{\Sigma f} \le \underset{\approx}{C}^{\mathrm{app}}_{\Sigma c} \le \underset{\approx}{C}^{\mathrm{eff}} \le \underset{\approx}{C}^{\mathrm{app}}_{Ec} \le \underset{\approx}{C}^{\mathrm{app}}_{Ef} \le \underset{\approx}{C}^{\mathrm{Voigt}} \tag{4.44}$$

$$\underset{\approx}{S}^{\mathrm{Voigt}} \le \underset{\approx}{S}^{\mathrm{app}}_{Ef} \le \underset{\approx}{S}^{\mathrm{app}}_{Ec} \le \underset{\approx}{S}^{\mathrm{eff}} \le \underset{\approx}{S}^{\mathrm{app}}_{\Sigma c} \le \underset{\approx}{S}^{\mathrm{app}}_{\Sigma f} \le \underset{\approx}{S}^{\mathrm{Reuss}} \tag{4.45}$$

$\underset{\approx}{C}^{\mathrm{Voigt}}$, $\underset{\approx}{S}^{\mathrm{Voigt}}$, $\underset{\approx}{C}^{\mathrm{Reuss}}$ and $\underset{\approx}{S}^{\mathrm{Reuss}}$ refer to the classical Voigt and Reuss bounds [172, 206]. The inequalities presented above can be used for verification of computational homogenisation results, as it was done for instance in [114, 115] for elastic and thermal properties. Moreover, the bounds $\underset{\approx}{C}^{\mathrm{app}}_{\Sigma}$ and $\underset{\approx}{C}^{\mathrm{app}}_{E}$ are usually far apart when the contrast of properties between phases is large. If the microstructure features a matrix phase, tighter bounds can be obtained by choosing elementary volumes including only the matrix at the boundary, as shown in [182, 183].

### 4.2.7 *Computational Homogenization Using the Finite Element Method*

In order to determine homogenized mechanical properties for a given microstructure, one has to solve boundary value problems in statics. The finite element (FE) method has proved to be quite an efficient technique to solve this kind of problems even in the case of highly nonlinear phenomena [29, 38, 81].

#### 4.2.7.1 FE Formulation of the Principle of Virtual Work

Galerkin's approach for continuum mechanics is implemented and used with the principle of virtual work. In each of the $n$ elements $e$, knowing the nodal displacements $\{u_e^*\}$, one can compute the virtual displacement field $\underline{u}^*$ and the strain tensor $\underset{\sim}{\varepsilon}$ as follows:

$$\underline{u}^* = [N]\{u_e^*\} \tag{4.46}$$

and,

$$\underset{\sim}{\varepsilon} = [B]\{u_e^*\} \tag{4.47}$$

with $[N]$, the shape function matrix and $[B]$, the matrix of shape function derivatives. Then, for all $\{u_e^*\}$ with prescribed body forces $\underline{f}$ and surface forces $\underline{F}$:

$$\sum_{e=1}^{n}\left(\int_{V_e} \underset{\sim}{\sigma}\left(\{u_e^*\}\right)[B]\{u_e^*\}\,dV\right)=\sum_{e=1}^{n}\left(\int_{V_e} \underline{f}\,[N]\{u_e^*\}\,dV+\int_{\partial V_e} \underline{F}\,[N]\{u_e^*\}\,dS\right) \tag{4.48}$$

Thus,

$$\sum_{e=1}^{n}\left(\{\mathcal{F}_e^{\text{int}}\}-\{\mathcal{F}_e^{\text{ext}}\}\right)\{u_e^*\}=0 \tag{4.49}$$

with $\{\mathcal{F}_e^{\text{int}}\}$ and $\{\mathcal{F}_e^{\text{ext}}\}$ respectively internal and external forces, in each element $e$, such that for the global problem:

$$\{\mathcal{F}^{\text{int}}\}=\int_{\Omega}[B]^T\underset{\sim}{\sigma}\left(\{u_i^*\}\right)dV \tag{4.50}$$

and,

$$\{\mathcal{F}^{\text{ext}}\}=\int_{\Omega}[N]^T\underline{f}\,dV+\int_{\partial\Omega}[N]^T\underline{F}\,dS \tag{4.51}$$

Balance between internal and external forces is achieved with a Newton iterative algorithm using the stiffness matrix $[K]$:

$$\begin{aligned}[K]&=\frac{\partial\{\mathcal{F}^{\text{int}}\}}{\partial\{u_i^*\}}\\&=\int_{\Omega}[B]^T\frac{\partial\{\sigma\}}{\partial\{\varepsilon\}}\frac{\partial\{\varepsilon\}}{\partial\{u_i^*\}}dV\\&=\int_{\Omega}[B]^T\frac{\partial\{\sigma\}}{\partial\{\varepsilon\}}[B]\,dV\end{aligned} \tag{4.52}$$

which yields, for linear elastic problems:

$$[K]=\int_{\Omega}[B]^T[\underset{\approx}{c}]\,[B]\,dV \tag{4.53}$$

#### 4.2.7.2 Application to Linear Elasticity

In the case of linear elasticity within a volume $V$ fulfilling RVE requirements, one can compute the effective elastic moduli $\underset{\approx}{\boldsymbol{C}}$ or compliances $\underset{\approx}{\boldsymbol{S}}$ using (4.6) by prescribing either the macroscopic strain $\underset{\sim}{\boldsymbol{E}}$ or macroscopic stress $\underset{\sim}{\boldsymbol{\Sigma}}$:

$$\begin{bmatrix} \Sigma_{11} \\ \Sigma_{22} \\ \Sigma_{33} \\ \Sigma_{23} \\ \Sigma_{31} \\ \Sigma_{12} \end{bmatrix} = \begin{bmatrix} C_{11} & C_{12} & C_{13} & C_{14} & C_{15} & C_{16} \\ - & C_{22} & C_{23} & C_{24} & C_{25} & C_{26} \\ - & - & C_{33} & C_{34} & C_{35} & C_{36} \\ - & - & - & C_{44} & C_{45} & C_{46} \\ - & - & - & - & C_{55} & C_{56} \\ - & - & - & - & - & C_{66} \end{bmatrix} \begin{bmatrix} E_{11} \\ E_{22} \\ E_{33} \\ 2E_{23} \\ 2E_{31} \\ 2E_{12} \end{bmatrix} \quad (4.54)$$

$$\begin{bmatrix} E_{11} \\ E_{22} \\ E_{33} \\ 2E_{23} \\ 2E_{31} \\ 2E_{12} \end{bmatrix} = \begin{bmatrix} S_{11} & S_{12} & S_{13} & S_{14} & S_{15} & S_{16} \\ - & S_{22} & S_{23} & S_{24} & S_{25} & S_{26} \\ - & - & S_{33} & S_{34} & S_{35} & S_{36} \\ - & - & - & S_{44} & S_{45} & S_{46} \\ - & - & - & - & S_{55} & S_{56} \\ - & - & - & - & - & S_{66} \end{bmatrix} \begin{bmatrix} \Sigma_{11} \\ \Sigma_{22} \\ \Sigma_{33} \\ \Sigma_{23} \\ \Sigma_{31} \\ \Sigma_{12} \end{bmatrix} \quad (4.55)$$

Linear relations thus appear between macroscopic stress and strain, and can readily be used to build up effective compliance and elastic moduli tensors for a given microstructure. The formalism is similar for any linear property, e.g. thermal conductivity. Such an approach has been successfully implemented for architectured materials in [31, 54, 56, 58, 87, 101, 104].

#### 4.2.7.3 The Element DOF Method for Periodic Problems

In the case of periodic boundary conditions (cf. Sect. 4.2.4.3), there is an alternative to the FE formulation presented in Sect. 4.2.7.1. It consists in adding global DOFs shared by all elements. These DOFs correspond to the macroscopic strain components $E_{ij}$ for displacements $v_i$, in addition to classical nodal DOFs. The balance equations can thus be written as follows:

$$\begin{aligned} \int_V \sigma_{ij} u_{i,j} dV &= \int_V \sigma_{ij} (E_{ik} x_k + v_i)_{,j} dV \\ &= \int_V \sigma_{ij} E_{ij} dV + \int_V \sigma_{ij} v_{i,j} dV \\ &= \int_V \sigma_{ij} E_{ij} dV + \int_V \left(\sigma_{ij} v_i\right)_{,j} dV \\ &= \int_V \sigma_{ij} E_{ij} dV + \underbrace{\int_{\partial V} \sigma_{ij} v_i n_j dS}_{=0} \\ &= \int_V \sigma_{ij} dV E_{ij} \\ &= V \Sigma_{ij} E_{ij} \\ &= RE_{ij} E_{ij} \end{aligned} \quad (4.56)$$

The FE problem left to solve concerns the homogeneous strain tensor $E_{ij}$ and its dual $RE_{ij}$, which corresponds to the macroscopic reaction stress. Prescribing $E_{ij}$ corresponds to the macroscopic strain approach, while prescribing $RE_{ij}$ leads to the macroscopic stress approach. In that way, mixed macroscopic problems, e.g. tension, can be solved with periodic boundary conditions. Implementation of additional degrees of freedom in the FE framework is done as follows:

$$\{\underset{\sim}{\varepsilon}\} = [B]\{u\} + \{\underset{\sim}{E}\} \tag{4.57}$$

Also,

$$\{\underset{\sim}{\varepsilon}\} = [B']\{u'\} \tag{4.58}$$

with

$$[B'] = \begin{bmatrix} 1 & & & & & & & \\ & 1 & & & & & & \\ & & 1 & & & & & \\ & & & 1 & & & & \\ & & & & 1 & & & \\ & & & & & 1 & & \\ & & & & & & N^i & \\ & & & & & & & \ddots \end{bmatrix} \tag{4.59}$$

and

$$\{u\} = \begin{Bmatrix} E11 \\ E22 \\ E33 \\ E23 \\ E31 \\ E12 \\ u^i \\ \vdots \end{Bmatrix} \tag{4.60}$$

### 4.2.8 *Case of Application: Periodic Auxetics*

#### 4.2.8.1 Auxetics

Auxetic materials are a type of architectured materials exhibiting a negative Poisson's ratio. They present interesting advantages for both functional and structural applications. In the case of isotropic elasticity, mechanical behavior is described by any couple of parameters among these: Young's modulus $E$, Poisson's ratio $\nu$, the bulk modulus $k$ and Lamé's coefficients $\lambda$ and $\mu$ (also referred to as the shear modulus). Poisson's ratio is defined as the ratio of the contraction in the transverse direction to the extension in the longitudinal direction. Material stability requires the tensor of elastic moduli to be positive definite, resulting in a positive Young's modulus $E$ and a Poisson's ratio $\nu$ ranging from $-1$, for unshearable materials, and 0.5 for incompressible or rubber-like materials. Most materials naturally present a positive Poisson's ratio, although negative Poisson's ratio materials, or auxetics [69], have been engineered since the mid-1980s with the pioneering works of [4, 93] and [124]. This new class of materials has been drawing more and more attention since then [2, 7, 23, 37, 55, 56, 58, 68, 80, 100, 113, 119, 122, 125, 144, 155, 169, 178, 193, 194, 212], as well as their potential applications [15, 28, 46, 67, 135]. Auxetic materials have also been expected to present enhanced acoustic damping [131]; this was shown experimentally in [43, 44]. The use of auxetics as building blocks for wave-guiding metamaterials has also been investigated in [13, 198]. Moreover, experiments on auxetic foams seem to provide evidence of better resistance to crash compared to conventional cellular materials [188].

Poisson's ratio is defined only for isotropic elastic materials. In the anisotropic case, negative apparent Poisson's ratio can be defined as the opposite of the ratio between transverse and longitudinal strains for a given specific direction. There is no restriction anymore on the values of the apparent Poisson ratio $\nu^*$. We call anisotropic auxetics, materials for which negative apparent Poisson's ratio is observed for a sufficiently large range of tensile directions.

#### 4.2.8.2 Computational Strategy

The computational framework used for homogenization has been presented in Sect. 4.2.7. For this application, effective mechanical properties are computed over a unit-cell (defined by its periodicity vectors $\underline{v_i}$) with periodic boundary conditions (PBC) using FE. Homogenization requires separation between micro and macro scales. In the case of periodic homogenization, the computed effective properties correspond to those of an infinite continuum made of periodic tiles.

Let us consider an elementary volume $V$ including a solid phase $V_s$ and porous one $V_p$. In the latter, the stress field is extended by setting $\underset{\sim}{\sigma} = \underset{\sim}{0}$ in $V_p$. Practically, the macroscopic strain $\underset{\sim}{E}$ and stress $\underset{\sim}{\Sigma}$ are computed by averaging the local fields $\underset{\sim}{\varepsilon}$ and $\underset{\sim}{\sigma}$. Within the porous phase, stresses are assumed to be equal to zero. Using periodic boundary conditions, and applying a homogeneous macroscopic strain field $\underset{\sim}{E}$ (cf. Sect. 4.2.7.3), it yields:

$$
\begin{aligned}
\underset{\sim}{\boldsymbol{\Sigma}} = \langle \underset{\sim}{\sigma} \rangle &= \frac{1}{V}\int_V \underset{\sim}{\sigma} dV \\
&= \frac{1}{V}\int_{V_s} \underset{\sim}{\sigma} dV + \underbrace{\frac{1}{V}\int_{V_p} \underset{\sim}{\sigma} dV}_{=0} \\
&= \frac{V_s}{V}\frac{1}{V_s}\int_{V_s} \underset{\sim}{\sigma} dV \\
&= V_V^s \langle \underset{\sim}{\sigma} \rangle_s
\end{aligned} \tag{4.61}
$$

with $V_V^s$, volume fraction of the solid phase. Let us now consider the case of a prescribed homogeneous macroscopic stress $\underset{\sim}{\boldsymbol{\Sigma}}$:

$$
\begin{aligned}
\underset{\sim}{\boldsymbol{E}} = \langle \underset{\sim}{\varepsilon} \rangle &= \frac{1}{V}\int_V \underset{\sim}{\varepsilon} dV \\
&= \frac{1}{V}\int_{V_s} \underset{\sim}{\varepsilon} dV + \frac{1}{V}\int_{V_p} \underset{\sim}{\varepsilon} dV \\
&= \frac{V_s}{V}\frac{1}{V_s}\int_{V_s} \underset{\sim}{\varepsilon} dV + \frac{V_p}{V}\frac{1}{V_p}\int_{V_p} \underset{\sim}{\varepsilon} dV \\
&= V_V^s \langle \underset{\sim}{\varepsilon} \rangle_s + V_V^p \langle \underset{\sim}{\varepsilon} \rangle_p
\end{aligned} \tag{4.62}
$$

with $V_V^p$, volume fraction of the porous phase. $\langle \underset{\sim}{\varepsilon} \rangle_p$ can be computed from the values at the boundary of the porous phase.

#### 4.2.8.3 Periodic Unit-Cell: Hexachiral Lattice

This chiral microstructure was first proposed by Lakes [125], then studied in [2, 55, 56, 58, 169]. Based on the parameters defined in [2], cell geometry can be described in this way: the circular nodes have radius r, the ligaments have length L, and both have in common wall thickness t (cf. Fig. 4.1a) as well as depth d, which in our case is considered infinite due to periodicity conditions along direction 3. Hence, three dimensionless parameters can be derived as shown in (4.63).

$$
\alpha = \mathrm{L/r} \qquad \beta = \mathrm{t/r} \qquad \gamma = \mathrm{d/r} \tag{4.63}
$$

On Fig. 4.1b, $\alpha = 5$, $\beta = 0.25$ and $\gamma \to +\infty$. These parameters correspond to a volume fraction of 15%. The 6-fold symmetry provides transverse isotropy.

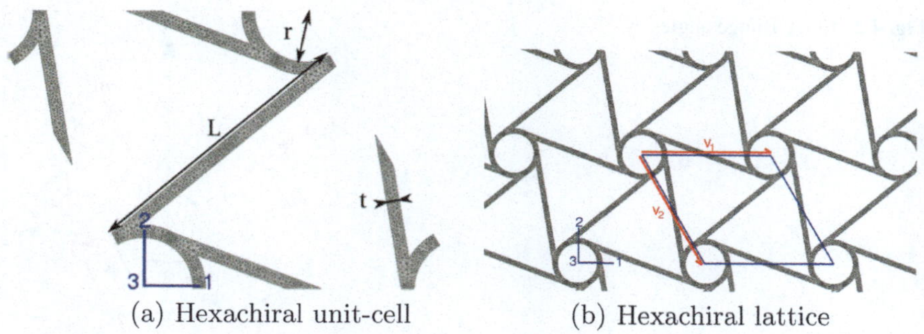


**Fig. 4.1** **a** Periodic cell with geometric parameters. **b** Hexachiral lattice with unit-cell (blue) and periodicity vectors $\underline{v_1}$ and $\underline{v_2}$ (red)

#### 4.2.8.4 Effective Elastic Properties

Elastic moduli tensor $\underset{\approx}{\boldsymbol{C}}$ is computed over a periodic unit-cell using Z-Set FE software.[1] Meshes are composed of volumic fully-integrated quadratic elements, such as 10-node tetrahedra and 20-node hexahedra, taking into account the finite thickness of the microstructure components. Using the Euler-Bunge [36] angles $\phi$, $\theta$ and $\psi$ as shown on Fig. 4.2, let us define 3 orthogonal vectors $\underline{\boldsymbol{l}}$, $\underline{\boldsymbol{m}}$ and $\underline{\boldsymbol{n}}$, such as:

$$[\underline{\boldsymbol{l}}] = \begin{bmatrix} \cos(\phi)\cos(\psi) - \sin(\phi)\sin(\psi)\cos(\theta) \\ \sin(\phi)\cos(\psi) + \cos(\phi)\sin(\psi)\cos(\theta) \\ \sin(\psi)\sin(\theta) \end{bmatrix} \tag{4.64}$$

$$[\underline{\boldsymbol{m}}] = \begin{bmatrix} -\cos(\phi)\sin(\psi) - \sin(\phi)\cos(\psi)\cos(\theta) \\ -\sin(\phi)\sin(\psi) + \cos(\phi)\cos(\psi)\cos(\theta) \\ \cos(\psi)\sin(\theta) \end{bmatrix} \tag{4.65}$$

$$[\underline{\boldsymbol{n}}] = \begin{bmatrix} \sin(\phi)\sin(\theta) \\ -\cos(\phi)\sin(\theta) \\ \cos(\theta) \end{bmatrix} \tag{4.66}$$

Using macroscopic strain and stress tensors $\underset{\sim}{\boldsymbol{E}}(\phi, \theta, \psi)$ and $\underset{\sim}{\boldsymbol{\Sigma}}(\phi, \theta, \psi)$, one can now define the Young modulus $E(\underline{\boldsymbol{l}})$ and effective Poisson ratio $\nu^*(\underline{\boldsymbol{l}}, \underline{\boldsymbol{m}})$ for tension along direction $\underline{\boldsymbol{l}}$:

$$E = \frac{\underline{\boldsymbol{l}} . \underset{\sim}{\boldsymbol{\Sigma}} . \underline{\boldsymbol{l}}}{\underline{\boldsymbol{l}} . \underset{\sim}{\boldsymbol{E}} . \underline{\boldsymbol{l}}} \tag{4.67}$$

$$\nu^* = -\frac{\underline{\boldsymbol{m}} . \underset{\sim}{\boldsymbol{E}} . \underline{\boldsymbol{m}}}{\underline{\boldsymbol{l}} . \underset{\sim}{\boldsymbol{E}} . \underline{\boldsymbol{l}}} \tag{4.68}$$

[1] http://www.zset-software.com/.

Fig. 4.2 Euler-Bunge angles

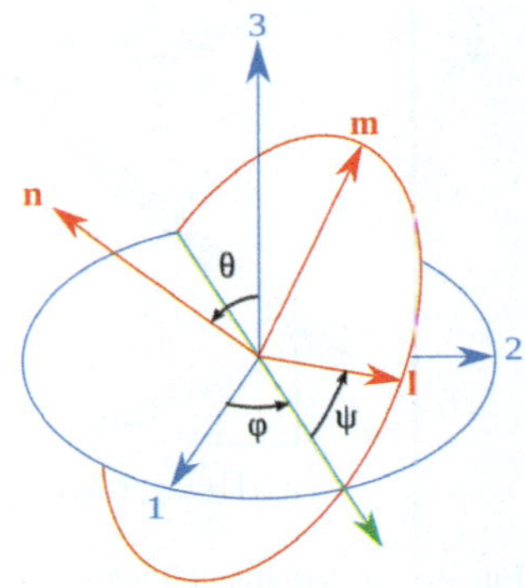


If we now consider simple shear in the plane $(\underline{l}, \underline{m})$, the shear modulus $\mu(\underline{l}, \underline{m})$ can be defined as follows:

$$\mu = \frac{\underline{l}.\underset{\sim}{\Sigma}.\underline{m}}{\underline{l}.\underset{\sim}{E}.\underline{m}} \tag{4.69}$$

For $\theta = \psi = 0$, elastic moduli and Poisson's ratio are obtained in the plane (1,2) of the microstructure as functions of $\phi$, we will refer to those as in-plane elastic properties. On the other hand, when $\phi = 0$ and $\theta = \frac{\pi}{2}$, one obtains moduli and Poisson's ratio within plane (1,3) as functions of $\psi$. These values will be considered as out-of-plane elastic properties.

For comparison purposes, normalized elastic moduli are defined using $f_V$, volume fraction of material, local constitutive isotropic elastic material parameters such as $E_0$ (Young's modulus) and $\mu_0$. Shear modulus $\mu_0$ is defined from $E_0$ and Poisson's ratio $\nu_0$ as follows:

$$\mu_0 = \frac{E_0}{2(1+\nu_0)}. \tag{4.70}$$

Thus, normalized Young's modulus $E^*$ is obtained as follows:

$$E^* = \frac{1}{E_0 f_V} E \tag{4.71}$$

Normalized shear modulus $\mu^*$ is defined in this way:

$$\mu^* = \frac{1}{\mu_0 f_V} \mu \tag{4.72}$$

In-plane elastic properties are shown in Table 4.1. The use of auxetic lattices in engineering applications might involve out-of-plane loading. Hence, $\nu^*$, $E^*$ and $\mu^*$ were plotted against $\psi$ in Fig. 4.3a (polar plot). For this work, $E_0 = 210{,}000$ MPa and $\nu_0 = 0.3$ are the isotropic elastic properties of the constitutive material. Resulting elastic moduli tensors are presented hereafter as (4.73). Components are expressed in MPa.

**Table 4.1** In-plane Poisson's ratio and normalized elastic moduli

| | Hexachiral |
|---|---|
| $\nu^*$ | −0.74 |
| $E^*$ | $2.3 \times 10^{-2}$ |
| $\mu^*$ | $2.3 \times 10^{-1}$ |

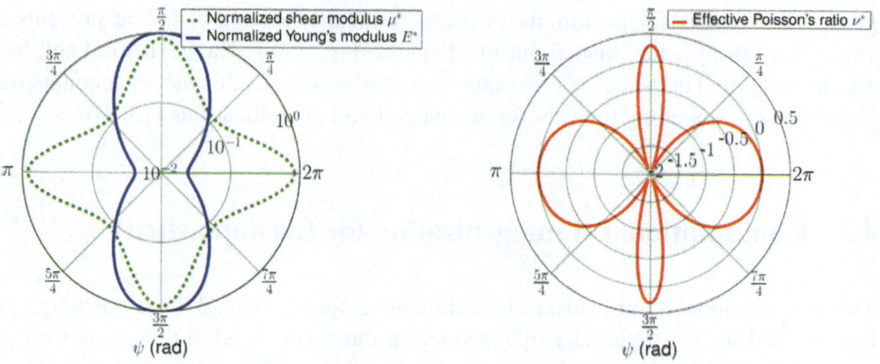


(a) $\mu^*, E^* = f(\psi)$ (b) $\nu^* = f(\psi)$

**Fig. 4.3** Results for hexachiral lattice ($\theta = \frac{\pi}{2}, \phi = 0$)

#### 4.2.8.5 Results

Elastic moduli tensor was computed for the hexachiral lattice as shown in (4.73). Transverse isotropy is verified since $\frac{C_{11}-C_{12}}{2} = C_{66}$.

$$[\underset{\approx}{C}] = \begin{bmatrix} 1650 & -1218 & 130 & 0 & 0 & 0 \\ -1218 & 1650 & 130 & 0 & 0 & 0 \\ 130 & 130 & 31968 & 0 & 0 & 0 \\ 0 & 0 & 0 & 5075 & 0 & 0 \\ 0 & 0 & 0 & 0 & 5075 & 0 \\ 0 & 0 & 0 & 0 & 0 & 1434 \end{bmatrix} \quad (4.73)$$

Components were used to obtain the in-plane properties gathered in Table 4.1. $\nu^*$ is underestimated compared to the value from [2], while our estimation of the normalized Young modulus $E^*$ is higher. This is discussed later. Figure 4.3a shows an increase of $E^*$ when the material is streched out-of-plane, while reaching its maximum value along direction 3 ($\psi = \frac{\pi}{2}$ or $\psi = \frac{3\pi}{2}$). Figure 4.3b shows that Poisson's ratio $\nu^*$ is always negative, except for $\psi = 0$ or $\psi = \pi$ where $\nu^*$ is close to 0, and $\psi = \frac{\pi}{2}$ or $\psi = \frac{3\pi}{2}$ where it takes the constitutive material value 0.3. Normalized shear modulus $\mu^*$ fluctuates in a decade around the in-plane value depending on angle $\psi$.

#### 4.2.8.6 Discussion

Values obtained for $E^*$ (cf. Table 4.1) exceed those from [2]. This is due to the boundary conditions of the FEM problem. With periodicity over displacements and nodal force loading, the prescribed loading in [2] corresponds to a static uniform boundary conditions (SUBC) micromechanical problem, which is known to lead to an underestimation of the elastic moduli [114]. On the other hand, the PBC problem gives exact results for an infinite medium. Both the hexachiral lattice presents a strong anisotropy when loaded out-of-plane (cf. Fig. 4.3a). The hexachiral cell has also been studied in the case of second-gradient elasticity, as this model accounts for chirality, hence opening the way for acoustic cloaking applications [13, 16].

## 4.3 Computational Homogenization for Elastoplasticity

The computational homogenization method developed in Sect. 4.2 for elasticity can be extended for elastoplaticity by considering the elastic moduli tensor, as well as an equivalent plasticity model to account for nonlinearities.

A case of application is made from the study of elastoplasticity [55] in the hexachiral auxetic lattice introduced in Sect. 4.2.8.3.

Let us consider the following yield function $f(\underset{\sim}{\sigma})$:

$$f(\underset{\sim}{\sigma}) = \sigma^{eq} - r \tag{4.74}$$

with the *von Mises* equivalent stress,

$$\sigma^{eq} = \sqrt{\frac{3}{2}\underset{\sim}{\sigma}^{dev} : \underset{\sim}{\sigma}^{dev}} \tag{4.75}$$

where $\underset{\sim}{\sigma}^{dev}$ is the deviatoric part of the stress tensor. Linear isotropic hardening rule is adopted:

$$r = r_o + hp \tag{4.76}$$

where $r_o$ is the yield stress, $h$ the hardening modulus and $p$ is the cumulative plastic strain variable.

### *4.3.1 Plastic Anisotropy*

Local material is now considered isotropic *von Mises*-elastoplastic. Plastic material parameters are shown in Table 4.2.

**Table 4.2** Local constitutive plastic parameters

| | |
|---|---|
| Yield stress (MPa) | 100 |
| Hardening modulus (MPa) | 1000 |

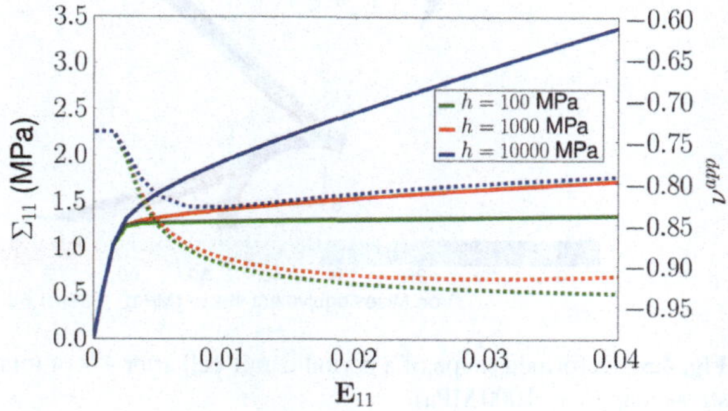


**Fig. 4.4** Stress (plain curves) and apparent Poisson's ratio (dashed curves) versus strain response for 3 different hardening moduli

First, the auxetic behavior is investigated. Although the parameters given in Table 4.2 will be used in the following sections, a short parametric study has been performed in order to assess the effect of the hardening modulus on the Poisson ratio. Uniaxial strain-controlled tensile test is performed along direction 1 until 4% of total macroscopic strain. The homogenized cell exhibits a nonlinear elastoplastic behavior (cf. Fig. 4.4). Now, if one considers the ratio of transverse over longitudinal macroscopic strains, an apparent Poisson ratio can be defined in the nonlinear regime as defined in (4.77) and plotted on Fig. 4.4. From these curves we observe that the auxetic nature of the lattice is kept with plasticity.

The effect is even stronger than in elasticity when the hardening modulus is in the range ($h$ = 100 MPa and $h$ = 1000 MPa). The auxetic effect is dependent on the size of the plastic zone in the unit cell. If the plastic zone is confined in a small domain around the junction between the rotating nodes and the connecting beams, as shown on Fig. 4.5 for low values of the hardening modulus, the auxetic deformation mechanism is strengthened. For $h$ = 10,000 MPa, the plastic zone spreads almost over the entire cell, thus fading the effect of plasticity on the auxetic behavior. The hardening modulus value, $h$ = 1000 MPa, is kept for the rest of this work since it is of the same order of magnitude of several common alloys.

$$\nu^{\text{app}} = -\frac{\mathbf{E}_{22}}{\mathbf{E}_{11}} \tag{4.77}$$

Now, anisotropy in the plastic regime is investigated. As a matter of fact, there is no guarantee for the 6-fold symmetric material to behave isotropically in the plastic regime. Polar plots shown on Fig. 4.6a–c are obtained from uniaxial tensile and shear tests in every direction of the plane (1,2). Each point corresponds to a test for a different direction with angle $\phi$ from the principal direction 1 of the structure defined

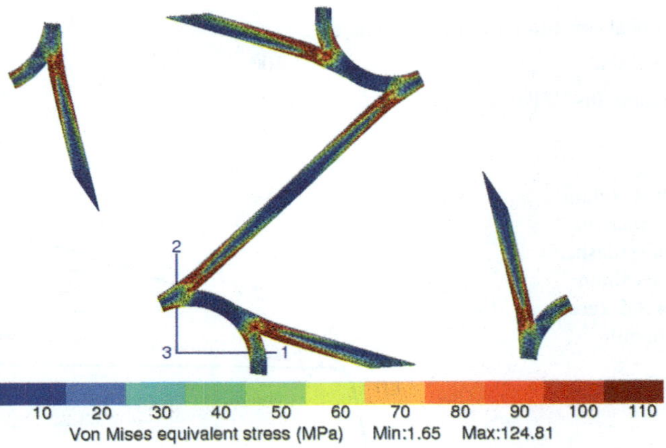


**Fig. 4.5** Deformed shape of a periodic unit-cell after 4% of total strain, with von Mises equivalent stress map ($h = 1000\,\text{MPa}$)

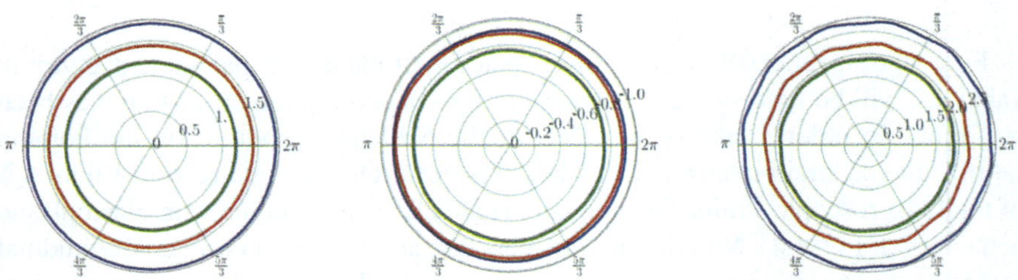

(a) Tensile stress (MPa) (b) Poisson's ratio (c) Shear stress (MPa)

**Fig. 4.6** **a** Tensile stress, **b** apparent Poisson's ratio, and **c** shear stress for 0.2% (green), 1% (red) and 4% (blue) total strain

on Fig. 4.5. Figure 4.6a, c shows stress level versus angle $\phi$ for three given strain states: respectively 0.2% (green), 1% (red) and 4% (blue) total strain for tension, and 0.1, 0.5 and 2% for shear. Figure 4.6b uses the same color code but for the apparent Poisson ratio versus angle $\phi$ for the same given tension states. The three plots show a quasi-transversely isotropic response for the hexachiral lattice with plasticity.

### *4.3.2 Macroscopic Modeling*

An additional upscaling can be performed. The mesoscopic elastoplastic behavior is now modelled as a constitutive behavior for further use in large structural computations. First, let us consider an isotropic compressible plasticity model such as those developed by Abouaf et al. [1] and Green [85] for porous metals, and by Deshpande and Fleck [52] and Miller [143] for cellular materials. An extension to the anisotropic case was proposed by Badiche et al. [17] and Forest et al. [76].

Let us now consider a yield function $f(\underset{\sim}{\boldsymbol{\Sigma}})$ such as,

$$f(\underset{\sim}{\boldsymbol{\Sigma}}) = \Sigma_{eq} - R \tag{4.78}$$

where $R$ is the macroscopic yield stress. Moreover, let us adopt the following equivalent yield stress:

$$\Sigma_{eq} = \sqrt{\frac{3}{2} C\, \underset{\sim}{\boldsymbol{\Sigma}}^{dev} : \underset{\sim}{\boldsymbol{\Sigma}}^{dev} + F \left(\mathrm{Tr}\, \underset{\sim}{\boldsymbol{\Sigma}}\right)^2} \tag{4.79}$$

where Tr $\underset{\sim}{\boldsymbol{\Sigma}}$ is the trace of the stress tensor. $C$ and $F$ are coefficients accounting for the relative influence of deviatoric and hydrostatic stress, they are usually expressed as functions of the porosity $\rho$ for isotropic materials.

Associated plasticity is assumed, such as the macroscopic plastic strain rate is:

$$\dot{\underset{\sim}{\mathbf{E}}}^p = \frac{\partial f}{\partial \underset{\sim}{\boldsymbol{\Sigma}}} = \frac{\dot{p}}{\sigma_{eq}} \left( \frac{3}{2} C\, \underset{\sim}{\boldsymbol{\Sigma}}^{dev} + F \left(\mathrm{Tr}\, \underset{\sim}{\boldsymbol{\Sigma}}\right) \underset{\sim}{\mathbf{1}} \right) \tag{4.80}$$

In the case of uniaxial tension, we define the in-plane plastic Poisson ratio:

$$\nu^p = -\frac{\dot{\mathbf{E}}^p_{22}}{\dot{\mathbf{E}}^p_{11}} = -\frac{F - \frac{C}{2}}{C + F} = \frac{\frac{C}{2} - F}{C + F} \tag{4.81}$$

When $F = 0$, incompressible plasticity is recovered. If $C = 1$, then $\nu^p < 0$ for $F > \frac{1}{2}$ and $\lim_{F \to +\infty} \nu^p = -1$. $\nu^p$ as a function of $F$ is plotted on Fig. 4.7.

Such a plasticity model is not fully capable of describing the anisotropic behavior of our microstructure along direction 3, especially transverse contraction when tension is applied in plane (1,2). In order to simplify the model, instead of using a fully anisotropic Hill tensor applied to the deviatoric stress tensor and a separate contri-

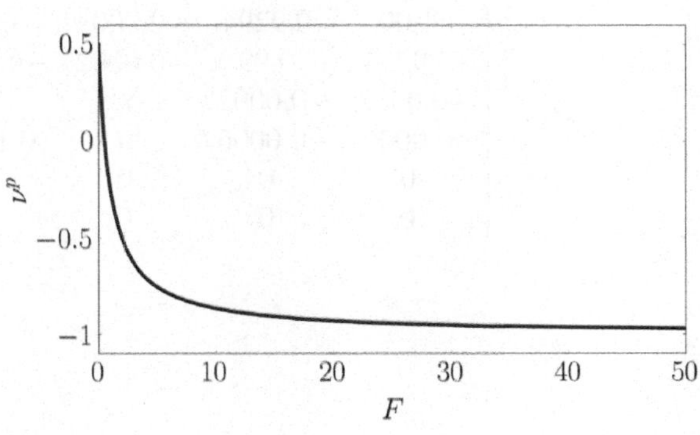


**Fig. 4.7** Plastic Poisson's ratio for an isotropic material as a function of parameter $F$, with $C = 1$

bution of the hydrostatic stress, we consider here a generalized Hill tensor applied to the Cauchy stress tensor.

We consider the same yield function $f(\underset{\sim}{\boldsymbol{\Sigma}})$ as in (4.78) with the following equivalent yield stress:

$$\Sigma^{eq} = \sqrt{\underset{\sim}{\boldsymbol{\Sigma}} : \underset{\approx}{\boldsymbol{H}} : \underset{\sim}{\boldsymbol{\Sigma}}} \tag{4.82}$$

where $\underset{\approx}{\boldsymbol{H}}$ is the applied generalized Hill fourth-rank tensor.

For the hardening rule, we consider an isotropic hardening function with a non-linear potential and a linear part:

$$\mathrm{R} = \mathrm{R}_0 + \mathrm{H}p + \mathrm{Q}(1 - e^{-\mathrm{b}p}). \tag{4.83}$$

### 4.3.3 Simulation and Identification

In order to determine parameters for the model, some of them are determined from reference curves obtained by periodic simulations of the unit-cell. Then comparison between reference data with results computed on a RVE is made and optimization of macroscopic material parameters is run using a simplex algorithm. The experimental database includes tensile, shear and Poisson's ratio curves. While loading in tension, we consider out-of-plane contraction. However, direction 3 and out-of-plane shear are not taken into account. Tensorial components of $\underset{\approx}{\boldsymbol{H}}$ (cf. 4.85) and parameters for the hardening rule (4.83) are thus identified:

$$R = 1.29 + 8.61p + 0.1\left(1 - e^{-140p}\right) \tag{4.84}$$

Comparison between curves from full-field simulations and the identified macroscopic model provides a good correlation as shown on the tensile stress and apparent Poisson's ratio versus strain curve (cf. Fig. 4.8), the shear stress versus strain curve (cf. Fig. 4.9) and the transverse strain vs. longitudinal strain curve (cf. Fig. 4.10).

$$[\underset{\approx}{\boldsymbol{H}}] = \begin{bmatrix} 1.00 & 0.9294 & -0.00031 & 0.0006 & 0 & 0 \\ 0.9294 & 0.99 & -0.00027 & -0.00067 & 0 & 0 \\ -0.00031 & -0.00027 & \times & 0 & 0 & 0 \\ 0.0006 & -0.00067 & 0 & 0.11554 & 0 & 0 \\ 0 & 0 & 0 & 0 & \times & 0 \\ 0 & 0 & 0 & 0 & 0 & \times \end{bmatrix} \tag{4.85}$$

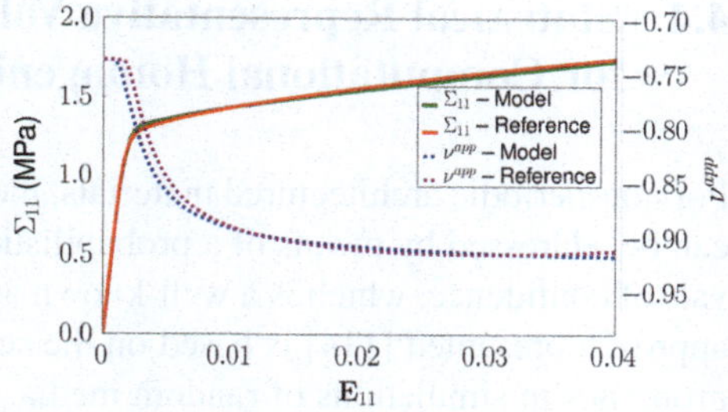


**Fig. 4.8** Stress and apparent Poisson's ratio versus strain for full-field simulation and macroscopic model for an uniaxial tensile test along direction 1

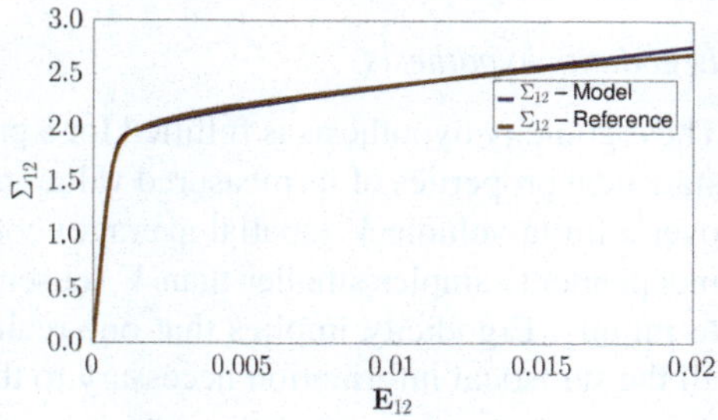


**Fig. 4.9** Stress versus strain for full-field simulation and macroscopic model for an pure shear test in plane (1,2)

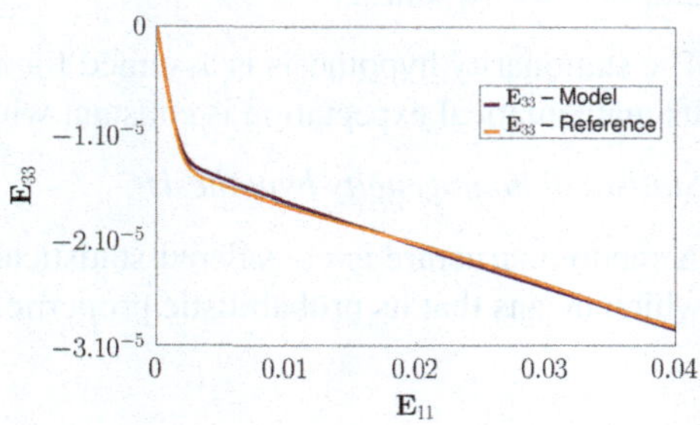


**Fig. 4.10** Transverse strain versus longitudinal strain for full-field simulation and macroscopic model for an uniaxial tensile test along direction 1

### *4.3.4 Conclusions*

Full-field simulations and macroscopic modelling using an anisotropic compressible plasticity framework have been performed for an auxetic microstructure: the hexachiral lattice. Plasticity of auxetics has been explored, showing that the auxetic effect persists and becomes even stronger with plastic yielding. It was also shown that the effect of plasticity on auxeticity fades with the expansion of the plastic zone. The plastic anisotropy for this 6-fold symmetric lattice is becoming weaker with plastic saturation. The proposed fully anisotropic Hill criterion seems to be suitable for modeling periodic architectured cellular materials as it was able to catch negative Poisson's ratio, transverse contraction, and volume change.

## 4.4 Statistical Representative Volume Element Size for Computational Homogenization

For non-periodic architectured materials, the problem of representativity of samples can be addressed by means of a probabilistic approach giving size-dependent intervals of confidence, which is a well-known approach used in geostatistics [137]. The approach presented [114] is based on the scaling effect on the variance of effective properties in simulations of random media. Several assumptions have to be considered regarding the statistics of the microstructures considered.

*Ergodicity hypothesis*

The ergodicity hypothesis is fulfilled for a property or a random function $Z$ when the statistical properties of its measured value (mathematical expectation, variance, etc.) over a finite volume $V$ (spatial average) converge to those estimated over series of independent samples smaller than $V$ (ensemble average), when the volume $V$ goes to infinity. Ergodicity implies that one realization of a volume $V \geq V_{\mathrm{RVE}}$ contains all the statistical information necessary to the description of its microstructure.

*Stationarity hypothesis*

The stationarity hypothesis is assumed for a property or a random function $Z$ when its mathematical expectation is constant with respect to time and space.

*Statistical homogeneity hypothesis*

A random structure is considered statistically homogeneous, when it is stationary, which means that its probabilistic properties are invariant by translation.

### 4.4.1 RVE Size Determination for Media with Finite Integral Range

Let us consider a microstructure that fulfills the ergodicity and stationarity conditions for a given physical quantity $Z(x)$ regarded as a random function with average $E\{Z(x)\}$ and point variance $D_Z^2$. The ensemble variance $D_Z^2(V)$ of its average value $\overline{Z}(V)$ over the domain $\Omega$ with volume $V$ can be obtained using the centered second-order correlation function $\overline{W}_2$ in this way:

$$D_Z^2(V) = \frac{1}{V} \iint_{\Omega} \overline{W}_2(x-y)dxdy \tag{4.86}$$

with

$$\overline{W}_2(h) = E\left\{\left(Z(x) - \overline{Z}\right)\left(Z(x+h) - \overline{Z}\right)\right\} \tag{4.87}$$

For determining the RVE size for the physical property $Z$ one can rely on the geostatistical notion of integral range [39, 109, 112, 126, 127, 138]. The integral range $A_n$ is homogeneous to a volume of dimension $n$ in $\mathbb{R}^n$. For $n = 3$, the integral range is given by:

$$A_3 = \frac{1}{D_Z^2} \int_{\mathbb{R}^3} \overline{W}_2(h) dh \tag{4.88}$$

The physical interpretation of the integral range is such that for a given volume $V$, one can define $n = \frac{V}{A_3}$ volume elements for which the $i$ average values $Z_i(V')$ over the $n$ sub-volumes $V' = \frac{V}{n}$ are uncorrelated random variables. Hence, for a large specimen, i.e. $V \gg A_3$, (4.86) can be rewritten introducing the point variance of $Z$, $D_Z^2$ as follows:

$$D_Z^2(V) = D_Z^2 \frac{A_3}{V} \tag{4.89}$$

Let us analyze this asymptotic relation. First, in general one has no guarantee on the finiteness of point variance $D_Z^2$ [137]: let us consider a large domain $\Omega$ and a smaller domain $V \subset \Omega$ that is attainable by means of experimentation or computation, one can then compute an experimental variance which is in fact a function of $\Omega$ supported by $V$, that will increase with $\Omega$. If the variance over $V$ is finite, it should be regarded as a limit of the experimental variance for $\Omega \to +\infty$. $D_Z^2$ can be computed over $V$ as follows:

$$\begin{aligned} D_Z^2 &= \frac{1}{V} \int_V \left(Z(x) - \overline{Z}\right)^2 dV \\ &= \frac{1}{V} \int_V Z^2(x) - \overline{Z}^2 dV \\ &= \frac{1}{V} \int_V Z^2(x) dV - \left(\frac{1}{V} \int_V Z(x) dV\right)^2 \end{aligned} \tag{4.90}$$

On the other hand, the ensemble variance $D_Z^2(V')$ is computed from the average values $\overline{Z_i}$ over $n$ sub-volumes:

$$\begin{aligned} D_Z^2(V') &= \frac{1}{n} \sum_{i=1}^{n} \left(Z_i(V') - \overline{Z_i}\right)^2 \\ &= \frac{1}{n} \sum_{i=1}^{n} Z_i^2(V') - \overline{Z_i}^2 \\ &= \frac{1}{n} \sum_{i=1}^{n} Z_i^2(V') - \left(\frac{1}{n} \sum_{i=1}^{n} Z_i(V')\right)^2 \end{aligned} \tag{4.91}$$

(4.91) uses the average value of the average values $\overline{Z_i}$ over $n$ sub-volumes $V'$, which is expected to converge towards the effective property $Z_{\text{eff}}$ when $V \to +\infty$. If $Z_{\text{eff}}$ is already known, it might be of interest to use it instead of $\overline{Z_i}$ in order to obtain a better estimate.

If $Z(x)$ is the indicator function of the stationary random set $A$, then one can obtain analytically the variance of the local volume fraction as a function of the point variance as follows:

$$D_Z^2 = p - p^2 = p(1-p) \tag{4.92}$$

with $p$, probability for a point $x$ to belong to the random set $A$, which is equivalent to the volume fraction of $A$ in $V$.

The asymptotic scaling law given in (4.89) can be used for any additive variable $Z$ over the domain $\Omega$. In the case of elastic properties for instance, average stress $\langle \underset{\sim}{\sigma} \rangle$ or strain $\langle \underset{\sim}{\varepsilon} \rangle$ fields have to be computed. For determining the RVE size for a given property $Z$, one thus has to know its integral range $A_3$. There is no theoretical covariance for mechanical fields. However, there are two ways to estimate it; first by assuming that $\overline{Z}$ is equal to the arithmetic average of properties (rule of mixtures) for a biphasic medium, hence (4.90) yields:

$$D_Z^2 = p(1-p)\,(Z_1 - Z_2)^2 \tag{4.93}$$

with $Z_1$ and $Z_2$, respectively property $Z$ of phase 1 and 2. $D_Z^2$ can also be estimated computationally on the largest virtual sample available, in order to minimize boundary layer effects and obtain a converged value. The approach proposed by di Paola [53] consists in taking only into account the inner part of the simulation volume. This could present an advantage for determining the point variance.

Once the point variance has been estimated for a given property, the integral range can be obtained using the procedure proposed by Matheron [139] for any random function: consider realizations of domains $\Omega$ with an increasing volume $V$ (or non-overlapping sub-domains of large simulations, with a wide range of sizes), the parameter $A_3$ can be estimated by fitting the obtained variance according to (4.89):

$$\log D_Z^2(V) = \log D_Z^2 + \log A_3 - \log V \tag{4.94}$$

Following the method proposed in [114], itself based on the approach developed in [39], considering a large number $n$ of realisations (or sub-volumes), the following sampling error in the estimation of the effective properties arises:

$$\epsilon_{\text{abs}} = \frac{2D_Z(V)}{\sqrt{n}} \tag{4.95}$$

From which the relative error $\epsilon_{\text{rel}}$ can be defined:

$$\epsilon_{\text{rel}} = \frac{\epsilon_{\text{abs}}}{\overline{Z}} = \frac{2D_Z(V)}{\overline{Z}\sqrt{n}} \Rightarrow \epsilon_{\text{rel}}^2 = \frac{4D_Z^2 A_3}{\overline{Z}^2 nV} \tag{4.96}$$

Hence a volume size that we will consider statistically representative can be defined for a prescribed property $Z$, number of realizations $n$ and relative error (e.g. 5%):

$$V_{\text{RVE}} = \frac{4D_Z^2 A_3}{\epsilon_{\text{rel}}^2 \overline{Z}^2 n} \tag{4.97}$$

This RVE size then depends on the point variance $D_Z^2$, integral range $A_3$ and mean value $\overline{Z}$, 3 parameters that are estimated from simulations.

### 4.4.2 *Generalization of the Statistical Approach to Microstructures with Non-finite Integral Range*

The method presented above is now adapted and generalized to the case of media with non-finite integral range, especially Poisson linear varieties and Boolean random models made of Poisson linear varieties, e.g. Poisson fibers, which will be used hereafter for modeling non-woven architectured materials. Since the integral range of linear Poisson varieties is not finite [107], (4.89) does not apply anymore. It was proposed in [126] to use a modified scaling law with exponent $\gamma \neq 1$. The variance can thus be rewritten as follows [110]:

$$D_Z^2(V) = D_Z^2 \left( \frac{A_3^*}{V} \right)^{\gamma} \tag{4.98}$$

which yields by linearization,

$$\log D_Z^2(V) = \log D_Z^2 + \gamma \log A_3^* - \gamma \log V \tag{4.99}$$

$A_3^*$ is not the integral of the centered second-order correlation function $\overline{W}_2(h)$ anymore, as defined before in (4.88). Nonetheless, it is homogeneous to a volume of material and can readily be used to determine RVE sizes which can then be obtained by updating the previous definition for relative error:

$$\epsilon_{\text{rel}} = \frac{\epsilon_{\text{abs}}}{\overline{Z}} = \frac{2D_Z(V)}{\overline{Z}\sqrt{n}} \Rightarrow \epsilon_{\text{rel}}^2 = \frac{4D_Z^2 A_3^{*\gamma}}{\overline{Z}^2 n V^{\gamma}} \tag{4.100}$$

Hence yielding an updated definition of the RVE size:

$$V_{\text{RVE}} = A_3^* \sqrt[\gamma]{\frac{4D_Z^2}{\epsilon_{\text{rel}}^2 \overline{Z}^2 n}} \tag{4.101}$$

The generalized integral range $A_3^*$ and scaling-law exponent $\gamma$ can be estimated from simulations as it was done in [6, 57, 114]. When considering statistical RVE sizes of microstructures with non-finite integral range for other properties than morphological ones, for which there is no information about the theoretical value of the point variance $D_Z^2$, it may be useful to reformulate (4.98) as follows:

$$D_Z^2(V) = KV^{-\gamma} \tag{4.102}$$

with $K = D_Z^2 A_3^{*\gamma}$, leaving only 2 parameters to identify from the statistical data obtained by simulation. We will adopt this formulation when studying Poisson fibers in Sect. 4.4.3. The method for determining statistical RVE sizes has been studied and used for media with finite integral range in the [6, 31, 105, 106, 114, 115, 133, 134, 152, 156, 199]. This approach is implemented for media with infinite integral range in [57], for the case of Poisson fibers.

In most papers, the authors resorted to periodic boundary conditions (PBC) since [114] showed from computational experiments that mean apparent properties obtained with PBC converge faster towards the effective properties than with the Dirichlet and Neumann-type BC. Nevertheless, KUBC and SUBC can be useful since they correspond to the Voigt and Reuss bounds in elasticity. They can thus be used for bounding the effective properties of random architectured materials. If the microstructure features a matrix phase, tighter bounds can be obtained by choosing elementary volumes including only the matrix at the boundary, as shown in [182].

### *4.4.3 Non-woven Architectured Materials*

A challenging candidate material was imagined for assessing the applicability of such methods to architectured materials: stochastic random networks made of infinite fibers, more specifically Poissonian fibrous networks. The determination of the effective properties of Poisson fibers is not trivial as it is a porous medium with infinite integral range. This pathological morphology could give rise to gigantic RVE sizes, maybe even no RVE at all, i.e. yield non-homogenizability in the sense of [14]. Although infinite fibers do not exist in nature, they can be considered a limit case representative of sintered long-fiber non-woven materials, such as those studied by Mezeix et al. [141]. This random fibrous architectured material can be modeled computationally. The generation of such virtual specimens relies on a tridimensional Poisson point implantation process. Due to the long-range correlation induced by the model of random structures chosen, the size of the representative volume element is *a priori* unknown. Moreover, periodization of these microstructures is generally impossible, making the periodic homogenization tools ineffective. Very large virtual samples computations have to be performed in order to compute their representative volume element sizes a posteriori based on statistical arguments [114] and finally predict their effective mechanical properties.

Many studies are available regarding finite-length fibrous media and strongly oriented infinite-fiber media. For instance, [50, 158] dealt with the morphology of 3D long-fiber media randomly oriented in-plane, [192] used a 3D random model of randomly oriented long-fibers for the design of an acoustic absorber, and [20, 21] generated virtual samples of long but finite fibers for modeling the mechanics of entangled materials. None of these studies accounted for the representativity of samples. Oumarou et al. [153] computed RVE sizes for 2D random arrays of fibers, using the statistical method of [114]. The works of [89, 161, 162, 196] on 2D fibrous fractal networks also deal with the representativity and homogenization of such, yet self-similar, fibrous media.

The non-woven material model used hereafter corresponds to a 3D stochastic network composed by randomly oriented and distributed infinitely-long interpenetrating rectilinear fibers. It exhibits an infinite integral range [107], i.e. an infinite morphological correlation length; this medium is non-periodizable without modifying its morphology, thus falling beyond the spectrum of periodic homogenization and the definition of RVE proposed by Sab and Nedjar [181].

### 4.4.4 *Case of Application: RVE Size of Random Fibrous Media*

The example is taken from [57] where the enquiring reader can find details and a full analysis of the problem of representativity in fibrous random media. The results presented hereafter are based on the equations introduced in Sect. 4.2.8.2 for the homogenization of porous media.

As shown on Fig. 4.11, the elementary volume considered in the computation is the volume $V$ that is composed of two complementary phases $V_f$ and $V_p$, respectively accounting for the Poisson fibers and the porous phase.

#### 4.4.4.1 Mixed Mechanical Boundary Conditions—MBC

A type of mixed BC, appropriate for porous random media, is presented here, instead of SUBC, in order to compute values that we can compare with those obtained using KUBC. Since only a minimal number of DOFs are prescribed on the boundary $\partial V$, mixed mechanical boundary conditions (MBC) are less constraining, or “softer”, than KUBC. They differ from the mixed uniform boundary conditions (MUBC) proposed in [90, 91] which are restricted, in general, to tensile loading of a volume element. They also differ from the normal mixed BC (NMBC) proposed by Gelebart et al. [82], since those are applied on the whole boundary $\partial V$, similarly to MUBC. The periodicity compatible mixed uniform boundary conditions (PMUBC) were proposed by Pahr and Zysset [154] as a generalization of MUBC, and implemented by Chateau et al. [42]. The BC proposed here are similar to PMUBC, but simpler

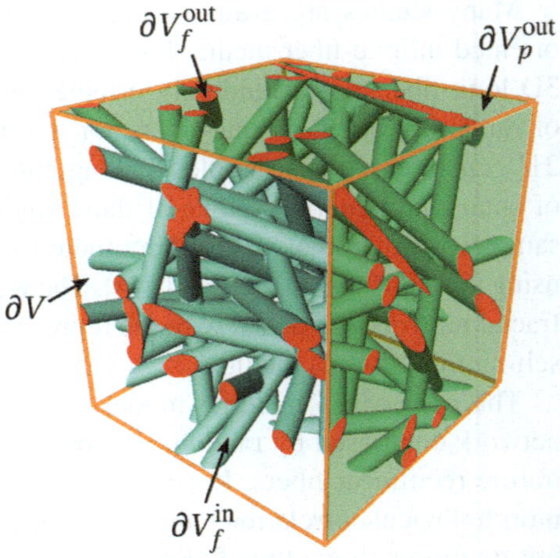


**Fig. 4.11** Example of an elementary volume of Poisson fibers used in the simulation

because only 2 loading configurations are considered. MBC are considered with a view to estimating the overall bulk and shear moduli of isotropic random porous media, thus being less general than PMUBC; the 2 loading cases are presented hereafter.

*Mixed triaxial loading*

Displacement $\underline{u}$ is prescribed along normals on $\partial V$, such that:

$$\begin{aligned} u_1 &= E_{11}x_1 \quad \forall \underline{x} \in F_1 \\ u_2 &= E_{22}x_2 \quad \forall \underline{x} \in F_2 \\ u_3 &= E_{33}x_3 \quad \forall \underline{x} \in F_3 \end{aligned} \tag{4.103}$$

The traction vector $\underset{\sim}{\sigma} \cdot \underline{n}$ is prescribed in this way:

$$\begin{aligned} \sigma_{21}n_1 = \sigma_{31}n_1 &= 0 \quad \forall \underline{x} \in F_1 \\ \sigma_{12}n_2 = \sigma_{32}n_2 &= 0 \quad \forall \underline{x} \in F_2 \\ \sigma_{13}n_3 = \sigma_{23}n_3 &= 0 \quad \forall \underline{x} \in F_3 \end{aligned} \tag{4.104}$$

From (4.62), it yields:

$$\langle \varepsilon_{11} \rangle = E_{11}; \quad \langle \varepsilon_{22} \rangle = E_{22}; \quad \langle \varepsilon_{33} \rangle = E_{33} \tag{4.105}$$

The macroscopic stress is computed using (4.61).

*Mixed shear loading*

Displacement $\underline{u}$ is prescribed along direction 1 on $F_2$ and direction 2 on $F_1$, such that:

$$u_2 = E_{12} x_1 \quad \forall \underline{x} \in F_1$$
$$u_1 = E_{12} x_2 \quad \forall \underline{x} \in F_2 \tag{4.106}$$

Two components of the traction vector $\underset{\sim}{\sigma} \cdot \underline{n}$ are prescribed in this way:

$$\sigma_{11} n_1 = \sigma_{31} n_1 = 0 \quad \forall \underline{x} \in F_1$$
$$\sigma_{22} n_2 = \sigma_{32} n_2 = 0 \quad \forall \underline{x} \in F_2. \tag{4.107}$$

The traction vector is fully prescribed on $F_3$:

$$\sigma_{33} n_3 = \sigma_{13} n_3 = \sigma_{23} n_3 = 0 \quad \forall \underline{x} \in F_3 \tag{4.108}$$

(4.62) gives:

$$\langle \varepsilon_{12} \rangle = E_{12} \tag{4.109}$$

Details of the calculation can be found in [57]. The boundary value problems for estimating the overall bulk and shear moduli are explicited hereafter for both KUBC and MBC.

#### 4.4.4.2 Overall Properties

The fibers are considered linear elastic following the generalized Hooke law:

$$\underset{\sim}{\sigma} = \underset{\approx}{c} : \underset{\sim}{\varepsilon} \quad \forall \underline{x} \in V_f \tag{4.110}$$

with $\underset{\approx}{c}$, fourth-order positive definite tensor of elastic moduli of the isotropic elastic fibers, and $V_f$ volume of the fibers.

*Overall bulk modulus*

For the case of linear elasticity, $\langle \underset{\sim}{\sigma} : \underset{\sim}{\varepsilon} \rangle$ is used as an estimate of $\left( \underset{\approx}{C}^{\text{eff}} : \underset{\sim}{E} \right) : \underset{\sim}{E}$, with $\underset{\approx}{C}^{\text{eff}}$, the tensor of effective elastic moduli of the homogenized medium, that can be rewritten in this way:

$$\underset{\approx}{C}^{\text{eff}} = 3k^{\text{eff}} \underset{\approx}{J} + 2\mu^{\text{eff}} \underset{\approx}{K} \tag{4.111}$$

with $\underset{\approx}{K}$ and $\underset{\approx}{J}$ respectively deviatoric and spherical projector on 2nd-order tensors.

Hence, for a hydrostatic macroscopic strain tensor,

$$\underset{\sim}{E} = \begin{bmatrix} 1 & 0 & 0 \\ 0 & 1 & 0 \\ 0 & 0 & 1 \end{bmatrix} \tag{4.112}$$

it yields:

$$2\mathcal{E}^{el} = \langle \underset{\sim}{\sigma} : \underset{\sim}{\varepsilon} \rangle = \underset{\sim}{\Sigma} : \underset{\sim}{E} = \left( \underset{\approx}{C}^{eff} : \underset{\sim}{E} \right) : \underset{\sim}{E} = 3k^{eff} \underset{\sim}{E} : \underset{\approx}{J} : \underset{\sim}{E} = 9k^{eff} \tag{4.113}$$

when computed on a given volume element, $k^{app} = \frac{1}{9} \langle \underset{\sim}{\sigma} : \underset{\sim}{\varepsilon} \rangle$ is an estimate of $k^{eff}$.

*Overall shear modulus*

For a deviatoric macroscopic strain tensor,

$$\underset{\sim}{E} = \begin{bmatrix} 0 & \frac{1}{2} & 0 \\ \frac{1}{2} & 0 & 0 \\ 0 & 0 & 0 \end{bmatrix} \tag{4.114}$$

it yields:

$$2\mathcal{E}^{el} = \langle \underset{\sim}{\sigma} : \underset{\sim}{\varepsilon} \rangle = \underset{\sim}{\Sigma} : \underset{\sim}{E} = \left( \underset{\approx}{C}^{eff} : \underset{\sim}{E} \right) : \underset{\sim}{E} = 2\mu^{eff} \underset{\sim}{E} : \underset{\approx}{K} : \underset{\sim}{E} = \mu^{eff} \tag{4.115}$$

When computed on a given volume element, $\mu^{app} = \langle \underset{\sim}{\sigma} : \underset{\sim}{\varepsilon} \rangle$ is an estimate of $\mu^{eff}$.

#### 4.4.4.3 Results

The estimation of elastic properties is now performed on hundreds of realizations. Fibers were attributed the constitutive material properties listed in Table 4.3. The variability of each realization gives rise to different results for apparent elastic properties. Mean values over $n$ realizations of the same size are considered in this section. The number of realizations $n$ fluctuates, depending on the size and BC. Ideally, $n$ should be different for each volume size in order to achieve a similar sampling error for all sizes considered. The fluctuation observed here is due to errors during the meshing step; the larger the virtual sample, the more defects in the geometry are likely to happen. Nevertheless, the values gathered in Table 4.4 are large enough for achieving statistical representativity.

**Table 4.3** Constitutive material parameters for Poisson fibers

| | |
|---|---|
| Young's modulus (GPa) | 210 |
| Poisson's ratio | 0.3 |
| Shear modulus (GPa) | 81 |
| Bulk modulus (GPa) | 175 |
| Volume fraction | 0.16 |

**Table 4.4** Number of realizations $n$ considered depending on boundary conditions and simulation size

| BC | Volume | | | | | | | | | |
|---|---|---|---|---|---|---|---|---|---|---|
| | $10^3$ | $20^3$ | $30^3$ | $40^3$ | $50^3$ | $60^3$ | $70^3$ | $80^3$ | $90^3$ | $100^3$ |
| $n$ – KUBC – $\overline{k^{app}}$ | 63 | 63 | 61 | 60 | 56 | 54 | 44 | 51 | 32 | 13 |
| $n$ – KUBC – $\overline{\mu^{app}}$ | 63 | 63 | 61 | 60 | 54 | 54 | 47 | 50 | 27 | 14 |
| $n$ – MBC – $\overline{k^{app}}$ | 40 | 60 | 53 | 55 | 45 | 48 | 37 | 46 | 28 | 13 |
| $n$ – MBC – $\overline{\mu^{app}}$ | 41 | 62 | 58 | 59 | 48 | 49 | 45 | 41 | 11 | 12 |

Results for the elastic properties are obtained over a large number of realizations (between 11 and 63, depending on their size). For the hydrostatic loading defined in (4.112), the ensemble averaged apparent bulk modulus $\overline{k^{app}}$ is computed in this way:

$$\overline{k^{app}} = \frac{1}{n}\sum_{i=1}^{n}(k^{app})_i = \frac{1}{n}\sum_{i=1}^{n}\left(\frac{1}{9}\langle \mathrm{Tr}\underset{\sim}{\sigma}\rangle_f\right)_i \tag{4.116}$$

On the other hand, the ensemble averaged apparent shear modulus $\overline{\mu^{app}}$ is obtained as follows for the shear loading defined in (4.114):

$$\overline{\mu^{app}} = \frac{1}{n}\sum_{i=1}^{n}(\mu^{app})_i = \frac{1}{n}\sum_{i=1}^{n}\left(\langle \sigma_{12}\rangle_f\right)_i \tag{4.117}$$

Results for the bulk modulus $\overline{k^{app}}$ and shear modulus $\overline{\mu^{app}}$ are shown respectively on Figs. 4.12 and 4.13 for the different volumes of simulation and corresponding number of fibers ($\overline{N}$) considered. The mean values obtained for the largest system size considered ($\overline{N} \simeq 800$) are given with the corresponding intervals of confidence $[\overline{Z} \pm 2D_Z]$ in Table 4.5 and compared with analytical bounds. Boundary layer effects are significant for small elementary volume sizes, for both types of BC. Similarly

**Fig. 4.12** Mean values for the bulk modulus depending on the number of fibers

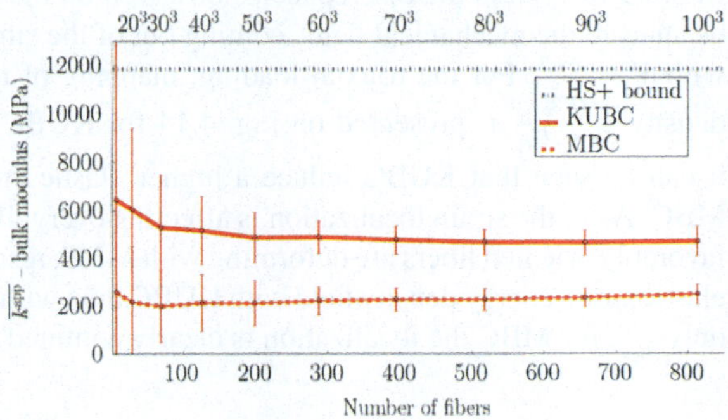

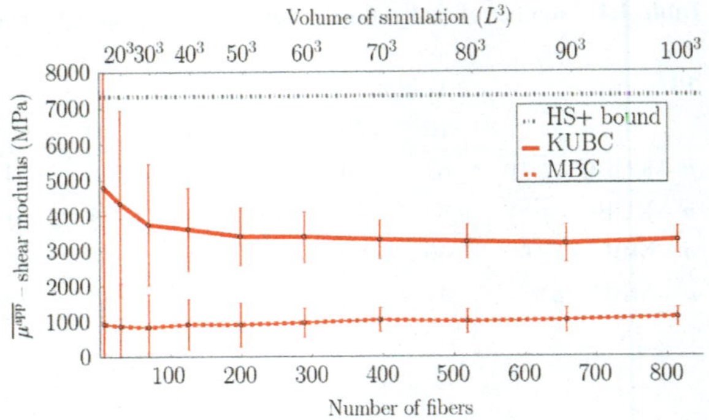


**Fig. 4.13** Mean values for the shear modulus depending on the number of fibers

**Table 4.5** Bounds and mean values for apparent thermal and elastic properties

| | $\overline{k^{app}}$ (MPa) | $\overline{\mu^{app}}$ (MPa) |
|---|---|---|
| FE-Uniform | $4763 \pm 624$ | $3270 \pm 394$ |
| FE-Mixed | $2527 \pm 413$ | $1097 \pm 272$ |
| Hashin–Shtrikman upper bound | 11,839 | 7328 |
| Voigt bound | 28,000 | 12,923 |

to the results observed for the thermal properties, mean apparent values for KUBC are decreasing with increasing volume size, whereas mean estimates obtained with MBC are slightly increasing with $V$. There are very strong discrepancies for mean values obtained with KUBC and MBC for both bulk and shear moduli, even for the largest volume considered: $\Delta\left(\overline{k^{app}}\right) = 88\%$ and $\Delta\left(\overline{\mu^{app}}\right) = 198\%$. Apparent properties obtained for KUBC are constantly above those computed with MBC; this is in accordance with the bounds stated in [91] for mixed boundary conditions, as well as the hierarchies of estimates given by Huet [99].

#### 4.4.4.4 Mechanical Fields

In order to explain the discrepancies observed on the mean apparent properties, let us analyze the mechanical fields coming out of the simulation done on a realization with $V = 50^3$. For the triaxial loading, mapping of the normalized elastic energy density $\frac{\underset{\sim}{\sigma}:\underset{\sim}{\varepsilon}}{V_f\langle\underset{\sim}{\sigma}:\underset{\sim}{\varepsilon}\rangle_f}$ is presented on Fig. 4.14 for KUBC and MBC. From this figure, it can be seen that KUBC induce a higher elastic energy density level than with MBC. Also, the strain localization is affecting every fiber for KUBC, while only the favorably oriented fibers are deforming with MBC, resulting in a more homogeneous elastic strain energy density field with KUBC than with MBC. On Fig. 4.15, mapping only $\frac{\varepsilon_{22}}{E_{22}}$, for MBC the localization is clearly confined to both ends of preferentially

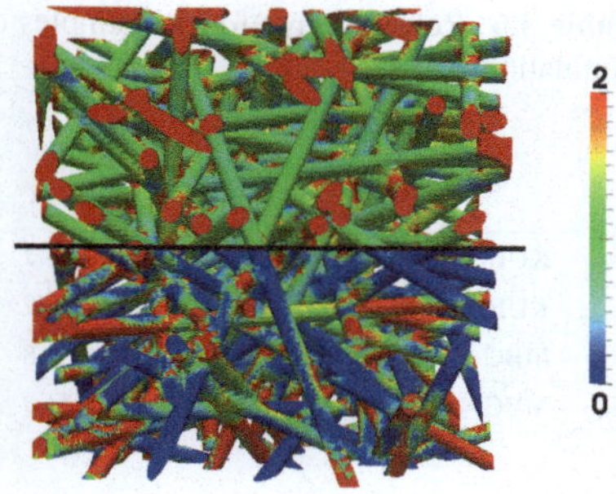


**Fig. 4.14** $\frac{\underset{\sim}{\sigma}:\underset{\sim}{\varepsilon}}{V_f\langle\underset{\sim}{\sigma}:\underset{\sim}{\varepsilon}\rangle_f}$ mapping for a $V = 50^3$ realization under hydrostatic load using KUBC (top) and MBC (bottom)

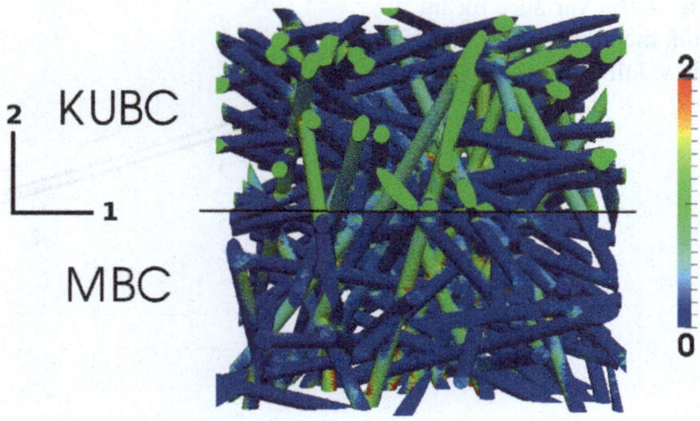


**Fig. 4.15** $\frac{\varepsilon_{22}}{E_{22}}$ mapping for a $V = 50^3$ realization under hydrostatic load using KUBC (top) and MBC (bottom)

oriented fibers, here along the vertical direction. On the other hand, for KUBC most of the deformation takes place all over the boundary $\partial V$, including fibers that are not preferentially oriented.

#### 4.4.4.5 Integral Range, Variance and RVE Size

The theory introduced in Sect. 4.4.2 is now applied. Using (4.101) and (4.102), it yields:

$$V_{\text{RVE}} = \left(\frac{4K}{\epsilon_{\text{rel}}^2 \overline{Z}^2 n}\right)^{\frac{1}{\gamma}} \tag{4.118}$$

Based on (4.100), the relative error associated with the number of realizations considered (cf. Table 4.4) is computed and presented in Table 4.6. As stated earlier, ideally, the error should be the same for all sample series by adapting the number of realizations. Although the relative error fluctuates, most values are very low, except for the smallest sample sizes.

Variance for the apparent bulk and shear moduli as functions of the volume are shown on Figs. 4.16 and 4.17.

The $\gamma$ exponents of the scaling-law for both moduli were estimated from the results of simulations, by fitting the slope of the variance curves. Only the data

**Table 4.6** Relative error for the samples considered $\epsilon_{rel}$ depending on boundary conditions and simulation size

| BC | Volume | | | | | | | | | |
|---|---|---|---|---|---|---|---|---|---|---|
| | $10^3$ (%) | $20^3$(%) | $30^3$ (%) | $40^3$ (%) | $50^3$ (%) | $60^3$ (%) | $70^3$ (%) | $80^3$ (%) | $90^3$ (%) | $100^3$ (%) |
| $\epsilon_{rel}$ – KUBC – $\overline{k^{app}}$ | 9.9 | 4.9 | 3.7 | 2.9 | 2.5 | 2.1 | 2.0 | 1.7 | 1.9 | 2.6 |
| $\epsilon_{rel}$ – KUBC – $\overline{\mu^{app}}$ | 11.9 | 6.2 | 4.7 | 3.7 | 3.3 | 2.7 | 2.6 | 2.2 | 2.7 | 3.3 |
| $\epsilon_{rel}$ – MBC – $\overline{k^{app}}$ | 25.6 | 11.7 | 8.8 | 6.0 | 5.3 | 4.1 | 3.9 | 3.0 | 3.3 | 4.2 |
| $\epsilon_{rel}$ – MBC – $\overline{\mu^{app}}$ | 51.6 | 21.8 | 14.7 | 9.8 | 8.8 | 6.8 | 5.6 | 5.4 | 8.9 | 7.1 |

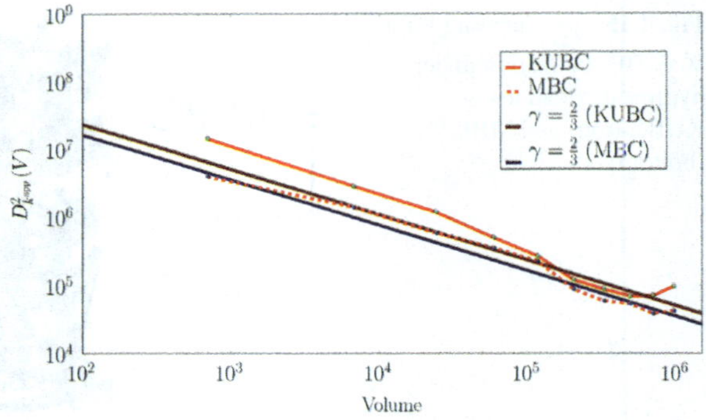


**Fig. 4.16** Variance for the bulk modulus depending on the volume of simulation

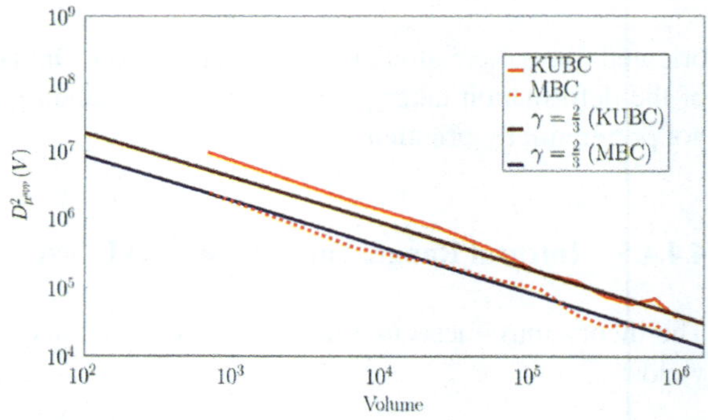


**Fig. 4.17** Variance for the shear modulus depending on the volume of simulation

points for volumes $V \geq 40^3$ have been considered, in order to mitigate any bias due to boundary layer effects with small volumes. The values obtained in this way are gathered in Table 4.7. For the case of Poisson fibers, [110, 111] gave the theoretical value of $\frac{2}{3}$ for $\gamma$ in (4.98) and (4.102). This value holds only for the indicator function of the Poisson fibers and its mean value, the volume fraction. The value estimated here ($\gamma = 0.67$) verifies the theoretical result of [110]. Results from [6], for 3D randomly distributed long-fibers, show a convergence of the variance on the volume fraction of fibers following a scaling law with exponent $\gamma = 0.8$; this value is in-between the values for infinite fibers ($\gamma = \frac{2}{3}$) and short-fibers ($\gamma = 1$), which could be compared to a random distribution of spheres in the extreme case of a shape factor

Table 4.7 Values for $\gamma$ exponent estimated from the simulation

| | $\gamma$ |
|---|---|
| $\overline{k^{app}}$-KUBC | 0.51 |
| $\overline{k^{app}}$-MBC | 0.77 |
| $\overline{\mu^{app}}$-KUBC | 0.64 |
| $\overline{\mu^{app}}$-MBC | 0.66 |
| $\overline{V_V^f}$ | 0.67 |

Table 4.8 RVE sizes estimated from computations with $\gamma = \frac{2}{3}$, $\epsilon_{rel} = 5\%$ and $n = 1$

| Z | BC | $K$ | $V_{RVE}$ | $\overline{N}$ |
|---|---|---|---|---|
| $V_V$ | – | $4.4 \times 10^{-1}$ | $164^3$ | $2.1 \times 10^3$ |
| $k^{app}$ | KUBC | $5.0 \times 10^8$ | $188^3$ | $2.8 \times 10^3$ |
| $k^{app}$ | MBC | $3.5 \times 10^8$ | $296^3$ | $6.9 \times 10^3$ |
| $\mu^{app}$ | KUBC | $4.0 \times 10^8$ | $245^3$ | $4.8 \times 10^3$ |
| $\mu^{app}$ | MBC | $1.8 \times 10^8$ | $489^3$ | $1.9 \times 10^4$ |

equal to 1, similarly to the simulations in [105]. For the apparent shear modulus, $\gamma$ fluctuates between 0.64 and 0.66. Hence, as a 1st-order approximation, these physical properties may depend on the indicator function of the fibers. The same conclusion can be drawn for the apparent bulk modulus, but with further variation since $\gamma$ varies between 0.51 and 0.77 depending on the BC.

Values for RVE size and relative error in Table 4.8 can be compared to the volume considered and associated relative error in Table 4.6: for instance, if one considers the experimental error associated with the estimate of $\overline{k^{app}}$ using KUBC for $V = 100^3$, with $n = 13$ realizations, $\epsilon_{rel} = 2.6\%$. For the same relative error and number of realizations, the asymptotic model, on which values in Table 4.8 are based, yields an RVE size of $V = 100.1^3$, which corresponds to an error of 0.1%. The model is thus appropriate for every data point considered, except in the case of volumes for which boundary layer effects are not negligible.

#### 4.4.4.6 Conclusions

This section questions the applicability of the concept of RVE for pathological cases such as Poisson fibers. The microstructural model was developed and implemented numerically. Hundreds of realizations were computed using FE, generating a large amount of data to be analyzed statistically. Overall, the method proposed in [114] and generalized in [57] for determining RVE sizes can be considered robust and effective, even for pathological microstructures. This is method is thus very useful for the homogenization of random architectured materials.

## 4.5 Conclusions and Outlook

The use of architectured materials in industrial applications is conditional upon the development of appropriate models for such materials. The present chapter covered the most critical aspects of computational homogenization for architectured materials. First, an introduction to classical computational homogenization was made and a description of a computational approach using the finite element method was given. This approach was applied for the specific case of periodic auxetic microstructures. Periodic homogenization elastoplastic constitutive materials was then explored. A systematic statistical approach for the computational homogenization of random architectured materials was presented. The problem of representativity is also critical when considering architectured materials in which the scale separation assumption becomes invalid, i.e. for finite structures made of architectured materials. Recent analytical results show that effective properties can indeed be determined independently of the number of unit cells [30, 160]. The polynomial approach was developed further in the works of [12, 102, 103]. This approach was applied for modeling the homogenized behavior of auxetics [16] and 3D woven textiles based on experiments [170]. However the dependence on the choice of the representative unit cell remains without solution. Approaches like multifield methods [186] and computational continua [73, 74] may be appropriate alternatives to the polynomial approach. For more details about the mechanics of generalized continua, see [5, 75, 180]. The study of structures made of architectured materials has been the object of intensive research in the last few years [168], and has resulted in promising results concerning the development of architectured metamaterials for acoustic [175, 177] and elastodynamic applications [174, 176].

## References

1. M. Abouaf, J.-L. Chenot, G. Raisson, P. Bauduin, Finite element simulation of hot isostatic pressing of metal powders. Int. J. Numer. Methods Eng. **25**, 191–212 (1988)
2. A. Alderson, K.L. Alderson, D. Attard, K.E. Evans, R. Gatt, J.N. Grima, W. Miller, N. Ravirala, C.W. Smith, K. Zied, Elastic constants of 3-, 4- and 6-connected chiral and anti-chiral honeycombs subject to uniaxial in-plane loading. Compos. Sci. Technol. **70**(7), 1042–1048 (2010)
3. G. Allaire, *Shape Optimization by the Homogenization Method* (Springer, Berlin, 2002)
4. R.F. Almgren, An isotropic three-dimensional structure with Poisson's ratio-1. J. Elast. **15**, 427–430 (1985)
5. H. Altenbach, V. Eremeyev (eds.), *Generalized Continua from the Theory to Engineering Applications* (Springer, Berlin, 2013)
6. H. Altendorf, D. Jeulin, F. Willot, Influence of the fiber geometry on the macroscopic elastic and thermal properties. Int. J. Solids Struct. **51**(23–24), 3807–3822 (2014)
7. J.C. Alvarez Elipe, A. Diaz Lantada, Comparative study of auxetic geometries by means of computer-aided design and engineering. Smart Mater. Struct. **21**, 105004 (2012)
8. E. Andreassen, B. Lazarov, O. Sigmund, Design of manufacturable 3d extremal elastic microstructure. Mech. Mater. **69**(1), 1–10 (2014)

9. U. Andreaus, F. dell'Isola, I. Giorgio, L. Placidi, T. Lekszycki, N.L. Rizzi, Numerical simulations of classical problems in two-dimensional (non) linear second gradient elasticity. Int. J. Eng. Sci. **108**, 34–50 (2016)
10. A. Asadpoure, M. Tootkaboni, L. Valdevit, Topology optimization of multiphase architected materials for energy dissipation. Comput. Methods Appl. Mech. Eng. **325**, 314–329 (2017)
11. M.F. Ashby, Y. Bréchet, Designing hybrid materials. Acta Mater. **51**, 5801–5821 (2003)
12. N. Auffray, Analytical expressions for odd-order anisotropic tensor dimension. C. R. Mécaniques **342**(5), 284–291 (2014)
13. N. Auffray, J. Dirrenberger, G. Rosi, A complete description of bi-dimensional anisotropic strain-gradient elasticity. Int. J. Solids Struct. **69–70**, 195–210 (2015)
14. J.-L. Auriault, Heterogeneous medium is an equivalent macroscopic description possible? Int. J. Eng. Sci. **29**(7), 785–795 (1991)
15. S. Babaee, J. Shim, J. Weaver, E. Chen, N. Patel, K. Bertoldi, 3d soft metamaterials with negative poisson's ratio. Adv. Mater. **25**(36), 5044–5049 (2013)
16. A. Bacigalupo, L. Gambarotta, Homogenization of periodic hexa- and tetrachiral cellular solids. Compos. Struct. **116**, 461–476 (2014)
17. X. Badiche, S. Forest, T. Guibert, Y. Bienvenu, J.-D. Bartout, P. Ienny, M. Croset, H. Bernet, Mechanical properties and non-homogeneous deformation of open-cell nickel foams: application of the mechanics of cellular solids and of porous materials. Mater. Sci. Eng. A **289**, 276–288 (2000)
18. E. Bafekrpour, A. Dyskin, E. Pasternak, A. Molotnikov, Y. Estrin, Internally architectured materials with directionally asymmetric friction. Sci. Rep. **5**, 10732 (2015)
19. E. Bafekrpour, A. Molotnikov, J.C. Weaver, Y. Brechet, Y. Estrin, Responsive materials: a novel design for enhanced machine-augmented composites. Sci. Rep. **4**, 3783 (2014)
20. C. Barbier, R. Dendievel, D. Rodney, Numerical study of 3D-compressions of entangled materials. Comput. Mater. Sci. **45**, 593–596 (2009a)
21. C. Barbier, R. Dendievel, D. Rodney, Role of friction in the mechanics of nonbounded fibrous materials. Phys. Rev. E **80**(1), 016115 (2009b)
22. S. Bargmann, B. Klusemann, J. Markmann, J.E. Schnabel, K. Schneider, C. Soyarslan, J. Wilmers, Generation of 3d representative volume elements for heterogeneous materials: a review. Prog. Mater. Sci. **96**, 322–384 (2018)
23. R.J. Bathurst, L. Rothenburg, Note on a random isotropic granular material with negative Poisson's ratio. Int. J. Eng. Sci. **26**(4), 373–383 (1988)
24. M. Bendsøe, O. Sigmund, *Topology Optimization* (Springer, Berlin, 2004)
25. I. Benedetti, F. Barbe, Modelling polycrystalline materials: an overview of three-dimensional grain-scale mechanical models. J. Multiscale Model. **5**(1), 1350002 (2013)
26. M.J. Beran, *Statistical Continuum Theories* (Wiley, Hoboken, 1968)
27. C. Berdin, Z.Y. Yao, S. Pascal, Internal stresses in polycrystalline zirconia: microstructure effects. Comput. Mater. Sci. **70**, 140–144 (2013)
28. K. Bertoldi, P. Reis, S. Willshaw, T. Mullin, Negative poisson's ratio behavior induced by an elastic instability. Adv. Mater. **22**(3), 361–366 (2010)
29. J. Besson, G. Cailletaud, J.-L. Chaboche, S. Forest, M. Blétry, *Non-Linear Mechanics of Materials, Volume 167 of Solid Mechanics and Its Applications* (Springer, Berlin, 2010)
30. D. Bigoni, W. Drugan, Analytical derivation of cosserat moduli via homogenization of heterogeneous elastic materials. J. Appl. Mech. **74**, 741–753 (2007)
31. A. Bironeau, J. Dirrenberger, C. Sollogoub, G. Miquelard-Garnier, S. Roland, Evaluation of morphological representative sample sizes for nanolayered polymer blends. J. Microsc. **264**(1), 48–58 (2016)
32. M. Bornert, T. Bretheau, P. Gilormini, *Homogénéisation en mécanique des matériaux, Tome 1 : Matériaux aléatoires élastiques et milieux périodiques*. (Hermès, 2001)
33. O. Bouaziz, Y. Bréchet, J.D. Embury, Heterogeneous and architectured materials: a possible strategy for design of structural materials. Adv. Eng. Mater. **10**(1–2), 24–36 (2008)
34. O. Bouaziz, J.P. Masse, S. Allain, L. Orgéas, P. Latil, Compression of crumpled aluminum thin foils and comparison with other cellular materials. Mater. Sci. Eng. Struct. Mater. Prop. Microstruct. Process. **570**, 1–7 (2013)

35. Y. Bréchet, J.D. Embury, Architectured materials: expanding materials space. Scripta Mater. **68**(1), 1–3 (2013)
36. H.J. Bunge, *Texture Analysis in Materials Science* (Butterworths, 1982)
37. B.D. Caddock, K.E. Evans, Microporous materials with negative Poisson's ratios: I. Microstructure and mechanical properties. J. Phys. D: Appl. Phys. **22**, 1877–1882 (1989)
38. G. Cailletaud, S. Forest, D. Jeulin, F. Feyel, I. Galliet, V. Mounoury, S. Quilici, Some elements of microstructural mechanics. Comput. Mater. Sci. **27**, 351–374 (2003)
39. G. Cailletaud, D. Jeulin, P. Rolland, Size effect on elastic properties of random composites. Eng. Comput. **11**(2), 99–110 (1994)
40. O. Caty, E. Maire, R. Bouchet, Fatigue of metal hollow spheres structures. Adv. Eng. Mater. **10**(3), 179–184 (2008)
41. V.J. Challis, A.P. Roberts, A.H. Wilkins, Design of three dimensional isotropic microstructures for maximized stiffness and conductivity. Int. J. Solids Struct. **45**, 4130–4146 (2008)
42. C. Chateau, L. Gélébart, M. Bornert, J. Crépin, D. Caldemaison, Multiscale approach of mechanical behaviour of sic/sic composites: elastic behaviour at the scale of the tow. Tech. Mechanik **30**(1–3), 45–55 (2010)
43. I. Chekkal, M. Bianchi, C. Remillat, F.-X. Becot, L. Jaouen, F. Scarpa, Vibro-acoustic properties of auxetic open cell foam: model and experimental results. Acta Acustica united Acustica **96**(2), 266–274 (2010)
44. C.P. Chen, R.S. Lakes, Micromechanical analysis of dynamic behavior of conventional and negative Poisson's ratio foams. J. Eng. Mater. Technol. **118**(3), 285–288 (1996)
45. Y. Chen, X.N. Liu, G.K. Hu, Q.P. Sun, Q.S. Zheng, Micropolar continuum modelling of bi-dimensional tetrachiral lattices. Proc. R. Soc. A Math. Phys. Eng. Sci. **470**(2165), 20130734 (2014)
46. J.B. Choi, R.S. Lakes, Design of a fastener based on negative Poisson's ratio foam. Cell. Polym. **10**(3), 205–212 (1991)
47. L. Courtois, E. Maire, M. Perez, D. Rodney, O. Bouaziz, Y. Bréchet, Mechanical properties of monofilament entangled materials. Adv. Eng. Mater. **14**(12), 1128–1133 (2012)
48. A.S. Dalaq, D.W. Abueidda, R.K.A. Al-Rub, I.M. Jasiuk, Finite element prediction of effective elastic properties of interpenetrating phase composites with architectured 3d sheet reinforcements. Int. J. Solids Struct. **83**, 169–182 (2016)
49. L. Decker, D. Jeulin, I. Tovena, 3d morphological analysis of the connectivity of a porous medium. Acta Stereologica **17**(1), 107–112 (1998)
50. C. Delisée, D. Jeulin, F. Michaud, Morphological characterization and porosity in 3D of cellulosic fibrous materials. C.-R. de l'Acad. des Sci. Serie IIb : Mécanique **329**(3), 179–185 (2001)
51. F. dell'Isola, A. Della Corte, I. Giorgio, Higher-gradient continua: The legacy of piola, mindlin, sedov and toupin and some future research perspectives. Math. Mech. Solids **22**(4), 852–872 (2017)
52. V.S. Deshpande, N.A. Fleck, Isotropic constitutive models for metallic foams. J. Mech. Phys. Solids **48**, 1253–1283 (2000)
53. F. di Paola, *Modélisation multi-échelles du comportement thermo-élastique de composites à particules sphériques*. Ph.D. thesis, Ecole Centrale Paris, 2010
54. J. Dirrenberger, *Effective properties of architectured materials*. Ph.D. thesis, MINES-ParisTech, Paris, 2012
55. J. Dirrenberger, S. Forest, D. Jeulin, Elastoplasticity of auxetic materials. Comput. Mater. Sci. **64**, 57–61 (2012)
56. J. Dirrenberger, S. Forest, D. Jeulin, Effective elastic properties of auxetic microstructures: anisotropy and structural applications. Int. J. Mech. Mater. Des. **9**(1), 21–33 (2013)
57. J. Dirrenberger, S. Forest, D. Jeulin, Towards gigantic RVE sizes for stochastic fibrous networks. Int. J. Solids Struct. **51**(2), 359–376 (2014)
58. J. Dirrenberger, S. Forest, D. Jeulin, C. Colin, Homogenization of periodic auxetic materials, in *11th International Conference on the Mechanical Behavior of Materials (ICM11) Procedia Engineering*, vol. 10, pp. 1847–1852 (2011)

59. L. Djumas, A. Molotnikov, G.P. Simon, Y. Estrin, Enhanced mechanical performance of bio-inspired hybrid structures utilising topological interlocking geometry. Sci. Rep. **6**, 26706 (2016)
60. L. Djumas, G.P. Simon, Y. Estrin, A. Molotnikov, Deformation mechanics of non-planar topologically interlocked assemblies with structural hierarchy and varying geometry. Sci. Rep. **7**(1), 11844 (2017)
61. W.J. Drugan, J.R. Willis, A micromechanics-based nonlocal constitutive equation and estimates of representative volume element size for elastic composites. J. Mech. Phys. Solids **44**(4), 497–524 (1996)
62. A.V. Dyskin, Y. Estrin, A.J. Kanel-Belov, E. Pasternak, Toughening by fragmentation–how topology helps. Adv. Eng. Mater. **3**, 885–888 (2001)
63. D. Embury, O. Bouaziz, Steel-based composites: driving forces and classifications. Ann. Rev. Mater. Res. **40**, 213–241 (2010)
64. J. Escoda, D. Jeulin, F. Willot, C. Toulemonde, Three-dimensional morphological modelling of concrete using multiscale poisson polyhedra. J. Microsc. **258**(1), 31–48 (2015)
65. Y. Estrin, A.V. Dyskin, E. Pasternak, Topological interlocking as a material design concept. Mater. Sci. Eng. C **31**, 1189–1194 (2011)
66. Y. Estrin, A.V. Dyskin, E. Pasternak, H.C. Khor, A.J. Kanel-Belov, Topological interlocking of protective tiles for the space shuttle. Philos. Mag. Lett. **83**, 351–355 (2003)
67. K.E. Evans, The design of doubly curved sandwich panels with honeycomb cores. Compos. Struct. **17**(2), 95–111 (1991)
68. K.E. Evans, A. Alderson, Auxetic materials: Functional materials and structures from lateral thinking!. Adv. Mater. **12**(9), 617–628 (2000)
69. K.E. Evans, M.A. Nkansah, I.J. Hutchinson, S.C. Rogers, Molecular network design. Nature **353**, 124 (1991)
70. A. Fallet, P. Lhuissier, L. Salvo, Y. Bréchet, Mechanical behaviour of metallic hollow spheres foam. Adv. Eng. Mater. **10**(9), 858–862 (2008)
71. A. Faure, G. Michailidis, G. Parry, N. Vermaak, R. Estevez, Design of thermoelastic multi-material structures with graded interfaces using topology optimization. Struct. Multi. Optim. **56**(4), 823–837 (2017)
72. Y. Feng, T. Siegmund, E. Habtour, J. Riddick, Impact mechanics of topologically interlocked material assemblies. Int. J. Impact Eng. **75**, 140–149 (2015)
73. J. Fish, V. Filonova, D. Fafalis, Computational continua revisited. Int. J. Numer. Methods Eng. **102**(3–4), 332–378 (2015)
74. J. Fish, S. Kuznetsov, Computational continua. Int. J. Numer. Methods Eng. **84**, 774–802 (2010)
75. S. Forest, *Mechanics of Generalized Continua and Heterogeneous Materials* (Les Presses de lEcole des Mines de Paris, 2005)
76. S. Forest, J.-S. Blazy, Y. Chastel, F. Moussy, Continuum modeling of strain localization phenomena in metallic foams. J. Mater. Sci. **40**, 5903–5910 (2005)
77. D. François, A. Pineau, A. Zaoui, *Mechanical Behaviour of Materials, Volume 1: Micro- and Macroscopic Constitutive Behaviour, volume 180 of Solid Mechanics and Its Applications* (Springer, Berlin, 2012)
78. A.J. Freeman, Materials by design and the exciting role of quantum computation/simulation. J. Comput. Appl. Math. **149**(1), 27–56 (2002)
79. F. Fritzen, S. Forest, D. Kondo, T. Böhlke, Computational homogenization of porous materials of green type. Comput. Mech. **52**(1), 121–134 (2013)
80. N. Gaspar, X.J. Ren, C.W. Smith, J.N. Grima, K.E. Evans, Novel honeycombs with auxetic behaviour. Acta Mater. **53**, 2439–2445 (2005)
81. M.G.D. Geers, J. Yvonnet, Multiscale modeling of microstructure property relations. MRS Bull. **41**(8), 610–616 (2016)
82. L. Gélébart, C. Chateau, M. Bornert, Conditions aux limites mixtes normales. In *19ème Congrès Français de Mécanique, 24–28 August 2009, Marseille* (2009)

83. A. Ghaedizadeh, J. Shen, X. Ren, Y.M. Xie, Tuning the performance of metallic auxetic metamaterials by using buckling and plasticity. Materials **9**(54), 1–17 (2016)
84. I.M. Gitman, H. Askes, L.J. Sluys, Representative volume: Existence and size determination. Eng. Fract. Mech. **74**, 2518–2534 (2007)
85. R.J. Green, A plasticity theory for porous solids. International J. Mech. Sci. **14**, 215–224 (1972)
86. J.K. Guest, J.H. Prévost, Optimizing multifunctional materials: design of microstructures for maximized stiffness and fluid permeability. Int. J. Solids Struct. **43**(22–23), 7028–7047 (2006)
87. L. Guiducci, P. Fratzl, Y. Bréchet, J. Dunlop, Pressurized honeycombs as soft-actuators: a theoretical study. J. R. Soc. Interface **11**, 20140458 (2014)
88. Z. Hashin, Analysis of composite materials–a survey. J. Appl. Mech. **50**, 481–505 (1983)
89. H. Hatami-Marbini, R.C. Picu, Heterogeneous long-range correlated deformation of semi-flexible random fiber networks. Phys. Rev. E **80**(4) (2009)
90. S. Hazanov, Hill condition and overall properties of composites. Arch. Appl. Mech. **68**, 385–394 (1998)
91. S. Hazanov, C. Huet, Order relationships for boundary conditions effect in heterogeneous bodies smaller than the representative volume. J. Mech. Phys. Solids **42**(12), 1995–2011 (1994)
92. J. Henry, S. Pimenta, Increasing damage tolerance in composites using hierarchical brick-and-mortar microstructures. J. Mech. Phys. Solids **118**, 322–340 (2018)
93. C.T. Herakovich, Composite laminates with negative through-the-thickness poisson's ratios. J. Compos. Mater. **18**(5), 447–455 (1984)
94. R. Hill, Elastic properties of reinforced solids: Some theoretical principles. J. Mech. Phys. Solids **11**, 357–372 (1963)
95. R. Hill, The essential structure of constitutive laws for metal composites and polycrystals. J. Mech. Phys. Solids **15**, 79–95 (1967)
96. J.B. Hopkins, L.A. Shaw, T.H. Weisgraber, G.R. Farquar, C.D. Harvey, C.M. Spadaccini, Design of nonperiodic microarchitectured materials that achieve graded thermal expansions. J. Mech. Rob. **8**(5), 051010 (2016)
97. A. Hor, N. Saintier, C. Robert, T. Palin-Luc, F. Morel, Statistical assessment of multiaxial hcf criteria at the grain scale. Int. J. Fatigue **67**, 151–158 (2014)
98. C. Huet, Application of variational concepts to size effects in elastic heterogeneous bodies. J. Mech. Phys. Solids **38**(6), 813–841 (1990)
99. C. Huet, An integrated micromechanics and statistical continuum thermodynamics approach for studying the fracture behaviour of microcracked heterogeneous materials with delayed response. Eng. Fracture Mech. **58**(5–6), 459–556 (1997)
100. T.P. Hughes, A. Marmier, K.E. Evans, Auxetic frameworks inspired by cubic crystals. Int. J. Solids Struct. **47**, 1469–1476 (2010)
101. A. Iltchev, V. Marcadon, S. Kruch, S. Forest, Computational homogenisation of periodic cellular materials: application to structural modelling. Int. J. Mech. Sci. **93**, 240–255 (2015)
102. R. Jänicke, H. Sehlhorst, A. Dster, S. Diebels, Micromorphic two-scale modelling of periodic grid structures. Int. J. Multiscale Comput. Eng. **11**, 161–176 (2013)
103. R. Jänicke, H. Steeb, Minimal loading conditions for higher-order numerical homogenisation schemes. Arch. Appl. Mech. **82**(8), 1075–1088 (2012)
104. A. Jean, G.C. Engelmayr, Finite element analysis of an accordion-like honeycomb scaffold for cardiac tissue engineering. J. Biomech. **43**, 3035–3043 (2010)
105. A. Jean, D. Jeulin, S. Forest, S. Cantournet, F. N'Guyen, A multiscale microstructure model of carbon black distribution in rubber. J. Microsc. **241**(3), 243–260 (2011a)
106. A. Jean, F. Willot, S. Cantournet, S. Forest, D. Jeulin, Large-scale computations of effective elastic properties of rubber with carbon black fillers. Int. J. Multiscale Comput. Eng. **9**(3), 271–303 (2011b)
107. D. Jeulin, Modèles de fonctions aléatoires multivariables. Sci. de la Terre **30**, 225–256 (1991)
108. D. Jeulin, Random texture models for material structures. Stat. Comput. **10**(2), 121–132 (2000)

109. D. Jeulin, *Caractérisation Morphologique et Modèles de Structures Aléatoires, volume 1 of Homogénéisation en Mécanique des Matériaux*, chapter 4, pp. 95–132. Hermès (2001)
110. D. Jeulin, Variance scaling of Boolean random varieties. Technical report, Centre de Morphologie Mathématique. N/10/11/MM (2011), hal-00618967, version 1 (2011)
111. D. Jeulin, Power laws variance scaling of Boolean random varieties. Methodol. Comput. Appl. Probab., pp. 1–15 (2015)
112. D. Jeulin, M. Ostoja-Starzewski, *Mechanics of Random and Multiscale Microstructures* (Springer, CISM Courses, 2001)
113. N. Kaminakis, G. Drosopoulos, G. Stavroulakis, Design and verification of auxetic microstructures using topology optimization and homogenization. Arch. App. Mech. **85**(9), 1289–1306 (2015)
114. T. Kanit, S. Forest, I. Galliet, V. Mounoury, D. Jeulin, Determination of the size of the representative volume element for random composites: Statistical and numerical approach. Int. J. Solids Struct. **40**, 3647–3679 (2003)
115. T. Kanit, F. N'Guyen, S. Forest, D. Jeulin, M. Reed, S. Singleton, Apparent and effective physical properties of heterogeneous materials: representativity of samples of two materials from food industry. Comput. Methods Appl. Mech. Eng. **195**, 3960–3982 (2006)
116. S. Khakalo, J. Niiranen, Isogeometric analysis of higher-order gradient elasticity by user elements of a commercial finite element software. Comput.-Aided Des. **82**, 154–169 (2017)
117. S. Khandelwal, T. Siegmund, R.J. Cipra, J.S. Bolton, Adaptive mechanical properties of topologically interlocking material systems. Smart Mater. Struct. **24**(4), 045037 (2015)
118. A. Kolopp, S. Rivallant, C. Bouvet, Experimental study of sandwich structures as armour against medium-velocity impacts. Int. J. Impact Eng. **61**, 24–35 (2013)
119. C. Körner, Y. Liebold-Ribeiro, A systematic approach to identify cellular auxetic materials. Smart Mater. Struct. **24**(2), 025013 (2015)
120. M. Kotani, S. Ikeda, Materials inspired by mathematics. Sci. Technol. Adv. Mater. **17**(1), 253–259 (2016)
121. N. Kowalski, L. Delannay, P. Yan, J.F. Remacle, Finite element modeling of periodic polycrystalline aggregates with intergranular cracks. Int. J. Solids Struct. **90**, 60–68 (2016)
122. V. Krasavin, A. Krasavin, Auxetic properties of cubic metal single crystals. Phys. Status Solidi b **251**(11), 2314–2320 (2014)
123. T. Krause, A. Molotnikov, M. Carlesso, J. Rente, K. Rezwan, Y. Estrin, D. Koch, Mechanical properties of topologically interlocked structures with elements produced by freeze gelation of ceramic slurries. Adv. Eng. Mater. **14**(5), 335–341 (2012)
124. R.S. Lakes, Foam structures with a negative Poisson's Ratio. Science **235**, 1038–1040 (1987)
125. R.S. Lakes, Deformation mechanisms in negative Poisson's ratio materials: structural aspects. J. Mater. Sci. **26**, 2287–2292 (1991)
126. C. Lantuéjoul, Ergodicity and integral range. J. Microsc. **161**, 387–403 (1991)
127. C. Lantuéjoul, *Geostatistical Simulation: Models and Algorithms* (Springer, Berlin, 2002)
128. L. Laszczyk, R. Dendievel, O. Bouaziz, Y. Br'echet, G. Parry, Design of architectured sandwich core materials using topological optimization methods. In *Symposium LL Architectured Multifunctional Materials, volume 1188 of MRS Proceedings* (2009)
129. A. Lebée, K. Sab, Homogenization of thick periodic plates: application of the bending-gradient plate theory to a folded core sandwich panel. Int. J. Solids Struct. **49**(19–20), 2778–2792 (2012)
130. M. Lewandowski, M. Amiot, A. Perwuelz, Development and characterization of 3D nonwoven composites. Mater. Sci. Forum **714**, 131–137 (2012)
131. A.W. Lipsett, A.I. Beltzer, Reexamination of dynamic problems of elasticity for negative poisson's ratio. J. Acoust. Soc. Am. **84**(6), 2179–2186 (1988)
132. J. Liu, T. Gu, S. Shan, S.H. Kang, J.C. Weaver, K. Bertoldi, Harnessing buckling to design architected materials that exhibit effective negative swelling. Adv. Mater. **28**(31), 6619–6624 (2016)
133. K. Madi, S. Forest, M. Boussuge, S. Gailliègue, E. Lataste, J.-Y. Buffière, D. Bernard, D. Jeulin, Finite element simulations of the deformation of fused-cast refractories based on x-ray computed tomography. Comput. Mater. Sci. **39**, 224–229 (2007)

134. K. Madi, S. Forest, P. Cordier, M. Boussuge, Numerical study of creep in two-phase aggregates with a large rheology contrast: implications for the lower mantle. Earth Planet. Sci. Lett. **237**(1–2), 223–238 (2005)
135. J. Martin, J.-J. Heyder-Bruckner, C. Remillat, F. Scarpa, K. Potter, M. Ruzzene, The hexachiral prismatic wingbox concept. Phys. Status Solidi (b) **245**(3), 570–577 (2008)
136. A. Mather, R. Cipra, T. Siegmund, Structural integrity during remanufacture of a topologically interlocked material. Int. J. Struct. Integrity **3**(1), 61–78 (2012)
137. G. Matheron, *The Theory of Regionalized Variables and its Applications* Les Cahiers du Centre de Morphologie Mathématique de Fontainebleau. (Ecole des Mines de Paris, 1971)
138. G. Matheron, *Random Sets and Integral Geometry* (Wiley, 1975)
139. G. Matheron, *Estimating and Choosing* (Springer, Berlin, 1989)
140. K. Matouš, M.G.D. Geers, V.G. Kouznetsova, A. Gillman, A review of predictive nonlinear theories for multiscale modeling of heterogeneous materials. J. Comput. Phys. **330**, 192–220 (2017)
141. L. Mezeix, C. Bouvet, J. Huez, D. Poquillon, Mechanical behavior of entangled fibers and entangled cross-linked fibers during compression. J. Mater. Sci. **44**(14), 3652–3661 (2009)
142. J.-C. Michel, H. Moulinec, P. Suquet, Effective properties of composite materials with periodic microstructure: a computational approach. Comput. Methods Appl. Mech. Eng. **172**, 109–143 (1999)
143. R.E. Miller, A continuum plasticity model for the constitutive and indentation behaviour of foamed metals. Int. J. Mech. Sci. **42**, 729–754 (2000)
144. G.W. Milton, Composite materials with Poisson's Ratios close to -1. J. Mech. Phys. Solids **40**(5), 1105–1137 (1992)
145. A. Molotnikov, Y. Estrin, A.V. Dyskin, E. Pasternak, A.J. Kanel-Belov, Percolation mechanism of failure of a planar assembly of interlocked osteomorphic elements. Eng. Fracture Mech. **74**, 1222–1232 (2007)
146. A. Molotnikov, R. Gerbrand, O. Bouaziz, Y. Estrin, Sandwich panels with a core segmented into topologically interlocked elements. Adv. Eng. Mater. **15**(8), 728–731 (2013)
147. A. Molotnikov, R. Gerbrand, Y. Qi, G.P. Simon, Y. Estrin, Design of responsive materials using topologically interlocked elements. Smart Mater. Struct. **24**(2), 025034 (2015)
148. G.B. Olson, Beyond discovery: design for a new material world. Calphad **25**(2), 175–190 (2001)
149. M. Osanov, J.K. Guest, Topology optimization for architected materials design. Ann. Rev. Mater. Res. **46**, 211–233 (2016)
150. M. Ostoja-Starzewski, Microstructural randomness versus representative volume element in thermomechanics. J. Appl. Mech. **69**(1), 25–35 (2002)
151. M. Ostoja-Starzewski, *Microstructural Randomness and Scaling in Mechanics of Materials* (Chapman & Hall/CRC, Mordern Mechanics and Mathematics, 2008)
152. M. Oumarou, D. Jeulin, J. Renard, Etude statistique multi-échelle du comportement élastique et thermique d'un composite thermoplastique. Rev. des Compos. et des matériaux avancés **21**, 221–254 (2011)
153. M. Oumarou, D. Jeulin, J. Renard, P. Castaing, Multi-scale statistical approach of the elastic and thermal behavior of a thermoplastic polyamid-glass fiber composite. Tech. Mechanik **32**(2–5), 484–506 (2012)
154. D.H. Pahr, P.K. Zysset, Influence of boundary conditions on computed apparent elastic properties of cancellous bone. Biomech. Model. Mechanobiol. **7**, 463–476 (2008)
155. E. Pasternak, A. Dyskin, Materials and structures with macroscopic negative poissons ratio. Int. J. Eng. Sci. **52**, 103–114 (2012)
156. C. Pelissou, J. Baccou, Y. Monerie, F. Perales, Determination of the size of the representative volume element for random quasi-brittle composites. Int. J. Solids Struct. **46**, 2842–2855 (2009)
157. X.L. Peng, E. Husser, G.Y. Huang, S. Bargmann, Modeling of surface effects in crystalline materials within the framework of gradient crystal plasticity. J. Mech. Phys. Solids **112**, 508–522 (2018)

158. C. Peyrega, D. Jeulin, C. Delisée, J. Malvestio, 3D morphological modelling of a random fibrous network. Image Anal. Stereol. **28**, 129–141 (2009)
159. C. Peyrega, D. Jeulin, C. Delisée, J. Malvestio, 3D morphological characterization of phonic insulation fibrous media. Adv. Eng. Mater. **13**(3), 156–164 (2011)
160. Pham, T. T.T, *Un modèle d'endommagement à gradient de déformation à partir de la méthode d'homogénéisation pour les matériaux fragiles*. Ph.D. thesis, Université Paris XIII, 2010
161. R.C. Picu, Mechanics of random fiber networks–a review. Soft Matter **7**, 6768–6785 (2011)
162. R.C. Picu, H. Hatami-Marbini, Long-range correlations of elastic fields in semi-flexible fiber networks. Comput. Mech. **46**, 635–640 (2010)
163. E. Piollet, *Amortissement non-linéaire des structures sandwichs à matériau d'âme en fibres enchevêtrées*. Ph.D. thesis, ISAE, Toulouse, 2014
164. E. Piollet, G. Michon, D. Poquillon, Nonlinear vibration behavior of sandwich beams with entangled fiber core material. In *ASME 2013 International Design Engineering Technical Conferences and Computers and Information in Engineering Conference* (American Society of Mechanical Engineers, 2013), pp. V008T13A013–V008T13A013
165. E. Piollet, D. Poquillon, G. Michon, Dynamic hysteresis modelling of entangled cross-linked fibres in shear. J. Sound Vibr. **383**, 248–264 (2016)
166. L. Placidi, E. Barchiesi, A. Della Corte, *Mathematical Modelling in Solid Mechanics, volume 69 of Advanced Structured Materials, chapter Identification of Two-Dimensional Pantographic Structures with a Linear D4 Orthotropic Second Gradient Elastic Model Accounting for External Bulk Double Forces* (Springer, Singapore, 2017), pp. 211–232
167. L. Placidi, A.R. El Dhaba, Semi-inverse method à la saint-venant for two-dimensional linear isotropic homogeneous second-gradient elasticity. Math. Mech. Solids **22**(5), 919–937 (2015)
168. M. Poncelet, A. Somera, C. Morel, C. Jailin, N. Auffray, An experimental evidence of the failure of Cauchy elasticity for the overall modeling of a non-centro-symmetric lattice under static loading. Int. J. Solids Struct. **147**, 223–237 (2018)
169. D. Prall, R.S. Lakes, Properties of a Chiral honeycomb with a Poisson's ratio of -1. Int. J. Mech. Sci. **39**(3), 305–314 (1997)
170. Y. Rahali, I. Goda, J.-F. Ganghoffer, Numerical identification of classical and nonclassical moduli of 3d woven textiles and analysis of scale effects. Compos. Struct. **135**, 122–139 (2016)
171. X. Ren, J. Shen, A. Ghaedizadeh, H. Tian, Y.M. Xie, A simple auxetic tubular structure with tuneable mechanical properties. Smart Mater. Struct. **5**(6), 065012 (2016)
172. A. Reuss, Berechnung der fließgrenze von mischkristallen auf grund der plastizitätsbedingung für einkristalle. Zeitschrift für angewandte Mathematik und Mechanik **9**(1), 49–58 (1929)
173. D. Rodney, B. Gadot, O.R. Martinez, S.R. Du Roscoat, L. Orgéas, Reversible dilatancy in entangled single-wire materials. Nat. Mater. **15**(1), 72 (2016)
174. G. Rosi, N. Auffray, Anisotropic and dispersive wave propagation within strain-gradient framework. Wave Motion **63**, 120–134 (2016)
175. G. Rosi, V.H. Nguyen, A. Loseille, S. Naili, Ultrasonic characterization of porous gyroid scaffolds for bone tissue engineering: mechanical modelling and numerical validation. J. Acoust. Soc. Am. **144**(3), 1854 (2018a)
176. G. Rosi, L. Placidi, N. Auffray, On the validity range of strain-gradient elasticity: a mixed static-dynamic identification procedure. Eur. J. Mech. A/Solids **69**, 179–191 (2018b)
177. G. Rosi, I. Scala, V. Nguyen, S. Naili, Wave propagation in strain gradient poroelastic medium with microinertia: closed-form and finite element solutions. Zeitschrift für angewandte Mathematik und Physik **63**(3), 58 (2017)
178. L. Rothenburg, A.A. Berlin, R.J. Bathurst, Microstructure of isotropic materials with negative Poisson's ratio. Nature **354**, 470–472 (1991)
179. K. Sab, On the homogenization and the simulation of random materials. Eur. J. Mech. A/Solids **11**(5), 585–607 (1992)
180. K. Sab, A. Lebée, *Homogenization of Heterogeneous Thin and Thick Plates* (Wiley, Hoboken, 2015)

181. K. Sab, B. Nedjar, Periodization of random media and representative volume element size for linear composites. C.-R. de l'Acad. des Sci. Serie IIb : Mécanique **333**, 187–195 (2005)
182. M. Salmi, F. Auslender, M. Bornert, M. Fogli, Apparent and effective mechanical properties of linear matrix-inclusion random composites: Improved bounds for the effective behavior. Int. J. Solids Struct. **49**, 1195–1211 (2012a)
183. M. Salmi, F. Auslender, M. Bornert, M. Fogli, Various estimates of representative volume element sizes based on a statistical analysis of the apparent behavior of random linear composites. C.-R. de l'Acad. des Sci. Serie IIb : Mécanique **340**, 230–246 (2012b)
184. K. Salonitis, D. Chantzis, V. Kappatos, A hybrid finite element analysis and evolutionary computation method for the design of lightweight lattice components with optimized strut diameter. Int. J. Adv. Manufact. Technol. **90**(9–12), 2689–2701 (2017)
185. E. Sanchez-Palencia, A. Zaoui, *Homogenization Techniques for Composite Media*, vol. 272. Lecture Notes in Physics (Springer, Berlin, 1987)
186. V. Sansalone, P. Trovalusci, F. Cleri, Multiscale modeling of composite materials by a multifield finite element approach. Int. J. Multiscale Comput. Eng. **3**, 463–480 (2005)
187. B. Sarac, J. Wilmers, S. Bargmann, Property optimization of porous metallic glasses via structural design. Mater. Lett. **134**, 306–310 (2014)
188. F. Scarpa, J.R. Yates, L.G. Ciffo, S. Patsias, Dynamic crushing of auxetic open-cell polyurethane foam. Proc. Inst. Mech. Eng. Part C-J. Mech. Eng. Sci. **216**(12), 1153–1156 (2002)
189. S. Schaare, W. Riehemann, Y. Estrin, Damping properties of an assembly of topologically interlocked cubes. Mater. Sci. Eng. A Struct. Mater. Prop. Microstruct. Process. **521–522**, 380–383 (2009)
190. T.A. Schaedler, W.B. Carter, Architected cellular materials. Ann. Rev. Mater. Res. **46**, 187–210 (2016)
191. T.A. Schaedler, A.J. Jacobsen, A. Torrents, A.E. Sorensen, J. Lian, J.R. Greer, L. Valdevit, W.B. Carter, Ultralight metallic microlattices. Science **334**(6058), 962–965 (2011)
192. K. Schladitz, S. Peters, D. Reinel-Bitzer, A. Wiegmann, J. Ohser, Design of acoustic trim based on geometric modeling and flow simulation for non-woven. Comput. Mater. Sci. **38**, 56–66 (2006)
193. I. Shufrin, E. Pasternak, A. Dyskin, Planar isotropic structures with negative poissons ratio. Int. J. Solids Struct. **49**(17), 2239–2253 (2012)
194. I. Shufrin, E. Pasternak, A. Dyskin, Negative poissons ratio in hollow sphere materials. Int. J. Solids Struct. **54**, 192–214 (2015)
195. P. Smith, S. Torquato, Computer simulation results for the two-point probability function of composite media. J. Comput. Phys. **76**(1), 176–191 (1988)
196. M.A. Soare, R.C. Picu, An approach to solving mechanics problems for materials with multiscale self-similar microstructure. Int. J. Solids Struct. **44**, 7877–7890 (2007)
197. C. Soyarslan, S. Bargmann, M. Pradas, J. Weissmuller, 3D stochastic bicontinuous microstructures: generation, topology and elasticity. Acta Mater. **149**, 326–340 (2018)
198. A. Spadoni, M. Ruzzene, S. Gonella, F. Scarpa, Phononic properties of hexagonal chiral lattices. Wave Motion **46**(7), 435–450 (2009)
199. J. Teixeira-Pinto, C. Nadot-Martin, F. Touchard, M. Gueguen, S. Castagnet, Towards the size estimation of a representative elementary domain in semi-crystalline polymers. *Mechanics of Materials* in press (2016)
200. K. Terada, M. Hori, T. Kyoya, N. Kikuchi, Simulation of the multi-scale convergence in computational homogenization approaches. Int. J. Solids Struct. **37**, 2285–2311 (2000)
201. S. Torquato, Morphology and effective properties of disordered heterogeneous media. Int. J. Solids Struct. **35**(19), 2385–2406 (1998)
202. S. Torquato, *Random Heterogeneous Materials* (Springer, Berlin, 2001)
203. D.K. Trinh, R. Jänicke, N. Auffray, S. Diebels, S. Forest, Evaluation of generalized continuum substitution models for heterogeneous materials. Int. J. Multiscale Comput. Eng. **10**(6), 527–549 (2012)

204. N. Vermaak, G. Michailidis, G. Parry, R. Estevez, G. Allaire, Y. Brechet, Material interface effects on the topology optimizationof multi-phase structures using a level set method. Struct. Mult. Optim. **50**(4), 623–644 (2014)
205. W.M. Vicente, Z.H. Zuo, R. Pavanello, T.K.L. Calixto, R. Picelli, Y.M. Xie, Concurrent topology optimization for minimizing frequency responses of two-level hierarchical structures. Comput. Methods Appl. Mech. Eng. **301**, 116–136 (2016)
206. W. Voigt, Ueber die beziehung zwischen den beiden elasticitätsconstanten isotroper körper. Ann. der Phys. und Chem. **38**, 573–587 (1889)
207. Y. Wang, H. Xu, D. Pasini, Multiscale isogeometric topology optimization for lattice materials. Comput. Methods Appl. Mech. Eng. **316**, 568–585 (2017a)
208. Z.P. Wang, L.H. Poh, J. Dirrenberger, Y. Zhu, S. Forest, Isogeometric shape optimization of smoothed petal auxetic structures via computational periodic homogenization. Comput. Methods Appl. Mech. Eng. **323**, 250–271 (2017b)
209. P.M. Weaver, M.F. Ashby, The optimal selection of material and section-shape. J. Eng. Des. **7**(2), 129–150 (1996)
210. B. Xu, X. Huang, S.W. Zhou, Y.M. Xie, Concurrent topological design of composite thermoelastic macrostructure and microstructure with multi-phase material for maximum stiffness. Compos. Struct. **150**, 84–102 (2016a)
211. S. Xu, J. Shen, S. Zhou, X. Huang, Y.M. Xie, Design of lattice structures with controlled anisotropy. Mater. Des. **93**, 443–447 (2016b)
212. W. Yang et al., Review on auxetic materials. J. Mater. Sci. **39**, 3269–3279 (2004)
213. C.L.Y. Yeong, S. Torquato, Reconstructing random media. Phys. Rev. E **57**(1), 495 (1998)

# Chapter 5
# Design Methods for Architectured Materials

**F. X. Kromm and H. Wargnier**

**Abstract** Architectured materials offer a great potential of performance for various applications, but they have to be tailored to fulfil each set of requirements. Designing an architectured material implies determining all its attributes: components, architecture, volume fractions, interfaces…Numerous methods have been developed for product design or for single material selection, but few deal with architectured materials. Because of the difficulty to determine all the parameters at the same time, studies have been carried out about specific tasks in architecture materials design. In this chapter, after having presented material selection methods and design or creativity methods that can be useful in this context, some results are detailed about methods for analysing the set of requirements, identify some incompatibilities between the functions, and select the components in the case where the architecture is defined.

## 5.1 Materials Selection and Architectured Materials

### *5.1.1 Definition of an Architectured Material*

In order to define clearly the domain of this study, it is necessary to give a precise determination of the notion of architectured material. Indeed, various works on this topic sometimes use different vocabularies, although some of the terms are synonymous.

The term "architectured material" refers to both notions of materials combination and structure, and as a consequence, it focuses the attention on observation scale.

An architectured material can be defined as the association of several non-miscible materials (or the association of one material and air), disposed following a predefined

F. X. Kromm (✉) · H. Wargnier
Université de Bordeaux, I2M, UMR5295, 33400 Talence, France
e-mail: francois-xavier.kromm@u-bordeaux.fr

H. Wargnier
e-mail: herve.wargnier@u-bordeaux.fr



Y. Estrin et al. (eds.), *Architectured Materials in Nature and Engineering*, Springer Series in Materials Science 282,
https://doi.org/10.1007/978-3-030-11942-3_5

morphology such that a representative elementary volume has at least one dimension that is very small in comparison with the dimensions of the part it composes [1].

This definition includes a great number of materials of very different types like composites, sandwich, foams… It must be mentioned that other terms are proposed in the literature to name these materials (hybrid materials [2], multi-materials [3], tailored materials [4] …).

The notion of scale is fundamental for this type of design because it outlines the limits of the definition. Indeed, applying a thermal treatment to a metal or polymer to obtain a wanted microstructure resulting in specific properties is a classical task in material science, but concern a too small scale to be considered in this study. On the other hand, the term architectured material doesn't include assemblies whose parts would be made of various materials, so the domain of the developed methods will be limited to an intermediate scale.

## *5.1.2 Materials Selection*

### 5.1.2.1 Selection Methods

One of the steps of the design process is the choice of the constitutive material of the product. Theoretically, the choice of the material, the geometry and the manufacturing process of a part are strongly interconnected, but no method allow the designer a simultaneous definition of all these characteristics.

The selection of a suitable material for an application implies a rigorous definition of the requirements, separating clearly:

- the constraints: criteria presenting a fixed required value the material has to reach;
- the objectives: design characteristics that the material has to optimise (searching for a maximum or minimum value);
- the free design variables (generally geometric parameters).

Several kinds of methods have been developed for materials selection as shown in Fig. 5.1. Each one defines an approach for elimination of unsuitable materials and ranking of potential solutions, using a systematic screening of the material database [5, 8], a research in a specific domain knowledge database [7], or questionnaire based method [8–10] for example.

Whatever the chosen approach, the selection procedure is sequenced in three steps:

(1) filtration to limit the number of candidate materials,
(2) ranking of the remaining candidates,
(3) expertise and documentation to ensure the validity of the solution.

Concerning the second step, an example of possible evaluation of the performance of the materials for an application is given in the next paragraph. Although the filtration step has been less developed, it can be interesting to add some remarks in a further paragraph.

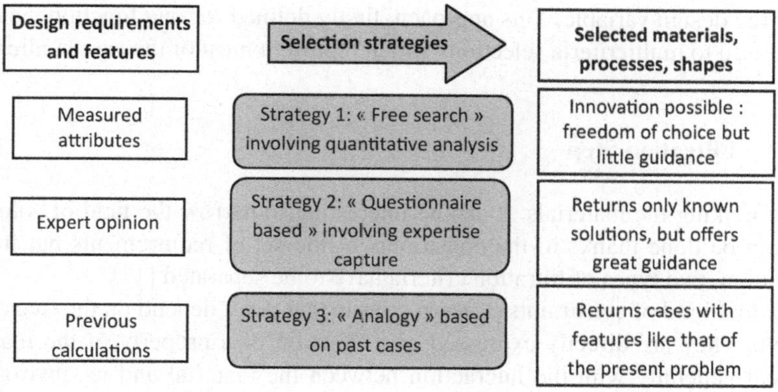


**Fig. 5.1** Selection strategies [11]

### 5.1.2.2 Performance Index

The materials selection methods developed by Ashby [5] are generally employed in the design process after the definition of the geometry of the product. These methods give the opportunity to rank materials of a database in order to choose the most suitable for an application.

Before selecting materials, it is necessary to define a criteria for their evaluation, and to determine the material properties influencing the product performance. This performance can be written as a function of the fixed parameters of the geometry (G), the constitutive material (M) the loading (F), and the free design variable (e), as it is shown in the following equation:

$$P = f(M, G, F, e) \tag{5.1}$$

The comparison of the materials can be made only if the expression of the performance doesn't depend on the free design variable (e). In order to obtain that, a constraint (C) depending on (G), (M), (F), and (e) has to be used:

$$C = h(M, G, F, e) \tag{5.2}$$

This equation can give the expression of (e) allowing to reach the threshold value $C_0$ of the constraint, so that the expression of P doesn't depend any more on (e) and can be shared in terms depending on M and other terms:

$$P = f_1(M) f_2(G, F) \tag{5.3}$$

Finally, the function $f_1$ is considered as a performance index of the material M, and that method can be applied for each triplet composed of an objective, a constraint and a free design variable. This approach, firstly defined for one function, has been generalised to multicriteria selections which represent most of the case studies [12].

#### 5.1.2.3 Filtration Step

Before ranking the materials, it can be interesting to narrow the field of solutions. This can be done thanks to the constraints of the set of requirements but in their application, two types of filtration criteria have to be separated [13].

The first kind of constraints concern criteria that don't depend on the free design variable. They are directly expressed as a function of a property of the material, and deal generally with the interaction between the material and its environment (maximum service temperature, resistance to chemical environment, wear conditions…) or with some of its attributes (possibility to obtain a particular shape, to use a manufacturing process…).

The second type of constraints, depending on the free variable, is also called soft requirements because they don't result in a binary choice on a material property, but in the definition of an allowable interval of the free geometric parameter to fulfil the requirement.

The approach developed for the problems with only one free design variable can be extended to the cases with several free parameters. Mathematical expression deduced from the constraints results in an inequation system where the number of inequations is the number of constraints and the number of unknown is the number of free parameters.

If the number of variables is equal to the number of inequations, the associated system of equations has a unique solution reaching exactly all the limit levels fixed by the selection criteria. If the system gives several solutions (for example when there are more free design variables than equations), the pertinence of the choice of the variables has to be verified because some of them may be linked, and so, the number of unknown may be decreased.

The following approach concerns the problems with a number of constraints bigger than the number of free parameters. It is supposed that for every problem, a linear system of equations can be obtained, which is a realistic assumption because the models employed in preliminary design are often simple.

Then, a classical method for system resolution gives the filtration criteria. A more complete description of the method is presented in the following example.

This example deals with the materials selection for a beam that mustn't be too expensive, and must have fixed minimum flexural stiffnesses in two different planes $(x_1, x_3)$ and $(x_2, x_3)$. The cross section of the beam is rectangular, and its dimensions a and b are free.

The requirements can be expressed as follows:

– constraint 1: stiffness in the $(x_1, x_3)$ plane

$$\frac{Ea^3b}{L^3} \geq S_{10} \tag{5.4}$$

– constraint 2: stiffness in the $(x_2, x_3)$ plane

$$\frac{Eab^3}{L^3} \geq S_{20} \tag{5.5}$$

– constraint 3: cost

$$\frac{1}{\rho C_m abL} \geq \frac{1}{C_0} \tag{5.6}$$

A linear system of equations associated to this system can be written as follows using the logarithms:

$$\begin{pmatrix} 3 & 1 \\ 1 & 3 \\ -1 & -1 \end{pmatrix} \begin{pmatrix} \log(a) \\ \log(b) \end{pmatrix} = \begin{pmatrix} \alpha \\ \beta \\ \gamma \end{pmatrix} \tag{5.7}$$

where $\alpha = \log\left(\frac{S_{10}L^3}{E}\right)$ $\beta = \log\left(\frac{S_{20}L^3}{E}\right)$ and $\gamma = \log\left(\frac{\rho C_m L}{C_0}\right)$.

Using for instance the first two equations, the values $a_0$ and $b_0$ of the free variables allowing the first two conditions are calculated. The level $\gamma*$ of the third constraint reached with the values $a_0$ and $b_0$ can be calculated by:

$$\begin{vmatrix} 3 & 1 & \alpha \\ 1 & 3 & \beta \\ -1 & -1 & \gamma^* \end{vmatrix} = 0 \quad \Leftrightarrow \quad \gamma^* = -\frac{(\alpha+\beta)}{4} \tag{5.8}$$

If $\gamma^* \geq \gamma$, all the constraints are fulfilled with the values $a_0$ and $b_0$, so the considered material can answer the set of requirements. On the contrary, if $\gamma^* < \gamma$, all the constraints aren't fulfilled with the values $a_0$ and $b_0$, so the considered material can't answer the set of requirements. Thus, the filtration can be expressed by the condition $-\frac{(\alpha+\beta)}{4} \geq \gamma$, that is to say:

$$\log(E) \leq 2\log(\rho\rho_m) + \log\left(\frac{C_0^2}{\sqrt{S_{10}S_{20}L^5}}\right) \tag{5.9}$$

This condition can be represented graphically as shown in Fig. 5.2 on a property chart.

This example shows that with a complex set of requirements presenting several soft constraints, a filtration criterion can be applied to the material database to reduce the number of candidates before ranking them.

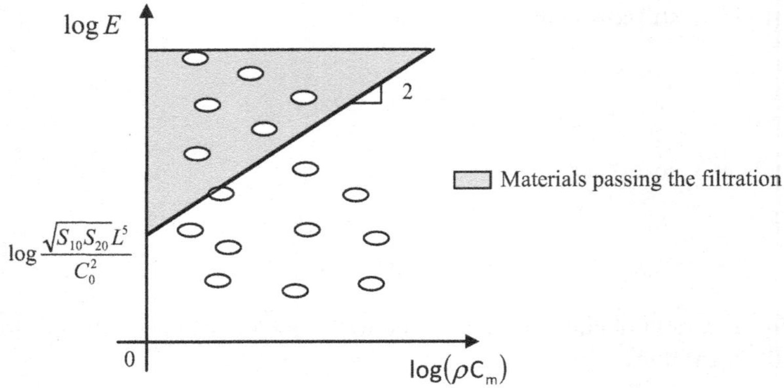


**Fig. 5.2** Property chart for filtration

### 5.1.3 *Why an Architectured Material?*

There can be several reasons for using an architectured material rather than a monolithic material. The first distinction that can be made is to note that in some cases, architectured materials are used to replace monolithic materials which until then have been satisfactory, whereas in other cases architectured materials are used as no monolithic material fulfils their specifications. We therefore differentiate between cases where the material selection process carried out beforehand provides improved solutions thanks to architectured material, and cases where the filtration stage is sufficient to show that no monolithic material can solve the problem.

There are several reasons that can be put forward for replacing a monolithic material by an architectured material:

(1) Improvement in the performances of a part: when the monolithic material used is the optimal material for the selection, any further improvement in performance can only come from new materials, hence from architectured materials.
(2) Reduction in implementation or operating costs: the main gain in terms of cost can be achieved by replacing a very expensive material by a combination of less costly materials, each of which fulfils part of the function required (e.g. differentiating surface and volume functions). A reduction in operating costs can mean a decrease in mass, resulting in easier and less energy-consuming transport.
(3) Easier manufacturing: redesigning a part using an architectured material can often result in a simplified assembly with a reduced number of parts. Even if some of the complexity is transferred to the material, long assembly operations can be avoided, along with any additional mass which in certain areas can be detrimental.
(4) Avoid oversizing: if several constraints are stipulated, when the number of free variables is small, using a monolithic material may lead to oversizing of the part

in relation to certain requirements. An architectured material, as highlighted in Sect. 5.1.1, intrinsically has more free morphological parameters, and thus the properties can be adapted as well as possible to match the requirements. By designing a custom-made material, additions of surplus material can be avoided.

The second type of case where an architectured material is appropriate is in applications with specifications that are so demanding that no monolithic material can meet all the requirements:

(1) integration of new functions: in order to reduce the number of parts in a system, designers sometimes want certain functions to be attributed to the material. An accumulation of such requirements may sometimes result in the monolithic materials no longer being able to meet these demands.
(2) Incompatibility of requirements: it can happen that failure to select a traditional material is not due to the number of constraints but to their nature. This is the case when despite being few in number, constraints are incompatible, for example, searching simultaneously for high thermal conductivity and high electrical resistivity.

This non-exhaustive list of situations justifying the use of architectured materials helps to identify the key design issues: number and nature of constraints, and adjusting properties to match. In what follows, we give an overview of the design of an architectured material.

## 5.2 Methodological Study of Product Design

The choice of material and the design of an architectured material must be looked at in the context of a general approach to design. In fact, the tools used will differ, depending on the stage the project has reached. In addition, it may be interesting to consider whether some existing general tools could be adapted for use with architectured materials.

### 5.2.1 *Product Design Methods*

To start up the design process for an industrial product of any kind (material, software, organisation, service, etc.), it is first essential to consider market need. The definition and specific expression of this need constitutes a fundamental step that determines the success of the design. A crucial step is where the designer translates the needs of the user into precise technical terms to which the design methods can be applied. This work, called functional analysis, should result in the drafting of the functional specifications of the product.

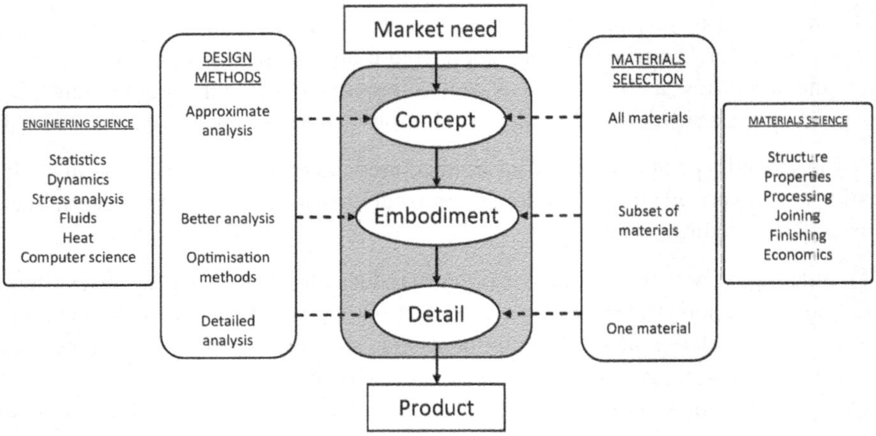


**Fig. 5.3** Materials selection and engineering science in design process [5]

Design is then divided into several phases in which a product is gradually described, by refining its attributes at each stage, and gradually reducing the scope of possible solutions.

The different phases described here can be represented in the form of an organisational chart [5] as shown in Fig. 5.3.

To try to systematise the way in which concepts of solutions are determined, various tools have been developed in an attempt to rationalise the design process. It will be interesting to examine these methods and pinpoint the principles that can be exploited when designing an architectured material.

### *5.2.2 Design and Creativity Methods*

The approach described in the previous section gives only a very global notion of design work, since it does not cover the way in which concepts and solutions are found. Yet this is a crucial aspect of design, even though it may perhaps be the one that is handled least well, since the mechanisms of creativity are so complex.

To systematise this operation, the designer has design tools at his disposal. These methods attempt to rationalise the design process, with each one having its own original approach. The aim here is not to produce an exhaustive list of these methods, but rather to give a brief description of those that are most commonly used, their advantages and disadvantages, especially regarding the design of architectured material.

#### 5.2.2.1 Concurrent Engineering

Concurrent engineering is a method based on the principle that all aspects associated with the life-cycle of a product can be taken into consideration during the design process. The result obtained with this method should ensure a balance between the requirements of each phase and will produce the most satisfactory compromise.

The advantage of this method is that it incorporates the multi-disciplinary nature of design which previously moved forward in successive and clearly distinct phases. However, it is difficult to determine whether concurrent engineering should be considered as a set of design techniques, a new approach, or indeed a philosophy to be applied at company level [14].

This approach is more and more appropriate for the choice of material or the design of architectured materials since in addition to the functions that are routinely requested, further criteria are often added concerning the supply of materials, implementation, dismantling, re-use or recycling [15]. For example, part of the work may consist in creating recyclability criteria or assessment of environmental impact [16].

#### 5.2.2.2 Decomposition into Functional Modules

A methodological approach consists in dividing the whole problem into several sub-problems, known as functional modules, which have minimum interaction between them [9, 17]. In fact, a product is often a fairly complex system that can be divided into several components with the result that functions are often better focused. A unique and complex problem is therefore transformed into a sum of small studies each with much simpler specifications.

The design of architectured materials can be approached in the same way. When confronted with specifications with many requirements, it is possible to search for a material that meets each one. The next step consists of either simplifying to see if the same material can effectively fulfil several functions, or assembling these materials in the hope that all their advantages will then be combined.

In practice, although this approach may guide the design process towards possible components, it is by no means certain that combining them will produce a satisfactory result. In fact, some requirements may concern all the components of the material (e.g. mass criterion) and can therefore not be processed in a simple fashion using this kind of deconstruction. In addition, considering the architectured material in this way gives only part of the answer because if we refer to the definition given in Sect. 5.1.1, clearly this work will only result in constituents being proposed, with no information on the morphology they are to have or the interface properties that will link them together.

Lastly, we see that although the design phase has indeed been made less restrictive, the proliferation of independent functional modules increases the complexity of the result.

#### 5.2.2.3 TRIZ Method

When a complex set of functions and constraints is induced by an application, these methods often give no solution, showing that "incompatible" required properties force the designer to use multimaterials. In front of this problem, it seems necessary to define a methodological approach to design multimaterials for that class of applications.

This notion of incompatibility between two functions already appeared in product design, particularly when TRIZ (TIPS) method was developed [18]. Indeed, this design method is based on the principle that insoluble design problems come from contradictory requirements. Pointing out various types of contradictory requirements from numerous examples, several general problems could be identified and listed. Besides, considering all the solutions operating in the examples, a solution has been proposed for each general problem. However, this method is hardly adaptable to multimaterials because the related given answers have been formulated in a very general way which leaves the engineer's creativity a great part in the result.

### *5.2.3 Strategy for a Toolbox*

The steps in the overall process of designing an architectured material can be summarised as shown in Fig. 5.4. In this process, design aid tools can be applied at different stages.

After drawing up the functional set of requirements and the material set of requirements, the design of the architectured material can be guided via a questionnaire to discover the reasons why a multi-material is necessary. This analysis of the specifications will either be used to identify the traditional problems for which solutions can be proposed, or to carry out preliminary work using a free solution search tool.

The other tools relate to the subsequent steps in the process: search for morphologies, for components, for interface properties, and simulation to assess the solutions. In this work, the large number of parameters that define the architectured material give rise to a huge number of combinations. To get around this problem, the proposed tools take into account only a small number of these parameters, with the others being determined by hypotheses, thus reducing the scope of the study. This can result in several scenarios, as shown in Fig. 5.5.

The presentation of the tools begins with this preliminary analysis of the specifications, then continues with the solution search methods developed for the possible scenarios, with the exception of the selection of adapted interface properties for which no tool has yet been studied.

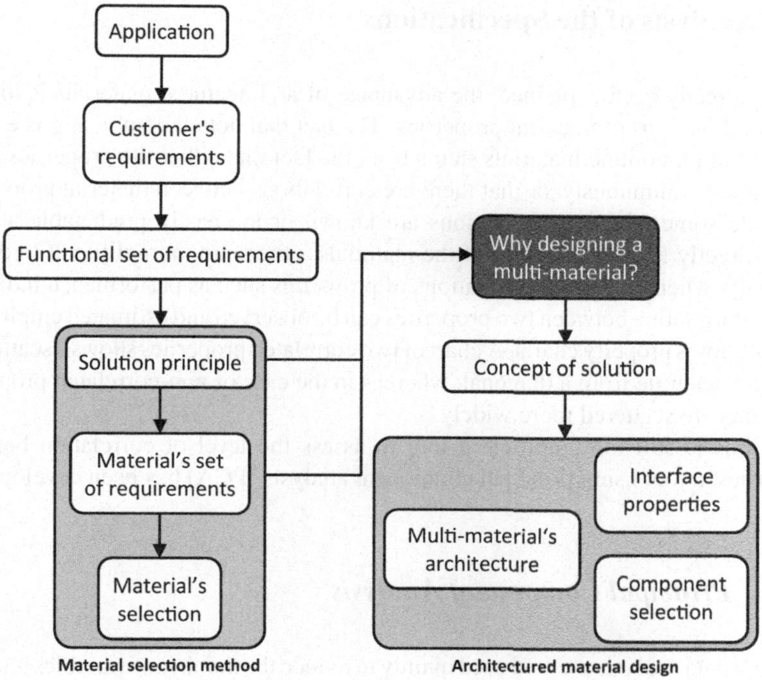


**Fig. 5.4** Architectured material design process

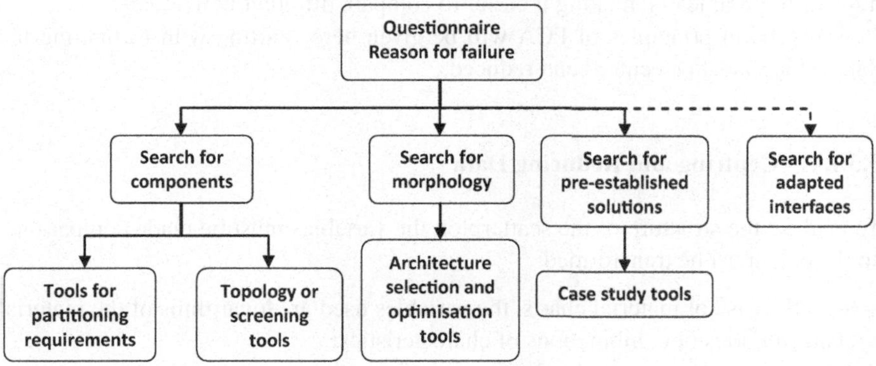


**Fig. 5.5** Methods and tools to search for an architectured material

## 5.3 Analysis of the Specifications

As has already been explained, the advantage of architectured materials is to offer new combinations of material properties. The fact that not all combinations exist in the field of monolithic materials stems from the fact that either the properties space is not filled continuously, or that there are correlations between material properties.

While some of these correlations are known or are easily predictable as they result directly from the physics of the materials, others are more difficult to detect, especially when they are combinations of properties such as performance indices.

The correlation between two properties can be observed and estimated empirically using Ashby's property charts. A chart of two correlated properties shows a scatterplot that deviates little from a diagonal, whereas in the case of non-correlated properties the points are scattered more widely.

As this is still too empirical a tool to assess the level of correlation between properties, a tool using principal component analysis (PCA) has been developed.

### *5.3.1 Principal Component Analysis*

PCA is a data analysis method used mainly to reduce the number of variables required to describe a scatterplot. By applying this method, a space with many dimensions (such as the material properties space) can be correctly represented by a smaller number of parameters, making it easier to compare different individuals.

The general principles of PCA will be given here, starting with formatting the data, which must be centred and reduced.

#### 5.3.1.1 Centring and Reducing Data

To analyse the structure of the scatterplot, the variables must be made comparable, so the data must be transformed:

- as in the case of material charts, the variables used are logarithms of the material characteristics or combinations of characteristics;
- we perform a centring and reduction transformation.

After transformation, a column vector $x$ of matrix $A$ is denoted $x^*$: the mean of $x^*$ is therefore 0 and standard deviation is 1. The transformed matrix is denoted $A^*$

$$x_i^* = \frac{x_i - m(x)}{\sqrt{\upsilon(x)}} \tag{5.10}$$

where $m(x) = \sum_i \frac{x_i}{n}$ is the mean of $n$ elements of $x$, $\upsilon(x) = \sum_i \frac{(x_i - m(x))^2}{n}$ is the standard deviation of $x$, and $n$ the number of materials.

#### 5.3.1.2 Correlation Coefficients

The study of links between variables or criteria is based on determining correlation coefficients between the column vectors of matrix $A^*$. Generally, the correlation coefficient $r(x, y)$ between vectors $x$ and $y$ is defined on the basis of the covariance of the centred variables and the standard deviations of the variables.

$$c(x, y) = \sum_i \frac{(x_i - m(x))(y_i - m(y))}{n} \tag{5.11}$$

$$r(x, y) = \frac{c(x, y)}{\sqrt{v(x)}\sqrt{v(y)}} \tag{5.12}$$

A correlation coefficient varies between $-1$ and 1. In the case of centred and reduced variables, the covariance between two vectors is also the correlation coefficient:

$$c(c^*, y^*) = \sum_i \frac{x_i^* y_i^*}{n} \tag{5.13}$$

$$r(x^*, y^*) = c(x^*, y^*) \tag{5.14}$$

As we saw earlier, it is possible either to look at individuals, i.e. the materials, and consider a scatterplot in the variable space, or to look at the variables. In the latter case, a variable is represented by a vector in a vector space where each dimension corresponds to a material: the $i$th component of this vector is the value of the variable for the $i$th material. As the data are centred and reduced, the ends of the vectors representing the variables are on a sphere of radius 1, if we limit the numbers to three individuals, and this is shown in graph form in Fig. 5.6: in practice, it is more common to talk about a hypersphere since the space being considered has more than 3 dimensions.

Note here that the correlation coefficient between two variables is also the cosine of the angle between the vectors representing these variables.

If the angle is close to 0°, the cosine of the angle is close to 1, the two vectors are almost equal, and the two variables are very positively correlated.

If the angle is close to 180°, the cosine is close to $-1$, the two vectors are almost opposite, and the two variables are very negatively correlated.

If the angle between two variables is close to 90°, the cosine of the angle is 0, hence the correlation coefficient is zero, and the variables represented by these vectors are not correlated

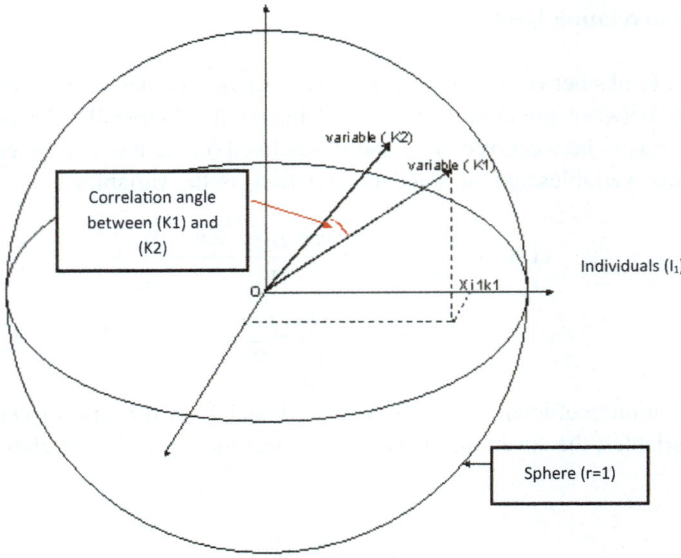


**Fig. 5.6** Set of variables represented in a 3 individuals space

#### 5.3.1.3 Factorial Axes

If we consider the scatterplot of the materials (or individuals) in the variable space, then we can construct a sequence of favoured directions, also called factorial axes. This sequence can be defined as follows:

- the first factorial axis minimises the inertia of the scatterplot in relation to the axis
- the second factorial axis is perpendicular to the first and such that the factorial design
- constructed from these two axes minimises the inertia of the scatterplot in relation to the design
- the other factorial axes are constructed using the same procedure.

There are several interpretations that can give a better appreciation of the notion of factorial axes:

- the factorial axes minimise the gap, in the sense of least squares, between the scatterplot of individuals and its projection on the axis.
- the factorial axes are the directions of maximal lengthening of the scatterplot. This lengthening is linked to the variability of the projection of the scatter.
- the directions of the factorial axes are such that the distance between the projections of individuals is the closest real distance between individuals.

The projected inertia of an axis is defined as the ratio of inertia to the origin of an axis of the projection of the scatterplot on the axis, and relative inertia is the ratio of projected inertia to total inertia. If we consider variable scatter, it is also possible to construct principal axes. The way they are constructed based on the inertia of the scatter is identical to the scatter of individuals but interpretation is different:

- the variable associated with the first factorial axis, also called the principal component, is the variable that best summarises all of the initial variables.

By denoting as $X^*$ the matrix representing the centred and reduced data, i.e. a line represents an individual and a column represents a variable, in order to carry out a PCA, we must calculate the correlation matrix $C$:

$$C = \frac{1}{n} X^{*t} X^* \tag{5.15}$$

The eigenvectors ($u_i$) of the correlation matrix define the factorial axes of the analysis of the individuals. The relative inertia (projected inertia/total inertia) of the scatterplot projected on this axis is proportional to the associated eigenvalue ($\lambda_i$).

The factorial axes of the analysis of variables are defined from these eigenvectors and eigenvalues. The principal component $F_i$, which is a new variable, is defined as:

$$F_i = X^* u_i \tag{5.16}$$

It is usual to classify eigenvalues in decreasing order, where the first factorial axis corresponds to the eigenvalue with the highest value.

#### 5.3.1.4 Reconstruction of the Scatterplot Using Factorial Axes

The scatterplot can be reconstructed from the factorial axes:

$$X^* = \sum_{i=1}^{p} (X^* u_i) u_{ii}^t \tag{5.17}$$

where $p$ is the number of initial variables

It is also possible to perform a partial reconstruction of the scatterplot without using all the factorial axes

$$\hat{X}_q = \sum_{i=1}^{q} (X^* u_i) u_{ii}^t \tag{5.18}$$

where $q < p$

We then show that the distance between the two matrices is directly linked to the sum of the relative inertias of the axes not used in the reconstruction

$$d^2\left(X^*, \hat{X}_q\right) = \sum_{i=1}^{q} \lambda_{ii} \tag{5.19}$$

A measurement of the quality of the representation of a plot on an axis can then be proposed from the relative inertia. This relative inertia is cumulative, in other words the relative inertia of a factorial design is the sum of the inertias of the design axes. The relative inertia of axis $i$ is determined by:

$$\frac{\lambda_i}{\sum_i \lambda_j} \tag{5.20}$$

### 5.3.2 *Study of Correlations Between Material Properties*

The principal component analysis tool that has been described can be applied to a space made up of material properties or combinations of these properties. In fact, by applying variable centring and reduction to two given properties, then calculating the factorial axes, all the materials in a database can be reconstructed along a factorial axis.

This procedure can be illustrated using a simple example stipulating that the density of a material must not exceed value $\rho_0$, and the Young's modulus must remain higher than $E_0$. Both of these criteria can be formulated in a new way, according to the position of the materials on factorial axis $F_1$. Thus each interval defined for the density and the Young's modulus is translated into an interval of validity of $F_1$ (see Fig. 5.7).

Two possible scenarios may then arise:

(1) if the intersection of these intervals is non-empty, the amplitude of this validity interval with respect to the total amplitude of the given database gives a measurement for the compatibility of the "studied" functions;
(2) if the intersection is empty, the "distance" separating the two intervals can be used to assess the degree of incompatibility of the functions, and also possibly to see if a small change in the level of these requirements could generate solutions.

### 5.3.3 *Study of the Density of a Materials Chart*

In the case where the requirements aren't correlated, the previous analysis becomes unsuitable because the first factorial axis isn't representative of the scatter plot. However, even if this scatter plot is very spread out and covers a bigger part of the space, the repartition of the materials may be non-uniform, so it can be interesting to study the plot density, particularly in the zones concerned by the requirements.

To evaluate this density, the variable space has to be discretised, and the number of materials in each elementary domain has to be calculated. Then, the density can be calculated with respect to the domain size or to the average density of the scatter plot. It is worthy of note that this method can be easily automatized.

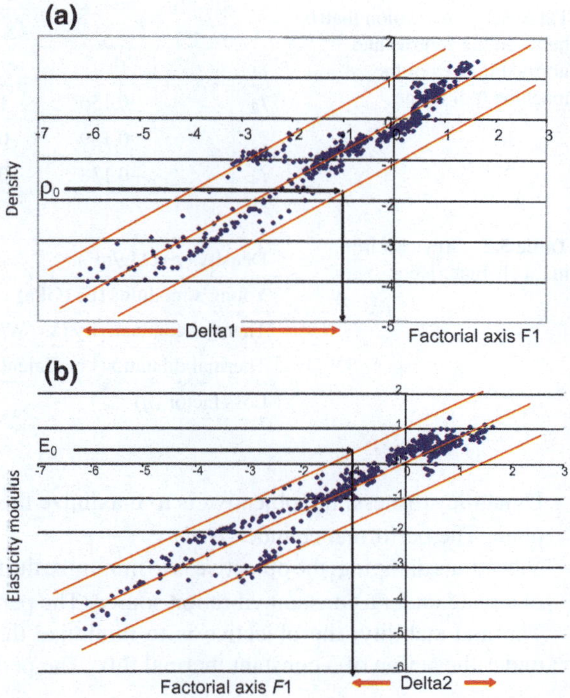


**Fig. 5.7** Relation between material properties and factorial axis: **a** density (centred and normed) versus first factorial axis, **b** Young's modulus (centred and normed) versus first factorial axis

## 5.3.4 *Application: Machine Tool Frame*

An improvement of the performance of a machine tool implies higher production rates while keeping precisions around 1–10 μm. As a consequence, the machining speed, the power of the drivers and the frame stiffness have to be increased.

The machining quality is partially related to the capability of the machine to maintain the relative position between the tool and the part to be machined, but the evaluation of this parameter is very complex because it is influenced by the machine tool architecture, the assembly of the different parts, the drivers, the frame… As these information are generally unkown in a preliminary design stage, general requirements for the frame are formulated to allow a material selection for this application:

- static and dynamic stiffness,
- vibration damping capability,
- thermal stability.

In order to select a material, performance indices are built for a plate under each solicitation:

- Static stiffness: the objective is to minimize the material cost with a fixed static stiffness. In the case of a bending plate the performance index is: $I_1 = \frac{E^{1/3}}{\rho C_m}$

**Table 5.1** Correlation matrix between the centred and normed indices of the machine tool frame

| | $I_1$ | $I_2$ | $I_3$ | $I_4$ |
|---|---|---|---|---|
| $I_1$ | 1 | | | |
| $I_2$ | 0.156 | 1 | | |
| $I_3$ | 0.139 | 0.41 | 1 | |
| $I_4$ | −0.138 | −0.233 | **−0.794** | 1 |

**Table 5.2** Grey cast iron main characteristics

| | |
|---|---|
| Density ($\rho$) (Mg/m$^3$) | 7.2 |
| Young's modulus (E) (GPa) | 150 |
| Thermal conductivity ($\lambda$) (W/m K) | 53.5 |
| Thermal dilatation coefficient ($\alpha$) ($10^6$ K$^{-1}$) | 12 |
| Loss factor ($\eta$) | 0.013 |

- Dynamic stiffness: the objective is to maximize the transverse eigenvalues of the plate. The performance index is: $I_2 = \frac{E^{1/3}}{\rho}$
- Vibrations damping: the objective is to minimize the transversal dynamic excitation of a plate under a constant vibration source. The performance index is: $I_3 = \eta \frac{E^{1/3}}{\rho}$
- Thermal stability: the objective is to minimize the thermal distortion of a part under the action of a constant thermal flux. The performance index is: $I_4 = \frac{\lambda}{\alpha}$

The correlation matrix is calculated for these indices (Table 5.1).

The only significant correlation in Table 5.1 is the negative correlation between the index $I_3$ (vibration damping) and $I_4$ (thermal stability) with a coefficient of −0.794. This means that minimizing simultaneously the transverse excitation of a plate under a constant vibration source and its thermal distortion caused by a heat flow are intrinsically conflicting requirements.

In order to apply the ACP method, these performance indices have to be turned into constraints, defining available intervals for each combination of properties. In this case study, the new material has to result in better performances than the currently used cast iron.

The improvement comparing to the cast iron should be of 5% of the variation of the index for all the materials database. From the cast iron properties (Table 5.2) the improved performance index $I_3 = 0.018$ and $I_4 = 10.2$ (in the same unities of Table 5.2). After centring and normalising, $I_3^* = 0.93$ and $I_4^* = 0.86$.

Figure 5.6a shows the evolution of the index $I_3$ as a function of the first main component, as well as the best linear correlation and the band representing the set of points (95% of materials inside). The result of the compatibility study depends on the selection made to fix the band: the minimum value for $F_1$, that here is 0.32, is related to this selection. The width of the band will be as little as the inertia of the set of points, from the factorial axis is little. Then the associated eigenvalue will be high.

We can say the same from Fig. 5.8b, that represents the evolution of the performance index related to the thermal distortion $I_4$ as a function of the first main

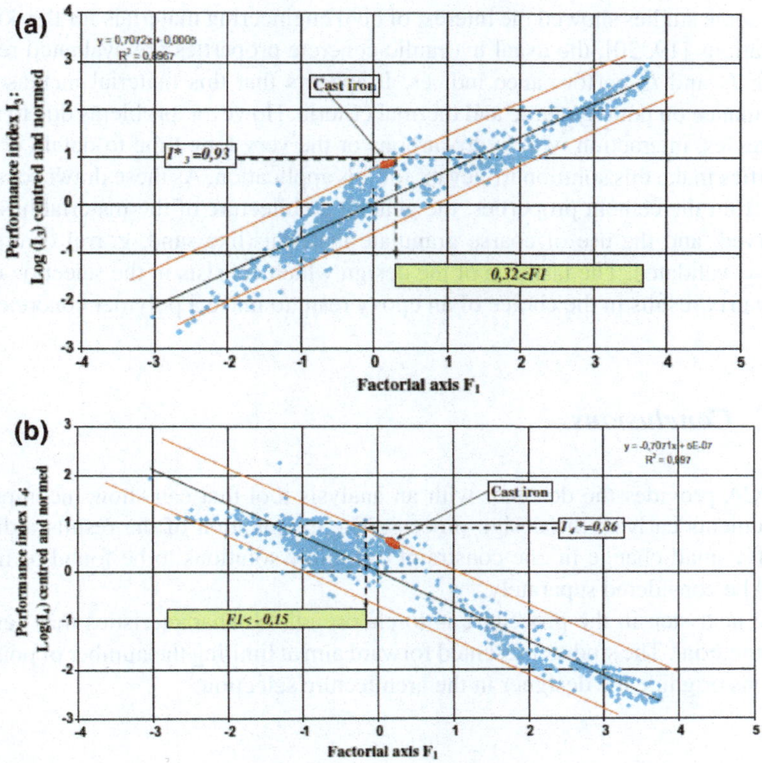


**Fig. 5.8** Compatibility analysis: **a** performance index $I_3$ versus first factorial axis; **b** performance index $I_4$ versus first factorial axis

**Table 5.3** Compatibility between indices $I_3$ and $I_4$

| $F_{1\,min}$ | $F_{1\,max}$ | Interval on $F_1$ | Allowable interval | Distance of incompatibility | Incompatibility index |
|---|---|---|---|---|---|
| −3 | 3.8 | $I_3$ (vibration): [0.32, 3.8]<br>$I_4$ (thermal): [−3, −0.15] | No interval | 0.32 − (−0.15) = 0.47 | 0.47/6.8 = 8.1% |

component. After the PCA application on the $I_3$ and $I_4$ indices and after the graphs Fig. 5.8a, b, Table 5.3 can be built.

The intersection of the allowable intervals is empty, and the distance of incompatibility is evaluated at 0.47, which corresponds to about 8% of the total range of value of $F_1$ on the database. This value seems too high to consider that solutions can be found near the requirements, so the incompatible functions have been studied specifically, and a multi-material solution has to be searched.

As some studies showed the interest of civil engineering materials for this kind of applications [19, 20], the usual hydraulic concrete properties are evaluated regarding the $I_3$ and $I_4$ performance indices. It appears that this material increases the performance on both dynamic and thermal criteria. However, problems due to residual stresses, interaction with hydrocarbons or the very long time to obtain de final properties make this solution irrelevant for this application. As these drawbacks only come from the cement properties, the general architecture of the material has been conserved, and the use of coarse granulate material (like sand, gravel or crushed stone) is validated. The last step of the design which consists in the selection of the new matrix results in the choice of an epoxy resin to make a polymer concrete.

### *5.3.5 Conclusions*

The PCA provides the designer with an analysis tool that can show incompatible requirements early in the design process. The exploitation of the results indicates then if a small change in one constraint can allow solutions to be found or if they should be considered separately.

The next step in the procedure is to choose all the characteristics that define a multi-material. The studies presented forward aim at limiting the number of potential solutions or guide the designer in the architecture selection.

## 5.4 Component Selection—Materials Database Filtration

The number of parameters in multi-material design increases greatly in comparison with a single material selection. Indeed, it has been noted previously that it implies the determination of:

- the components,
- the volume fractions of the component,
- the architecture, or morphology of the multi-material,
- the coupling mode, the nature of the interface.

In order to reduce the study domain and the number of candidates, a concept of solution (i.e. the functional mechanism) and some parameters have to be predetermined. For example, we firstly consider that the architecture of the multi-material and the nature of the interface are fixed. In this case, the only parameters remaining free are the constitutive materials and their volume fraction. It appears that the number of possible combinations is still very important, even if the material database is reduced, so it seems necessary to operate a filtration on the database to determine the materials that can be a component of the multi-material before ranking the solutions.

This filtration method for a multi-materials can be divided in three steps:

- elimination of materials that can't be a component,
- searching possible combinations of materials,
- searching possible volume fractions or design parameters.

### 5.4.1 Elementary Example

This study aims at selecting the components of a material with two layers to answer a set of requirements concerning a tie with a fixed length L and whose cross section area A is free. The tie is submitted to constraints $C_1$ on mass (lower than $m_0$), $C_2$ mechanical (tensile stiffness greater than $S_0$) and $C_3$ electrical behaviour (electrical resistance greater than $R_0$).

These three constraints can be treated separately, or can be used to define two filters. Each filter results from the compatibility between two constraints because, as seen previously for a single material selection (at Sect. 5.1.2.3), it is important to focus on the intersection of the intervals of possible values of the free variable imposed by each constraint. Using these filters allows a filtration on materials property chart.

In the example of the tie, the method can be applied as follows:

- Mass constraint $C_1$:

$$\rho LA \geq m_0 \quad \text{so } A \leq \frac{m_0}{\rho L} \tag{5.21}$$

- Mechanical constraint $C_2$:

$$\frac{EA}{L} \geq S_0 \quad \text{so } A \geq \frac{S_0 L}{E} \tag{5.22}$$

- Electrical constraint $C_3$:

$$\frac{rL}{A} \geq R_0 \quad \text{so } A \leq \frac{rL}{R_0} \tag{5.23}$$

where $\rho$, $r$ and $E$ are respectively the density, the electrical resistivity and the Young's modulus of the material.

The first filter $F_1$, defined by the compatibility of $C_1$ and $C_2$, leads to the following in (5.24) and can be graphically represented by Fig. 5.9:

$$E \geq \frac{A_0 L^2}{m_0} \rho \tag{5.24}$$

The second filter $F_2$, defined by the compatibility of $C_2$ and $C_3$, leads to the following in (5.25) and can be graphically represented by Fig. 5.10:

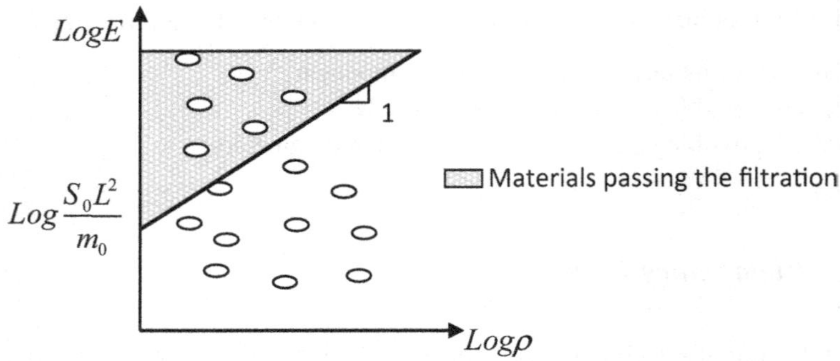


**Fig. 5.9** Property chart for $F_1$ filter

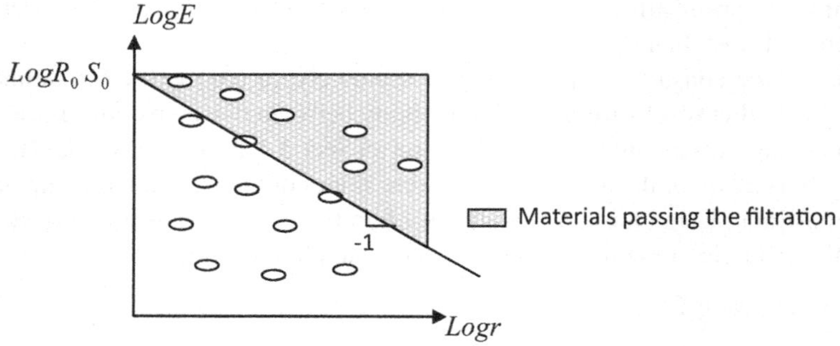


**Fig. 5.10** Property chart for $F_2$ filter

$$E \geq \frac{R_0 S_0}{r} \tag{5.25}$$

These graphical representations show the realised filtration by defining different domains on property charts, but another illustration can be given considering this filtration for one material—Fig. 5.11). Indeed, this material passes the first filter $F_1$ (resp. $F_2$) if the $I_1$ (resp. $I_2$) interval is not void.

This kind of filtration can be applied to any material, showing the possible intervals for the free variable imposed by each constraint. This method, available for single material selection, has to be adapted to work on multi-materials, taking into account the different steps presented previously.

- *First and second steps: component selection*

If the set of materials passing all the filters (that is to say if the intersection of the two intervals) is void, the only solution to fill the set of requirements is to design a multi-material.

The set of materials passing the filter $F_1$ (respectively $F_2$) is noted $M_1$ (respectively $M_2$). A combination of materials to satisfy the three constraints should be constituted

Fig. 5.11 Representation of both filters for a given material

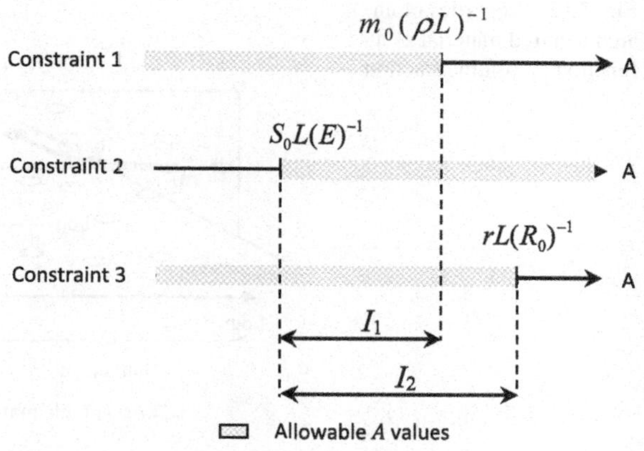


of one component from each set ($M_1$ and $M_2$), so couples of materials have to be selected by this way in the two sets.

Two types of constraints can be added to those ones. First, possible incompatibilities between the components must be taken into account. Then, the two components are not concerned by the same constraints because of their position (for example environment resistance or surface treatment). These supplementary constraints have to be considered at this stage.

- *Third step: searching for acceptable combinations and multi-materials*

As the morphology of the multi-materials is already determined (two layers), its behaviour can be modelled easily thanks to the equations:

$$m = LA(\rho_1 V_1 + \rho_2(1 - V_1)) \tag{5.26}$$

$$S_0 = \frac{A}{L}(E_1 V_1 + E_2(1 - V_1)) \tag{5.27}$$

$$\frac{1}{R_0} = \frac{A}{L}\left(\frac{V_1}{r_1} + \frac{1 - V_1}{r_2}\right) \tag{5.28}$$

where $\rho_i$, $E_i$, $r_i$, $V_i$ are the density, the Young's modulus, the electrical resistivity and the volume fraction of the $i$th component.

Then, the variations of the values of these properties as a function of the volume fractions of the materials can be represented on property charts by lines joining two components.

The application of the filters $F_1$ and $F_2$ defines limits on property charts that enhance limit values of the volume fraction $V$ of material 1 as shown in Fig. 5.12. The fulfilment of the set of requirements can be done only if $V_a < V_b$ because if this condition is not verified, no volume fraction can allow both filter to be passed.

An approach of the filtration for the selection of components of an architecture material has been exposed on an example consisting in a problem with two materials

Fig. 5.12 Properties of an architectured material as a function of volume fraction

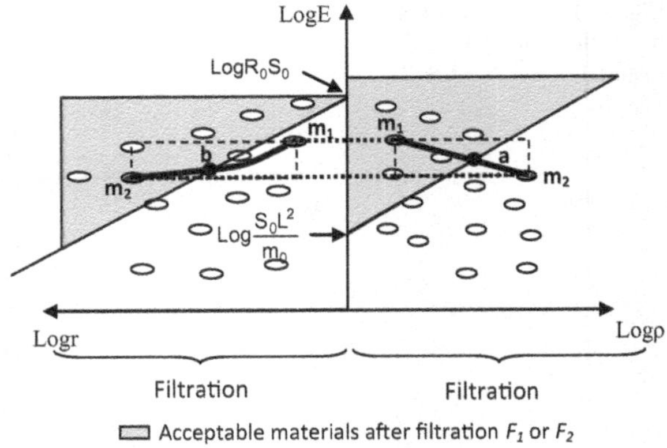


and three constraints depending on the free variables. This approach can be generalised to problems with two materials and more constraints first, and problems with more than two materials then.

### 5.4.2 *Design of an Architectured Material with Two Components and* n *Constraints*

In the previous paragraph, each filter was defined thanks to two constraints, one constraint imposing a minimum value to the free variable, the other one imposing a maximum value. If there are $p^-$ constraints imposing a maximum value and $p^+$ constraints imposing a minimum value, the number of filters for this problem is $p = p^- p^+$.

The application of each filter (noted $F_i$) determines a set of materials $M_i$ passing the filter. If the intersection of all the sets $M_i$ is void, there is no monolithic material that can be a solution for the problem.

For a material $m_k$, it is possible to represent the filtering results by a $p$-uplet $a_{k1}$, $a_{k2}, \ldots, a_{kp}$ where $a_{ki} = 1$ if the material $m_k$ passes the filter $F_i$, and $a_{ki} = 0$ otherwise.

Thanks to this coding, the condition for a couple of materials ($m_k$, $m_l$) to be a solution is expressed as: $\forall i \in [1, p], \ a_{ki} + a_{li} \neq 0$. Then, two cases have to be considered:

$a_{ki} + a_{li} = 2$, both materials belong to the $M_i$ set, so this filter will be passed whatever the volume fractions of the components,

$a_{ki} + a_{li} = 1$, one of the materials doesn't pass the filter $F_i$, so an interval of acceptable volume fractions will be defined in the form $[V_{ki}, 1]$ if the material $m_k$ passes the filter or $[0, V_{ki}]$ if the material $m_k$ doesn't pass the filter.

As a result, a multi-materials whose components are $m_k$ and $m_l$ is a solution if a volume fraction allows to pass all the filters, that is to say if the intersection of the available volume fractions intervals is not void.

It is important to note that the couples of materials are ordered. Indeed, a couple ($m_k$, $m_l$) differs from a couple ($m_l$, $m_k$) because each component plays a different role in most multi-materials (matrix/reinforcement, core/skin ...).

### 5.4.3 *Design of an Architectured Material with m Components and n Constraints*

The proposed coding for the selection of components of a bi-materials can be generalised to a greater number m of components, so the *p*-uplet representing the *m* materials is the sum of the *m* *p*-uplets of the components. As it was done in the previous paragraph, two cases have been considered:

$\sum_{k=1}^{m} a_{ki} = m$, all the components pass the *i*th filter, so the multi-materials will pass it too whatever the volume fractions of the components,

$\sum_{k=1}^{m} a_{ki} \langle m$, so the intervals of acceptable volume fractions have to be determined.

In a last part, a case study concerning a pipeline has been carried out to illustrate the approaches developed in this section.

### 5.4.4 *Application*

The case study concerns the development of new materials for oil extraction pipes lying on the marine ground at great depths. The pipe is supposed to be multi-layered and axisymmetric, with two or three layers (either inner and outer layers, or inner, medium and outer layers), so the free variables are the layer's thicknesses.

The functional analysis of such a pipe during its installation and its exploitation results in a set of requirements of the materials whose constraints can be divided in four groups detailed in Table 5.4: three groups don't depend on the free variable and concern either inner or outer or medium layer; the fourth one depends on the free variable.

The first step of the filtration consists in the elimination of the materials that don't fill the constraints of the first or the second group because they can't be used as an internal or external layer of the multi-materials. In order to be more easily treated, the database has been restricted to 62 monolithic materials. After the filtration, there are only 18 materials left for the internal layer and 14 for the external one.

The second step consists in applying the criteria fixed by the fourth group of constraints. There are four constraints imposing a minimum thickness and two imposing a maximum thickness, so eight filters can be defined. Each material is represented by a vector (whose terms are 0 or 1), and the sum of two vectors represents the properties of a two layer materials, that can be easily evaluated with respect to the requirements. After this step, there are only 8 couples of materials left.

**Table 5.4** Set of requirements by layers and types of constraints

| Group | Implied layer | Relation with the free variable | Description of the functions |
|---|---|---|---|
| 1 | Inner | No | – Resistance to service temperature |
| | | | – Resistance to corrosion due to the effluent (hydrocarbon) |
| 2 | Outer | No | – Resistance to service temperature |
| | | | – Resistance to corrosion due to sea water |
| | | | – Wear resistance |
| 3 | Medium | No | – Resistance to service temperature |
| 4 | Inner/outer or inner/medium/outer | Yes | – Resistance to mechanical loads during installation |
| | | | – Maximum force on the installation ship |
| | | | – Resistance to local buckling due to external pressure during exploitation |
| | | | – Limited thermal losses to ensure the effluent flow |
| | | | – Buoyancy: the pipe mustn't float |
| | | | – External diameter must be adapted to installation system |

Finally, for each couple of materials, the interval of acceptable volume fractions to satisfy the requirements has been calculated, leading to a set of only 5 possible bi-materials for this application. However, all these solutions must be rejected because it appears that polymer foams should be used in every case, but the contact with sea water or extracted fluid will probably result in a decrease of the foam properties.

In order to avoid this problem, a three layers solution was searched. The groups of constraints include the three groups presented in the two-layer pipe, and the group concerning the medium layer (see Table 5.4).

The first filtration step determines the materials which can be a component of the multi-materials: the constraints of the groups 1, 2 and 3 are applied, leading to a supplementary set of 45 materials left for the medium layer.

**Table 5.5** Results of the selection for a three layer material

| Type of combination (inner layer/medium layer/outer layer) | Number of solutions |
|---|---|
| Metal/polymer foam/metal | 4 |
| Metal/polymer foam/composite | 2 |
| Composite/polymer foam/metal | 8 |
| Metal/polymer foam/polymer | 10 |
| Polymer/polymer foam/metal | 24 |

The previous procedure is applied to this multi-materials, and results in 48 possible combinations which are summarised in Table 5.5.

It is important to note that currently used structures of materials can be found with these combinations (metal/polymer foam/metal), but in this case, the classical carbon steel is replaced by stainless steels or other metal alloys.

The combinations including composite materials as a component would require further developments, because several parameters (number of plies, orientations, volume fractions…) have to be determined.

The filtration method allowed the determination of a set of solutions, near of the existing ones but employing new materials. The important number of solutions can be reduced by focusing on the manufacturability of each multi-materials, which should be the next step of this design case study.

## 5.5 Architecture Selection

### 5.5.1 *First Approach*

The selection of the best material for an application consists in determining the optimal value of a combination of material properties called performance index. This indicator is derived from the expression of the performance of the product (mass or cost for example) in which all the parameters except the materials properties and one free design variable (i.e. one geometrical or functional parameter) are fixed.

There could be several ways to define the performance of an architecture. For example, the improvement on a property (or combination of properties) could be calculated comparing it with a law of mixture with the same components and volume fraction. This principle is close to the shape factor [21, 22] that is used to select the shape of the cross section of a beam. However, the calculation of these shape factors always rests on the separation of geometric parameters and material properties in the expression of the performance, which is generally impossible to manage when several materials are involved. So, it seems that this technique is suitable when trying to add an architecture to a single material, but becomes inapplicable for multi-material design.

Another method consists in using shape optimisation algorithms like level set for example [23]. With this method, the architecture isn't picked in a database, but evolves at each step of the calculation. The drawbacks of this method are that the final morphology depends greatly on the initial one, and that the calculation times are very important.

### 5.5.2 *Simultaneous Selection of Components and Architecture*

Trying to develop a free search method for the design of an architectured material, the simultaneous selection of the architecture and the components can follow the principle shown in Fig. 5.13. In order to calculate the properties of the multi-material, the architecture database must include homogenisation models, and must be linked to a materials database for the components properties.

The selection in a database allows a quick comparison of very different morphologies. Then, when a first ranking of the candidates is made, the parameters of the chosen architecture can be optimised thanks to classical topology optimisation methods in a final step.

The database has to be representative of the various possibilities of the multi-materials, but doesn't have to be exhaustive because it mustn't lead to important calculation times.

Trying to make a collection of the most classical architectured materials, it appears fundamental to create a hierarchy in the database. Indeed, multi-material morphologies can sometimes appear quite complex, but they can be considered as a combination of elementary patterns [24, 25], so, the database has to be organized as a function of the length scale.

For example, at macroscopic scale, all the multi-materials can be supposed to be multi-layered, so the only morphology that is considered in this category is a division of the material through the thickness.

Then, each layer can be filled with a predefined morphology. The possible elementary patterns can be separated in different types as:

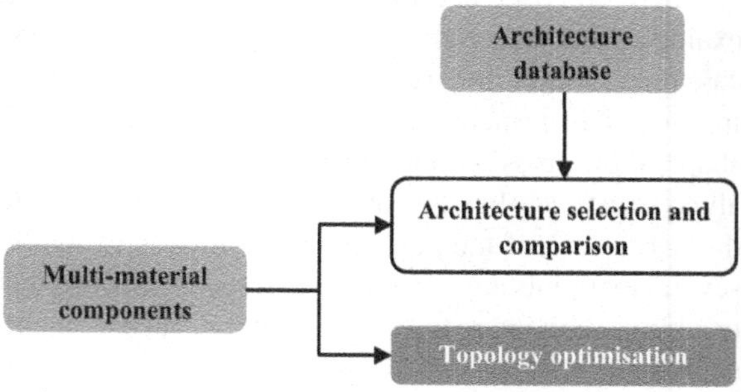


**Fig. 5.13** General method for architecture selection

- monolithic material,
- composite material: matrix reinforced by particles, short or continuous fibers,
- cellular material like foams or honeycombs.

This combinatorial problem gives rise to a huge number of potential solutions, even if material and architecture databases contain rather small numbers of items. Indeed, each architecture is characterised by several specific parameters describing the morphology. As a consequence, the numerical method for the selection of the architecture and components has to deal with a lot of different parameters to evaluate the homogenized properties and the performance. Some of these parameters are quantitative (volume fractions, particle dimensions…), but others are qualitative (type of architecture, constitutive materials…) so different kinds of variable have to be manipulated. As a complete screening of the solution space is impossible, the numerical method has to be chosen to give solutions in a reasonable time (for example genetic algorithms are already frequently used for the definition of stacking sequence or the choice of the components of composites [26]).

### *5.5.3 Architecture and Interface*

In addition to the coupling between architecture and components, the interest can be focused on the role of the interface between the components. Indeed, in the previous case, the properties of the architecture materials depended only on geometrical parameters or material properties because the interface between components was supposed to be perfectly adhesive.

Among the approaches of the definition of the attributes of the architecture materials, interface can be considered separately from the others, but often revealed coupled with architecture. For example, requiring a high tensile stiffness and a low flexural one for a beam can result in the design of a cable. This solution can be considered as an architectured material because even if it is made of only one material, it is characterised by a segmented morphology and a sliding interface between the monofilaments. Some parameters of this cable are clearly related to the architecture (number and diameter of filaments, twisted or not…) but the requirements can be fulfilled only thanks to this specific interface.

To illustrate this, the importance of the interface properties and its coupling with the architecture has been studied on a structure with two components [27]. A level set method was used to optimise the arrangement of the materials. The different types of interfaces that were tried resulted in very different architectures, showing that the efficiency of an architecture can't be separated from the interface properties.

## 5.6 Conclusions

Designing architectured materials fulfilling a set of requirements is a very complex task, and methods allowing a rational design are often still to develop. The first difficulty comes from the number and diversity of factors the designer has to deal with. In this chapter, some tools have been presented, showing that taking assumptions narrowing the field of solutions, some specific methods can be applied.

A statistical approach allows to answer the questions "why designing an architectured material?". The result of this study provides the designer with the possibility to separate incompatible requirements, and split the set of requirements in several more consistent ones.

Then, trying to extend selection methods already available for a single material, some approaches can be defined for the component selection, making an important filtration eliminating the biggest part of the possible materials combination.

Other approaches could consist in searching a solution to the set of requirements in terms of coupling mechanisms directly, which would involve the components, architecture, and interface. This method can be based on an analogy research, but implies that a database of standard solutions answering standard problems. Although this kind of approach is less innovative than free search, it can be interesting to give robust solutions.

Finally, all these methods have been illustrated with industrial cases, but didn't deal with one of the most stringent requirements for the future that is recycling. Indeed, the final step of the life cycle of the products has to be taken into account in the design process, and becomes even more complicated when several components are involved like in most of the architectured materials.

## References

1. H. Wargnier, F.X. Kromm, M. Danis, Y. Brechet, Mater. Des. **56**, 44–49 (2014)
2. M.F. Ashby, Y. Brechet, Acta Mater. **51**, 5801–5821 (2003)
3. F.X. Kromm, J.M. Quenisset, T. Lorriot, R. Harry, H. Wargnier, Mater. Des. **28**, 2641–2646 (2007)
4. S. Gasser, Y. Brechet, F. Paun, Adv. Eng. Mater. **6**(1–2), 97–102 (2004)
5. M.F. Ashby, Mat. Sci. Tech. **5**, 517–525 (1989)
6. M. Farag, in *Handbook of Materials Selection*, ed. by M. Kutz (2002), pp. 3–24
7. S.M. Sapuan, Mater. Des. **22** (2001)
8. M. Buggy, C. Conlon, J Mater. Process. Technol. **153–154**, 213–218 (2004)
9. A.M. Lovatt, H.R. Shercliff, Mater. Des. **19**, 205–215 (1998)
10. K.L. Edwards, Mater. Des. **26**, 469–473 (2005)
11. M.F. Ashby, Y. Brechet, D. Cebon, L. Salvo, Mater. Des. **25**, 51–67 (2004)
12. M.F. Ashby, Acta Mater. **48**, 359–369 (2000)
13. S. Giaccobi, F.X. Kromm, H. Wargnier, M. Danis, Mater. Des. **31**(4), 1842–1847 (2010)
14. J.R. Wilson, S.M. Grey-Taylor, Int. J. Ind. Ergonom. **16**, 353–365 (1995)
15. N. Perry, A. Bernard, F. Laroche, S. Pompidou, CIRP Ann. - Manuf. Technol. **61**(1), 151–154 (2012)
16. M.F. Ashby, *Materials and the Environment* (Butterworth-Heinemann/Elsevier, 2009)

17. G. Pahl, W. Beitz, *Engineering Design* (Springer, 1984)
18. D. Cavallucci, Techniques de l'ingénieur **A5**(211), 1–18 (1999)
19. M. Rahman, M.A. Mansur, Z. Feng, Mater. Des. **16**(4) (1995)
20. H. Sugishita, H. Nishiyama, O. Nagayasu, T. Shin-nou, H. Sato, M. O-hori, CIRP Ann. Manuf. Technol. **37**(1) (1988)
21. P.M. Weaver, M.F. Ashby, J. Eng. Des. **7**(2), 129–150 (1996)
22. D. Pasini, Mater. Des. **28**(7) (2007)
23. L. Laszczyk, Ph.D. thesis, Grenoble University, 2011
24. J.W.C. Dunlop, P. Fratzl, Scripta Mater. **68**, 8–12 (2013)
25. P. Bollen, N. Quievy, I. Huynen, C. Bailly, C. Detrembleur, J.M. Thomassine, T. Pardoen, Scripta Mater. **68**, 50–54 (2013)
26. L. Duratti, L. Salvo, D. Landru, Y. Brechet, Adv. Eng. Mater. **4**(6) (2002)
27. N. Vermaak, G. Michailidis, G. Parry, R. Estevez, G. Allaire, Y. Brechet, Struct. Multidisc Optim. **50**, 623–644 (2014)

# Chapter 6
# Topological Optimization with Interfaces

**N. Vermaak, G. Michailidis, A. Faure, G. Parry, R. Estevez, F. Jouve, G. Allaire and Y. Bréchet**

**Abstract** Design of architectured materials and structures, whether in nature or in engineering, often relies on forms of optimization. In nature, controlling architecture or spatial heterogeneity is usually adaptive and incremental. Naturally occuring architectured materials exploit heterogeneity with typically graded interfaces, smoothly transitioning across properties and scales in the pursuit of performance and longevity. This chapter explores an engineering tool, topology optimization, that is at the frontier of designing architectured materials and structures. Topology optimization offers a mathematical framework to determine the most efficient material layout for prescribed constraints and loading conditions. In engineering, topology optimization is identifying designs with interfaces, materials, manufacturing meth-

N. Vermaak (✉)
Department of Mechanical Engineering and Mechanics,
Lehigh University, Bethlehem, PA 18015, USA
e-mail: vermaak@lehigh.edu

G. Michailidis · A. Faure · G. Parry · R. Estevez · Y. Bréchet
SIMaP, CNRS, Université Grenoble, 38000 Grenoble, France
e-mail: georgios.michailidis@simap.grenoble-inp.fr

A. Faure
e-mail: alexis.faure@simap.grenoble-inp.fr

G. Parry
e-mail: guillaume.parry@simap.grenoble-inp.fr

R. Estevez
e-mail: Rafael.Estevez@simap.grenoble-inp.fr

Y. Bréchet
e-mail: Yves.Brechet@simap.grenoble-inp.fr

F. Jouve
Laboratoire J.L. Lions (UMR 7598), University Paris Diderot,
Batiment Sophie Germain. 5 rue Thomas Mann, 75205 Paris, CEDEX 13, France
e-mail: jouve@ljll.univ-paris-diderot.fr

G. Allaire
CMAP, Ecole Polytechnique, 91128 Palaiseau, Cedex, France
e-mail: gregoire.allaire@polytechnique.fr



Y. Estrin et al. (eds.), *Architectured Materials in Nature and Engineering*, Springer Series in Materials Science 282,
https://doi.org/10.1007/978-3-030-11942-3_6

ods, and functionalities unavailable to the natural world. The particular focus is on the variety of roles that interfaces may play in advancing architectured materials and structures with topology optimization.

## 6.1 Introduction

Designing architectured materials and structures introduces a scale for materials organization: this is the "architecture" between the microstructure and macroscopic shape [1–3]. In this approach, "spatial heterogeneity" is controlled such that combinations of materials or of materials and space are arranged in configurations and with connectivities that offer enhanced performance. In this way, interfaces are integral to architectured materials/structures. Interfaces are dynamic, diverse, sometimes dangerous frontiers. The idea of an interface implies the presence of some kind of transition, difference, change, discontinuity, and/or heterogeneity. Within the context of engineering applications, the interfaces of present interest may be mathematical and numerical tools or physically-based constructs. Physically, interfaces occur within and between the primary physical states of matter (solid, liquid, and gas).[1] Mathematically and numerically, interfaces are surfaces, infinitely sharp, that define the boundaries of regions.

Design of architectured materials/structures, whether in nature or in engineering, often relies on forms of optimization. In nature, controlling spatial heterogeneity is usually adaptive and incremental. Naturally occuring architectured materials exploit heterogeneity with typically graded interfaces, smoothly transitioning across properties and scales in the pursuit of performance and longevity. In engineering, topology optimization is identifying architectured designs with interfaces, materials, manufacturing methods, and functionalities unavailable to the natural world. Topology optimization provides a mathematical framework to determine the most efficient material layout for prescribed constraints and loading conditions [4].

This chapter focuses on engineering applications within structural topology optimization and explores the roles that interfaces can play. Within the classification of physical solid or bulk interfaces, several types of interfaces can occur. These include bi-material or hybrid, dissimilar interfaces, heterophase boundaries, localization of defects, grain boundaries, interphases, complexions, interlayers, coatings, and joints. In many of these cases, lattice-parameter changes in the interfacial region, induced by interfacial stresses, may have a pronounced effect on the physical properties and chemical composition at or near the interface [5]. Note that liquid, gas, or mixed, solid/liquid, and liquid/gas interfaces are also of engineering interest but beyond the scope of this chapter. In many structural applications, it is critical to design these material or solid/solid interfaces for system performance. For example, material interfaces often dictate tolerances and processing choices, lifetime and failure characteristics. Recently, the tool of topology optimization has begun to consider

[1] Plasma and other states of matter that occur under extreme conditions are not considered.

more and more numerical and physical roles for interfaces in design. This chapter explores some of these trends and provides perspectives and suggestions for future developments in the context of architectured materials and structures.

## 6.2 Structural Optimization

The phrase "structural optimization" is frequently used in engineering fields to describe processes and methodologies which aim to improve some characteristics of a structural part related to its mechanical performance. Historically, mechanical design has been primarily based either on the experience and intuition of designers, or on very simplistic analytical mechanical models. Extensive progress in computation over the last decades has enabled engineers to capitalize on mathematical optimization methods and algorithms to solve increasingly complex structural design problems in an automated manner.

Several categories of structural optimization exist; they depend on the kind of objectives or optimization variables of interest [6–8]. These differences in optimization methods can have dramatic effects on the types of design problems that can be considered and the way that interfaces can be accounted for in the design of architectured materials/structures. The following sections give an overview of the main methods of structural optimization. The focus is on problems of "Topology Optimization" (T.O.), i.e. problems where the design variables define the shape and the connectivity of a structure. A general mathematical formulation of a T.O. problem reads:

$$\inf_{\Omega \in \mathscr{U}_{ad}} J(\Omega), \tag{6.1}$$

where $\Omega$ is the domain occupied by the structure, $J$ is the objective function to be minimized and $\mathscr{U}_{ad}$ is a set of admissible shapes constraining the problem and to which $\Omega$ shall belong. In general, this problem (6.1) lacks a solution, unless the admissible set, $\mathscr{U}_{ad}$, is adequately constrained.

### 6.2.1 Categories of Topology Optimization

A Topology Optimization method can be characterized by two choices: (i) how the structural shape is described and (ii) how the shape evolves during the optimization process. How the shape evolves during the optimization process is directly related to the way a descent direction is calculated, i.e. how a direction that causes a decrease in the objective function to be optimized is found. The majority of T.O. applications use gradient-based methods. Using the calculus of variations, and depending on the nature of the optimization parameters, a descent direction is identified that guarantees the decrease of the objective function, at least for a small change of the design variables.

Despite the fact that calculating such a descent direction can be quite cumbersome, very efficient methods have been developed (MMA, CONLIN, MFD, SQP, etc.) that allow one to quickly identify an optimal solution, even for problems with a large number of optimization variables.

In terms of how the structural shape is described, the majority of numerical applications of topology optimization in the literature, as well as commercial software codes, are based on what is called a "density approach" to topology optimization. In these density approaches, the T.O. problem, (6.1), is translated into a problem of finding an optimal density distribution in a design domain, i.e.

$$\inf_{\theta \in \mathscr{U}_{ad}} J(\theta). \tag{6.2}$$

The first density approaches to appear in the topology optimization literature used homogenization theory [9–15]. The goal of these approaches was to overcome the difficulty of the non-existence of solutions for the general T.O. problem (6.1) by including "composite" materials as admissible solutions. That is, within a design domain, every point in the domain has an assigned density: 0 for material "A", 1 for material "B", and any value between 0 and 1 represents a kind of "composite" or mixture of "A" and "B" at that point. Note that material "A" or "B" may be physical materials or void (empty space).

Following the early work based on homogenization theory, several simplified methods have been proposed in order to force classical "0–1" shapes (shapes that include only solid or void) as solutions to the T.O. problem (6.2). The most well-known among the density approaches is the "Solid Isotropic Material with Penalization" or SIMP method [4]. SIMP uses a simple penalization scheme to suppress the formation of intermediate densities (density values between 0 and 1 with little physical meaning). With the advance of additive manufacturing, the motivation for penalizing these intermediate densities is being reevaluated. Multi-material printers continue to be developed with advanced capabilities. For example, one goal is to mix materials like painters mix red and yellow to achieve any desired orange at any point of interest. A great benefit of these density methods is the simplicity of the description and the ease of computing derivatives with respect to the density field. On the other hand, when the exact position of boundaries of or within the structural shape plays an important role in the mechanics or physics of the problem, density methods are not, in general, well-suited.

Alternatively, geometric methods, as the name implies, use a geometric description of the shape, $\Omega$, and constrain the admissible domain, $\mathscr{U}_{ad}$, to ensure the existence of optimal solutions. In the past, geometric methods have mostly been based on mesh deformation. They were often considered inadequate for performing T.O. due to the difficulty of enacting topological changes while moving the mesh. Recently, the use of level-set methods [16] for T.O. [17–19] made it possible to preserve a geometric description of the shape while performing topological changes in a simplified way. In addition to level-set approaches, advanced mesh deformation methods have also been developed [20] that handle topological changes with mesh evolution. Level-set methods for T.O. will be described in greater detail in Sect. 6.2.2.1. Geometric methods are usually more complicated than density methods, but they also offer two major benefits. First, they do not inherently introduce intermediate densities into designs. Intermediate densities have limited physical interpretation and 0–1 designs are usually preferred. Thus, geometric methods do not require further post-processing penalizations and re-interpretations of the design results like most density methods apply for 0–1 designs. Second, geometric methods can be applied in any mechanical framework without the need to modify the mechanics of the problem to accommodate intermediate densities.

Other approaches to T.O. include phase-field, evolutionary, and stochastic methods. The phase-field approach combines characteristics from both density and geometric methods [21, 22]. These methods are particularly suitable for multi-material problems, however optimized solutions depend strongly on the penalization parameters used to obtain 0–1 solutions (without intermediate densities). In the so-called "evolutionary methods," the update of optimization parameters is based on a heuristic criterion. For structural problems, the heuristic criterion is usually a stress-based criterion. This criterion is used to decide whether to add material, to remove material, or to advect a geometric shape [23, 24]. This process is similar to the way that a tree branch will adapt to changing mechanical loads, slowly adding or removing material [25]. Evolutionary methods are very simple to develop, however they are also only effective for a limited range of problems where a heuristic criterion is relevant and readily formulated. Lastly, stochastic methods are well-known for solving general structural optimization problems, mainly involving integer values. However, they are limited to a small number of optimization parameters and their use in Topology Optimization problems that are representative for engineering applications is impractical [26].

### *6.2.2 Topology Optimization for Interface Problems*

The choice of the T.O. method to be used for structural design depends strongly on the complexity and the nature of the problem. For example, simplified density methods perform very well for two-dimensional compliance minimization problems and there is no reason to develop an elaborate geometric method in order to obtain similar results. However, for problems where the position of the structural boundary or the precise definition of shape is important, classical density methods may not be sufficient due to the lack of geometric information. There is no explicit definition of an interface for a shape described via a density field varying continuously between 0 and 1. Instead, for T.O. implementations that penalize intermediate density values, the interface is recognized as a region of rapid density variation from 0 to 1. Despite the recent combination of density methods with projection filters [27, 28] in T.O., the position of the interface cannot be described as accurately as when a geometric method is used. In addition, although geometric methods are more difficult to implement numerically and require more complexity to allow topological changes to occur than density approaches, they also present an inherent benefit when it comes to problems where interfaces are important. Namely, the positions of interfaces or shape boundaries are always explicitly defined. This makes all operations related to the interface or shape boundary significantly easier.

#### 6.2.2.1 Level-Set Method for T.O.

Since the first publications on the topic of level-set-based T.O. [17–19], there has been rapid growth in this field that is reflected in the numbers of related publications and industrial projects [29]. The combination of the level-set method with the shape sensitivity approach [18, 30, 31] allows one to obtain a gradient-based geometric method that easily and elegantly performs topological changes. The following provides a brief description of the main elements of a level-set based T.O. algorithm.

**Shape description**

Using the level-set method for shape description, the boundary of a domain $\Omega$, denoted $\partial\Omega$, is defined as the zero level-set of an auxiliary function $\phi$ (see Fig. 6.1). Thus, discretizing the level-set function, $\phi$, which is part of a larger working domain, $D$, and defining $\phi$ as:

$$\begin{cases} \phi(x) = 0 \leftrightarrow x \in \partial\Omega \cap D, \\ \phi(x) < 0 \leftrightarrow x \in \Omega, \\ \phi(x) > 0 \leftrightarrow x \in \left(D \setminus \overline{\Omega}\right), \end{cases} \tag{6.3}$$

one ensures an immediate and precise knowledge of the interface or shape boundary. It should be noted that, based on the definition in (6.3), a domain, $\Omega$, can be equivalently described by an infinity of level-set functions. A "good" choice of level-set

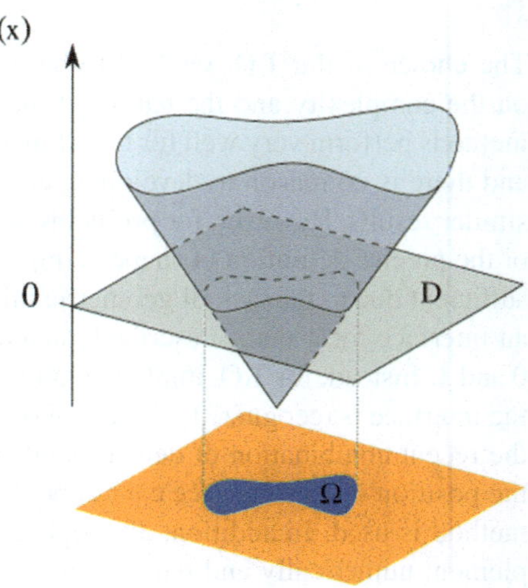


**Fig. 6.1** Level-set based definition of a domain $\Omega$

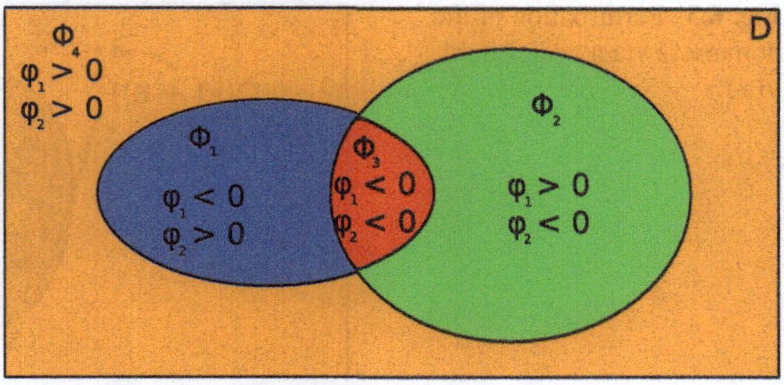


**Fig. 6.2** Multiphase representation in the level-set framework

function is one that is smooth enough to guarantee sufficient accuracy for all of the necessary numerical approximations: finding the exterior normal vector, the mean curvature, etc. Such a "good" choice is the **signed-distance function** to the domain, defined as:

$$d_\Omega(x) = \begin{cases} -d(x, \partial\Omega) & \text{if } x \in \Omega, \\ 0 & \text{if } x \in \partial\Omega, \\ d(x, \partial\Omega) & \text{if } x \in \overline{^c\Omega}, \end{cases} \tag{6.4}$$

where $d(x, \partial\Omega)$ denotes the standard Euclidean distance from a point, $x \in \Omega$, to the boundary. Beyond the smoothness of the signed-distance function, its use allows one to obtain valuable geometric information about the shape in order to define geometric characteristics, like a zone of prescribed thickness.

Another benefit of the level-set framework for T.O. is the ability to capture and describe multiple phases or materials in a natural way [32–34]. By defining $n$ level-set functions in the same design domain and combining their values, one can describe up to $m = 2^n$ different materials (see Fig. 6.2).

### Shape derivative

To employ a gradient-based optimization method, one needs to compute a derivative for the functionals contained in the optimization problem with respect to the design variables considered. Using the level-set framework for the description of the shape requires no parameterization of the shape, i.e. the shape is implicitly defined through the values of the discretized level-set function, $\phi$, and no design variables need "to" be considered? Instead, the notion of a **shape derivative**, i.e. a derivative of a functional with respect to variations of the shape in a direction $\theta(x)$, which dates back to Hadamard, can be defined as follows [31, 35, 36].

Starting from a domain $\Omega$, one considers perturbations by a smooth enough vector field $\theta(x)$, such that the new domain, $\Omega_\theta$, is described by (see Fig. 6.3):

$$\Omega_\theta = \big(Id + \theta\big)\Omega.$$

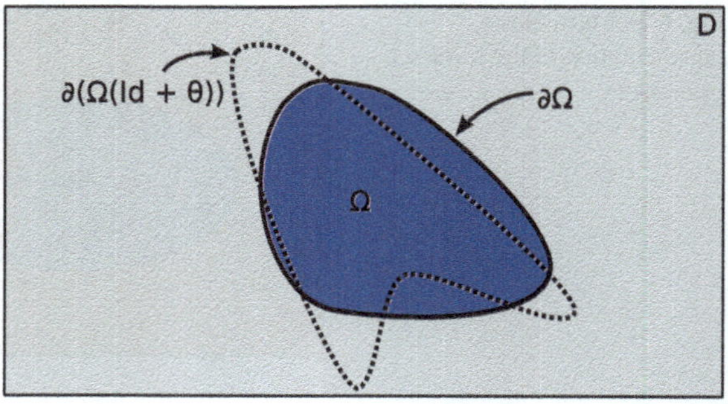


**Fig. 6.3** Perturbation of the domain $\Omega$ via a vector field $\theta(x)$

Then, the shape derivative $J'(\Omega)(\theta)$ of the functional $J(\Omega)$ in a direction $\theta(x)$ is obtained through an asymptotic expansion formula of the type:

$$J\big((Id+\theta)\Omega\big) = J(\Omega) + J'(\Omega)(\theta) + o(\theta) \quad \text{with} \quad \lim_{\theta\to 0} \frac{|o(\theta)|}{\|\theta\|} = 0\,.$$

Once calculated, a descent direction can be found by advecting the shape in the direction $\theta = -tJ'(\Omega)$ for a small enough descent step $t > 0$. For the new shape $\Omega_t = (\,\mathrm{Id} + t\theta)\,\Omega$, if $J'(\Omega)$, one can formally write:

$$J\,(\Omega_t) = J\,(\Omega) - t(J'(\Omega))^2 + \mathcal{O}(t^2) < J\,(\Omega)\,,$$

which guarantees a descent direction for small positive $t$.

**Shape evolution**

During the optimization process the shape is advected with a velocity $V(x)$ obtained from shape differentiation, as presented above. The advection is described in the level-set framework by introducing a pseudo-time, $t \in \mathbb{R}^+$, and solving the well-known Hamilton-Jacobi transport equation:

$$\frac{\partial \phi}{\partial t} + V(x)|\nabla\phi| = 0, \tag{6.5}$$

using an explicit second order upwind scheme [37, 38].

With this general description of the components of the level-set T.O. method, the following sections focus on how level-set T.O. can account for interfaces from numerical and physical perspectives. In both perspectives, this chapter considers interfaces as either sharp or smooth. Sharp interfaces represent discontinuous jumps. Smooth interfaces represent regions of gradation between distinct regions. Physically and in real materials, examples of both sharp and smooth interfaces can be found. Smooth or graded interfaces are commonly found in naturally occuring architectured materials. Sharp interfaces are representative of the atomically sharp boundaries that can be engineered in modern devices. If one looks at metals at the atomic scale, one can see a physical motivation for both kinds of interface models (sharp or smooth/graded). At an atomic scale it is possible to get a nearly perfectly sharp boundary; this is typically achieved in the semiconductor industry by molecular beam epitaxy. However, in many cases one gets (by design or not) smooth or graded interface zones that are a result of interdiffusion and surface reactions. These graded interface zones can be controlled and tuned by parameters like temperature. All of these types of interfaces serve functions but it is commonly thought that sharp interfaces sacrifice lifetime for performance. Thus, accounting for engineering interfaces in material/structural design is of great interest.

## 6.3 Interfaces as Numerical Tools in Topology Optimization

In T.O. one needs to handle criteria that depend on the solution of partial differential equations (PDEs) that represent the physics and mechanics of the problem of interest. These PDEs are defined with respect to a domain, $\Omega$, which is populated by one or more materials and that is a part of a larger working domain, $D$. The complementary of $D$, i.e. $D \setminus \Omega$, is filled with a weak material ("ersatz material") representing void or the absence of material. When an interface region between two phases or two materials is present, then the PDE must contain boundary conditions describing compatibility conditions across the interface. In linear elasticity, compatibility conditions typically refer to the displacement fields and the normal component of the stress tensor.

In T.O., treating a sharp interface between different materials or phases is a delicate issue, due to discontinuities in the material properties. This is because discontinuities are not differentiable and the calculation of a shape derivative becomes more tedious. As such, historically, smooth interfaces were initially introduced and used in T.O. as numerical tools to circumvent difficulties related to the differentiation of functionals rather than as models of real interface properties. More specifically, the principal goal of introducing smooth interfaces was to avoid difficulties related to shape differentiation or to avoid limitations of the numerical methods, rather than to represent material realities.

To the best of the authors' knowledge, smooth interfaces in T.O. first appeared in [19] and were followed by a long series of publications that follow this strictly numerical interpretation. In this legacy treatment: (i) mechanical properties are smoothly interpolated across the whole design domain using the level-set function, $\phi$, (ii) all of the criteria in the optimization problem are expressed in terms of $\phi$, and (iii) classical variational calculus is used to calculate a descent direction, i.e. the derivative of a functional, $J(\phi)$, in a direction, $\xi$, reads:

$$\langle J'(\phi), \xi \rangle = \lim_{\delta \to 0} \frac{J(\phi + \delta\xi) - J(\phi)}{\delta},$$

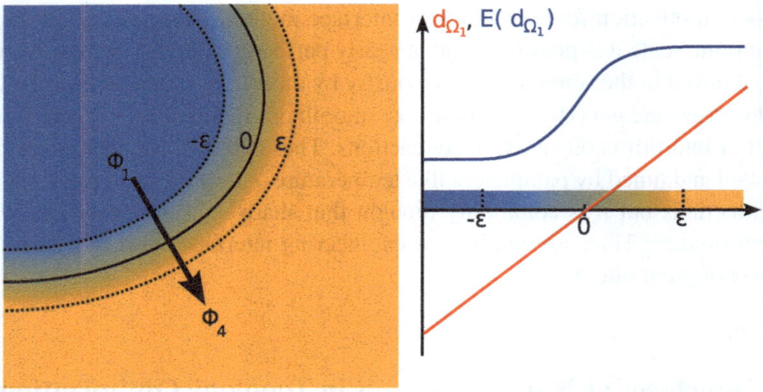


**Fig. 6.4** Interpolation between two phases $\Phi_1$ and $\Phi_4$ using the signed-distance function $d_{\Omega_1}$

which usually includes much simpler calculations compared to computing a classical shape derivative.

Similar interface modeling was proposed by Allaire et al. in [34] using the signed-distance function (see Fig. 6.4), coupled with a shape derivative. As it is explained in [34], the shape derivative in the multi-material setting contains the jumps of discontinuous stress and strain components across the interface. However, these jumps cannot be computed using Lagrange finite elements on a fixed mesh. Instead of employing techniques that accurately discretize the material interface [20, 39, 40], a remedy to this numerical limitation consists in smoothing the problem using a shape differentiable function (the signed-distance function) and constructing an interpolation scheme that converges to the sharp interface framework when the interpolation width tends to zero.

Beyond smoothing discontinuities or facilitating the derivation, interfaces can also be used numerically to ensure the existence of a minimizer. A minimizer refers to the case where a sequence of shapes during the optimization process (minimizing sequence) converges to a shape (the minimizer) in $\mathcal{U}_{ad}$. As previously mentioned, the general T.O. problem (6.1) lacks a solution unless the admissible set is adequately constrained. For example, the minimizing sequence may converge to a "composite" design that does not belong to the $\mathcal{U}_{ad}$ that is composed of 0–1 shapes. Controlling the complexity of the shapes in the sequence can be achieved by penalizing the perimeter of the shape [41], choosing smooth vector fields to advect smooth domains [6], or imposing manufacturing constraints like minimum thicknesses [42, 43].

Another approach to stop a minimizing sequence is to add a fictitous interface of constant width. This interface is a numerical tool to control the complexity of design features using a projection scheme [44, 45]. Figure 6.5 presents a typical example of a bi-material minimizing sequence obtained using a level-set based topology optimization algorithm that imposes an equality volume constraint at each iteration [44, 45]. From left to right, the microstructural features increase in complexity and a highly-interconnected distribution of materials (white and black) is observed. These

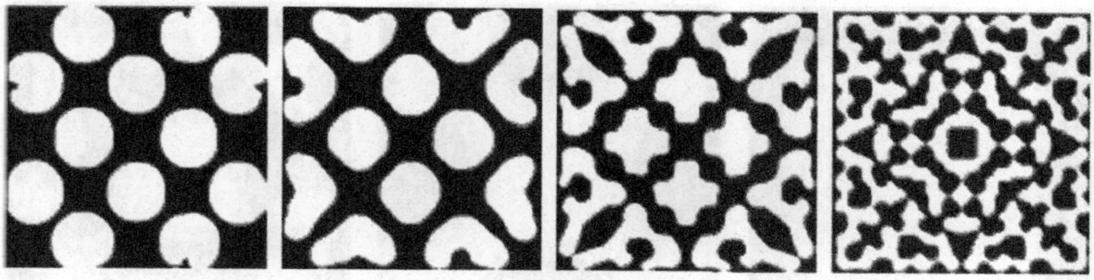

**Fig. 6.5** A minimizing sequence of a bi-material distribution converging weakly to a constant that respects a resource constraint ratio of 50% of each material at each iteration

kinds of minimizing sequences are converging to the theoretical limit in which every point in the domain is occupied by the prescribed volume fraction of materials. For example, if a resource constraint with a volume fraction of 50% of one material (white) and 50% of another material (black) was specified, the optimal solution would be a domain in which every material point is comprised of 50% of each material (the domain would look completely grey). Numerically, the minimizing sequences exhibit features that become vanishingly small while still respecting the required fixed volume fractions of materials at every point in the domain.

Figure 6.6 illustrates schematically how introducing a fictitous interface of constant width (in grey) can stop a minimizing sequence for the bi-material problem (black and white) with a 50–50 resource constraint. Figure 6.6 presents, from left to right, an initial design, three intermediate profiles, and the optimal design obtained at convergence with the fictitious interface removed. In this case, the interface is understood as a minimal distance constraint for the members of the projected phase (in black) [43]. As a result, significant oscillations of the boundary are prevented and the optimization process converges smoothly. The role of the interface here is purely geometric and it is assumed to have the same properties as the weaker material (in white). In this way, an extremely simple, although indirect, way to control manufacturing complexity is achieved.

## 6.4 Considering Material Interfaces in Topology Optimization

Beyond numerical reasons, interfaces can be considered in T.O. to represent real material observations. As depicted in Fig. 6.7, the interaction of atoms close to the interface of two phases or materials can result in a region that exhibits properties that are vastly different from the bulk phases or materials. The transition from one phase to another can be monotonic or not, also having a great impact on the optimal configuration.

In order to account for interface effects in T.O., one has to find an appropriate mathematical formulation. The mathematical background presented in [34] for han-

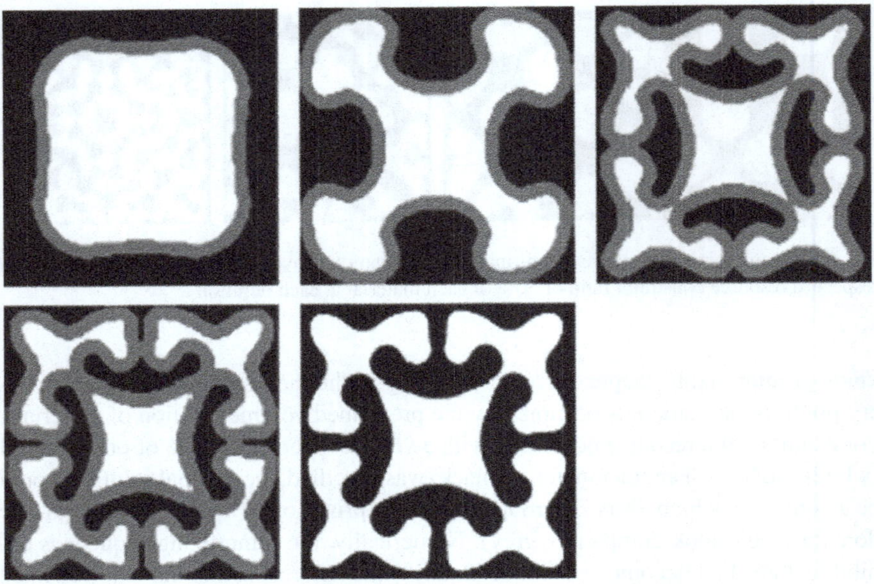

**Fig. 6.6** Using a fictitious interface of constant thickness (in grey) to avoid a minimizing sequence

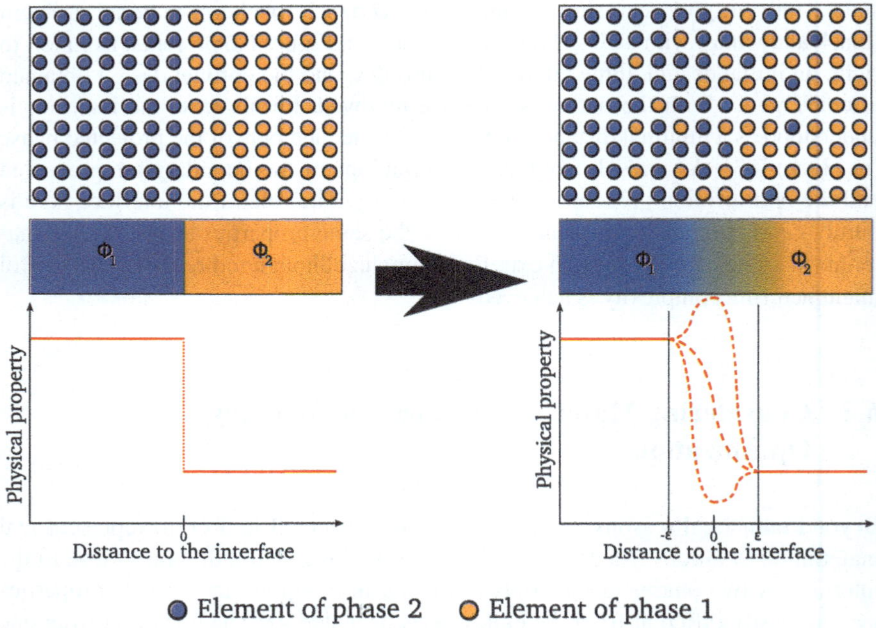


**Fig. 6.7** Left: sharp interface; right: smooth interface with various interpolation schemes

dling smooth interpolation profiles provided the formalism to efficiently treat any kind of regular interpolation scheme. The two key ingredients in this formalism are:

- the use of the signed-distance function in the interface interpolation scheme and
- the use of the co-area formula to obtain a shape derivative.

Using the signed-distance function to interpolate a physical quantity like the Young's modulus, $E(x)$, follows the general form:

$$E(x) = E_0 + h_\varepsilon(d_\Omega)(E_1 - E_0), \text{ where } h_\varepsilon(d_\Omega) = \begin{cases} 0 & \text{if } d_\Omega < -\varepsilon \\ h(d_\Omega) & \text{if } -\varepsilon \le d_\Omega \le \varepsilon \\ 1 & \text{if } d_\Omega > \varepsilon, \end{cases} \tag{6.6}$$

where $h(d_\Omega)$ is a smooth and differentiable function. This formulation requires that the interpolation width (width of the interface zone) remains constant during the optimization process (equal to $2\varepsilon$). More specifically, if the level-set function, $\phi$, is used instead of $d_\Omega$ in (6.6), the algorithm could try to enlarge or shrink the interface zone during the shape evolution (see Fig. 6.8) and optimization process in order to improve the objective. This is undesirable for two reasons. First, a great part of the final shape may contain intermediate values for the mechanical properties, eliminating the benefit of using a geometric method. Second, re-initializing the level-set function, $\phi$, to the signed-distance function to the domain changes the values of the functionals. As a result, a descent direction is not guaranteed. For more detailed information about this issue, see Allaire et al. [34] and Sect. 6.3.

The choice of $d_\Omega$ in (6.6) comes at the cost of complexity in the shape derivation [34]. This key point has been neglected in previous publications using similar interpolations [19], leading to erroneous formulations. Intuitively, the interface profile considered should have an impact on the advection velocity, and therefore also on the shape derivative. This intuition is substantiated and guaranteed when the correct shape derivation is applied. As explained in [34], using the co-area formula, one can

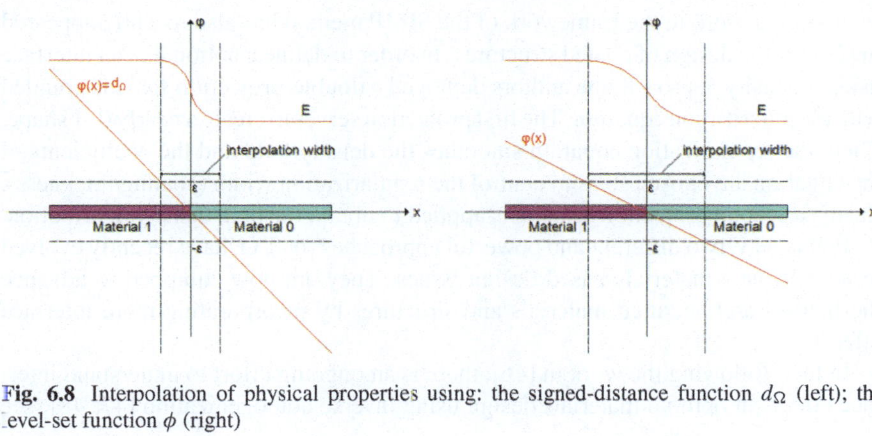


**Fig. 6.8** Interpolation of physical properties using: the signed-distance function $d_\Omega$ (left); the level-set function $\phi$ (right)

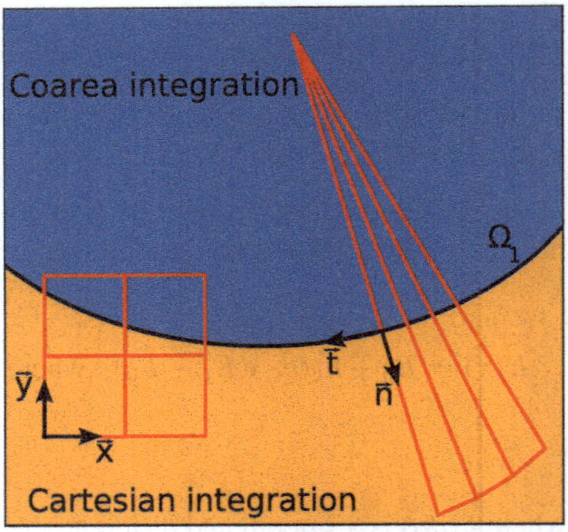


**Fig. 6.9** Rays emerging from the internal intermediate interface, $\partial\Omega_1$

split a volume integral defined in a domain, $D$, into a surface integral on the internal interface, $\partial\Omega$, and two one-dimensional integrals along the normal lines emerging from every point $y \in \partial\Omega$ (the so-called rays, see Fig. 6.9). Thus, all of the information associated with the interface zone is taken into consideration in the final shape derivative.

### *6.4.1 Some Applications*

The influence of incorporating internal material interface zones with distinct material properties in T.O. was first examined in [46] for macroscopic structures. The authors presented the impact of various interface zone interpolation profiles with elastic and thermoelastic example problems. The tendency to augment the total interface perimeter, whenever it is beneficial for the optimization problem, was highlighted and directions for further work to resolve manufacturing issues were proposed.

A similar work in the framework of the SIMP method has also recently appeared in [28] for the design of coated structures. In order to define a notion like an interface using a density approach, the authors deployed a double-projection scheme coupled with a regularization equation. The first projection serves to create a nearly 0–1 shape. Then, the regularization equation smoothes the density field and the coefficients of the equation are chosen so as to control the regularization width (coating thickness). Lastly, a second projection scheme is applied to obtain the final density distribution.

In this way, two different and powerful approaches for T.O. have recently evolved to address new materials-based design issues. They are now equipped to advance the field of architectured materials and structures by incorporating more interface effects.

In fact, following the work in [46], there is an ongoing effort to understand interface effects in optimal materials design using inverse homogenization [47, 48] (see

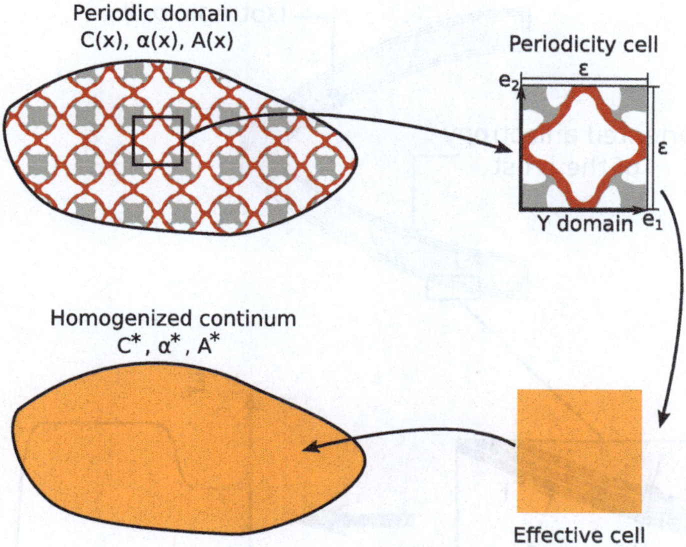


**Fig. 6.10** Inverse homogenization concept: design of a microstructure for target effective properties

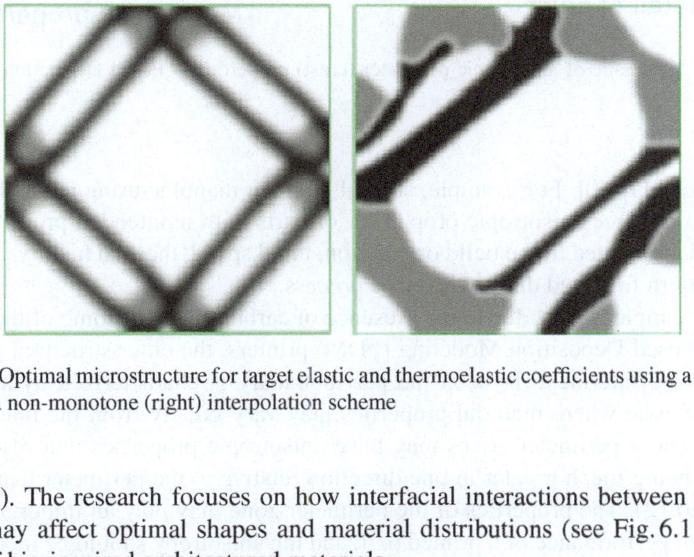

**Fig. 6.11** Optimal microstructure for target elastic and thermoelastic coefficients using a monotone (left) and a non-monotone (right) interpolation scheme

Fig. 6.10). The research focuses on how interfacial interactions between different phases may affect optimal shapes and material distributions (see Fig. 6.11) in the design of bio-inspired architectured materials.

In concert with the advance of T.O. methods, recent progress in additive manufacturing (AM) is expected to significantly broaden the range of T.O. applications. This is due to the extreme design flexibility that additive manufacturing methods afford. The capability to realize increasingly complex geometries allows designers to take full advantage of T.O. and impose fewer manufacturing constraints. As more research is focused on AM techniques and understanding the control of processing features, it is becoming increasingly necessary to tailor T.O. methods to specific AM

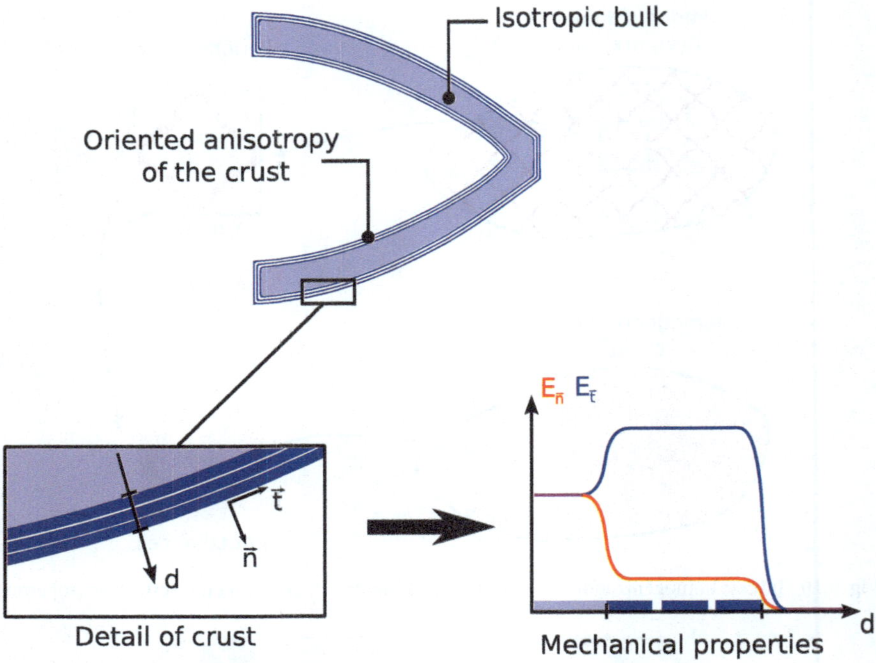


**Fig. 6.12** Schematic of anisotropic perimeter (crust) properties in Fused Deposition Modelling (FDM)

techniques [49, 50]. For example, several additive manufacturing technologies are known to produce anisotropic properties, or parts with unintended property gradations that are related to the build orientation, build speed, thermal history, and to the in-fill pattern followed during the build process.

One example is that, due to the presence of carbon fibers in some of the printing inks for Fused Deposition Modeling (FDM) printers, the outer structural perimeter or solid/void interfacial layer at the part boundary is characterized by a physical and finite zone where material properties may vary greatly from the interior infill regions. These perimeter zones may have anisotropic properties that result in the stiffness being much greater in one direction relative to the perimeter than another (see Fig. 6.12). The properties of the perimeter zone may play an important role in the overall performance of a printed part, and the anisotropy should be modeled and considered in the optimal design process. This topic is also related to coatings in materials and structures and is the subject of ongoing research.

## 6.5 Discussion

The above examples illustrate some of the rich opportunities for including interface effects in topology optimization. In addition to offering a better model of many physical or actual multi-material problems, including interface effects also offers new functionality by leading optimal designs in new directions. Including interface effects may allow optimizers to identify configurations that differ from those identified with classical sharp or monotonic interface modeling. Certainly, for a fixed design domain, these differences become more significant as the presence of the interface zone increases.

One could argue that the area occupied by the interface zone could also be modeled as a third material. The main difference between the interface zone and a third material is that the interface zone, by definition, is always surrounded or in contact with both of the other two materials. Simply introducing a third material does not guarantee or adequately represent this physical feature of interface zones. Nevertheless, this concern is valid and especially relevant when the existence of the interface zone is beneficial for the problem at hand. In this beneficial case, the tendency to promote and create tortuous paths for interface zones puts into question the notion of calling this an interface or interface zone at all. Interfaces, by definition, should also have an associated length scale that is much smaller than the dominant structural feature of the design domain of interest. For example, in Fig. 6.13, one can see how a region of fictitiously high Young's modulus can be assembled by small material islands that approach each other. These islands effectively create an interface network. This artifact is shown in more detail in Fig. 6.14. A non-monotonic interface transition of fixed width is considered between the phase in blue and the one in the grey color. However, since the size of the feature in grey is smaller than the interface width, one can argue that there is no longer an interface between the two phases, but instead a third material with elastic properties defined via the interpolation function.

However, since the position of the interface zone depends on the intermediate interface between the two bulk phases or materials, it is not correct to claim that the interface is simply a third material. Of course, in order to be physically meaningful,

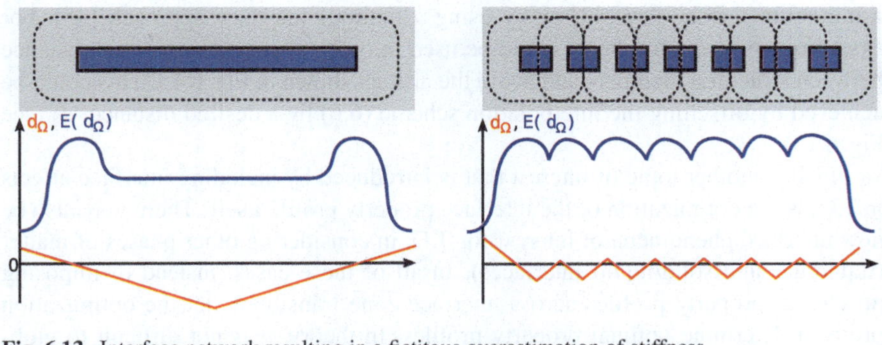


**Fig. 6.13** Interface network resulting in a fictitous overestimation of stiffness

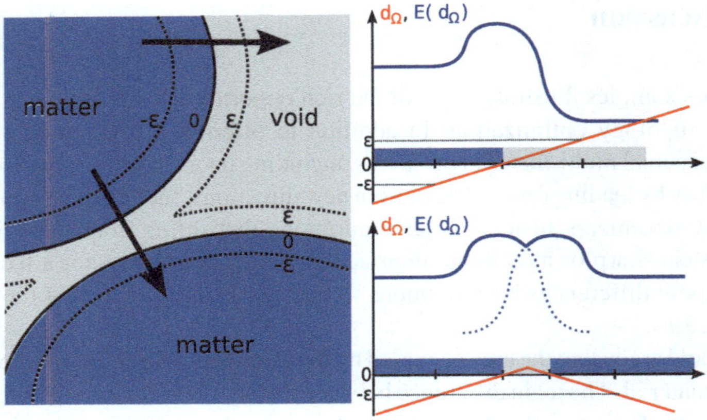


**Fig. 6.14** Another interface network resulting in a fictitous overestimation of stiffness

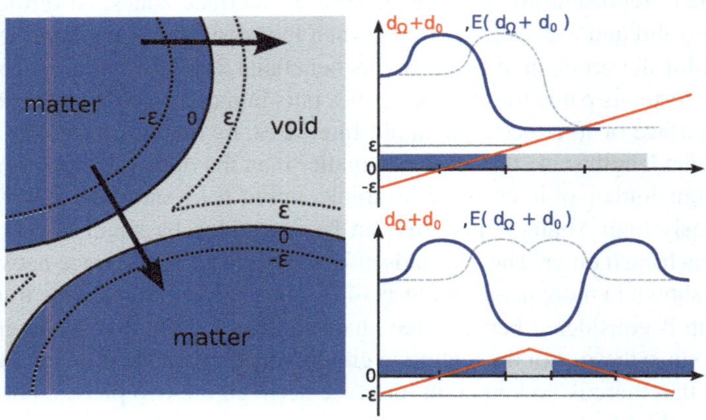


**Fig. 6.15** Imposing a minimal distance $d_0$ between structural features via a projection method (offsetting the interpolation scheme)

geometric constraints are required. Geometric constraint could be applied through a minimum feature size [43] or by using a fictitous interface approach [45]. For example, a projection scheme could be used in order to ensure a minimal distance between structural members and avoid the artifact shown in Fig. 6.13. This could be achieved by offsetting the interpolation scheme (6.6) by a desired distance, $d_0$ (see Fig. 6.15).

Finally, another topic of interest that is introduced by including interface effects in T.O. is the optimization of the interface property profile itself. There may also be new interface phenomena of interest for T.O. in considering other phases of matter (liquid/gas and solid/liquid interfaces). In all of these cases, instead of imposing prescribed property profiles across interface zone transitions, let the optimization protocol determine optimal property profiles. In theory, it is not difficult to globally parametrize the interface profile and to proceed with a parametric optimization derivation. One could also consider the interface thickness, as well as the intermediate property value at the center of the interface zone as optimization variables while employing a combined shape and parametric optimization. However, implementation becomes far more complicated when one is interested in locally optimizing the interface profile and a detailed explanation is out of the scope of this chapter.

## 6.6 Conclusions

By accounting for interfaces, topology optimization is poised to advance the design and aid the understanding of architectured materials and structures. Whether in nature or in engineering, architecture is allowing designs to capitalize on property differences to extract new or improved performance. Capturing these interface frontiers and transitions across boundaries offers many opportunities and challenges for materials scientists, mathematicians, designers, and engineers. It is suggested that all of these issues are best understood in a holistic context that takes inspiration from nature, manufacturing processes, and performance targets into account.

**Acknowledgements** This material is, in part, based upon work supported by the National Science Foundation under Grant No. 1538125.

## References

1. M. Ashby, Designing architectured materials. Scr. Mater. **68**(1), 4–7 (2013)
2. M. Ashby, Y. Brechet, Designing hybrid materials. Acta Mater. **51**(19), 5801–5821 (2003)
3. S. Torquato, Optimal design of heterogeneous materials. Ann. Rev. Mater. Res. **40**, 101–129 (2010)
4. M. Bendsoe, O. Sigmund, in *Topology Optimization: Theory, Methods and Applications* (Springer, 2004)
5. D. Wolf, J. Jaszczak, in *Materials Interfaces: Atomic-level Structure and Properties* (1992)
6. G. Allaire, Conception optimale de structures, vol. 58, in *Mathématiques & Applications* (Springer-Verlag, Berlin, 2007)
7. P. Christensen, A. Klarbring, in *An Introduction to Structural Optimization*, vol. 153. (Springer, 2009)
8. J.D. Deaton, R.V. Grandhi, A survey of structural and multidisciplinary continuum topology optimization: post 2000. Struct. Multi. Optim. **49**(1), 1–38 (2014)
9. G. Allaire, E. Bonnetier, G. Francfort, F. Jouve, Shape optimization by the homogenization method. Numer. Math. **76**(1), 27–68 (1997)
10. M. Bendsøe, N. Kikuchi, Generating optimal topologies in structural design using a homogenization method. Comput. Methods Appl. Mech. Eng. **71**(2), 197–224 (1988)
11. L. Gibiansky and A. Cherkaev, Design of composite plates of extremal rigidity, in *Topics in the Mathematical Modelling of Composite Materials* (Springer, 1997), pp. 95–137
12. R. Kohn, G. Strang, Optimal design and relaxation of variational problems, I. Commun. Pure Appl. Math. **39**(1), 113–137 (1986a)

13. R. Kohn, G. Strang, Optimal design and relaxation of variational problems, II. Commun. Pure Appl. Math. **39**(2), 139–182 (1986b)
14. R. Kohn, G. Strang, Optimal design and relaxation of variational problems, III. Commun. Pure Appl. Math. **39**(3), 353–377 (1986c)
15. F. Murat, L. Tartar, Calcul des variations et homogénéisation. Les méthodes de lhomogénéisation: théorie et applications en physique **57**, 319–369 (1985)
16. S. Osher, J. Sethian, Fronts propagating with curvature-dependent speed: algorithms based on hamilton-jacobi formulations. J. Comput. Phys. **79**(1), 12–49 (1988)
17. J. Sethian, A. Wiegmann, Structural boundary design via level set and immersed interface methods. J. Comput. Phys. **163**(2), 489–528 (2000)
18. G. Allaire, F. Jouve, A.-M. Toader, Structural optimization using sensitivity analysis and a level-set method. J. Comput. Phys. **194**(1), 363–393 (2004)
19. M. Wang, X. Wang, D. Guo, A level set method for structural topology optimization. Comput. Methods Appl. Mech. Eng. **192**(1), 227–246 (2003)
20. A. Christiansen, M. Nobel-Jørgensen, N. Aage, O. Sigmund, J. Bærentzen, Topology optimization using an explicit interface representation, in *Structural and Multidisciplinary Optimization* (2013), pp. 1–13
21. L. Blank, M. Farshbaf-Shaker, H. Garcke, C. Rupprecht, V. Styles, Multi-material phase field approach to structural topology optimization, in *Trends in PDE Constrained Optimization* (Springer, 2014), pp. 231–246
22. S. Zhou, M. Wang, Multimaterial structural topology optimization with a generalized cahn-hilliard model of multiphase transition. Struct. Multi. Optim. **33**(2), 89–111 (2007)
23. O. Querin, G. Steven, Y. Xie, Evolutionary structural optimisation (eso) using a bidirectional algorithm. Eng. Comput. **15**(8), 1031–1048 (1998)
24. A. Baumgartner, L. Harzheim, C. Mattheck, Sko (soft kill option): the biological way to find an optimum structure topology. Int. J. Fatigue **14**(6), 387–393 (1992)
25. C. Mattheck, Design and growth rules for biological structures and their application to engineering. Fatigue Fract. Eng. Mater. Struct. **13**(5), 535–550 (1990)
26. O. Sigmund, On the usefulness of non-gradient approaches in topology optimization. Struct. Multi. Optim. **43**(5), 589–596 (2011)
27. J. Guest, Topology optimization with multiple phase projection. Comput. Methods Appl. Mech. Eng. **199**(1), 123–135 (2009)
28. A. Clausen, N. Aage, O. Sigmund, Topology optimization of coated structures and material interface problems. Comput. Methods Appl. Mech. Eng. **290**, 524–541 (2015)
29. T. Abballe, M. Albertelli, G. Allaire, A. Caron, P. Conraux, L. Dall'Olio, C. Dapogny, C. Dobrzynski, B. Jeannin, F. Jouve, et al., *Rodin Project, Topology Optimization 2.0?* (2015). HAL preprint: https://hal.archives-ouvertes.fr/hal-01237051
30. F. Murat, J. Simon, Etude de problèmes d'optimal design. Optim. Tech. Model. Optim. Serv. Man Part **2**, 54–62 (1976)
31. J. Simon, F. Murat, in *Sur le contrôle par un domaine géométrique*. Publication 76015 du Laboratoire d'Analyse Numérique de l'Université Paris VI, (76015):222 pages (1976)
32. B. Merriman, J.K. Bence, S.J. Osher, Motion of multiple junctions: a level set approach. J. Comput. Phys. **112**(2), 334–363 (1994)
33. M. Wang, X. Wang, Color level sets: a multi-phase method for structural topology optimization with multiple materials. Comput. Methods Appl. Mech. Eng. **193**(6), 469–496 (2004)
34. G. Allaire, C. Dapogny, G. Delgado, G. Michailidis, Mutli-phase structural optimization via a level-set method. ESAIM Control Optim. Calc. Var. **20**(2), 576–611 (2014). https://doi.org/10.1051/cocv/2013076
35. O. Pironneau, *Optimal Shape Design for Elliptic Systems*, Springer Series in Computational Physics. (Springer-Verlag, New York, 1984)
36. J. Sokołowski and J.-P. Zolésio. *Introduction to Shape Optimization*, vol. 16, Springer Series in Computational Mathematics. (Springer-Verlag, Berlin, 1992). Shape sensitivity analysis
37. S. Osher, R. Fedkiw, Level set methods and dynamic implicit surfaces, in *Applied Mathematical Sciences*,vol. 153 (Springer-Verlag, New York, 2003)

38. J. Sethian, *Level Set Methods and Fast Marching Methods: Evolving Interfaces in Computational Geometry, Fluid Mechanics, Computer Vision, and Materials Science* (Cambridge University Press, Cambridge, 1999)
39. G. Allaire, C. Dapogny, P. Frey, A mesh evolution algorithm based on the level set method for geometry and topology optimization. Struct. Multi. Optim. **48**(4), 711–715 (2013)
40. Q. Xia, T. Shi, S. Liu, M. Wang, A level set solution to the stress-based structural shape and topology optimization. Comput. Struct. **9091**, 55–64 (2012)
41. L. Ambrosio, G. Buttazzo, An optimal design problem with perimeter penalization. Calc. Var. Partial. Differ. Equ. **1**(1), 55–69 (1993)
42. C. Dapogny, *Optimisation de formes, méthode des lignes de niveaux sur maillages non structurés et évolution de maillages*. Ph.D. thesis, Université Pierre et Marie Curie-Paris VI, 2013. Available at http://tel.archives-ouvertes.fr/tel-00916224
43. G. Allaire, F. Jouve, G. Michailidis, Thickness control in structural optimization via a level set method. Struct. Multi. Optim. **53**(6), 1349–1382 (2016)
44. F. Feppon, *Design and Optimization for Wear of Bi-material Composite Surfaces*. Master's thesis, Ecole Polytechnique, Promotion X2012, 2015
45. F. Feppon, G. Michailidis, M. Sidebottom, G. Allaire, B. Krick, N. Vermaak, Introducing a level-set based shape and topology optimization method for the wear of composite materials with geometric constraints. Struct. Multi. Optim. **55**(2), 547–568 (2017). https://doi.org/10.1007/s00158-016-1512-4
46. N. Vermaak, G. Michailidis, G. Parry, R. Estevez, G. Allaire, Y. Bréchet, Material interface effects on the topology optimizationof multi-phase structures using a level set method. Struct. Multi. Optim. **50**(4), 623–644 (2014)
47. O. Sigmund, Tailoring materials with prescribed elastic properties. Mech. Mater. **20**(4), 351–368 (1995)
48. A. Faure, G. Michailidis, G. Parry, N. Vermaak, R. Estevez, Design of thermoelastic multi-material structures with graded interfaces using topology optimization. Struct. Multi. Optim. **56**(4), 823–837 (2017). https://doi.org/10.1007/s00158-017-1688-2
49. A. Clausen, N. Aage, O. Sigmund, Exploiting additive manufacturing infill in topology optimization for improved buckling load. Engineering **2**(2), 250–257 (2016)
50. P. Zhang, J. Liu, A. To, Role of anisotropic properties on topology optimization of additive manufactured load bearing structures. Scripta Mater. **135**, 148–152 (2016)

# Chapter 7
# Friction Stir Processing for Architectured Materials

**Aude Simar and Marie-Noëlle Avettand-Fènoël**

**Abstract** Friction Stir Processing (FSP) is a solid-state process derived from Friction Stir Welding (FSW). FSP may be applied for the efficient manufacturing of metallic alloys based architectured materials. Indeed, the FSP tools allow locally modifying the microstructure of alloys or assembling dissimilar materials. The architectured materials that were or may be manufactured by friction stir processing will be discussed in this chapter. FSP may improve the mechanical performances of cast alloys, process metal matrix composites (MMC), make sandwiches, foams or additively manufactured structures. The aim is to process materials with improved lightweight performances, static or fatigue properties, crack resistance, toughness or wear resistance.

## 7.1 Friction Stir Processing for Architectured Materials

Friction stir processing (FSP) is a solid-state process derived from an innovative welding process patented by TWI in 1991 called Friction stir welding (FSW) [1–3]. It uses a non-consumable tool with complex features favoring the mixing of material in the solid state. A rotating tool is mounted on a machine similar to a milling machine. This tool is composed of a pin and a shoulder (Fig. 7.1). It is introduced into the material along the joint axis until the shoulder gets in contact with the upper surface of the workpiece to process. The tool is then travelling forward with the shoulder remaining in contact with the upper surface of the workpiece. The side where the rotational tool movement is in the same direction as the advancing movement is called the advancing side (AS) and the opposite side is called the retreating side

A. Simar (✉)
Institute of Mechanics, Materials and Civil Engineering (iMMC),
Université catholique de Louvain, Louvain-la-Neuve, Belgium
e-mail: aude.simar@uclouvain.be

M.-N. Avettand-Fènoël
Université de Lille, Unité Matériaux Et Transformations, UMR CNRS 8207,
59655 Villeneuve d'Ascq, France
e-mail: Marie-Noelle.Avettand-Fenoel@univ-lille.fr



Y. Estrin et al. (eds.), *Architectured Materials in Nature and Engineering*, Springer Series in Materials Science 282,
https://doi.org/10.1007/978-3-030-11942-3_7

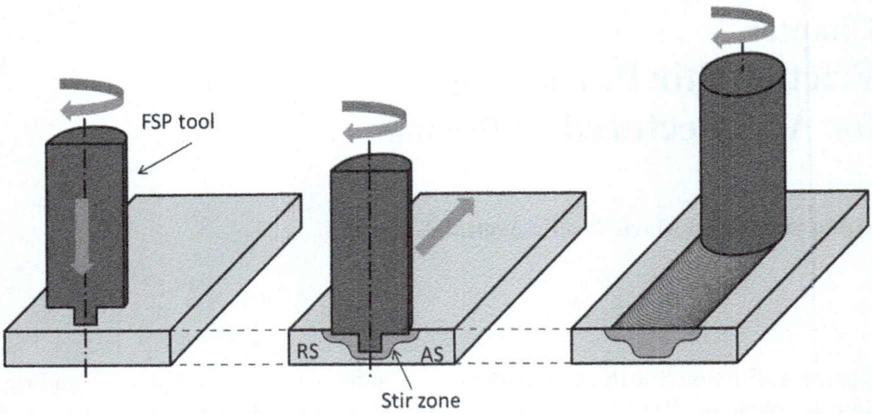


**Fig. 7.1** Principle of Friction Stir Processing (FSP). RS = Retreating Side, AS = Advancing Side (Schematic by N. Jimenez-Mena)

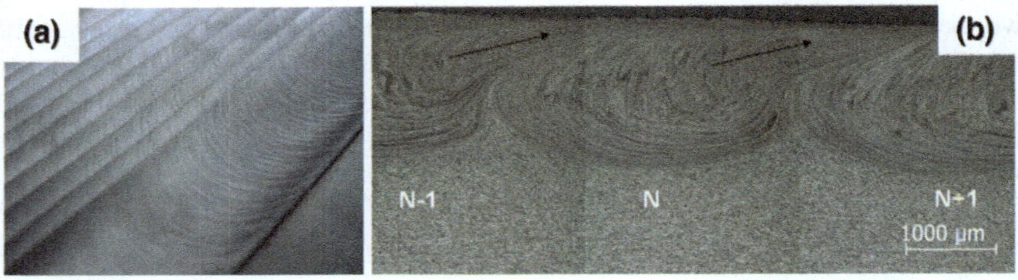


**Fig. 7.2** Multipass surface modification of an Al5083-H111 by FSP. **a** Surface finish. **b** Multiple adjacent stir zones (or nugget zone). From Gandra et al. [6]

(RS) (Fig. 7.1). Heat generated by friction and deformation brings the material into a malleable state that promotes the material flow from the front to the back of the tool and around the pin where it cools down. The temperatures reached are typically of 0.6–0.9 times the melting temperature for aluminum alloys [4]. This process may be applied on large surfaces by shifting the tool over a distance that is a fraction (called the overlapping ratio) of the modified region in the material [5, 6]. Figure 7.2 presents the surface finish and stir zone (or nugget zone) shape of a 5083 aluminum alloy surface processed by FSP.

FSP is particularly well suited for low melting point metals, such as Al, Mg or even Cu alloys, since the tool can then easily be machined from hard steel. However, FSW, and thus possibly FSP, has also been applied to higher melting point metallic alloys such as steel and titanium making use of much harder at high temperature and wear resistant tools such as PCBN or W-Re [3].

FSP may be applied for the efficient manufacturing of metallic alloys based architectured materials. Indeed, the FSP tools allow locally modifying the microstructure of metal based alloys or assembling dissimilar materials. As largely discussed in the other chapters of this book, architectured materials often involve dissimilar material assemblies or gradients of properties that could be reached by the local modification of a bulk material. In addition, improving the mechanical performances of an entire structural component is usually not necessary. In practice, plasticity, damage and cracking is only initiated in regions presenting high stress concentration. A FSP treatment can thus be locally applied to these regions to make overall higher performance structural parts.

The present chapter will discuss the possible architectured materials that may be manufactured by friction stir processing. Most of the work that is presented here did not specifically aim at making architectured materials but some of the ideas can be exploited to make such new structural materials and definitely aim at improved overall performances by the smart combination of materials and processing. FSP may locally modify the microstructure of cast or wrought materials, process metal matrix composites (MMC), make sandwiches, foams or additively manufactured structures. The ability of making these modifications at the local scale make the nature of the material architectured. The aim is to process materials with improved lightweight performances, static or fatigue properties, crack resistance, toughness or wear resistance.

## 7.2 Local Microstructure Modifications and Graded Materials

### *7.2.1 Grain Refinement*

Friction stir processing is commonly known to produce a significant refinement of the grain size in the stir zone. This is due to the dynamic recrystallization process that takes place in the zone mixed by the tool shoulder and pin. The type of recrystallization mechanism likely depends on the processing parameters and material nature [7]. FSP may cause discontinuous [8] or continuous [9] dynamic recrystallization.

Figure 7.3 shows the grain refinement observed in a twin-roll cast FSPed Al-Mg-Sc alloy [10]. In that particular case, the grain size is reduced by a factor close to 40. The effect of the rotational speed on the grain size is also evidenced in Fig. 7.3. A clear decrease of the grain size is observed as the rotational speed decreases [10, 11]. This is associated to a decrease of the maximum temperature as the rotational speed decreases due to less frictional and deformation heat generated by the process. Thus, grain growth is less activated after recrystallization. Other research teams have managed to further reduce the grain size in Al7075 [8, 12, 13] by hindering the grain growth by fast cooling using dry ice on the surface with an additionally cooled support anvil (100–400 nm [8], 25–40 nm [12]) or by liquid nitrogen circulating in the support anvil [13] (210–380 nm). Liquid nitrogen cooling of the FSP anvil may also refine grains in magnesium alloys, e.g. 300 nm in AZ61 [14] and 500 nm in AZ91 [15]. Dry ice surface cooling also allowed to refine the microstructure to a 50 nm mean grain size in pure Cu [16] near the exit hole. These grains then grew to a

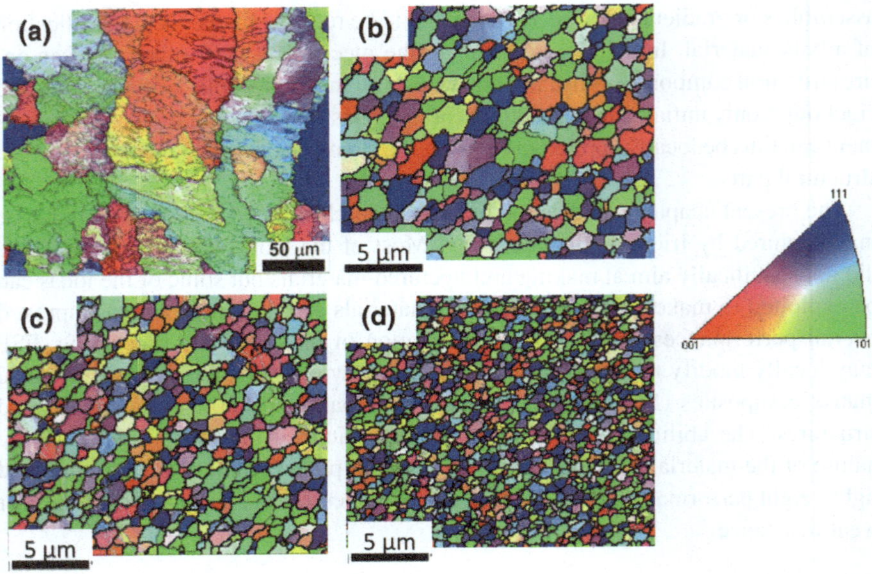


**Fig. 7.3** Grain size and texture evolution due to FSP at various rotational speeds in a twin-roll cast Al-4Mg-0.8Sc-0.08Zr alloy (EBSD map). The advancing speed is kept constant at 3.4 mm/s. **a** Base cast material (mean size = 19 μm), **b** FSP at 800 rpm (mean size = 0.89 μm), **c** FSP at 400 rpm (mean size = 0.73 μm) and **d** FSP at 325 rpm (mean size = 0.49 μm). From Kumar et al. [10]

mean size of 300 nm in the final FSP stir zone, emphasizing the effect of the cooling stage on grain growth.

The grain refinement is not always homogeneous in the processed zone. Indeed, the grain size is generally larger on the top surface of the processed zone [8, 17, 18]. This is associated to the extra heating due to the tool shoulder favoring grain growth after recrystallization. In the high entropy alloy Al0.1CoCrFeNi a graded increase of grain size is observed in the processed zone from the advancing (AS) to the retreating side (RS) [19]. This is due to both a difference of strain accumulation and a sluggish diffusion which controls and retards grain growth during cooling, entailing a hardness increase in the processed zone from the AS to the RS.

The importance of the pinning action of second phase particles on hindering grain growth and favoring particle stimulated nucleation during recrystallization has been particularly evidenced by Morishige et al. [20]. They have compared the grain size after FSP for pure aluminum presenting various degrees of impurities. They clearly show that smaller grains after FSP can only be maintained if second phase particles are present, i.e. in 99% purity Al compared to 99.999% purity Al. The alloying elements may also impact on the formation of extra barriers to the movement of dislocations like twins. Xue et al. [21] have evidenced more twin formation during FSP in copper containing 15%Al than in Cu-5%Al due to the lower stacking fault energy associated with the extra Al in solid solution. The grain size was similar in both alloys, i.e. 500–600 nm. The Cu-15%Al material presented significantly enhanced strength and ductility associated to the formation of these extra twin boundaries.

Thus, reaching ultra-fine grains or extra twin boundaries by FSP requires low maximum temperatures, fast cooling rates, second phase particles or alloying elements. During the FSP process, the strains are expected to be in the 5–10 range and the strain rates in the order of 10–400 $s^{-1}$ [22]. FSP allows reaching locally, where needed, grain sizes that are typical of severe plastic deformation processes. Su et al. [8] reached a grain size and misorientation distribution typical of 10 ECAP (Equal Channel Angular Pressing) passes with a single pass FSP. Kulitskiy et al. [23] showed that FSPed Al-Mg-Sc presented only slightly larger grain size after 1 pass of FSP than after 12 ECAP passes (2.1 μm after FSP vs. 1.2 μm after ECAP). More than 80% of high angle grain boundaries are found in an Al-Mg-Sc FSPed alloy [23, 24]. FSP can be easily performed locally contrarily to ECAP.

An issue related to the fine grains generated by FSP is that they are generally not very stable during a solution treatment above typically 450 °C in Al7010 [25] or above 490 °C in Al5083 [26]. Indeed, abnormal grain growth is likely to occur because the grains are free of dislocations or free of pinning phases at high temperature [25, 27]. Grains typically of a few millimeters have been reported [25, 27, 28]. Abnormal grain growth will inevitably lead to a loss in strength and ductility [29]. This may be problematic as sometimes a solution heat treatment[1] is necessary to fully dissolve the large precipitates prior to an aging treatment aiming at restoring the fine strengthening precipitation in aluminum alloys.

Many authors have evidenced increases in hardness and strength associated to the grain refinement by FSP in Al and Mg alloys [11, 14, 30–32]. The local modification of the mechanical properties is thus possible by the FSP process and associated grain refinement. However, not many authors have attempted to only partially process materials to take advantage of the possibility to make a multilayer. Mansoor et al. [33] have processed a ZK60 magnesium alloy on the top and bottom surface and report grains of 800 nm in the processed zone and a higher strength "multilayer" with only limited loss in ductility. Figure 7.4 shows the multilayered structure they processed.

### 7.2.2 *Cast Materials: Porosity Elimination and Microstructure Homogenization*

Cast materials have the particularity to contain porosities due to gas entrapping or gas dissolution in the molten state. Volume shrinkage associated to solidification is also a source of large porosities [34]. In addition, often the microstructure is inhomogeneous due to dendritic growth [34]. This may lead to low strength and premature failure. However, casting is an interesting process to manufacture geometrically complex parts. Thus, it may be interesting to be able to locally modify the microstructure

[1]The solution heat treatment is typically performed at temperatures larger than 500 °C in Al alloys.

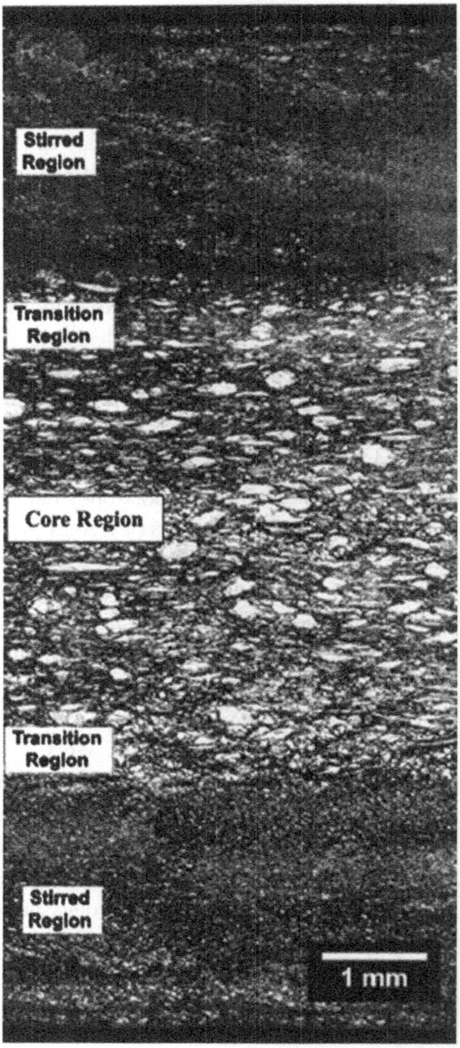


**Fig. 7.4** Multilayered microstructure obtained by FSP a ZK60 magnesium alloy on top and bottom surface. From Mansoor et al. [33]

of cast materials in order to improve the mechanical performances in regions of high stress concentrations where fatigue cracks are likely to initiate and propagate. Friction stir processing has largely proven to eliminate casting porosities and refine the microstructure [35–42]. This process may be applied locally and is thus an alternative to the hot isostatic pressing of the entire part [35]. Eutectic modifications are possible, e.g. by adding Sr in cast Al alloys [43] but are not likely to have a positive effect on casting porosities. The volume fraction of porosities might even be increased due to longer time spent at high temperature needed to ensure the eutectic modification [44].

Friction stir processing allows the elimination of casting porosities both in aluminum [35, 39, 41] and magnesium [39, 45] alloys. Ma et al. [36] have reported 0.1% of porosities in as-cast A356 and 0.02% when FSPed in the best conditions. Hannard et al. [28] have even reported the full elimination of porosities in wrought alloy 6056 after 3 passes of FSP. This is due to the large pressure under high temperature applied by the tool shoulder on the surface of the material. Prior porosities significantly decrease the ductility of cast materials [34] while many authors have reported a significant increase in ductility by FSP [36, 41, 46].

The cast microstructure is also homogenized and refined by FSP. Figure 7.5 shows the breakage and homogenization by FSP of the inhomogeneous acicular distribution of Si rich particles in an Al-7Si-0.6Mg alloy [41]. The large plastic deformations of the FSP process cause the redistribution, fracture and erosion of coarse particles [3, 36] as will be further discussed for metal matrix composites (Sect. 7.3). The aspect ratio of particles can be significantly reduced [36]. The homogenization is observed at a macroscopic scale through hardness measurements [35]. If performing overlapping passes, Ma et al. showed that the homogeneity of the hardness is maintained [37]. However, banded structure with coarse Si particles segregation have been reported for low rotational speeds [38]. A more homogenous and finer second phase distribution is generally observed at higher rotational speed [38, 41]. Chen et al. [42] have reported the alignment of the particles in the transition deformed region close to the stir zone (called the thermomechanically affected zone, TMAZ) in partially overlapping FSP leading to preferential failure at that location. In magnesium alloys, Feng et al. [40] show bands of higher content in $Mg_{17}Al_{12}$ particles in AZ80 alloy single pass FSPed material. In the double pass material this problem was solved. Hannard et al. [28] have quantified the microstructure homogenization due to FSP in Al6056. After one pass of FSP, the fragments of large particles are not yet homogeneously distributed, three passes are needed to ensure homogeneity.

In what concerns the mechanical properties, as already mentioned, the ductility of FSPed material is generally much higher than the ductility as cast [36, 41, 46].

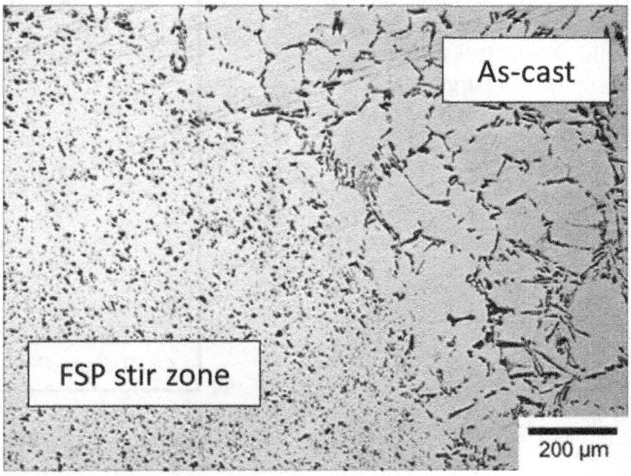


**Fig. 7.5** Interface between the cast and FSP region in an Al-7Si-0.5Mg alloy (F357) showing extended microstructure refinement and homogenization in the stir zone. From Jana et al. [41]

In addition to porosity elimination, the redistribution and refinement of the second phases also significantly improves the strength and the ductility [41, 47]. Jana et al. [41] have compared the fracture surface of as cast Al-7Si-0.6Mg (1.1% elongation at fracture) with the samples FSPed (20% elongation at fracture). The fracture surface presents a more ductile character with fine dimples in the FSPed samples while cleavage along the acicular second phases is observed on the fracture surface of the as-cast material.

The fatigue performances of FSPed cast materials are also impressively improved compared to their as-cast counterpart [45, 48–52]. This is mainly attributed to the elimination of porosities and to the particle refinement avoiding easy path for fatigue crack growth [48, 49, 52]. The fine second phase particles distribution favors crack tip blunting by the ductile matrix [50]. Figure 7.6 shows the increase by 80% of the fatigue stress threshold by FSP of an A356 cast Al alloy.

In what concerns fracture toughness, very limited results are available. Aktarer et al. [53] have measured the Charpy impact energy in as-cast and FSPed Al-12Si. It increased by a factor 7 from 1.2 to 8.3 J/cm$^2$. Here again, acicular Si particles where found to be preferential path for fracture in as-cast Al-12Si leading to a facetted fracture surface, while a large amount of small dimples were found on the fracture surface of FSPed Charpy tested samples.

The wear resistance of FSPed cast materials may also be improved compared to their as-cast counterpart thanks to the finely distributed second-phase particles. A lower wear rate (typically divided by a factor 2) is found for FSPed A356 Al alloy compared to the as-cast A356 [54, 55]. The wear rate was found to decrease if the rotational speed is very large (1250 rpm) [54]. This is consistent with a better distribution and refinement of the second-phase particles at higher rotational speeds [38, 41].

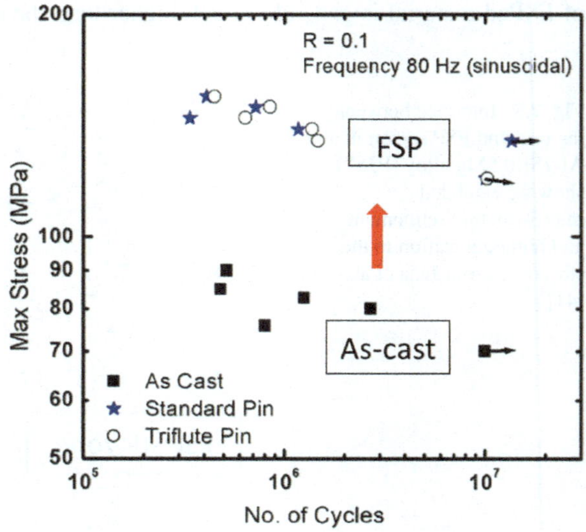


**Fig. 7.6** Improved fatigue life of a A356 Al alloy by FSP under a stress ratio of R = 0.1. Two tool pin geometries (standard and triflute) were tested showing no impact on fatigue life. Adapted from Sharma et al. [48]

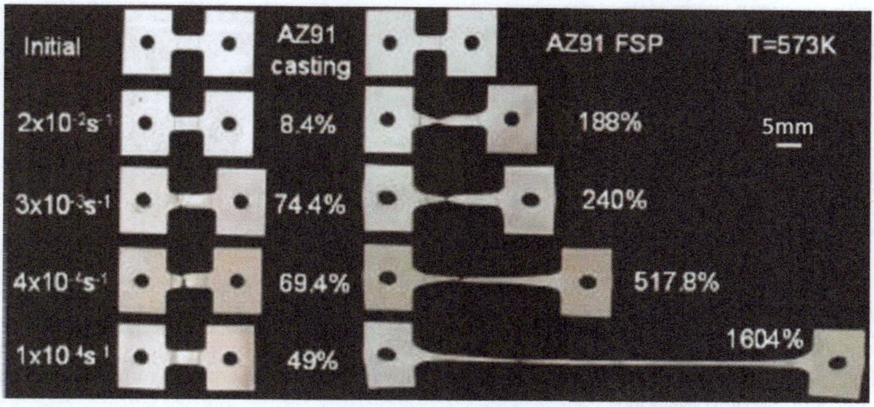


**Fig. 7.7** Comparison of the superplastic behavior of as-cast and FSPed AZ91 magnesium alloy at various strain rates for a temperature of 573 K. From Zhang et al. [72]

### 7.2.3 *Superplasticity*

The superplastic potential at relatively low temperatures and high strain rates compared to other processing methods has been largely reported for both aluminum [2, 56–70] and magnesium [71–75] friction stir processed alloys. Figure 7.7 shows the exceptional improvement of the plastic behavior of the FSPed AZ91 magnesium alloy compared to the as-cast material. High strain rate superplasticity allows to form sheets in a reasonable amount of time compared to materials presenting low strain rate superplasticity. Low strain rate superplasticity is rather classical for other severe plastic deformation processes but high strain rate superplasticity is exceptional. The main deformation mechanism leading to superplastic forming in FSPed materials is grain boundary sliding which is thus greatly enhanced in ultra-fine grain materials [56, 57, 61, 62, 70, 73] and in the presence of high angle grain boundaries [69] as typically found in FSPed materials. The appearance of the abnormal grain growth phenomenon is the upper temperature limit for superplastic forming [62, 65, 67, 68].

In the aim to process an architectured material with local ability for superplasticity, FSP could be applied locally. Mishra [59] reports the interest of making local FSP modifications to a part to ensure superplastic forming where needed, see Fig. 7.8.

### 7.2.4 *Precipitation State and Phase Modifications*

The high temperature reached during FSW and FSP generally affect the precipitation and phase distribution in metallic materials. Dissolution is enhanced in FSPed samples compared to just heated samples due to the shear driven effect [76]. Precipitate

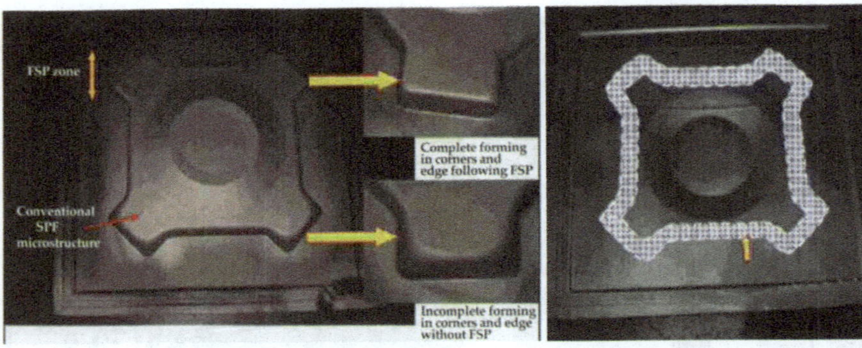


**Fig. 7.8** Local modification by FSP of an Al7xxx alloy for superior superplastic formability. From [59], courtesy of M. Mahoney

evolution may lead to gradients in mechanical properties that might be much more significant than what is resulting from grain refinement [77].

In aluminum alloys, the strengthening precipitates tend to dissolve in the stir zone leading to substantial softening [77–82]. Natural ageing may lead to the re-precipitation of GP zones that will slightly increase the hardness compared to a supersaturated state [82, 83]. The softening associated to the heating in the stir zone is also extended to the heat affected zones (HAZ). The HAZ generally present a lower strength than the stir zone due to the lack of dissolution of alloying elements in this zone that is rather subjected to precipitate coarsening [77, 82]. Figure 7.9 shows the gradient in hardness between the various zones of an Al6005A FSW in cold (advancing speed equal to 1000 mm/min) and hot (advancing speed equal to 200 mm/min) conditions. The hardness ranges from 65 to 105 HV. Figure 7.9 also shows that a post-welding heat treatment may modify the hardness gradient enhancing the strength in the stir zone due to formation of fine strengthening precipitates (here $\beta''$ precipitates). In thick FSPed Al alloys, the precipitate state may vary from the top to the bottom of the stir zone due to gradients in peak temperature [84, 85]. This will lead to hardness gradient as shown in Fig. 7.10 for an Al2050 alloy. In cast Al-Si-Mg alloys, $Mg_2Si$ precipitate dissolution in the stir zone has also been reported, and re-precipitation is possible during a post-welding aging treatment [36, 37].

In magnesium alloys, precipitate dissolution also occurs in the stir zone. For example, Xiao et al. [86] observed the dissolution of $\beta$-Mg(Gd,Y) in a FSP Mg-Gd-Y-Zr alloy. Jamili et al. [87] concluded that the $Mg_{24}Y_5Nd$ lamellar eutectic phase dissolves during FSP of WE43 alloy.

For details about the microstructure evolutions and resulting mechanical properties in high melting temperature materials, the reader is referred to [3]. The main limitation in processing these high melting point materials is the need for very hard tools, which might render FSP uneconomical for these materials. But, when cost is not part of the optimization criteria, FSP may also be considered for these materials. Seraj et al. [88] performed FSP on a AISI 52100 steel where the ferritic-pearlitic mi-

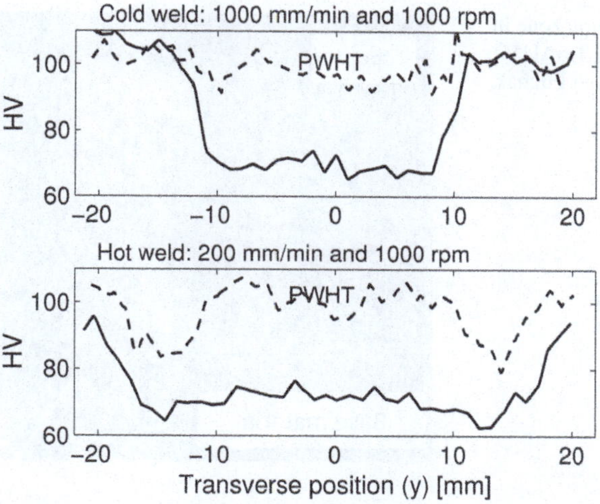


**Fig. 7.9** Hardness gradient in the stir zone and heat affected zone of Al6005A-T6 friction stir weld performed in cold conditions at 1000 mm/min and 1000 rpm and in hot conditions at 200 mm/min and 1000 rpm. PWHT = Post-Welding Heat Treatment at 185 °C for 6h. From Simar et al. [81]

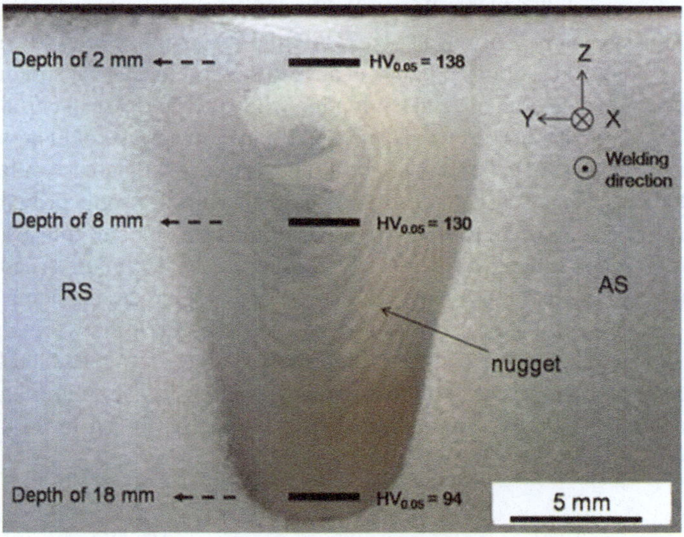


**Fig. 7.10** Hardness gradient in the stir zone of thick Al2050 FS weld. From Avettand-Fènoël et al. [84]

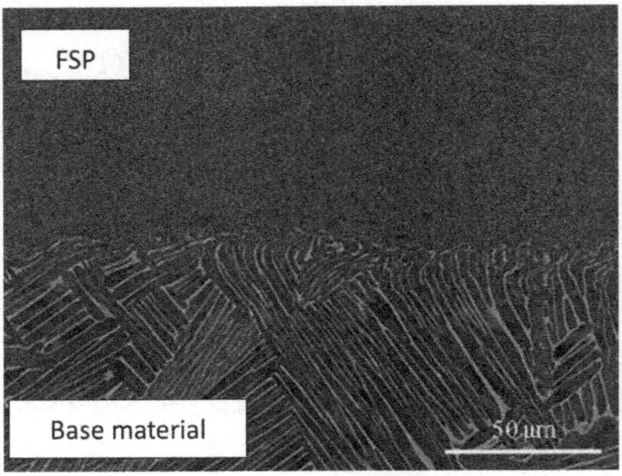


**Fig. 7.11** Transition zone in a FSPed zone of a Ti6Al4V titanium alloy. From Pilchak et al. [89]

crostructure was transformed to a martensitic microstructure with retained austenite. This leads to a significant increase in hardness (by a factor 2) and a reduction of the wear rate (by a factor 15).

In Ti6Al4V titanium alloy, phase transformations are observed during FSP. Figure 7.11 shows the modification of the microstructure between the base material and the stir zone in a cast Ti6Al4V alloy [89]. The base material presents $\alpha$ lamellae in prior $\beta$ grains typically 1.5 mm in size while the recrystallized stir zone presents 1–2 μm size equiaxed $\alpha$ grains. Pilchak et al. [89] suggest that a FSP structure on the part surface could delay fatigue crack initiation while the core base material structure favors a lower fatigue crack growth. Thus, a multilayer material would present better fatigue performances [90]. Pilchak et al. [90] have shown an increase in fatigue life from 58,000 cycles in as-cast conditions to 300,000 cycles for some FSPed samples. In the work by Pilchak et al. [89], FSP was performed above the $\beta$ transus. The FSP parameters were modified to perform FSP on a biphased $\alpha + \beta$ microstructure below the $\beta$ transus [91] with a lower rotational speed and/or a larger advancing speed. In that case, regions of transformed $\beta$ phase may be observed in-between equiaxed $\alpha$ grains [91]. Li et al. [92] have shown that faster cooling rates generated by higher advancing speeds lead to the formation of an $\alpha$ martensitic structure that is more wear resistant and may thus be interesting to generate on the surface of Ti6Al4V parts.

### 7.2.5 *Texture Modifications*

In Al alloys the texture after FSP is usually close to random (see Fig. 7.3). The texture after FSP is not affected by the initial texture even when FSP severely deformed Al alloys [93].

In magnesium alloys, FSP clearly affects the texture [94–100]. Generally a strong basal texture is observed with an alignment of (0001) basal plane with the tool shoulder surface in the upper stir zone [95, 96, 99, 100]. Vargas et al. [100] have shown that this crystallographic texture is enhanced by a second pass of FSP in a ZKX50 alloy. Xin et al. [96] have shown that this basal texture evolves from the processing direction to the transverse direction respectively from the center of the stir zone to its periphery. Yuan et al. [97] have shown that the tilt of the basal plane depends on the depth. The angle between the basal plane and the normal direction of the plate surface varies from 37 at 0.5 mm depth to 86 at 1.5 mm depth, see Fig. 7.12. The strong texture impacts the anisotropy of plasticity in the FSPed samples with significantly higher ductility in the transverse direction compared to the processing direction and harder (respectively softer) material than the base material in the transverse direction (respectively in the processing direction) [97].

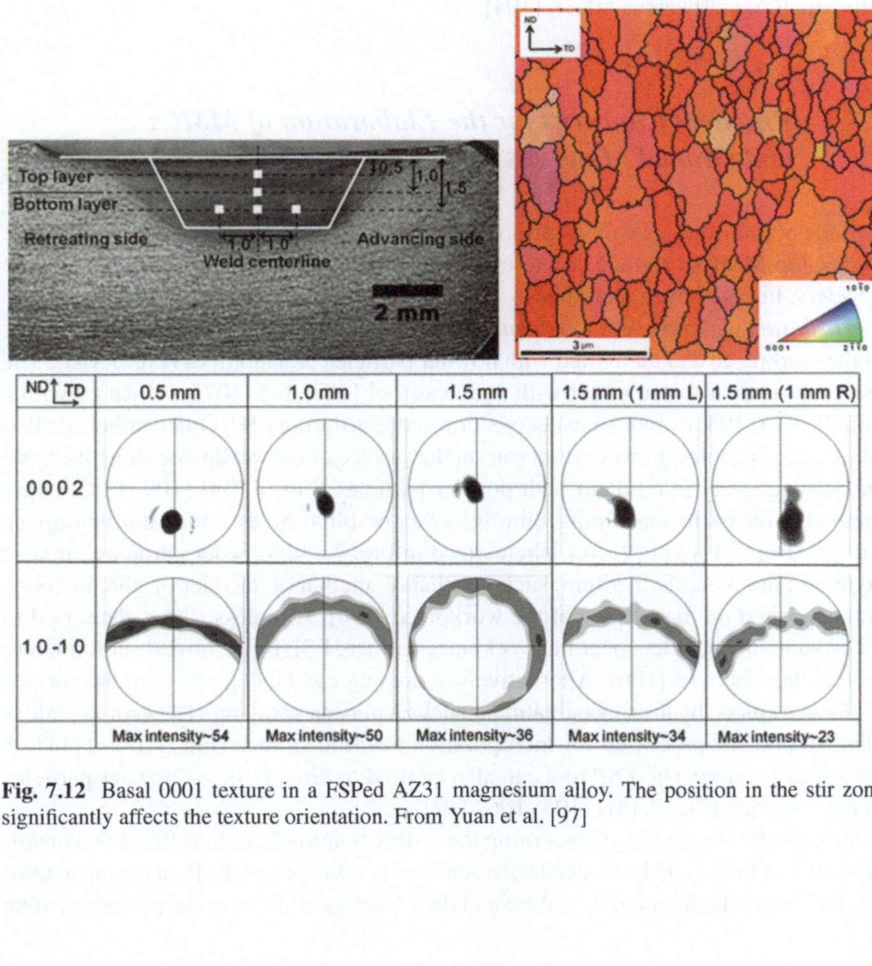


**Fig. 7.12** Basal 0001 texture in a FSPed AZ31 magnesium alloy. The position in the stir zone significantly affects the texture orientation. From Yuan et al. [97]

## 7.3 Metal Matrix Composites (MMC) and Graded Materials

Friction stir processing is also used to process metal matrix composites (MMCs) and graded materials. The reader interested by the FSW of metal matrix composites can also refer to the review by Avettand-Fènoël and Simar [101]. Indeed, FSW of MMC presents some common features with the use of FSP to make these materials.

The architectured materials presented in Sect. 7.2 generally refer to an evolution of the microstructure in the bulk or at the surface of the materials, affecting in particular the grain size. In the case of MMCs, and as previously mentioned in Sect. 7.2.1, the grain boundaries pinning by second phases can enhance the grain refinement effect of dynamic recrystallization. This is for instance shown in a FSPed $Al_2O_3$ reinforced ferritic stainless steel [102], in a FSPed TiC reinforced ferritic steel [103] or in a FSPed $Al_2O_3$ reinforced pure titanium, particularly if the latter reinforcements are very small, i.e. 20 versus 80 nm [104].

### *7.3.1 Practical Concerns for the Elaboration of MMCs and Graded Materials*

Various possibilities as well as disposals of reinforcements are reported in literature to develop MMCs or graded materials by FSP (Fig. 7.13). Reinforcements may be powders, fibers or even nanotubes.

*Disposals with powders or nanotubes*—Grooves can be machined at the surface of the workpiece and then filled with powder particles or nanotubes (Fig. 7.13a). The assembly is then sealed either with a pinless tool [102, 105–107] or covered with a thin sheet [5, 108] to avoid powder ejection and loss during FSP. Other authors drilled some periodic holes (powder reservoir) at the surface of the workpiece along its depth and subsequently filled them with powder particles (Fig. 7.13b) [109, 110]. Other methods consist in machining blind-grooves or blind-holes under the workpiece surface (Fig. 7.13c) [109]. As schematized in Fig. 7.13d, powder particles can also be mixed in a volatile medium, such as ethanol, methanol, lacquer or glue in spray, and deposited on the surface of the workpiece [106]. A pinless tool is then used to incorporate the powders near the workpiece surface. FSP is performed subsequently with a classical tool [106]. Alternatively, a coating can be deposited on the surface of the workpiece by a surface treatment such as plasma spraying. The coated plate is then friction stir processed for incorporation near the surface (Fig. 7.13e) [111]. A consumable composite FSP tool can also be used to directly insert powder particles in the substrate (Fig. 7.13f) [105, 106, 112].

*Disposals with fibers*—Concerning the setting to introduce short fibers in a metallic matrix, a fabric can be stacked between two metallic plates. FSP on the lap assembly will break the fibers of the fabric and distribute short fibers in the processed zone

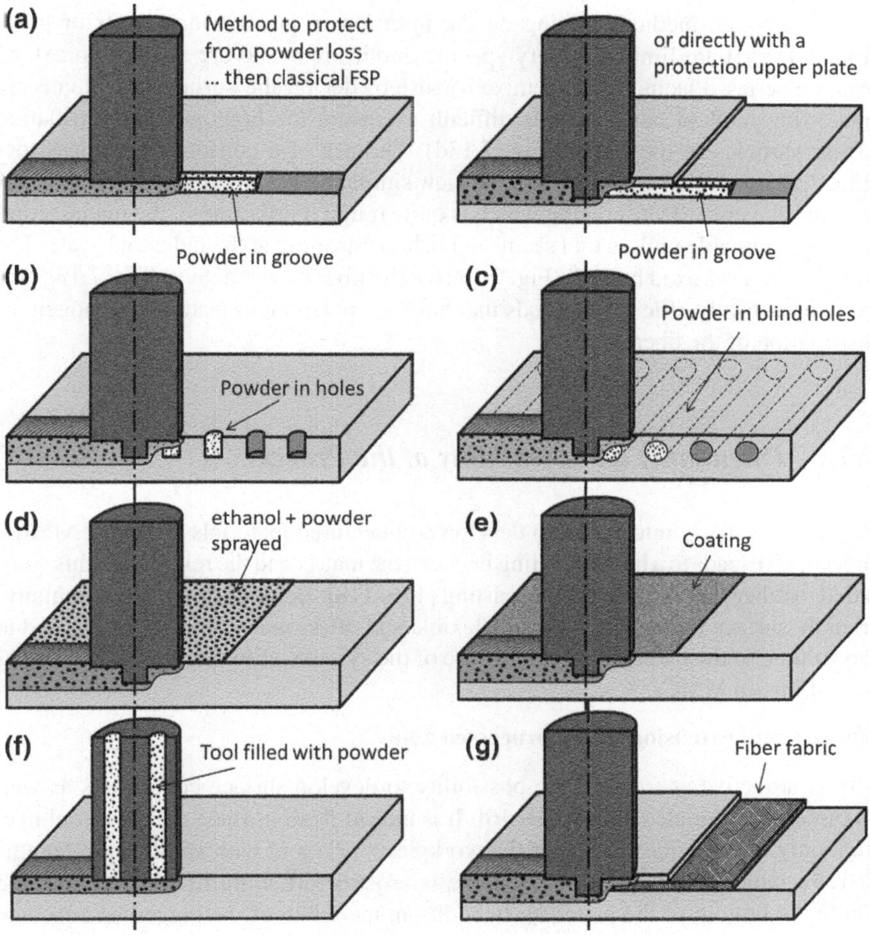


**Fig. 7.13** Schematic of various reinforcement disposal methods: **a** powder placed in grooves machined near the surface; **b** powder placed in holes drilled near the surface; **c** powder placed in blind holes; **d** powder dispersed in ethanol and sprayed on the surface; **e** coating prior to FSP; **f** consumable composite pinless tool; **g** fabric stacked between two plates (Schematic by N. Jimenez-Mena)

(Fig. 7.13g). Mertens et al. [113] have applied this process to reinforce Mg alloys with carbon fibers.

*Disposals with strips*—Liu et al. [114] directly placed a strip of an amorphous alloy in a groove machined at the surface of the Al alloy workpiece.

*Disposal with gas*—In this method, the first FSP pass is performed under reactive nitrogen blown in front of the advancing tool in order to introduce nitrogen gas into the stir zone of a workpiece machined with a groove. This was, for example, applied to develop in-situ TiN reinforced Ti. A TiN quasi uniform distribution was obtained after three subsequent FSP passes under argon [115].

The disposal methods relying on the insertion of powder in grooves or holes (Fig. 7.13a–c) are limited to very specific applications and are difficult to extend to large series. Placing ethanol mixed with powder on the surface is certainly an interesting method but it sounds difficult to ensure the homogenous distribution of the particles in the MMC (Fig. 7.13d). The use of a consumable pinless tool (Fig. 7.13f), which is in some points of view similar to additive manufacturing, leads to a multilayered MMC coating which is quite rough. Furthermore, the manufacture of the consumable pinless tool seems ticklish to transpose at the industrial scale. The use of coatings mixed by FSP (Fig. 7.13e) or the fiber fabric method (Fig. 7.13g) are more industrially efficient methods that have the potential to lead to a homogenous distribution of the fibers.

### *7.3.2 Potentiality and Flexibility of the Process*

FSP is particularly interesting to develop architectured materials involving MMCs. Indeed, FSP leads to a better bonding between the matrix and the reinforcements compared to other processes involving casting [116–118]. Besides, this environmentally-friendly surface treatment process is flexible and offers many possibilities related to the volume of the treated zone, the nature of the systems of materials and the formation of in-situ MMCs.

**Volume and extension of the processed zone**

FSP is attractive as it offers the possibility to develop surface composites as well as functionally graded composites [6]. It is indeed flexible since a pinless tool may treat only the extreme surface of the workpiece, or a tool with variable pin lengths or retractable pins may architecture the microstructure with differences from the top to the bottom of the plate [6]. In addition, the width of the processed zone can be adapted in order to tailor the volume fraction of reinforcement in the MMC. This is done by using a suitable setting of process parameters. For instance, the volume of the processed zone can be increased using a low advancing speed [113] or a large rotational speed [106]. Indeed, such processing conditions contribute to an enhancement of the malleability of the base material workpiece due to a higher processing temperature. An increase of both the pin and the shoulder diameters as well as a reduction of the rotational speed with increasing number of overlapping passes can also enable to adjust the size of the processed zone. This method was tested during FSP of Al5059 with 50 vol.% carbon nanotubes [119]. The use of a tool design with three pins under the shoulder [120] and of successive passes partially overlapping [5] are other ways to extend the width of the processed zone. FSP assisted by a supplementary electrical heat input was further shown to enlarge the processed zone size [106].

The powder disposal method (grooves, holes, etc.) also affects the local volume fraction of reinforcements [109]. For instance, according to the size (width or depth) of the machined grooves [104, 121] or according to the number and the size of the holes drilled in the workpiece [109], the volume fraction of reinforcements packed into them and finally dispersed in the MMC changes. In order to increase the volume fraction of reinforcements in the composite, it is also possible to machine new grooves at the surface of the workpiece following a first FSP treatment. These new grooves are then filled with powders before new FSP passes [122]. However, this last technique sounds labor intensive but is effective in increasing the volume fraction of reinforcements.

**Variety of systems**

On the point of view of the nature of matrix/reinforcement couples, various systems have been tested by FSP as detailed in Table 7.1. In the case of in-situ MMCs elaborated during FSP by the introduction of elementary components, intermetallic compounds were formed during the process and play the role of reinforcements. As mentioned previously, the initial reinforcements can further present various shapes namely particles [5], which can be coated to improve their bonding with the matrix [132], strips [114], coatings [144], fiber fabrics [113] or nanotubes [107, 119, 131, 151]. As one may see, the variety of couple reinforcement/metallic matrix is very large and allows considering any system that may lead to enhanced performances that may be mechanical or functional as will be discussed in Sect. 7.3.4.

The initial reinforcement can have micrometric or nanometric sizes. However, getting a uniform distribution of nanosized particles is difficult. Asadi et al. [118] noted an agglomeration of nanometric $Al_2O_3$ in Mg based composite because of the bad wetting between the reinforcements and the matrix, but obtained a uniform distribution of the micrometric $Al_2O_3$. Similarly, nanoparticles of TiC in mild steel were heterogeneously distributed after one pass of FSP because of their high cohesive energy [103]. By comparison, in the case of Mg based composite reinforced with micrometric SiC, although SiC particles do not agglomerate, the good wetting between Mg and SiC reinforcement leads to a heterogeneous distribution of SiC particles [118]. Moreover, at another size scale, Bauri et al. [128] proved that with an initial powder particle size of 70 $\mu$m of Ni, FSP does not lead to a uniform distribution of Ni reinforcements in the Al matrix whereas their distribution becomes uniform during FSP of 10 $\mu$m sized particles. In the same way, during FSP of AA5083 Al alloy with SiC particles of 119 or 37 or 12 $\mu$m mean size, the smallest sized reinforcements lead to higher concentration along the bead surface and to a smooth gradient in volume fraction both in depth and along the transverse direction [6]. Some examples of different MMCs obtained from various reinforcements shapes are displayed in Fig. 7.14.

**Ex-situ or in-situ MMCs and graded materials**

Ex-situ or in-situ MMCs may be processed by FSP. Ex-situ composites contain reinforcements produced separately and then introduced in the matrix. In-situ composites contain reinforcements produced during the process. In-situ composites may be formed via:

**Table 7.1** Nature of matrix and reinforcements of MMC systems processed by FSP. IMC = Intermetallic compounds

| | Systems | Relevant references |
|---|---|---|
| Base materials | Al based | [6, 105, 108, 110–112, 114, 116, 119, 123–141] |
| | Cu based | [5, 121, 122, 145–149] |
| | Mg based | [107, 113, 117, 118, 150–157] |
| | Ti based | [104, 109, 115, 156, 158] |
| | Steel based | [102, 103] |
| | Bronze | [159] |
| Amorphous reinforcements | $SiO_2$ | [133, 150, 153] |
| | $Al_{84.2}Ni_{10}La_{2.1}$ | [114] |
| Metallic reinforcements | Ni | [128, 130, 154] |
| | Mo | [127] |
| | Fe | [134] |
| | Cu | [144] |
| | Stainless steel | [136] |
| | W | [139, 160] |
| Ceramic reinforcements | AlN | [148] |
| | Hydroxyapatite | [156] |
| | SiC | [6, 102, 105, 112, 117, 118, 123, 132, 135, 145–147, 149, 158] |
| | TiC | [103, 109, 147] |
| | $B_4C$ | [121, 129, 147] |
| | $Al_2O_3$ | [102, 104, 105, 116, 118, 147, 152] |
| | $Y_2O_3$ | [5] |
| | $TiO_2$ | [151, 152] |
| | TiN | [115] |
| | $TiB_2$ | [116] |
| | $ZrB_2$ | [141] |
| | WC | [147] |
| | $SiO_2$ | [137, 155] |
| | $ZrO_2$ | [157] |
| | $Cr_2O_3$ | [111] |
| IMCs | NiTi | [110, 124, 140] |
| | $Al_2Cu$ | [126] |
| | $Al_3Ti$ | [126] |
| Other | Graphite | [122, 161] |
| | Carbon | [107, 113, 119, 131, 151] |
| Hybrid reinforcements | SiC + $MoS_2$ | [125, 144] |
| | SiC + $Al_2O_3$ | [108] |
| | $Al_2O_3$ + C nanotubes | [161] |
| | $Al_2O_3$ + $TiB_2$ | [138] |
| In-situ made reinforcements | $Al_{12}Mo$ + $Al_5Mo$ | [127] |
| | $Al_{13}Cr_2$ (4 passes) | [111] |
| | $Al_{11}Cr_2$ (6 passes) | [111] |
| | $Al_3Ni$ | [130] |
| | $Al_3Ni_2$, $Mg_2Ni$, AlNi, $MgNi_2$ | [154] |

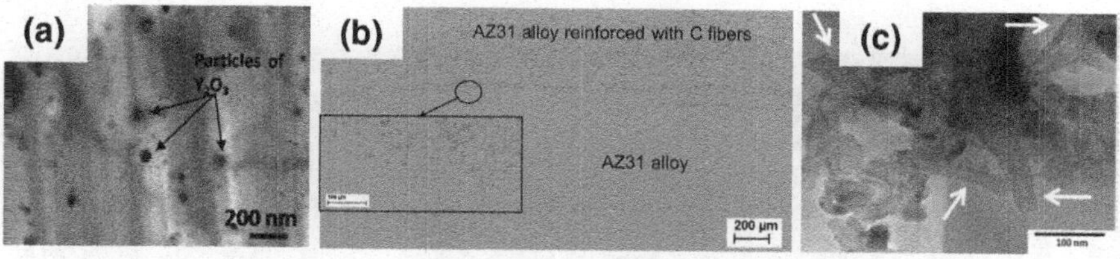


**Fig. 7.14** **a** Cu reinforced with $Y_2O_3$ particles (from Avettand-Fènoël et al. [5]). **b** AZ31 Mg alloy reinforced with C fibers (for research project, see Mertens et al. [113]). **c** Al 5059 reinforced with C nanotubes (indicated by white arrows) after 2 FSP passes (from Izadi et al. [119])

- a phase transformation like crystallization of an amorphous reinforcement [114];
- a chemical reaction between the matrix and initial solid elements inserted within it [111, 126, 127, 130, 133, 134, 137, 144, 154, 162], e.g. Al + $SiO_2$ → Al/$Al_2O_3$+$SiO_2$ [133];
- a reaction between the matrix and the reactive gas, such as nitrogen, flowing in the zone treated by the FSP tool [115].

The reinforcements produced in situ via chemical reactions are very fine (for instance about 100nm [111]), thermodynamically stable and generally uniformly distributed in the matrix [126]. Besides, they present a good interface with the matrix.

Some undesirable or unexpected chemical reactions or phase transformations can also occur during FSP due to the presence of the reinforcements. This may have the positive effect of further increasing the reinforcement effect. During FSP of Mg with $SiO_2$ particles [153], the good affinity between Mg and Si leads to the formation of additional $Mg_2Si$ reinforcements in the Mg composite. These $Mg_2Si$ are in complement to the amorphous initial $SiO_2$ particles still maintained after FSP. In the case of FSP of a ferritic stainless steel with SiC particles [102], SiC reacted with the matrix to form $Cr_3Si$ and $(Cr,Fe)_3C_7$ precipitates and further induces the matrix phase transformation into $Fe_3Si$ IMC and solid solutions of Si and C in a ferritic matrix. During FSP of CP-Ti under nitrogen, the formation of in-situ TiN precipitates (see Fig. 7.15) occurs together with the martensitic phase transformation of $\alpha$-Ti matrix because of the formation of an N interstitial solid solution in Ti [115].

### *7.3.3 Effect of Process Parameters on the Architectured Microstructure of the MMCs and Graded Materials*

Various process parameters, which directly govern the material flow during the process, can have an effect on the architecture of the MMCs and the graded materials. The homogeneity of reinforcement distribution and their structural stability may be affected. Both will directly govern the properties of MMCs. The following will discuss the effect of the groove location on the workpiece, the number of passes, the

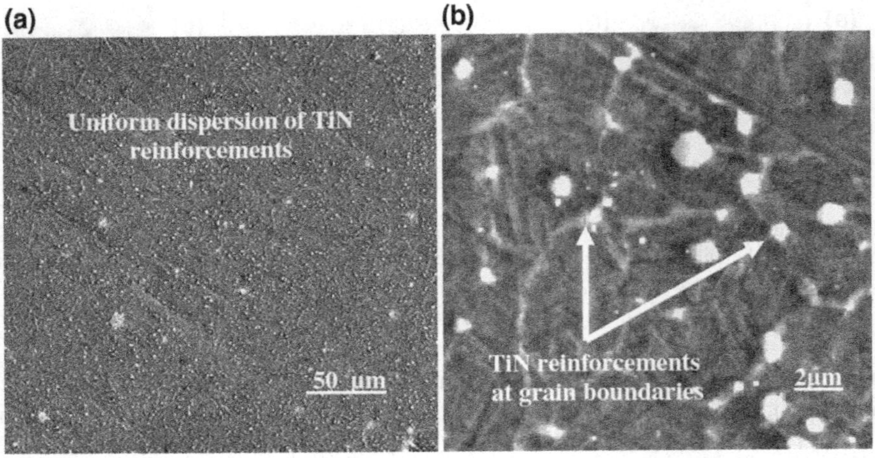


**Fig. 7.15** TiN/Ti MMC processed in situ by FSP (from Shamsipur et al. [115])

rotational and advancing speeds, the change of the tool direction sense at each new pass, the heat input and the tool geometry.

*Groove location and size*—Gandra et al. [6] showed that a good dispersion of reinforcements is ensured provided the groove machined at the surface of the workpiece is located in the pin flow zone, i.e. in the core of the stir zone, and not in the shoulder flow zone, i.e. near the top surface. In addition, a reduction of the groove width leads to a better distribution of reinforcements [147].

*Number of passes*—Several passes are generally required to disperse the reinforcements, in particular if they are nanometric [155]. An increase of the number of passes was indeed shown to reduce the clustering of powdered reinforcements [155, 158] and to improve their fineness and homogenize their distribution [103, 116, 118, 129, 130, 146, 152–155, 158, 162]. X-ray tomography analyses showed that fiber reinforcements follow the material flow induced during each pass [113]. One may thus question the uniform and random distribution of such reinforcements which present directionality by opposition to the particle reinforcements.

Besides, by increasing the number of passes, the porosity at the matrix-reinforcement interface is reduced and the interfacial bonding is enhanced. For example, after one FSP pass, some porosities are observed at the SiC/Cu interfaces of the MMC but are not observed after 2 passes [145, 149]. Thus FSP improves the densification of the MMC as shown by Khayyamin [155] and already presented in Sect. 7.2.2 for cast alloys.

The multiplication of passes may however also entail a structural instability of reinforcements and a change of their morphology by fracture and blunting. For instance, Bauri et al. [128] evidenced a fracture of initial Ni 70 µm sized powder particles during FSP of Al based MMC. Li et al. [109] also noticed a refinement during FSP of initial 5 µm sized TiC introduced in Ti-6Al-4V alloy reaching the nanometric scale. A refinement of hydroxyapatite particles introduced in Mg was also observed during FSP due to their breaking. Their morphology changed from flaked to spherical [156]. Izadi et al. [119] observed the fracture of some carbon nanotubes, then their disappearance which leads to spherical shelled carbon structures and formation of $Al_4C_3$. Three passes were required to homogenize the dispersion of reinforcements at the expense of the maintenance of the nanotubes shape. Mertens et al. [113] detected an erosion of carbon fibers in Mg based MMC obtained by FSP. Abdollahi et al. [154] showed an evolution of the nature of the reinforcements produced in situ according to the FSP passes number, namely a transformation of $Al_3Ni_2 + Mg_2Ni$ formed in situ after the first FSP pass of AZ31 with Ni powder, into $AlNi + MgNi_2$ after subsequent passes.

*Change of the direction of the tool advance at each new pass*—The asymmetric feature of the FSP process due to a difference of material flow on advancing (AS) and retreating (RS) sides leads to the agglomeration of reinforcements on the AS [146]. Some authors changed the direction of the tool advance at each new overlap pass to get an homogeneous distribution and a similar density of reinforcements on AS and RS [108, 152, 153].

*Rotational and advancing speeds*—Like the increase of the number of passes, a greater rotational speed [116, 147, 152, 162] or a reduction of the advancing speed [116, 145, 147, 151] leads to a refined distribution of reinforcements. Indeed, they contribute to an enhancement of the malleability of the base material workpiece and thus improve the stirring effect [106]. As a consequence, since the processed zone is expected to be larger, the volume fraction of reinforcements is reduced [113]. Nevertheless, a significant increase of the rotational speed may lead to the tool fracture [158].

*Friction stir vibration processing*—In friction stir vibration processing, a normal vibration is applied to the processed line during the process while $SiO_2$ particles are incorporated in an Al alloy [137]. Vibrations decrease the grain size in the processed zone, improve the homogeneity of the distribution of reinforcements by contributing to the breaking up of agglomerated particles. This finally leads to enhanced hardness, strength and ductility [137].

*Supplementary heat input*—The use of an electrical current which passes through the shoulder into the substrate can thermally assist the FSP by improving the substrate viscoplasticity [106, 140]. This increases the volume of the processed zone although it reduces the volume fraction of reinforcements within it. It then promotes the homogeneousness of the reinforcements dispersion and avoids the formation of clusters. This alternative which may easily be industrialized could avoid to increase the number of passes to get a suitable, uniform and fine distribution of reinforcements in the MMCs [106].

*Tool wear*—FSP tools are generally made of hard steels like H13 steel for light alloys (with eventually surface nitriding and oxidizing treatments [106]). WC/Co, PCBN, InRe, WRe, Co alloys and WC are preferred for processing MMCs and graded materials [144]. These tools are however expensive and their major concern is their wear which increases with both the abrasive feature and the volume fraction of the reinforcements [105]. The dimensions of a WC tool are for instance reduced by more than 20% during FSP of a mild steel with nanosized TiC reinforcements after four passes, which leads to the insertion of WC tool debris in the MMC processed zone [103]. It is worth noting that similar concerns, occur during friction stir welding of MMCs [101]. Various solutions were consequently proposed to reduce the tool wear during FSW. The process could be assisted by an electrical current in order to improve the stirring and to favor the viscoplasticity of the base material surface [106]. Besides, although some authors improved the reinforcement distribution with a threaded pin [152], an unthreaded pin is generally recommended to reduce tool wear otherwise the threads are abraded by the reinforcements during the process [163–165].

*Tool geometry*—The pin shape is an important factor since it governs the material flow and thus the quality of the reinforcements distribution [122]. The ideal pin would be square shaped, which entails a pulsating action leading to a uniform distribution of the reinforcements [166, 167] or a smooth frustum like tool shape which promotes a uniform material flow [163]. The use of three-fluted tool shape during FSP of $Al_2O_3$ reinforced Mg MMCs indeed led to cavities and microvoids [152]. Another study concerned by FSP of Cu with graphite particles showed that a triangular pin leads to a larger processed zone, to a better dispersion of the reinforcements than other tool shapes [122].

### *7.3.4 Mechanical and Functional Properties of FSP MMCs and Graded Materials*

The use of several passes and of a judicious and suitable setting of process parameters was shown to improve the dispersion of the reinforcements (Fig. 7.13) and thus are expected to enhance the surface and bulk properties.

**Surface composites**

*Hardness*—Like in most MMCs, the hardness of the MMC is generally larger than its matrix [111, 112, 115]. Shamsipur et al. [115] processed a TiN/Ti surface composite by FSP. Particles were embedded 6 mm deep[2] in a commercially pure Ti workpiece machined with a 3 mm deep groove subjected to reactive nitrogen. The surface composite reinforced with only 2.3 vol.% TiN (Fig. 7.15) presents a promising 972 HV hardness which is about 6 times larger than that of the commercially pure Ti (Fig. 7.16). Comparing reinforcement size effects, the hardness of the processed zone of $Al_2O_3$ reinforced Ti based MMC with a 20 nm size of reinforcements is larger than when they are 80 nm in size [104].

[2]The latter surface composite is significantly deeper than those obtained by classical diffusion or remelting nitriding methods.

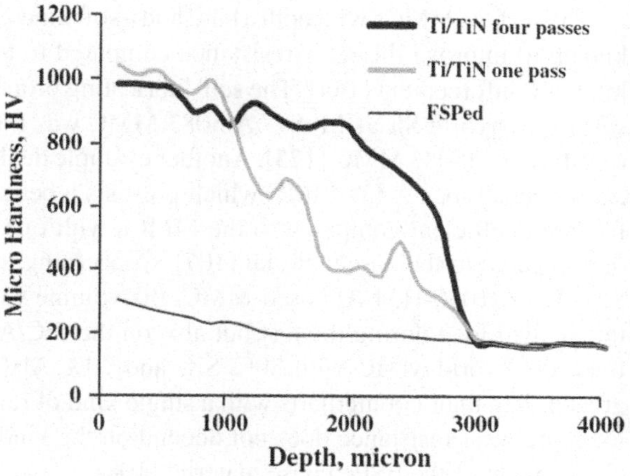


**Fig. 7.16** Hardness of the TiN/Ti MMC processed in-situ by FSP (from Shamsipur et al. [115])

*Ductility*—Compared to the TiC reinforcements, the stainless steel reinforcements in Al lead to a better ductility of the metal matrix composite due to the deformable nature of stainless steel and low Al work hardening around the reinforcements [136]. Besides, an increase of the number of passes from 1 to 5 was shown to increase the elongation from 12.5 to 24.5% for the W reinforced Al based composite [160].

*Wear resistance*—Many surface composites processed by FSP aim at improving wear resistance [103, 112, 118, 122, 129, 149]. Barmouz et al. [149] obtained a better wear resistance (Table 7.2), mechanical resistance and lower ductility by reinforcing Cu with 30 nm than with 5 μm sized SiC particles by FSP. Sarmadi et al. [122] showed that a 22 vol.% graphite reinforced Cu based MMC developed by 4 passes of FSP presents a friction coefficient 79% lower and a resistance to wear 65% greater than the pure annealed Cu because of the lubricating property of graphite. The TiC reinforced mild steel FSPed surface nanocomposite presents a wear resistance against steel disk two to three times greater than that of the steel substrate [103]. This behavior is attributed to a lower coefficient of friction and a greater hardness value of the surface composite. These are favored by a fine distribution of nanosized TiC and the refinement of the grain size.

**Table 7.2** Dry wear rate in pure Cu and in FSPed samples (data from Barmouz et al. [149])

| Sample | Wear rate ($mm^3/Nm$) |
|---|---|
| Pure Cu | $9.366 \times 10^{-5}$ |
| FSPed Cu—6vol%SiC (5 μm) | $8.107 \times 10^{-5}$ |
| FSPed Cu—18vol%SiC (5 μm) | $7.293 \times 10^{-5}$ |
| FSPed Cu—6vol%SiC (30 nm) | $5.950 \times 10^{-5}$ |
| FSPed Cu—18vol%SiC (30 nm) | $3.931 \times 10^{-5}$ |

The surface MMCs with both a hard and a soft kind of reinforcements are generally known to improve the wear resistance compared to their counterpart with a single kind of reinforcement [144]. The self-lubricating property and the wear resistance of surface of hybrid SiC + $MoS_2$ / Al5083 MMC was for instance higher than for SiC (or $MoS_2$)/ Al5083 MMC [125]. Another example deals with the 0.1% $Al_2O_3$ + 0.2% carbon nanotubes AZ31 MMC which presents a better wear resistance and a lower friction coefficient compared to the MMCs with one single kind of reinforcement or compared to the base material [107]. Concerning the wear resistance of a FSPed SiC-$Al_2O_3$/1050-H24 Al based MMC, the volume wear loss depends of course on the applied load during the test but also on the SiC/$Al_2O_3$ ratio. For a 5 N applied load, the hybrid MMC with 80% SiC and 20% $Al_2O_3$ presents a wear resistance greater than their counterparts with a single kind of reinforcement. For 10 N applied load, the wear resistance does not depend on the kind of reinforcement and is very close to that of the FSPed base material [108].

*Corrosion*—An addition of hydroxyapatite (HA) to Mg by FSP would weaken the basal texture of Mg thereby deteriorating the corrosion resistance to less than 20% of that in FSPed Mg. In the composite, the protective mixed layer was more easily generated on the surface since HA particles can act as nucleation sites. Although the accumulation of corrosion products is high, these products are porous and cannot prevent corrosion efficiently [156]. The 5A06 Al alloy based MMC with initial amorphous Al84.2Ni10La2.1 reinforcement was shown to present a lower corrosion density, a higher passivation current than the base material and a passivation zone [114].

*Optical properties*—Gudla et al. [142, 143] studied the optical properties of an aluminum alloy containing finely distributed $TiO_2$ particles after surface anodization. The color of the $TiO_2$ containing sample was much lighter. The reflectance of the surface is decreased as the anodizing time increases due to $TiO_2$ degradation.

**Graded bulk MMCs**

*Thermal properties*—The shape memory of the NiTi reinforcements decrease the coefficient of thermal expansion of the Al alloy in the 10 vol.% NiTi 6061 Al MMC which leads to a good combination of adjustable damping and thermal physical properties through a proper heat-treatment process [124].

*Electrical properties*—The FSP of a billet of carbon nanotubes (CNT) reinforced AA6061 MMC previously obtained by powder metallurgy leads to a slight increase of its electrical conductivity because of the Mg and Si segregation at the CNT/Al interface after heat treatment which led to the formation of precipitates free zones near the CNTs. This increase in electrical conductivity to a lower extent may also be attributed to the random distribution of the nanotubes in the matrix due to FSP [131]. Oppositely, the addition of SiC particles to Cu by FSP led to an enhancement of the electrical resistivity of pure Cu. But the electrical resistivity of the MMC did not evolve significantly with an increase of the number of passes [168]. The same observation was reported in the case of Al based composite reinforced with W particles [160].

## 7.4 Lap Welding Metallic Sandwiches

Friction stir welding and processing allow assembling dissimilar materials in an easy way in butt or lap welding configurations [3]. In the present chapter, the main issues related to the dissimilar metal lap welding by FSW will be discussed. Indeed, if one wishes to process architectured metal sandwiches, FSP might be an interesting option, see Fig. 7.17. FSP will ensure reconstitution of the metallic continuity at the interface between the two metals.

Sandwich-like architectured materials may be processed using friction stir lap welding. The interested reader is referred to the review on FSW of dissimilar metals by Simar and Avettand-Fènoël [169]. These structures could be processed in order to improve surface properties or combine efficiently properties of different materials. For example, the wear resistance of soft and ductile materials could be improved by placing a hard material sheet on top. The corrosion resistance may also be improved with a cover sheet, e.g. steel plates covered with Al or stainless steel sheets, or magnesium plates covered with Al. Li et al. [170] have processed an Al/Ti-6Al-4V bi-layer by friction stir lap welding, see Fig. 7.18. They claim that these assemblies may present interest for the good combination of high specific strength titanium alloy with high thermal conductivity aluminum alloy. Liu et al. [171] processed an Al-Si laser cladded coating by FSP and reached better corrosion resistance compared to the non-FSPed samples due to the suppression of coating cracks and pores. Finally, another interest of this type of assembly might be the processing of complex and brittle intermetallic phases already in the right shape and composition. The formation of the final material is ensured by a diffusion thermal treatment [172].

Dissimilar metal FS welding leads to the formation of intermetallic compounds at the interface. For example, $Fe_2Al_5$, $Fe_4Al_{13}$, $FeAl_3$ and FeAl intermetallic compounds have been reported in aluminum to steel lap welds [173–176]. $TiAl_3$ is systematically reported in aluminum to titanium welds [177] as well as magnesium alloys containing aluminum to titanium welds [178]. In the Al-Cu dissimilar FS welds many intermetallic compounds can be formed but $Al_2Cu$ and $Al_4Cu_9$ are systematically reported [179]. These intermetallics are generally brittle leading to premature failure compared to the base materials [176].

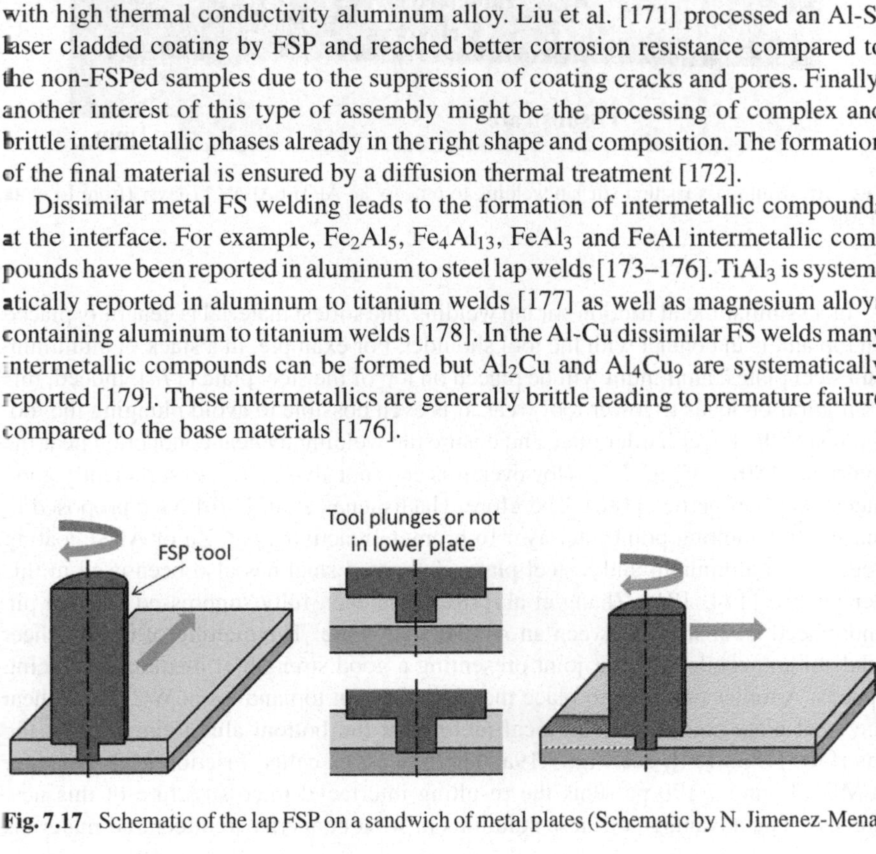


**Fig. 7.17** Schematic of the lap FSP on a sandwich of metal plates (Schematic by N. Jimenez-Mena)

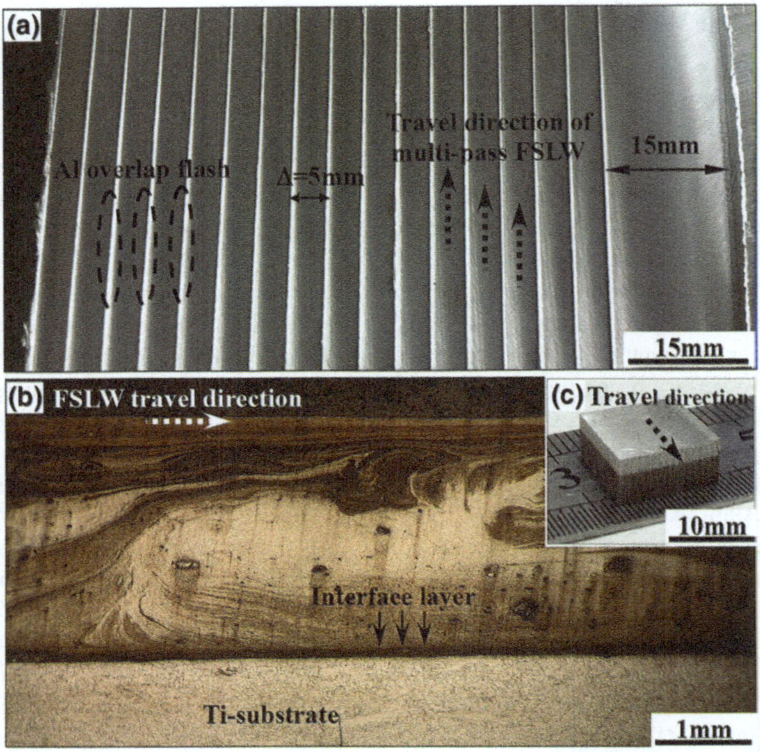


**Fig. 7.18** Multipass friction stir lap welding to process an Al/Ti-6Al-4V bi-layer (from Li et al. [170])

In dissimilar metal friction stir lap welding, the softest material is generally placed on top and is in contact with the tool shoulder. For example, in a stack of aluminum and steel plates, aluminum will be placed on top of the steel plate [174]. Indeed, this configuration leads to lower tool wear. It is even possible to avoid plunging the tool pin inside the lower harder plate and ensure the welding by heat conduction near the interface [180], see Fig. 7.17. However, this does not always lead to sufficiently good mechanical properties [180]. Therefore, Haghshenas et al. [176] have proposed to place a low melting point interlayer to promote reactivity, e.g. Zn or Al-Si coating between an aluminum and a steel plate. Zn coated steel has also been used in this perspective [174, 181]. Zhang et al. [182] have even fully suppressed the tool pin and placed a Zn sheet between an Al and steel plate. The melting of the Zn sheet leads to an excellent planar joint presenting a good strength at medium advancing speeds. Another option is to place the steel plate on top and use a WC tool to heat the steel plate and ensure the local melting of the bottom aluminum plate at the interface [183, 184], see Fig. 7.19a. The process is called Friction Melt Bonding (FMB). Figure 7.19b presents the resulting interfacial microstructure of this new process. The advantage of these solutions in what concerns architectured materials

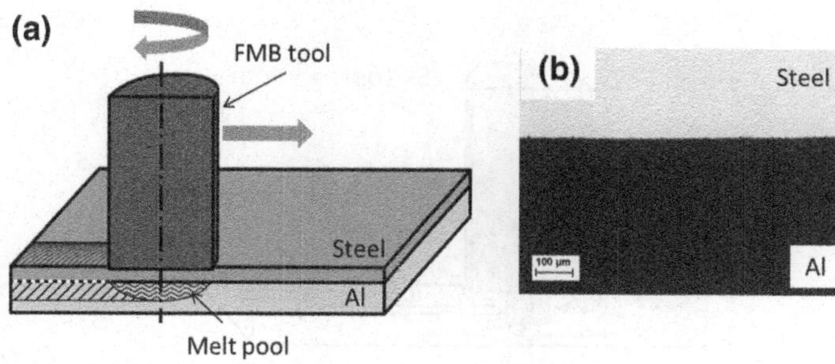


**Fig. 7.19** **a** Principle of the friction melt bonding (FMB) process. **b** Planar interface between ULC steel and 1050 Al alloy in a friction stir melting joint (Courtesy of N. Jimenez-Mena)

is that the interface is then much more planar than when the FSP tool pin penetrates the bottom material (Fig. 7.18b). Indeed, the penetration of the tool inside the lower plate leads to material flowing upwards and thus a hooking effect [185, 186] sometimes degrading the mechanical properties due to local stress concentration [185] or, oppositely, improving the mechanical properties thanks to interlocking [186]. The assembly of large panels by classical friction stir lap welding is expected to lead to inhomogeneous microstructures, however this could be acceptable.

Constitutional liquation has been reported at the interface between two dissimilar metals when they present a low temperature eutectic among themselves. This is particularly the case for the Al-Mg system [187, 188] and has also been reported for Al-Cu butt welds [189]. The Al-Mg phase diagram presents a eutectic phase at 450 °C while the FS temperatures can be much larger in particular for high rotational speeds [188]. This constitutional liquation may lead to the formation of solidification cracks that are evidently detrimental to the mechanical properties of the structure [188].

## 7.5 Porous Architectured Materials

### *7.5.1 Foam Based Structures*

Hangai and co-workers [190–196] have extensively studied the possibility to make porous aluminum alloy based foam architectured structures by FSP. They have used FSP to finely distribute blowing agent ($TiH_2$) and pore stabilizing agent powder ($Al_2O_3$) inside an aluminum plate [190, 191]. The distribution of the foaming agent is ensured by passing over the stack with a FSP tool, see Fig. 7.20. The porosities are formed by a heat treatment after FSP. The foaming agent (here $TiH_2$) presents explosion hazards. Thus, foam was manufactured in cast Al alloys with no foaming agent, only the pore stabilizing agent was inserted by FSP in the Al plates [192, 194]. The foaming effect during the subsequent heat treatment is ensured by the gas pores and dissolved gas blown during casting. More finely distributed porosities were

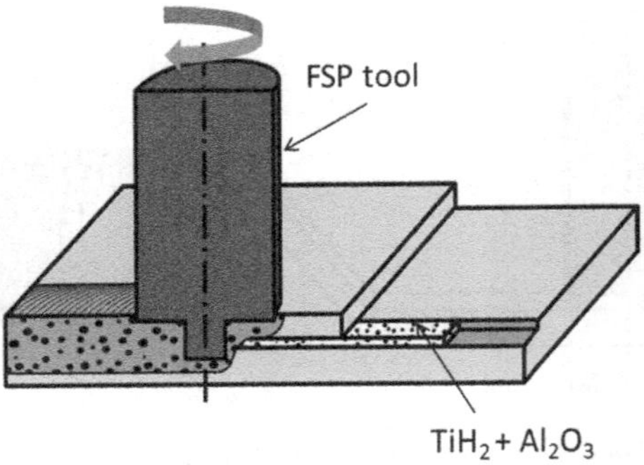


**Fig. 7.20** Manufacturing process to make a foam by FSP (Schematic by N. Jimenez-Mena, inspired from [190])

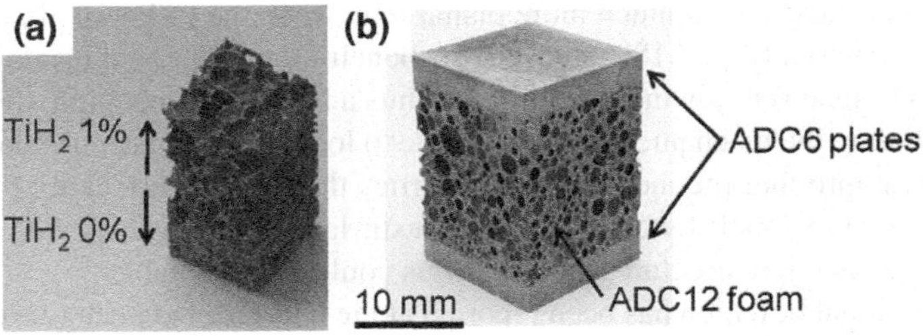


**Fig. 7.21** Architectured structures based on aluminum foam made by FSP. **a** Graded porosities structure (from Hangai et al. [193]) and **b** sandwich structure fully assembled by FSP (no adhesive) (from Hangai et al. [194])

resulting from this modification of the process. In all cases the porosity content can be controlled by changing the holding time of the heat treatment after FSP [194]. The FSP step is needed to distribute homogenously the pore stabilizing agent.

Many graded or architectured structures have been processed using this technology. Figure 7.21 shows some examples of these structures. Figure 7.21a shows an architectured material presenting a graded proportion of porosities [193]. This architectured material shows a good combination of compressive strength due to the low porosity part and compressive strain due to the high porosity part. Figure 7.21b shows a sandwich panel made of ADC12 foam in between two dense ADC6 cast sheets [194]. Once more, the assembly is performed by FSP without the use of adhesive. This sandwich can thus be used for higher temperature applications. The formation of the sandwich rather than the foam alone provides higher tensile and bending strength. Hangai et al. [196] also developed by FSP graded foams composed of Al1050 and Al6061 foams produced by FSP separately and then assembled together by FSP. They showed that this assembly presents better compression and impact strain than the individual foams exploiting the prior deformation of the Al1050 foam followed by the deformation of the Al6061 foam.

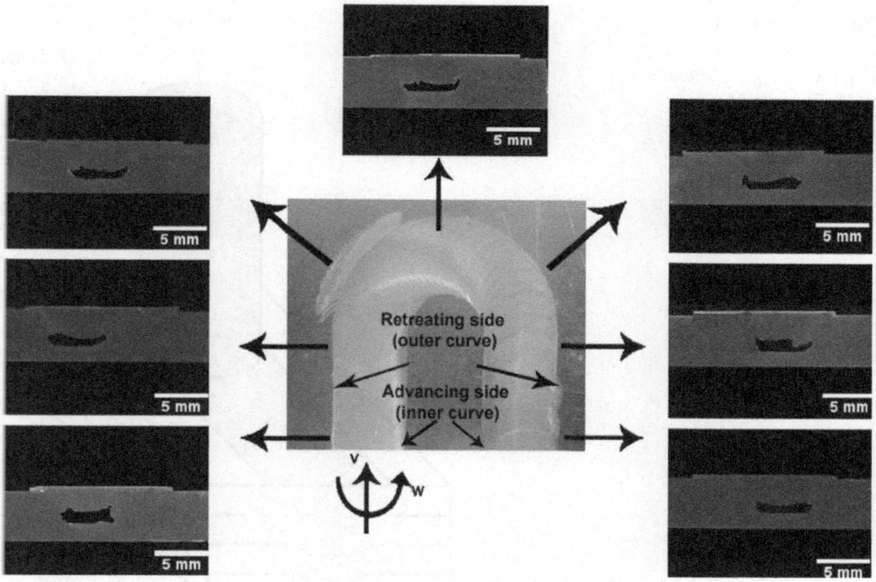


**Fig. 7.22** Curvilinear channel processed by friction stir channeling. From Balasubramanian et al. [197]

### 7.5.2 *Friction Stir Channeling*

Balasubramanian et al. [197, 198] have invented a process derived from FSP called Friction Stir Channeling (FSC). FSC uses the large defect called wormhole formed for specific friction stir processing parameters to create channels with a path dictated by the FSC tool. This may, for example, be used to manufacture compact heat exchangers. Figure 7.22 shows the typical shape of the channel that can be manufactured with a curvilinear path. The tool geometry, the process parameters and the applied force have to be correctly tailored in order for the channel to present a given shape and be continuous [197, 198].

## 7.6 Friction Stir Additive Manufacturing

The additive manufacturing technology to process architectured materials is largely discussed in Chap. 9. Here, we will just shortly discuss the possibility of using the friction stir technology to perform additive manufacturing.

Friction Stir Additive Manufacturing (FSAM) is a very simple process using friction stir welding to assemble parts to build a 3D structure [199–201]. Figure 7.23 presents the principle of this process. Small plates are lap jointed by FSW and may thus create a structure presenting the properties of the stir zone. Similarly to the laser

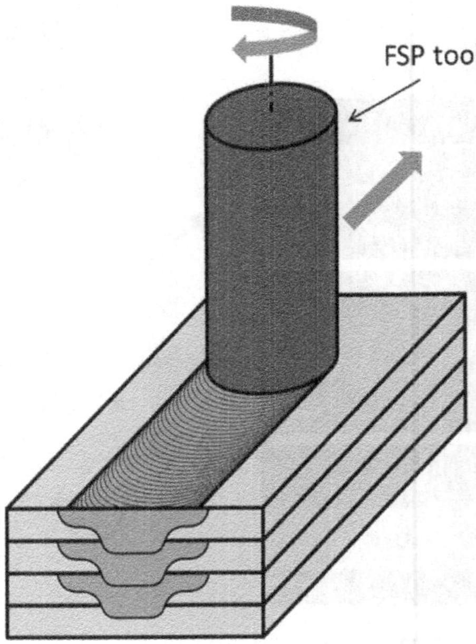


**Fig. 7.23** Principle of friction stir additive manufacturing (Schematic by N. Jimenez-Mena, inspired from [201])

cladding additive manufacturing process, this may help manufacture any structure with however limitations due to the need of a backing plate to support the load applied by the tool on the workpiece. Palanivel et al. [200] have shown that this technology may be applied to process stringers for the skin panel of an airplane fuselage. The stringer is thus presenting the specific microstructure of an FSP material, contrarily to the skin panel.

Palanivel et al. [199, 200] have made FSAM on a Mg-4Y-3Nd magnesium alloy. They found a high strength equal to 400 MPa with an excellent elongation of 17% due to the fine precipitates (2–7 nm) formed during a post-processing heat treatment. Yuqing et al. [201] show that the built microstructure presents some strong heterogeneities in an 7075 aluminum alloy.

## References

1. W.M. Thomas, E.D. Nicholas, J.C. Needham, M.G. Murch, P. Templesmith, C.J. Dawes, GB Patent Application no. 9125978.8, December 1991, US Patent No. 5460317, Oct 1995
2. R.S. Mishra, M.W. Mahoney, S.X. McFadden, N.A. Mara, A.K. Mukherjee, Scripta Mater. **42**, 163 (2000)
3. R.S. Mishra, P.S. De, N. Kumar, *Friction Stir Welding and Processing, Science and Engineering* (Springer, New York, 2014)
4. R.S. Mishra, Z.Y. Ma, Mater. Sci. Eng. R **50**(1), 1 (2005)
5. M.-N. Avettand-Fènoël, A. Simar, R. Shabadi, R. Taillard, B. De Meester, Mater. Des. **60**, 343 (2014)

6. J. Gandra, R. Miranda, P. Vilaa, A. Velhinho, J.P. Teixeira, J. Mater. Proc. Technol. **211**, 1659 (2011)
7. F. Khodabakhshi, A. Simchi, A.H. Kokabi, A.P. Gerlich, M. Nosko, Mater. Sci. Eng. A **642**, 204 (2015)
8. J.Q. Su, T.W. Nelson, C.J. Sterling, Scripta Mater. **52**, 135 (2005)
9. A.G. Rao, R.R. Ravi, B. Ramakrishnarao, V.P. Deshmukh, A. Sharma, N. Prabhu, B.P. Kashyap, Metall. Mater. Trans. A **44**, 1519 (2013)
10. N. Kumar, R.S. Mishra, Metall. Mater. Trans. A **44**, 934 (2013)
11. X. Feng, H. Liu, J.C. Lippold, Mater. Charact. **82**, 97 (2013)
12. C.G. Rhodes, M.W. Mahoney, W.H. Bingel, M. Calabrese, Scripta Mater. **48**, 1451 (2003)
13. A. Orozco-Caballero, P. Hidalgo-Manrique, C.M. Cepeda-Jimnez, P. Rey, D. Verdera, O.A. Ruano, F. Carreo, Mater. Charact. **112**, 197 (2016)
14. X.H. Du, B.L. Wu, Trans. Nonferrous Met. Soc. China **18**, 562 (2008)
15. J.A. del Valle, P. Rey, D. Gesto, D. Verdera, J.A. Jimnez, O.A. Ruano, Mater. Sci. Eng. A **628**, 198 (2015)
16. J.-Q. Su, T.W. Nelson, T.R. McNelley, R.S. Mishra, Mater. Sci. Eng. A **528**, 5458 (2011)
17. A. Shafiei-Zarghani, Int. J. Modern Phys. B **22**(18–19), 2874 (2008)
18. G.R. Cui, Z.Y. Ma, S.X. Li, Acta Mater. **57**, 5718 (2009)
19. M. Komarasamy, N. Kumar, Z. Tang, R.S. Mishra, P.K. Liaw, Mater. Res. Lett. **3**(1), 30 (2015)
20. T. Morishige, T. Hirata, M. Tsujikawa, K. Higashi, Mater. Lett. **64**, 1905 (2010)
21. P. Xue, B.L. Xiao, Z.Y. Ma, Scripta Mater. **68**, 751 (2013)
22. T. Long, W. Tang, A.P. Reynolds, Sci. Technol. Weld. Joining **12**, 311 (2007)
23. V. Kulitskiy, S. Malopheyev, S. Mironov, R. Kaibyshev, Mater. Sci. Eng. A **674**, 480 (2016)
24. N. Kumar, R.S. Mishra, C.S. Huskamp, K.K. Sankaran, Mater. Sci. Eng. A **528**, 5883 (2011)
25. K.A.A. Hassan, A.F. Norman, D.A. Price, P.B. Prangnell, Acta Mater. **51**, 1923 (2003)
26. M.A. Garcia-Bernal, R.S. Mishra, R. Verma, D. Hernandez-Silva, Mater. Sci. Eng. A **670**, 9 (2016)
27. I. Charit, R.S. Mishra, Scripta Mater. **58**, 367 (2008)
28. F. Hannard, S. Castin, E. Maire, R. Mokso, T. Pardoen, A. Simar, Acta Mater. **130**, 121 (2017)
29. S. Jana, R.S. Mishra, J.A. Baumann, G. Grant, Mater. Sci. Eng. A **528**, 189 (2010)
30. C.M. Hu, C.M. Lai, P.W. Kao, N.J. Ho, J.C. Huang, Scripta Mater. **60**, 639 (2009)
31. R. Kapoor, N. Kumar, R.S. Mishra, C.S. Huskamp, K.K. Sankaran, Mater. Sci. Eng. A **527**, 5246 (2010)
32. N. Kumar, R.S. Mishra, Mater. Sci. Eng. A **580**, 175 (2013)
33. B. Mansoor, A.K. Ghosh, Acta Mater. **60**, 5079 (2012)
34. J.G. Kaufman, E.L. Rooy, *Aluminum Alloy Castings, Properties, Processes, and Applications* (ASM International, U.S.A., 2004)
35. M.L. Santella, T. Engstrom, D. Storjohann, T.Y. Pan, Scripta Mater. **53**, 201 (2005)
36. Z.Y. Ma, S.R. Sharma, R.S. Mishra, Metall. Mater. Trans. A **37**, 3323 (2006)
37. Z.Y. Ma, S.R. Sharma, R.S. Mishra, Scripta Mater. **54**, 1623 (2006)
38. Z.Y. Ma, S.R. Sharma, R.S. Mishra, Mater. Sci. Eng. A **433**, 269 (2006)
39. Z.Y. Ma, A.L. Pilchak, M.C. Juhas, J.C. Williams, Scripta Mater. **58**, 361 (2008)
40. A.H. Feng, B.L. Xiao, Z.Y. Ma, R.S. Chen, Metall. Mater. Trans. A **40**, 2447 (2009)
41. S. Jana, R.S. Mishra, J.A. Baumann, G.J. Grant, Metall. Mater. Trans. A **41**, 2507 (2010)
42. Z.W. Chen, F. Abraham, J. Walker, Mater. Sci. Forum **706–709**, 971 (2012)
43. S.S. Shin, E.S. Kim, G.Y. Yeom, J.C. Lee, Mater. Sci. Eng. A **532**, 151 (2012)
44. T.J. Hurley, R.G. Atkinson, Trans. Am. Foundry Soc. **91**, 291–96 (1985)
45. D.R. Ni, D. Wang, A.H. Feng, G. Yao, Z.Y. Ma, Scripta Mater. **61**, 568 (2009)
46. L. Karthikeyan, V.S. Senthilkumar, V. Balasubramanian, S. Natarajan, Mater. Des. **30**, 2237 (2009)
47. K. Nakata, Y.G. Kim, H. Fujii, T. Tsumura, T. Komazaki, Mater. Sci. Eng. A **437**, 274 (2006)
48. S.R. Sharma, Z.Y. Ma, R.S. Mishra, Scripta Mater. **51**, 237 (2004)
49. P. Cavaliere, Mater. Charact. **57**, 100 (2006)
50. S.R. Sharma, R.S. Mishra, Scripta Mater. **59**, 395 (2008)

51. S. Jana, R.S. Mishra, J.B. Baumann, G. Grant, Scripta Mater. **61**, 992 (2009)
52. S. Jana, R.S. Mishra, J.B. Baumann, G. Grant, Acta Mater. **58**, 989 (2010)
53. S.M. Aktarer, D.M. Sekban, O. Saray, T. Kucukomeroglu, Z.Y. Ma, G. Purcek, Mater. Sci. Eng. A **636**, 311 (2015)
54. S.A. Alidokht, A. Abdollah-Zadeh, S. Soleymani, T. Saeid, H. Assadi, Mater. Charact. **63**, 90 (2012)
55. S.K. Singh, R.J. Immanuel, S. Babu, S.K. Panigrahi, G.D. Janaki Ram, J. Mater. Proc. Technol. **236**, 252 (2016)
56. Z.Y. Ma, R.S. Mishra, M.W. Mahoney, Acta Mater. **50**, 4419 (2002)
57. I. Charit, R.S. Mishra, Mater. Sci. Eng. A **359**, 290 (2003)
58. Z.Y. Ma, R.S. Mishra, M.W. Mahoney, R. Grimes, Mater. Sci. Eng. A **351**, 148 (2003)
59. R.S. Mishra, Adv. Mater. Process. **162**, 45 (2004)
60. Z.Y. Ma, R.S. Mishra, M.W. Mahoney, R. Grimes, Metall. Mater. Trans. A **36**, 1447 (2005)
61. Z.Y. Ma, R.S. Mishra, Scripta Mater. **53**, 75 (2005)
62. I. Charit, R.S. Mishra, Acta Mater. **53**, 4211 (2005)
63. F.C. Liu, Z.Y. Ma, Scripta Mater. **59**, 882 (2008)
64. Z.Y. Ma, R.S. Mishra, F.C. Liu, Mater. Sci. Eng. A **505**, 70 (2009)
65. M.A. Garcia-Bernal, R.S. Mishra, R. Verma, D. Hernandez-Silva, Scripta Mater. **60**, 850 (2009)
66. F.C. Liu, B.L. Xiao, K. Wang, Z.Y. Ma, Mater. Sci. Eng. A **527**, 4191 (2010)
67. Z.Y. Ma, F.C. Liu, R.S. Mishra, Acta Mater. **58**, 4693 (2010)
68. M.A. Garcia-Bernal, R.S. Mishra, R. Verma, D. Hernandez-Silva, Mater. Sci. Eng. A **534**, 186 (2012)
69. F.C. Liu, Z.Y. Ma, F.C. Zhang, J. Mater. Sci. Technol. **28**(11), 1025 (2012)
70. A. Orozco-Caballero, M. Alvarez-Leal, P. Hidalgo-Manrique, C.M. Cepeda-Jimenez, O.A. Ruano, F. Carreno, Mater. Sci. Eng. A **680**, 329 (2017)
71. P. Cavaliere, P.P. De Marco, J. Mater. Process. Technol. **184**, 77 (2007)
72. D. Zhang, S. Wang, C. Qiu, W. Zhang, Mater. Sci. Eng. A **556**, 100 (2012)
73. F. Chai, D. Zhang, Y. Li, W. Zhang, Mater. Sci. Eng. A **568**, 40 (2013)
74. Q. Yang, B.L. Xiao, Q. Zhang, M.Y. Zheng, Z.Y. Ma, Scripta Mater. **69**, 801 (2013)
75. S.R. Babu, V.S.S. Kumar, L. Karunamoorthy, G.M. Reddy, Mater. Des. **53**, 338 (2014)
76. S. Palanivel, A. Arora, K.J. Doherty, R.S. Mishra, Mater. Sci. Eng. A **678**, 308 (2016)
77. A. Simar, Y. Brechet, B. de Meester, A. Denquin, T. Pardoen, Acta Mater. **55**, 6133 (2007)
78. J.D. Robson, Acta Mater. **52**, 1409 (2004)
79. C. Genevois, A. Deschamps, A. Denquin, B. Doisneau-Cottignies, Acta Mater. **53**(8), 2447 (2005)
80. C. Gallais, A. Simar, D. Fabregue, A. Denquin, G. Lapasset, B. de Meester, Y. Brechet, T. Pardoen, Metall. Mater. Trans. A **38**(5), 964 (2007)
81. A. Simar, Y. Brechet, B. de Meester, A. Denquin, T. Pardoen, Mater. Sci. Eng. A **486**, 85 (2008)
82. A. Simar, Y. Brechet, B. de Meester, A. Denquin, C. Gallais, T. Pardoen, Progress. Mater. Sci. **57**, 95 (2012)
83. C.D. Marioara, S.J. Andersen, J. Janssen, H.W. Zandbergen, Acta Mater. **49**, 321 (2001)
84. M.-N. Avettand-Fènoël, R. Taillard, Metall. Mater. Trans. A **46**(1), 300 (2015)
85. M.-N. Avettand-Fènoël, R. Taillard, Mater. Des. **89c**, 348 (2016)
86. B.L. Xiao, Q. Yang, J. Yang, W.G. Wang, G.M. Xie, Z.Y. Ma, J. Alloy. Compd. **509**, 2879 (2011)
87. A.M. Jamili, A. Zarei-Hanzaki, H.R. Abedi, P. Minarik, R. Soltani, Mater. Sci. Eng. A **690**, 244 (2017)
88. R.A. Seraj, A. Abdollah-Zadeh, M. Hajian, F. Kargar, R. Soltanalizadeh, Metall. Mater. Trans. A. **47**, 3564 (2016)
89. A.L. Pilchack, M.C. Juhas, J.C. Williams, Metall. Mater. Trans. A **38**, 401 (2007)
90. A.L. Pilchack, D.M. Norfleet, M.C. Juhas, J.C. Williams, Metall. Mater. Trans. A. **39**, 1519 (2008)

91. J.C. Lippold, J.J. Livingston, Metall. Mater. Trans. A **44**, 3815 (2013)
92. B. Li, Y. Shen, W. Hu, L. Luo, Surf. Coat. Technol. **239**, 160 (2014)
93. M. Sarkari Khorrami, M. Kazeminezhad, Y. Miyashita, A.H. Kokabi, Metall. Mater. Trans. A **48**, 188 (2017)
94. G. Bhargava, W. Yuan, S.S. Webb, R.S. Mishra, Metall. Mater. Trans. A **41**, 13 (2010)
95. S. Mironov, Q. Yang, H. Takahashi, I. Takahashi, K. Okamoto, Y.S. Sato, H. Kokawa, Metall. Mater. Trans. A **41**, 1016 (2010)
96. R. Xin, B. Li, A. Liao, Z. Zhou, Q. Liu, Metall. Mater. Trans. A **43**, 2500 (2012)
97. W. Yuan, R.S. Mishra, B. Carlson, R.K. Mishra, R. Verma, R. Kubic, Scripta Mater. **64**, 580 (2011)
98. W. Yuan, R.S. Mishra, Mater. Sci. Eng. A **558**, 716 (2012)
99. V. Jain, R.S. Mishra, A.K. Gupta, Gouthama. Mater. Sci. Eng. A **560**, 500 (2013)
100. M. Vargas, S. Lathabai, P.J. Uggowitzer, Y. Qi, D. Orlov, Y. Estrin, Mater. Sci. Eng. A **685**, 253 (2017)
101. M.-N. Avettand-Fènoël, A. Simar, Mater. Charact. **120**, 1 (2016)
102. Y. Kimoto, T. Nagaoka, H. Watanabe, M. Fukusumi, in *Proceedings of 1st International Joint Symposium on Joining and Welding* (2013), p. 389
103. A. Ghasemi-Kahrizsangi, S.F. Kashani-Bozorg, Surf. Coatings Technol. **209**, 15 (2012)
104. A. Shafiei-Zarghani, S.F. Kashani-Bozorg, A. Gerlich, Mater. Sci. Eng. A **631**, 75 (2015)
105. R.M. Miranda, T.G. Santos, J. Gandra, N. Lopes, R.J.C. Silva, J. Mater. Proc. Technol. **213**, 1609 (2013)
106. T.G. Santos, N. Lopes, M. Machado, P. Vilaa, R.M. Miranda, J. Mater. Proc. Technol. **216**, 375 (2015)
107. D. Lu, Y. Jiang, R. Zhou, Wear **305**(1–2), 286 (2013)
108. E.R.I. Mahmoud, M. Takahashi, T. Shibayanagi, K. Ikeuchi, Wear **268**, 1111 (2010)
109. B. Li, Y. Shen, L. Luo, W. Hu, Mater. Sci. Eng. A **574**, 75 (2013)
110. M. Dixit, J.W. Newkirk, R.S. Mishra, Scripta Mater. **56**, 541 (2007)
111. S.R. Anvari, F. Karimzadeh, M.H. Enayati, J. Alloy. Compd. **562**, 48 (2013)
112. J. Gandra, P. Vigarinho, D. Pereira, R.M. Miranda, A. Velhinho, P. Vilaa, Mater. Des. **52**, 373 (2013)
113. A. Mertens, A. Simar, J. Adrien, E. Maire, H.M. Montrieux, F. Delannay, J. Lecomte-Beckers, Mater. Charact. **107**, 125 (2015)
114. P. Liu, Q.Y. Shi, Y.B. Zhang, Compos. Part B **52**, 137 (2013)
115. A. Shamsipur, S.F. Kashani-Bozorg, A. Zarei-Hanzaki, Surf. Coat. Technol. **218**, 62 (2013)
116. H. Eskandari, R. Taheri, Proc. Mater. Sci. **11**, 503 (2015)
117. Y. Morisada, H. Fujii, T. Nagaoka, M. Fukusumi, Mater. Sci. Eng. **433**, 50 (2006b)
118. P. Asadi, G. Faragi, A. Masoumi, M.K.B. Givi, Metall. Mater. Trans. **42A**, 2820 (2011)
119. H. Izadi, A.P. Gerlich, Carbon **50**, 4744 (2012)
120. W. Yuan, R.S. Mishra, S. Webb, Y.L. Chen, B. Carlson, D.R. Herling, J. Mater. Process. Technol. **211**, 972 (2011)
121. R. Sathiskumar, N. Murugan, I. Dinaharan, S.J. Vijay, Mater. Charact. **84**, 16 (2013)
122. H. Sarmadi, A.H. Kokabi, S.M.S. Reihani, Wear **304**, 1 (2013)
123. V. Sharma, U. Prakash, B.V.M. Kumar, Mater. Today Proc. **2**, 2666 (2015)
124. D.R. Ni, J.J. Wang, Z.Y. Ma, J. Mater. Sci. Technol. **32**(2), 162 (2016)
125. S. Soleymani, A. Abdollah-Zadeh, S.A. Alidokht, Wear **278–279**, 41 (2012)
126. C.J. Hsu, P.W. Kao, N.J. Ho, Mater. Lett. **61**, 1315 (2007)
127. I.S. Lee, P.W. Kao, C.P. Chang, N.J. Ho, Intermetallics **35**, 9 (2013)
128. R. Bauri, D. Yadav, C.N.S. Kumar, G.D.J. Ram, Data Brief **5**, 309 (2013)
129. N. Yuvaraj, S. Aravindan, Vipin. J. Mater. Res. Technol. **4**(4), 398 (2015)
130. M. Golmohammadi, M. Atapour, A. Ashrafi, Mater. Des. **85**, 471 (2015)
131. Z.Y. Liu, B.L. Xiao, W.G. Wang, Z.Y. Ma, J. Mater. Sci. Technol. **30**, 649 (2014)
132. J.N. Aoh, C.W. Huang, W.J. Cheng, Mater. Sci. Forum **783–786**, 1721 (2014)
133. G.L. You, N.J. Ho, P.W. Kao, Mater. Charact. **80**, 1 (2013)

134. M. Sarkari-Khorrami, S. Samadi, Z. Janghorban, M. Movahedi, Mater. Sci. Eng. A **641**, 380 (2015)
135. M. Sarkari-Khorrami, M. Kazeminezhad, A.H. Kokabi, Metall. Mat. Trans. A **46**, 2021 (2015)
136. S. Selvakumar, I. Dinaharan, R. Palanivel, B.G. Babu, Mater. Sci. Eng. A **685**, 317 (2017)
137. S. Fouladi, M. Abbasi, J. Mater. Proc. Technol. **243**, 23 (2017)
138. H. Eskandari, R. Taheri, F. Khodabakhshi, Mater. Sci. Eng. A **660**, 84 (2016)
139. C.N.S. Kumar, R. Bauri, D. Yadav, Tribol. Int. **101**, 284 (2016)
140. J.P. Oliveira, J.F. Duarte, P. Infacio, N. Schell, R.M. Miranda, T.G. Santos, Mater. Des. **113**, 311 (2017)
141. N. Kumar, G. Gautam, R.K. Gautam, A. Mohan, S. Mohan, Tribol. Int. **97**, 313 (2016)
142. V.C. Gudla, F. Jensen, A. Simar, R. Shabadi, R. Ambat, Appl. Surf. Sci. **324**, 554 (2015)
143. V.C. Gudla, K. Bordo, F. Jensen, S. Canulescu, S. Yuksel, A. Simar, R. Ambat, Surf. Coat. Technol. **277**, 67 (2015)
144. V. Sharma, U. Prakash, B.V. Manoj Kumar, J. Mater. Process. Technol. **224**, 117 (2015)
145. M. Barmouz, M.K. Besharati Givi, J. Seyfi, Mater. Charact. **62**, 108 (2011)
146. Y.F. Sun, H. Fujii, Mater. Sci. Eng. A **528**, 5470 (2011)
147. R. Sathiskumar, N. Murugan, I. Dinaharan, S.J. Vijay, Mater. Des. **55**, 224 (2014)
148. T. Thankachan, K.S. Prakash, Mater. Sci. Eng. A **688**, 301 (2017)
149. M. Barmouz, P. Asadi, M.K. Besharati Givi, M. Taherishargh, Mater. Sci. Eng. A **528**, 1740 (2011)
150. Y. Jiang, X. Yang, H. Miura, T. Sakai, Rev. Adv. Mater. Sci. **33**, 29 (2013)
151. Y. Morisada, H. Fujii, T. Nagaoka, M. Fukusumi, Mater. Sci. Eng. **419**, 344 (2006)
152. M. Azizieh, A.H. Kokabi, P. Abachi, Mater. Des. **32**, 2034 (2011)
153. C.J. Lee, J.C. Huang, P.J. Hsieh, Scripta Mater. **54**, 1415 (2006)
154. S.H. Abdollahi, F. Karimzadeh, M.H. Enayati, J. Alloy. Compd. **623**, 335 (2015)
155. D. Khayyamin, A. Mostafapour, R. Keshmiri, Mater. Sci. Eng. A **559**, 217 (2013)
156. D. Ahmadkhaniha, M. Fedel, M.H. Sohi, A.Z. Hanzaki, F. Deflorian, Corros. Sci. **104**, 319 (2016)
157. M. Navazani, K. Dehghani, J. Mater. Proc. Technol. **229**, 439 (2016)
158. A. Shamsipur, S.F. Kashani-Bozorg, A. Zarei-Hanzaki, Surf. Coat. Technol. **206**, 1372 (2011)
159. S. Thapliyal, D.K. Dwivedi, J. Mater. Proc. Technol. **238**, 30 (2016)
160. G. Huang, Y. Shen, R. Guo, W. Guan, Mater. Sci. Eng. A **674**, 504 (2016)
161. Z. Du, M.J. Tan, J.F. Guo, G. Bi, J. Wei, Mater. Sci. Eng. A **667**, 125 (2016)
162. D. Ahmadkhaniha, P. Asadi, Mechanical alloying by friction stir processing, Chap. 9, in *Advances in Friction Stir Welding and Processing* (2014), p. 387
163. R.A. Prado, L.E. Murr, K.F. Soto, J.C. McClure, Mater. Sci. Eng. A **349**, 156 (2003)
164. H.J. Liu, J.C. Feng, H. Fujii, K. Nogi, Int. J. Mach. Tools Manuf. **45**, 1635 (2005)
165. R.A. Prado, L.E. Murr, D.J. Shindo, K.F. Soto, Scripta Mater. **45**, 75 (2001)
166. K. Kalaiselvan, N. Murugan, Procedia Eng. **38**, 49 (2012)
167. S. Gopalakrishnan, N. Murugan, Mater. Des. **32**, 462 (2011)
168. M. Barmouz, M.K. Besharati Givi, Compos. Part A **42**, 1445 (2011)
169. A. Simar, M.-N. Avettand-Fènoël, Sci. Technol. Weld. Joining **22**, 389–403 (2016). https://doi.org/10.1080/13621718.2016.1251712
170. B. Li, Y. Shen, L. Luo, W. Hu, J. Alloy. Compd. **658**, 904 (2016)
171. F. Liu, Y. Ji, Q. Meng, Z. Li, Vacuum **133**, 31 (2016)
172. C. van der Rest, *Dissertation* (Université catholique de Louvain, Belgium, 2015)
173. A. Elrefaey, M. Takahashi, K. Ikeuchi, J. Jpn. Weld. Soc. **23**, 186 (2005)
174. Y.C. Chen, K. Nakata, Metall. Mater. Trans. A **39**, 1985 (2008)
175. R.S. Coelho, A. Kostka, S. Sheikhi, J.F. dos Santos, A.R. Pyzalla, Adv. Eng. Mater. **10**, 961 (2008)
176. M. Haghshenas, A. Abdel-Gwad, A.M. Omran, B. Gke, S. Sahraeinejad, A.P. Gerlich, Mater. Des. **55**, 442 (2014)
177. Y.C. Chen, K. Nakata, Mater. Des. **30**, 469 (2009)
178. M. Aonuma, K. Nakata, Mater. Sci. Eng. B **173**, 135 (2010)

179. M.N. Avettand-Fènoël, R. Taillard, G. Ji, D. Goran, Metall. Mater. Trans. A **43**, 4655 (2012)
180. J.T. Xiong, J.L. Li, J.W. Qian, F.S. Zhang, W.D. Huang, Sci. Technol. Weld. Joining **17**, 196 (2012)
181. C. Schneider, T. Weinberger, J. Inoue, T. Koseki, N. Enzinger, Sci. Technol. Weld. Joining **16**, 100 (2011)
182. G. Zhang, W. Su, J. Zhang, Z. Wei, Metall. Mater. Trans. A **42**, 2850 (2011)
183. C. van der Rest, P.J. Jacques, A. Simar, Scripta Mater. **77**, 25 (2014)
184. S. Crucifix, C. van der Rest, N. Jimenez-Mena, P.J. Jacques, A. Simar, Sci. Technol. Weld. Joining **20**, 319 (2015)
185. S. Jana, T. Hovanski, Sci. Technol. Weld. Joining **17**, 141 (2012)
186. J. Mohammadi, Y. Behnamian, A. Mostafaei, H. Izadi, T. Saeid, A.H. Kokabi, A.P. Gerlich, Mater. Charact. **101**, 189 (2015)
187. Y.C. Chen, K. Nakata, Scripta Mater. **58**, 433 (2008)
188. J. Yan, Z. Xu, Z. Li, L. Li, S. Yang, Scripta Mater. **53**, 585 (2005)
189. J. Ouyang, E. Yarrapareddy, R. Kovacevic, J. Mater. Process. Technol. **172**, 110 (2006)
190. Y. Hangai, M. Hasegawa, T. Utsunomiya, J. Mater. Process. Technol. **210**, 288 (2010)
191. Y. Hangai, Y. Oba, S. Koyama, T. Utsunomiya, Metall. Mater. Trans. A **42**, 3585 (2011)
192. Y. Hangai, H. Kato, T. Utsunomiya, S. Kitahara, O. Kuwazuru, N. Yoshikawa, Mater. Trans. **53**(8), 1515 (2012)
193. Y. Hangai, K. Takahashi, T. Utsunomiya, S. Kitahara, O. Kuwazuru, N. Yoshikawa, Mater. Sci. Eng. A **534**, 716 (2012)
194. Y. Hangai, H. Kamada, T. Utsunomiya, S. Kitahara, O. Kuwazuru, N. Yoshikawa, J. Mater. Process. Technol. **214**, 1928 (2014)
195. Y. Hangai, K. Saito, T. Utsunomiya, O. Kuwazuru, N. Yoshikawa, Mater. Sci. Eng. A **613**, 163 (2014)
196. Y. Hangai, N. Kubota, T. Utsunomiya, H. Kawashima, O. Kuwazuru, N. Yoshikawa, Mater. Sci. Eng. A **639**, 597 (2015)
197. N. Balasubramanian, R.S. Mishra, K. Krishnamurthy, J. Mater. Process. Technol. **209**, 3696 (2009)
198. N. Balasubramanian, R.S. Mishra, K. Krishnamurthy, J. Mater. Process. Technol. **211**, 305 (2011)
199. S. Palanivel, P. Nelaturu, B. Glass, R.S. Mishra, Mater. Des. **65**, 934 (2015)
200. S. Palanivel, H. Sidhar, R.S. Mishra, JOM **67**(3), 616 (2015)
201. M. Yuqing, K. Liming, H. Chunping, L. Fengcheng, L. Qiang, Int. J. Adv. Manuf. Technol. **83**, 1637–1647 (2015). https://doi.org/10.1007/S00170-015-7695-9

# Chapter 8
# Severe Plastic Deformation as a Way to Produce Architectured Materials

**Yan Beygelzimer, Roman Kulagin and Yuri Estrin**

**Abstract** In this chapter, a group of processing techniques leading to desired materials architectures is discussed. They are based on severe plastic deformation (SPD) by shear combined with high hydrostatic pressure. Originally, these techniques were developed for imparting to the material an ultrafine grained (UFG) microstructure thus improving its mechanical performance characteristics. An added benefit of SPD processing in the context of architectured materials is its ability to tune the inner makeup of a hybrid material at a macroscopic scale. After a brief introduction to the available SPD processing techniques, we provide an analysis of architectured multi-scale structures with UFG constituents they can produce. A target of this research is development of materials with a high specific strength and low overload sensitivity. Specific designs enabling a favourable combination of these properties are considered. An emphasis is put on structures that include soft layers whose presence delays strain localisation and failure of the hybrid material.

Y. Beygelzimer (✉)
Donetsk Institute for Physics and Engineering named after A.A. Galkin,
National Academy of Sciences of Ukraine, Nauki av. 46, Kiev 03680, Ukraine
e-mail: yanbeygel@gmail.com

R. Kulagin
Institute of Nanotechnology (INT), Karlsruhe Institute of Technology (KIT),
Hermann-von-Helmholtz-Platz 1, 76344 Eggenstein-Leopoldshafen, Germany
e-mail: rkulagin@gmail.com

Y. Estrin
Department of Materials Science and Engineering, Monash University,
Clayton 3800, Australia
e-mail: yuri.estrin@monash.edu

Department of Mechanical Engineering, The University of Western Australia,
Crawley 6009, Australia



Y. Estrin et al. (eds.), *Architectured Materials in Nature and Engineering*, Springer Series in Materials Science 282,
https://doi.org/10.1007/978-3-030-11942-3_8

## 8.1 Introduction

In a quest for improving the mechanical performance of solids, primarily metallic ones, a range of techniques based on severe plastic deformation (SPD) have been developed by materials scientists over the recent decades. A defining feature these techniques have in common is the use of giant shear deformation in conjunction with a very high, gigapascal-range, hydrostatic pressure. Despite the extreme cumulative equivalent strains imparted to the workpiece in these processes, its dimensions remain practically unchanged, which distinguishes the SPD techniques from traditional metal forming operations [1, 2].

The consequences for materials processed by SPD in terms of their mechanical properties are remarkable, as their initial grain structure is reduced in a spectacular way—down to deep submicron scale, and in some cases to nanoscale [3, 4]. Not only are such techniques suitable for providing *bulk* materials with an ultrafine-grained (UFG) structure, but they can also be efficient for consolidation and 'cold sintering' of metal powders [5], solid-state modification of polymers [6, 7], recycling of machining chips [8, 9], and other engineering applications.

This chapter aims at highlighting the possibilities of obtaining architectured materials offered by various SPD methods. The attractiveness of this approach is its ability to impart to the material a desired inner architecture while at the same time providing it with an UFG structure, thus improving the mechanical properties at both macro- and microstructural levels. We have been promoting this approach from its inception [8, 10–12] and feel that it has a great future. Indeed, the available experience with producing nanostructured gradient materials [13–15], nanostructured lattices [16], as well as layered [17] and spiral-shaped [10, 11] structures is encouraging. This chapter builds upon these developments. It comprises two sections. In the first one, various approaches to producing multiscale structures by SPD methods are considered. The second one provides an analysis of architectured multiscale structures with UFG constituents, which do not suffer from plastic instability under tensile loading—the Achilles verse of uniform UFG materials produced by SPD methods.

## 8.2 Producing Multiscale UFG Structures by SPD Techniques

### *8.2.1 Major SPD Techniques*

To date, more than 30 SPD techniques have been proposed. Among those, five principal methods are most prominent. These include equal-channel angular pressing (ECAP), high pressure torsion (HPT), accumulative roll-bonding (ARB), multi-directional forging (MDF), and twist extrusion (TE) [4]. Surface treatment techniques, such as surface mechanical attrition treatment (SMAT) [18], the cone-cone method (CCM) [19], and high pressure tube twisting (HPTT) suggested by Toth et al.

(2009) and further developed by Lapovok et al. [20], are becoming more and more popular for applications where there is no need for grain refinement throughout the thickness of a workpiece. These techniques are well suited for producing gradient structures across the workpiece.

There exists a rich literature on SPD, which is summarised in several review articles and monographs devoted to individual SPD processes [8, 21–23] and their comparative analysis [4, 24, 25]. Therefore, instead of going into the detail of SPD processing, we shall only dwell on the methods which in our view are most interesting with regard to architecturing the bulk or the surface of solids. The concepts underlying these methods are presented schematically in Fig. 8.1. The HPT, ECAP, ARB, TE, and SMAT processes mentioned above are complemented here with high pressure torsion-extrusion (HPTE), a recently developed technique permitting continuous HPT [26].

Most of these methods, viz. HPT, ECAP, ARB, TE, and HPTE, aim at uniform grain refinement throughout the bulk of a workpiece. SMAT is included to represent a range of techniques for plastic deformation of a near-surface region of the material, which also include surface mechanical grinding [29], broaching [30], burnishing [31], CCM, HPTT, etc.

### *8.2.2 Characteristic Strain Distributions Produced by SPD Processes*

The specific strain distributions associated with the deformation in each of the processes illustrated in Fig. 8.1 are shown in Fig. 8.2. It is seen that under HPT, TE, HPTE, and SMAT strain is non-uniform within a specimen cross-section, as distinct from the ECAP and ARB processes, for which, to a first approximation, the strain distribution is uniform. The non-uniformity of strain is not necessarily a negative, as it can be utilised to create various tailor-made gradient structures. Indeed, strain non-uniformity translates to non-uniformity in the grain structure, the smallest grain size occurring in the regions with the greatest cumulative strain [3, 4]. Thus, plates and rods with a nanostructured surface layer were obtained by SMAT [18, 32]. They were found to possess remarkably good strength and ductility characteristics.

Another example of the formation of a gradient structure are cylindrical copper or magnesium rods processed by repetitive three-roll planetary milling, also known as screw rolling, to large overall strain [33, 34]. An axisymmetric gradient microstructure obtained by this technique is characterised by a gradual decrease of the grain size with the radial distance from the billet axis to its periphery, where it attains a submicron level. It was shown that such a gradient structure gives rise to increased tensile elongation of the processed material.

By sharpening strain non-uniformity deliberately, e.g. by TE, one can manufacture rods of arbitrary cross-sectional shape and dimensions that possess a UFG structure in a near-surface layer and a coarse-grained structure near the billet axis [8]. The TE

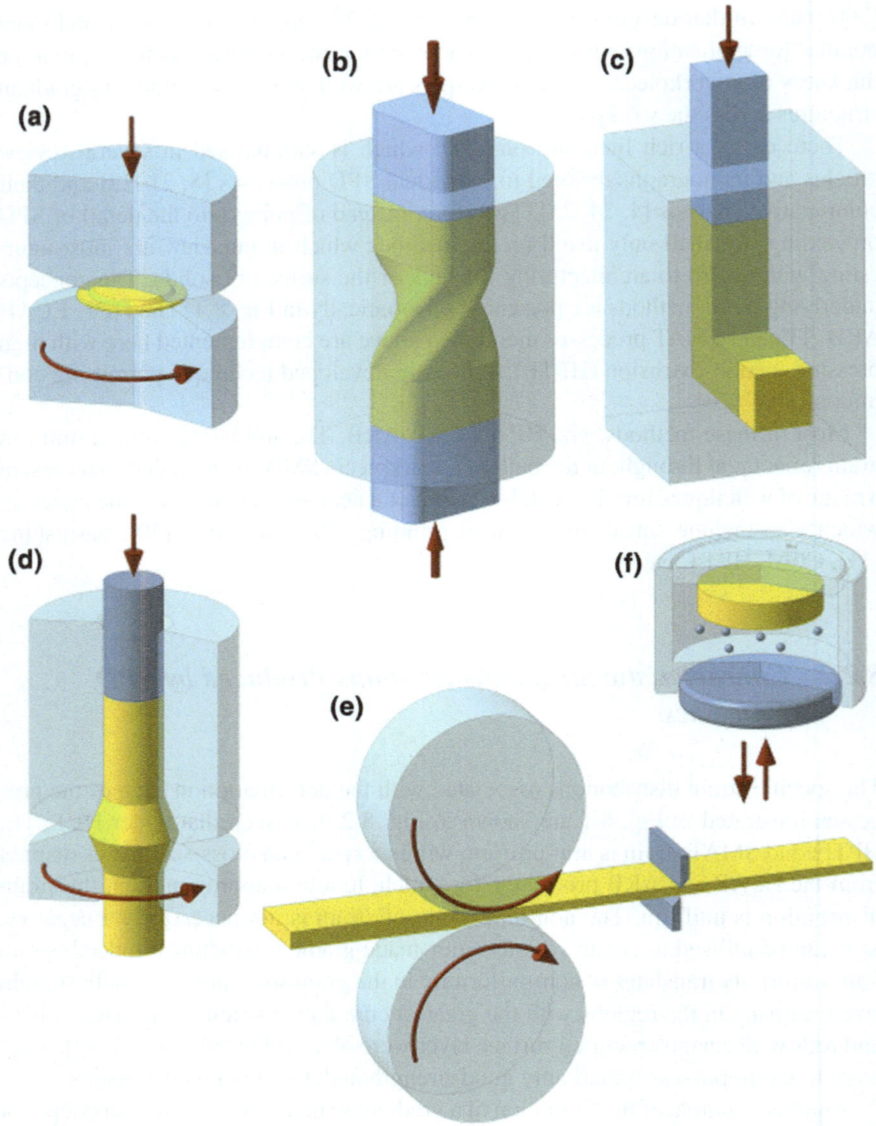


**Fig. 8.1** Schematics of selected SPD processes: HPT (**a**) [1], TE (**b**) [27], ECAP (**c**) [2], HPTE (**d**) [26], ARB (**e**) [28], SMAT (**f**) [18]

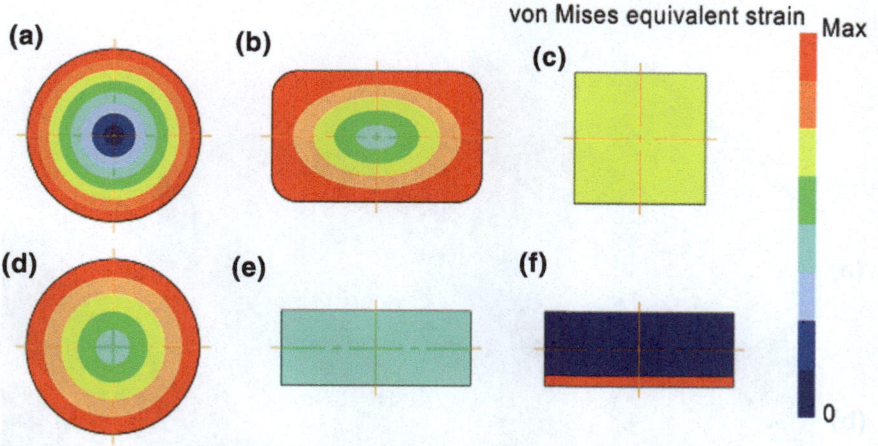


**Fig. 8.2** Schematic pictures of the distribution of the von Mises equivalent strain in a cross-section normal to the die axis for HPT (**a**), TE (**b**), ECAP (**c**), HPTE (**d**), and to the workpiece surface for ARB (**e**) and SMAT (**f**)

parameters for producing a gradient microstructure within copper rods with a hexagonal profile were determined in [35]. The mechanical properties of such gradient materials under uniaxial tensile deformation studied by computational simulations are presented in the following section.

Rods made from light alloys based on magnesium, aluminium, and titanium, which have a UFG structure at the surface and a coarse-grained crystalline structure near the axis are akin to various structural members used in the engineering practice, such as I-bars, tubes with various cross-sections, channel beams, L-angles, etc., cf. Fig. 8.3.

In addition to bulk materials with gradient structure, the SMAT technique can also be used to produce various nanostructured lattices on the surfaces of light alloys [16]. Such architectured materials are analogous to steel mesh structures.

### 8.2.3 Bulk Materials with Spiral-Shaped and Layered Armour

Prior to SPD, various kinds of inclusions (in form of solid particles or powders) can be introduced in a workpiece. Controlled material fluxes during the SPD process will transform them to produce a desired inner structure of the workpiece. The initial distribution of the introduced particles that generates the desired final architecture can be found computationally, by solving an inverse problem. Under certain SPD conditions, large strains under high pressure will lead to the occurrence of a strong metallic bond between the constituents of the workpiece similar to the one

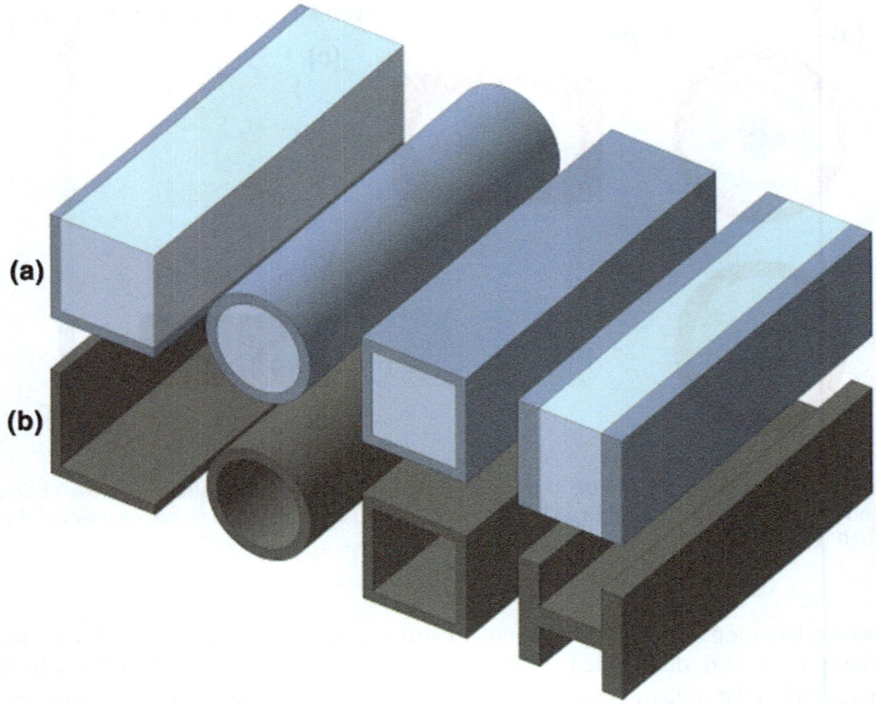


**Fig. 8.3** Rods made from light alloys with a gradient structure (**a**) and their structural member analogues (**b**)

that occurs during pressure welding. In this way, architectured materials with good mechanical characteristics can be produced. Some examples of creating a desired inner architecture in a bulk material by virtue of material fluxes generated by SPD are presented. A particular type of an architectured material is a spiral hybrid material with a disc-shaped geometry of the constituents produced by HPT [10]. UFG bars with embedded spiral fibres (helical armour) can be obtained by means of TE [11], free end torsion [36], and HPTE [26]. An interest in spiral-shaped architecture of reinforcing elements of a hybrid material goes back to the work of Bouaziz [37] who suggested that such a geometry may improve the tensile ductility of the matrix material due to artificial strain hardening associated with a 'spring' embedded in the matrix. It acts to postpone the onset of the Considère instability thus enhancing tensile ductility, while the way in which the spiral armour was produced by SPD provides the material with increased strength.

Another SPD process, HPTE, makes it possible to fabricate a UFG material with embedded spiral-shaped coarse-grained layers. The schematics of this process are seen in Fig. 8.4a.

The embedded fibres are to be made from a metal whose melting temperature is substantially lower than that of the billet, such as in the considered case of Al fibres in

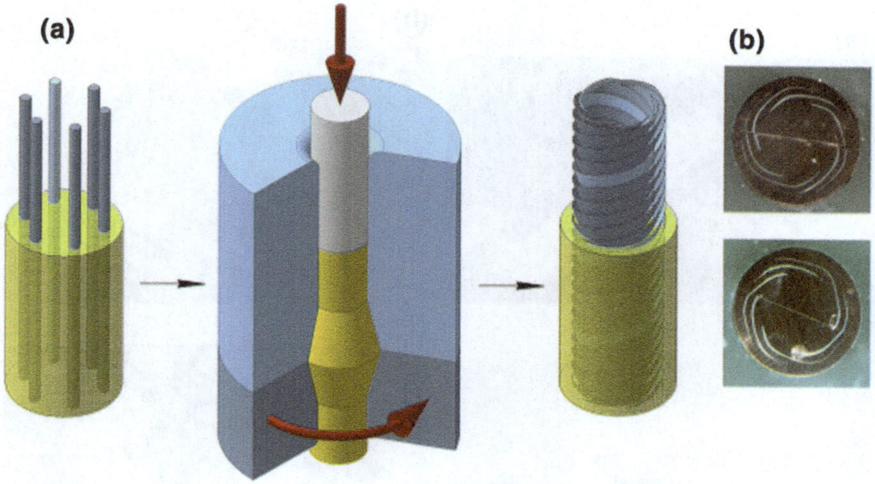


**Fig. 8.4** Formation of a spiral-shaped layered architecture by means of HPTE: initial arrangement of embedded fibres and schematics of the HPTE process (**a**); an example of a spiral layer structure in a copper/aluminium system (**b**). Here the Cu matrix is represented by yellow material and aluminium fibres (transformed to spiral Al sheets in the process of HPTE) are shown in blue

the Cu matrix. After HPTE, the hybrid material produced (the right picture in Fig. 8.4) needs to be subjected to a heat treatment that would preserve the UFG structure of the matrix, while causing recrystallisation of the spiral sheets to which the initial fibres have been transformed. A high strength of the resulting architectured material is ensured by that of the UFG matrix. The coarse-grained spiral sheets enable the necessary toughness by acting as traps for microcracks under monotonic or cyclic loading.

HPT of layered specimens can give rise to a plethora of architectured materials with interesting inner structures. Such structures can be formed as a result of a loss of stability of a non-homogenous layered material under shear and compression. Structures of this kind are also known in geology [38], Fig. 8.5.

In [39] the formation of vortices in a material deforming plastically under simple shear was associated with local blocking of shear deformation. It was shown by computational modelling [40] that in a sample undergoing HPT such blocking leads to twists and turns of the obstacles. We regard this as the physical cause for the turbulent flow during HPT.

The results of finite element (FE) calculations for a layer of a plastic material placed between two plates moving in the opposite directions are seen in Fig. 8.6. Obstacles to plastic flow within the layer are seen to block shear deformation locally. In the simulations a cylindrical obstacle normal to the plane of the figure was associated with a material with the yield stress ten times higher than that of the layer itself.

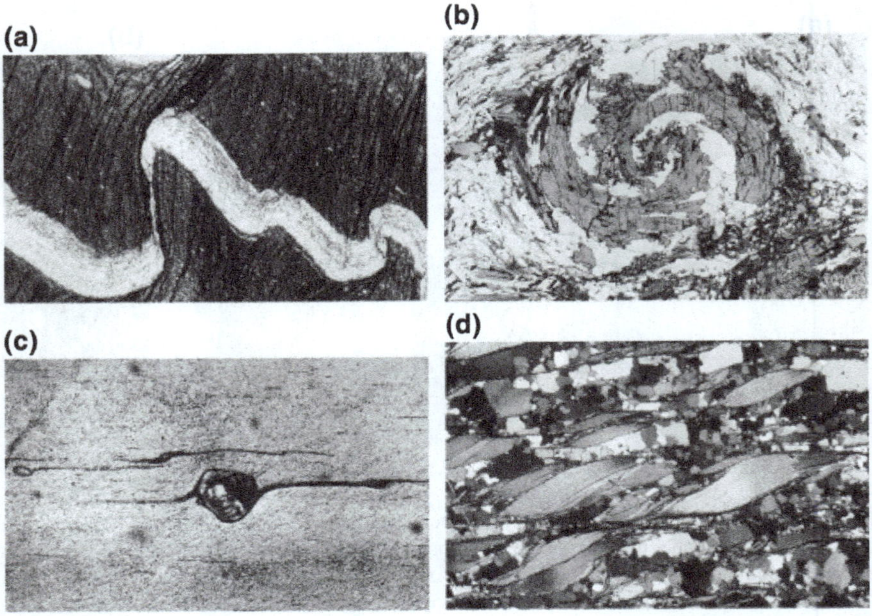


**Fig. 8.5** Geological structures stemming from instability of non-homogenous layered formations under shear and compression: geological fold (**a**), spiral (**b**), δ-type (**c**), shear bands structure (**d**) [38]

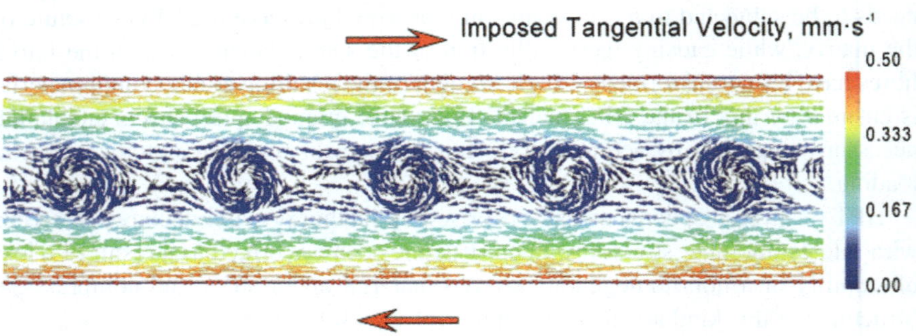


**Fig. 8.6** Simple shear blocked locally by cylindrical obstacles from a material that is ten times stronger than the layer undergoing shear deformation (FE simulation)

Figure 8.6 demonstrates that local blocking of shear does cause rotation of the obstacles, which leads to the formation of vortices at sufficiently large shear strain. In layered materials, the stronger layers play the role of such obstacles. As a result of twists, rotations and fracturing they undergo during the HPT process, these layers get transformed to a chain of vortices [41, 42]. This is illustrated schematically in Fig. 8.7a, c. Figure 8.7b, d show the transformation of the structure of a layered Al/Cu composite by HPT, which confirms the above scenario.

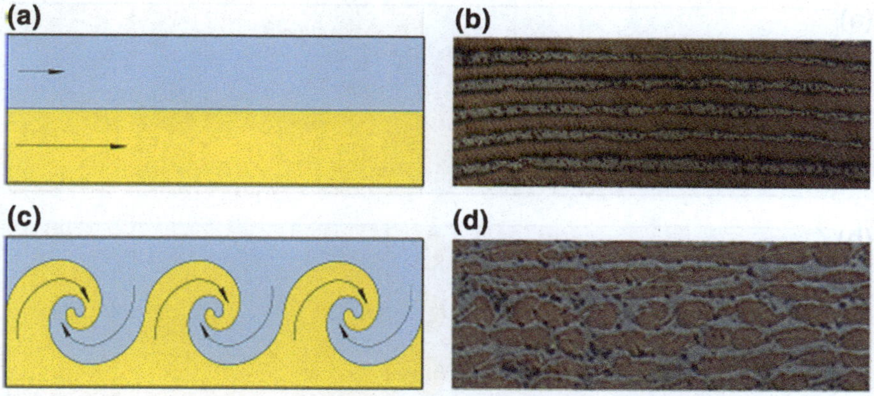


**Fig. 8.7** Transformation of a layered composite as a result of the occurrence of a chain of vortices under simple shear: **a**, **c** schematic pictures; **b**, **d** transformation of the structure of a real layered Al/Cu composite under HPT

A possible application of an architectured material in the form of a layered disc with inner vortices is as follows. Consider a shaft to be designed to operate in torsion under a large torque, which consists of two different materials (Fig. 8.8a). For many material couples, especially dissimilar ones, joining poses a great challenge. HPT makes it possible to join materials, which are normally not weldable, by high pressure and large shear strain [1]. This can be used to fabricate a layered disc whose surfaces are formed by two different materials (Fig. 8.7c). The disc can then be employed as a connector between the two parts of the shaft welded to their respective sides (Fig. 8.8b). This welding can be done in a conventional way, as the pieces to be welded on would face the same metal. The SPD-induced vortex structure of the disc reinforces the connection through mechanical locking of the dissimilar contact surfaces that comes on top of diffusion welding.

SPD processing of layered polymer specimens can lead to interesting new effects. As shown in a recent work [43], processing of a multi-layer billet with alternating amorphous and partially crystalline layers by a variant of ECAP, referred to as multiple angle ECAP, results in a controlled change of the inclination of the layers to the billet axis, densification of the polymers, an increase of their hardness and elastic moduli, as well as a better cohesion between the layers. This can be used to create architectured materials whose modulus of elasticity can be varied in a broad range. Figure 8.9a shows a representative volume of such a material: a plate from a rigid polymer (r) with a large modulus of elasticity with embedded layers of a compliant flexible polymer (f) whose modulus of elasticity is low.

Using the common methods of the elasticity theory [44], one obtains the following expression for the effective modulus of elasticity $E_{\text{eff}}$ of the composite plate:

$$\frac{1}{E_{\text{eff}}} = \frac{1}{E_{\text{r}}} + \frac{2}{E_{\text{f}}}\frac{h}{L}\cos\alpha\big((1+\nu_{\text{f}})(1-2\nu_{\text{f}})\cos^2\alpha + 2(1+\nu_{\text{f}})\sin^2\alpha\big) \tag{8.1}$$

**Fig. 8.8** Layered disc with a vortex structure that can be used as a connector between two pieces of a shaft designed to operate in torsion: **a** schematics of a shaft comprising two pieces made from different materials; **b** connection of the two pieces provided by a layered disc produced by high-pressure torsion

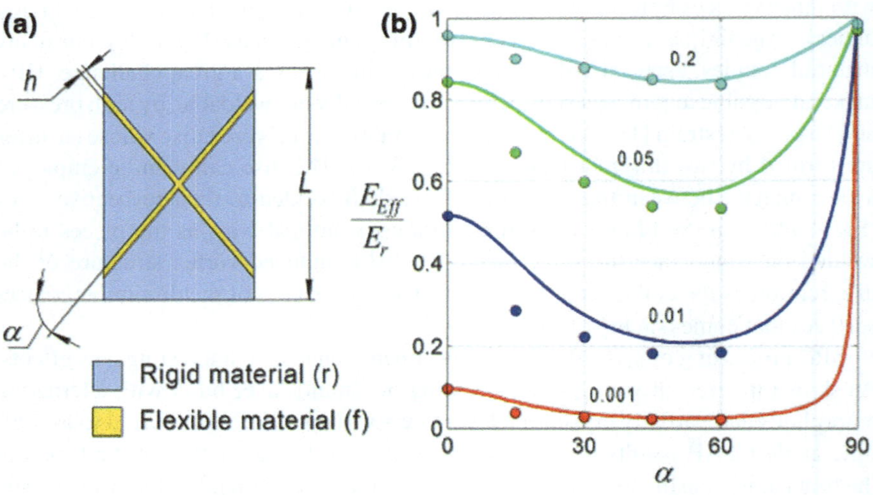


**Fig. 8.9** A representative volume of an architectured material with a variable effective modulus of elasticity depending on the inclination angle of compliant flexible layers to the vertical axis: schematics of the material architecture (**a**), effective modulus of elasticity as a function of the inclination angle $\alpha$ (**b**). The solid lines correspond to the calculated dependence according to the analytical formula (8.1); the symbols represent the results of finite element calculations in DEFORM 2D; the numbers above the curves indicate the magnitude of the ratio $E_f/E_r$

where $E_r$ and $E_f$ are, respectively, the elasticity moduli of the rigid and the flexible materials and $\nu_f$ is the Poisson's ratio of the flexible (compliant) material. The results of computer simulations using the finite element package DEFORM 2D support the validity of (8.1), cf. Fig. 8.9b. The diagrams presented in this figure indicate the range within which the effective modulus of elasticity can be varied by the choice of the material of the flexible layer and its inclination angle.

### *8.2.4 Mechanochemical Synthesis*

In an earlier publication [12], general principles of fabricating architectured materials by SPD-enabled mechanochemical synthesis were established. The factors that make SPD processing to a promising tool for mechanochemical synthesis include (i) the possibility to control the mutual positions of material points within a deforming solid during SPD processing, (ii) extremely fast mass transfer in the bulk of a deforming solid, and (iii) absorption of substances through the surface of a billet.

An interplay of these factors promotes efficient delivery of external substances to the sites within the workpiece where they can participate in a chemical reaction leading to the intended formation of a compound. In this way, a pattern of regions chemically different from the matrix material or a conceived geometry of new structural features can be produced within the bulk of the workpiece.

Chemically different ingredients can be introduced into a workpiece either prior to SPD (by embedding them in the bulk) or *during* the process (by infiltration of the bulk though the outer surface). The latter strategy can be realised by conducting SPD processing in a liquid or gaseous medium, thus enabling the absorption of atoms of the medium through the surface, with subsequent diffusive or, more importantly, convective transport within the bulk. Infiltration of the bulk through the surface during SPD operations can also be achieved if the workpiece or the tool is provided with an appropriate coating. Delivery of atoms to the bulk through the outer surface is promoted by SPD processes in which the surface area is cyclically varied, such as HPT, TE, and ARB. For example, in the case of twist extrusion, the surface of the material in contact with the die or the medium is 50% larger than before or after the billet passes through the die. Production of multiscale architectured materials with composition and grain size gradients presented in [13] provides an example of mechanochemical synthesis by means of HPT. A further example of architecturing of a material by HPT is the application of this technique to a layered Al/Mg specimen. In this case, vortices were formed at interfaces and cold fusion of the constituents occurred [45].

A promising route to producing materials with engineered inner structure is offered by SPD combined with powder metallurgy [12]. A particular example are the so called 'harmonic' materials produced by ball milling of spherical particles followed by spark plasma sintering (SPS) proposed in [46–48]. In the course of ball milling, a UFG shell is formed on the surface of powder particles, the shell thickness depending on the duration of ball milling. Through subsequent sintering, an interesting type of architectured material is produced. It is comprised by a hard cellular skeleton formed by the UFG shells and a softer coarse-grained material 'filling'. The skeleton structure of harmonic materials can be controlled by varying the SPS parameters, the thickness of the UFG shells of powders, and also through mixing of ball milled powders with unprocessed ones before SPS, pre-straining, etc. In this way, both monolithic and porous materials can be produced, with various applications being envisaged for them.

### 8.2.5 Combining SPD with Additive Manufacturing to Produce Architectured Materials

The great potential of SPD techniques for producing high-strength architectured materials discussed in the foregoing sections can be extended immensely by combining them with nascent additive technologies [49]. A review of the current state of such technologies focusing on 3D printing with polymers and metals can be found in Chap. 9 of this monograph.

For metallic materials, 3D printing by direct metal laser sintering (MLS) is the technology of choice. MLS with two or more alloys permits fabrication of monolithic parts with any inner architecture, including that shown, by way of example, in Fig. 8.10a. For a number of reasons, structural parts produced by additive manufacturing do not show good mechanical properties [49–51]. This drawback of the otherwise attractive additive manufacturing technologies can be overcome by means of post-processing by SPD, which can enable high mechanical characteristics to an MLS-produced material by inducing an UFG structure.

As discussed above, SPD processing causes a distortion of the inner architecture of a workpiece. In Fig. 8.10 this is illustrated by comparing the initial 3D printed pattern with the one transformed by TE (Fig. 8.10b) or ECAP (Fig. 8.10c). The distortions can be calculated using the simple shear model [21] in the case of ECAP and the experiment-based computed velocity field [8] in the case of TE. If the goal of the processing chain (printing + SPD) is to obtain the geometry depicted in Fig. 8.10a, the distorted pattern needs to be printed in the first place. Thus, in the case of post-processing by ECAP, the initial printed structure should be that of Fig. 8.10c. Then, ECAP in the direction opposite to that transforming Fig. 8.10a to Fig. 8.10c would restore the desired pattern. In the case of post-processing by TE, the initial printed structure should be that of Fig. 8.10b, while the screw geometry of the twist die should have the chirality opposite to the one used in the calculation of the transformation of Fig. 8.10a to Fig. 8.10b.

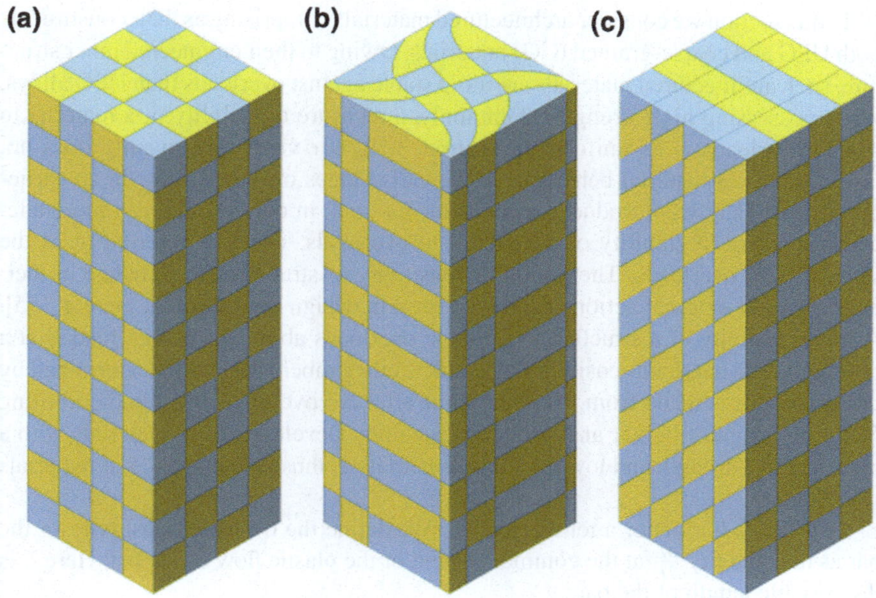


**Fig. 8.10** Producing an architectured material by combining SPD with 3D printing: desired pattern (**a**), pattern transformed by TE (**b**), pattern transformed by ECAP (**c**)

## 8.3 Multiscale UFG Structures with Low Sensitivity to Tensile Overloads

### *8.3.1 Definition of Overload Sensitivity*

UFG materials produced by means of SPD possess high strength characteristics, which is their unquestionable advantage [3, 4]. Unfortunately, they commonly lose their load bearing capacity once the yield stress is reached in tension. This is associated with necking or localised shear banding terminating in failure of the tensile specimen [52]. This is referred to as loss of stability of the material against tensile loading [53]. The loss of stability of a critical structural member, such as steering gear of an aircraft or an automobile, can be catastrophic. To avoid this, a large safety margin needs to be considered in the design with materials unstable against overloads.

In terms of their mechanical performance in critical structures, UFG materials can be categorised as high-strength, yet brittle, materials, for which a safety factor in excess of 3–4 is used. For materials with lower strength stable against tensile loading, a safety factor of 1.4–1.5 is sufficient [54]. As a result, the admissible design stress for structural members made from UFG materials may turn out to be lower than for conventional materials. This makes UFG materials less competitive.

In this section we consider architectured materials comprising as their constituents both UFG and coarse-grained (CG) materials. Owing to their engineered inner structure, such architectured materials are more stable against overloads than UFG alloys, while possessing high strength. Commonly, a measure of stability of a material to tensile overloads is its uniform elongation, $\delta$. In our view, this quantity does not reflect the local material behaviour in the early stages of plastic flow. In an earlier publication [12], we introduced a characteristic that, in our opinion, is a more adequate measure of stability of a material to overloads, which is referred to as the sensitivity to overloads. The notion of sensitivity of structures to a change in their parameters or external actions is broadly used in design optimisation, cf., e.g. [55]. A low sensitivity of a structure to possible overloads above the design load is at a premium. The overload sensitivity of a structural member is primarily determined by the material it is made from. An obnoxious effect of overloads is often the limiting factor in design of lighter and stronger structures. Development of materials with a high specific strength and low overload sensitivity is thus on the agenda of materials scientists.

Consider a bar under a tensile load $F$. We define the overload sensitivity of the bar as the quantity $\frac{\partial L}{\partial F}$ at the commencement of the plastic flow of the bar. Here $L$ is the variable length of the bar.

Neglecting the elastic volume change and using the relation $F = \sigma_s S$ (where $\sigma_s$ is the flow stress of the material (close to the yields strength) and $S$ is the cross-sectional area of the bar) one can easily show that

$$\frac{\partial L}{\partial F} = \frac{L}{F} \cdot \frac{\eta}{1-\eta}, \tag{8.2}$$

holds. Here $\eta$ is defined as

$$\eta = \sigma_s \left( \frac{\partial \sigma_s}{\partial e_M} \right)^{-1} \tag{8.3}$$

$e_M$ denoting the von Mises plastic strain of the bar.

As seen from (8.2), the overload sensitivity of the bar is determined by the quantity $\eta$, which depends solely on the intrinsic properties of the material. It is therefore reasonable to refer to this parameter as the material's sensitivity to overloads under uniaxial tension. It further follows from (8.2) that in the limit of $\eta$ approaching unity from below the derivative diverges: $\frac{\partial L}{\partial F} \to \infty$. That is to say, small overloads give rise to a dramatic irreversible elongation of the bar. For $\eta > 1$, the derivative is negative, $\frac{\partial L}{\partial F} < 0$, which means that the bar keeps elongating even when the load is dropping. This is tantamount to loss of stability of the bar. Finally, for $\eta < 0$ softening of the material in the course of deformation occurs, which gives rise to strain localisation in deformation bands culminating in loss of stability [56]. It follows from this brief analysis that only materials satisfying the condition

$$0 \le \eta < 1 \tag{8.4}$$

are suitable for structural applications.

It is easily recognised that at the point $\eta = 1$ the Considère condition, $\frac{\partial \sigma_s}{\partial e_M} = \sigma_s$, is fulfilled. The above considerations provide a physical interpretation for condition (8.4) as a requirement that only those materials can be employed in structural members subjected to tensile deformation that do not lose stability by either necking or shear banding.

It should be noted that in [57] the quantity $\eta$ was regarded as an order parameter that governs changes of the mechanisms of plastic deformation of a material. The parameter $\eta$ is related to the strain hardening rate $\Theta = \frac{\partial \sigma_s}{\partial e_M}$ via a simple formula [58]:

$$\eta = \sigma_s \Theta^{-1} \tag{8.5}$$

In annealed materials $\eta \ll 1$ $(\eta \sim 0.01)$ [57] holds for early stages of plastic deformation. UFG materials have large values of $\sigma_s$ and small $\Theta$ values, which is tantamount to a large overload sensitivity ($\eta \gg 1$, [57]). That is why it is dangerous to employ UFG materials in engineering structures that may be prone to overloads.

As foreshadowed above, this drawback can be combated by designing architectured materials combining high strength with low overload sensitivity. Two kinds of UFG structures whose high strength is achieved without sacrificing overload sensitivity are proposed in the following section.

### 8.3.2 'Artificial Crystals'—A Particular Kind of Architectured Materials

One of the promising designs of an architectured material is suggested by the biomimetics-driven idea that embedding soft layers in an otherwise strong material enables a good balance of strength and ductility [59]. In an earlier publication, we discussed a simple implementation of this idea [12], which is as follows. Thin planar layers of CG material are embedded in a UFG matrix. Upon a slight overload over the design load, these soft layers deform by shear and experience strain hardening, while the UFG matrix remains elastic. This picture suggests an analogy with a crystal, which also deforms on specific shear planes, and the term 'artificial crystal'. The coarse-grained layers within the artificial crystal are responsible for its strain hardening, thus reducing its sensitivity to possible tensile overloads.

A material architectured to contain CG layers embedded in a UFG plate thus forming an 'artificial crystal' can be fabricated by employing the following two-stage technology. First, a UFG strip is produced by SPD. Then, in a next stage, the strip is locally heated (e.g. by laser beam) according to a prescribed line pattern. The heat input is chosen in such a way as to cause local recrystallisation.

As demonstrated below, a specific feature of the thin layers is their high load bearing capacity under tension in the normal direction and the ease of shear along

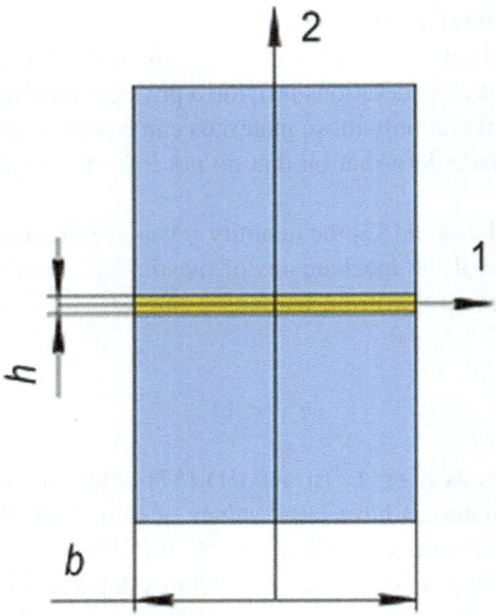


**Fig. 8.11** Tension of a plate with an embedded thin layer normal to the direction of tensile load

the layers. We start by considering a plate with a single thin layer, Fig. 8.11. For simplicity, plane strain condition is assumed. Outside of the thin layer (assumed to be ductile), the material is considered to be rigid. Its flow stress is denoted by $\sigma_s$. For the thin layer the condition $\frac{h}{b} \ll 1$ is used, where $b$ is the width of the plate. The flow stress $\sigma_s^*$ of the composite (matrix + soft thin layer) was calculated in [12]. For the case when the thin layer is normal to the tensile direction along axis 2, it is given by

$$\frac{\sigma_s^*}{\sigma_s} = \frac{2}{\sqrt{3}}\left(1 + \frac{1}{2}\frac{b}{h}\right) \tag{8.6}$$

Obviously, it exceeds by far the flow stress $\sigma_s$ of the base material of the matrix.

An interesting corollary following from this result is that a softer layer produced, for example, by welding that softens the material locally is not detrimental to the tensile strength of the structure.

The flow stress for an 'artificial crystal' with a thin layer inclined to axis 2 at non-zero angle $\alpha$ (see Fig. 8.12) was calculated to be [12]:

$$\sigma_s^* = \frac{2}{\sqrt{3}} \frac{\sigma_s}{\sin 2\alpha} \tag{8.7}$$

For the 'artificial crystal' to be able to deform exclusively by shear along the layers, the condition $\sigma_s^* < \frac{2}{\sqrt{3}}\sigma_{UFG}$ needs to be fulfilled. Here $\sigma_{UFG}$ is the flow stress

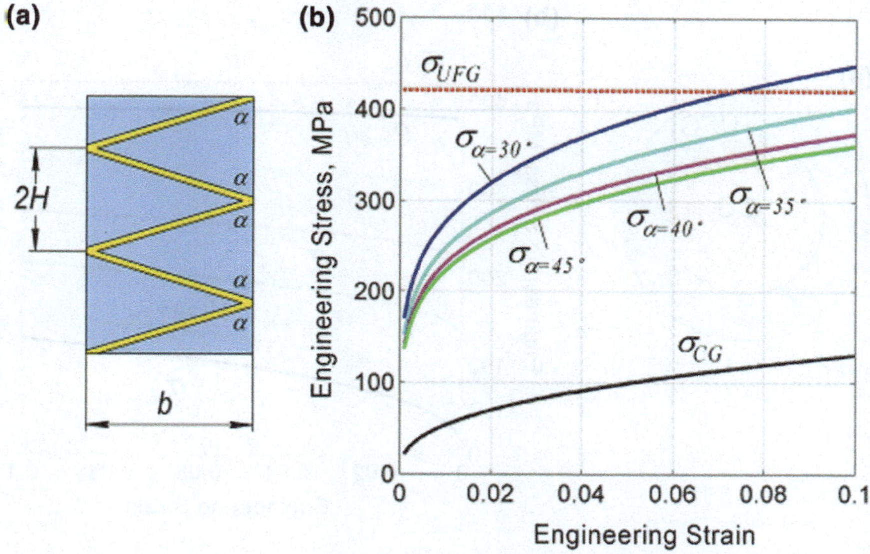


**Fig. 8.12** An architectured material of the 'artificial crystal' type: schematic drawing (**a**), engineering stress-strain dependence under uniaxial tension for several values of the inclination angle of the CG layers (**b**)

of the UFG matrix; the coefficient $\frac{2}{\sqrt{3}}$ accounts for the system to be a planar one (cf., e.g. [60]). Equation (8.7) then reduces to

$$\frac{\sigma_s}{\sin 2\alpha} < \sigma_{\mathrm{UFG}} \tag{8.8}$$

To calculate the overload sensitivity parameter $\eta^*$ for the 'artificial crystal', the relative extension of the composite, $\varepsilon^*$, can be expressed in terms of the shear strain $\gamma$ in an individual layer. As the plastic elongation of the 'artificial crystal' is furnished solely by shear in the embedded layers, it follows, purely from geometry considerations:

$$\varepsilon^* = \frac{h}{H}\gamma \cdot \cos\alpha, \tag{8.9}$$

where the geometry parameter $H$ is defined in Fig. 8.12 and $h$ is the thickness of the coarse-grained layer.

To characterise the deformation of the 'artificial crystal', we define the quantity $e_M^*$ analogous to the von Mises equivalent strain by relating it to the relative elongation of the material in the plane strain condition [60]:

$$e_M^* = \frac{2}{\sqrt{3}}\ln\left(1 + \varepsilon^*\right) \tag{8.10}$$

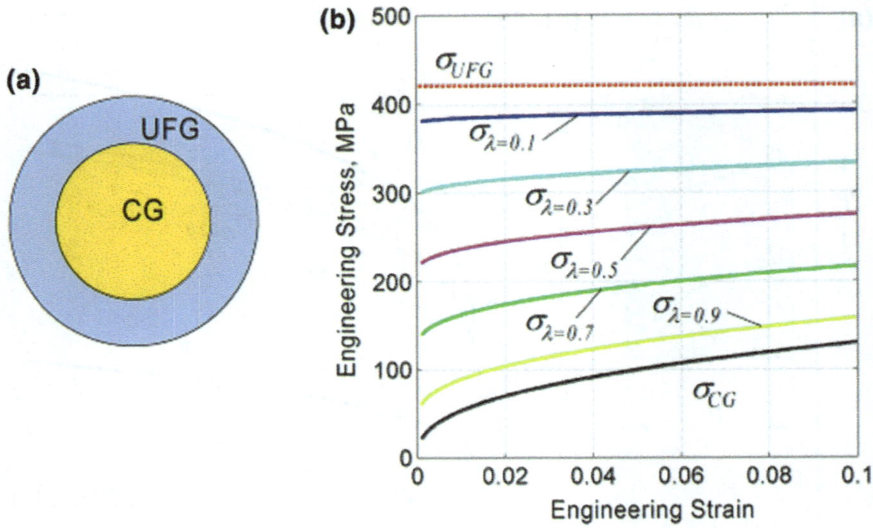


**Fig. 8.13** Architectured material of the 'soft core/hard sheath' type: **a** a cross-section showing a coarse grained (CG) core and an ultra-fine grained (UFG) sheath, **b** stress-strain curves for various values of the volume fraction of the core material

For $\varepsilon^* \ll 1$, this equation reduces to

$$e_M^* \approx \frac{2}{\sqrt{3}}\varepsilon^* \tag{8.11}$$

Substitution of (8.11) into (8.9) yields

$$e_{\mathrm{M}}^* = 2\frac{h}{H}e_{\mathrm{M}} \cdot \cos\alpha \tag{8.12}$$

(Here the well-known relation $e_{\mathrm{M}} = \frac{\gamma}{\sqrt{3}}$ (cf, e.g. [60], was used.)
The overload sensitivity parameter for the 'artificial crystal' is defined as

$$\eta^* = \sigma_{\mathrm{s}}^* \left(\frac{\partial \sigma_{\mathrm{s}}^*}{\partial e_{\mathrm{M}}^*}\right)^{-1} \tag{8.13}$$

It can be expressed in terms of the similar quantity for the base material, $\eta$, by combining (8.7), (8.12), and (8.3):

$$\eta^* = 2\eta\frac{h}{H}\cos\alpha \tag{8.14}$$

The deformation behaviour of the 'artificial crystal' can be discussed by way of the following example. The soft layer material is considered to be coarse grained copper with known mechanical properties [61]. As the matrix material UFG copper is taken. We assume that it was produced by simple shear deformation, e.g. by means of ECAP. The flow stress $\sigma_{UFG}$ is set at 421 MPa—a value that corresponds to saturation flow stress for copper pre-deformed by simple shear to very high strain [61].

Under uniaxial tensile loading of the 'artificial crystal' the embedded soft layers deform by simple shear. Under this loading mode, at least up to the equivalent von Mises strain of 3, the stress-strain curve for the layer material is well represented by the power law

$$\sigma_s = Ae_M^n, \tag{8.15}$$

with the parameters $A = 269$ MPa, $n = 0.21$ [61].

The engineering stress-strain curves for the 'artificial crystal' considered as obtained from (8.7), (8.11), (8.12) and (8.15) for $\frac{h}{H} = 0.1$ are shown in Fig. 8.12b. For comparison, the same figure contains the stress-strain curve for CG copper and marks the level of $\sigma_{UFG}$ [61].

For the matrix material with the power-law strain dependence of stress, it follows from substitution of (8.15) in (8.3):

$$\eta = \frac{e_M}{n}, \tag{8.16}$$

Using (8.14), (8.12), and (8.16) it is easy to show that in this case the same relation applies for the 'artificial crystal' as well:

$$\eta^* = \frac{e_M^*}{n}, \tag{8.17}$$

That is to say, the 'artificial crystal' exhibits the same sensitivity to tensile overloads as the coarse-grained matrix material. At the same time, according to (8.7), the yield strength of the 'artificial crystal' can by far exceed that of the soft layer material and in a limit case approach the flow stress of the UFG matrix. A more detailed discussion of this result will be given in Sect. 8.3.4.

### 8.3.3 *Architectured Materials with 'Soft Core/Hard Sheath' Design*

A further recipe for enhancing the tensile overload stability of a UFG material is to create a structure in which a soft core is encased in a hard sheath, for example in a cylindrical geometry shown in Fig. 8.13a.

This kind of structure mimics that of a bone, with a soft trabecular tissue 'encased' in a hard cortical bone. It can be produced by severe plastic deformation through a number of processes, including ECAP of a bimetallic rod [62]. A gradient structure with a soft CG core and a hard UFG outer skin can be produced by such processes as direct torsion, high pressure tube twisting, tree-roll milling, etc. Figure 8.13b shows that this architecture is capable of delivering a high strength and a reasonably high tensile stability.

The tensile overload sensitivity of this kind of architectured material, which can be regarded as a hybrid material, can be assessed in the following way. We assume that the connection between both constituents of the structure is fully cohesive and that the strains they develop during tensile deformation of the hybrid structure are the same. Its flow stress can then be expressed by a rule of mixtures:

$$\sigma_s^* = \lambda\sigma_s + (1 - \lambda)\sigma_{UFG}, \tag{8.18}$$

where $\lambda$ denotes the volume fraction of the core material. The subscripts 1 and 2 refer to the soft core and the hard skin, respectively.

Engineering tensile deformation curves for materials with different values of $\lambda$ calculated using (8.18) are presented in Fig. 8.13b. The experimental stress-strain curve for tensile deformation of copper [61] was employed as a basis for these calculations. The experimental curve was approximated by a power-law, (8.15), with the parameters $A = 320\,\text{MPa}$, $n = 0.39$. As in Fig. 8.12b, the engineering stress-strain curve for CG copper and the level of the flow stress for the UFG matrix, $\sigma_{UFG}$, are shown in Fig. 8.13b.

From (8.3) and (8.18) the overload sensitivity of the hybrid is easily obtained:

$$\eta^* = \frac{\lambda\kappa + (1 - \lambda)}{\lambda\kappa\eta^{-1} + (1 - \lambda)\eta_{UFG}^{-1}}, \tag{8.19}$$

where $\kappa = \frac{\sigma_s}{\sigma_{UFG}}$ denotes the ratio of the flow stresses in the two constituents.

Making use of the condition $\eta_{UFG} \gg 1$, one obtains from (8.19) the following relation:

$$\eta^* = \left(1 + \frac{1 - \lambda}{\lambda\kappa}\right)\eta \tag{8.20}$$

Equations (8.18) and (8.20) show that the 'soft core/hard sheath' type architecture makes it possible to tune the material properties to a desired property profile. In the following section we propose a material map in the 'overload sensitivity *versus* flow stress' space as a convenient tool for comparing such architectured materials with other materials.

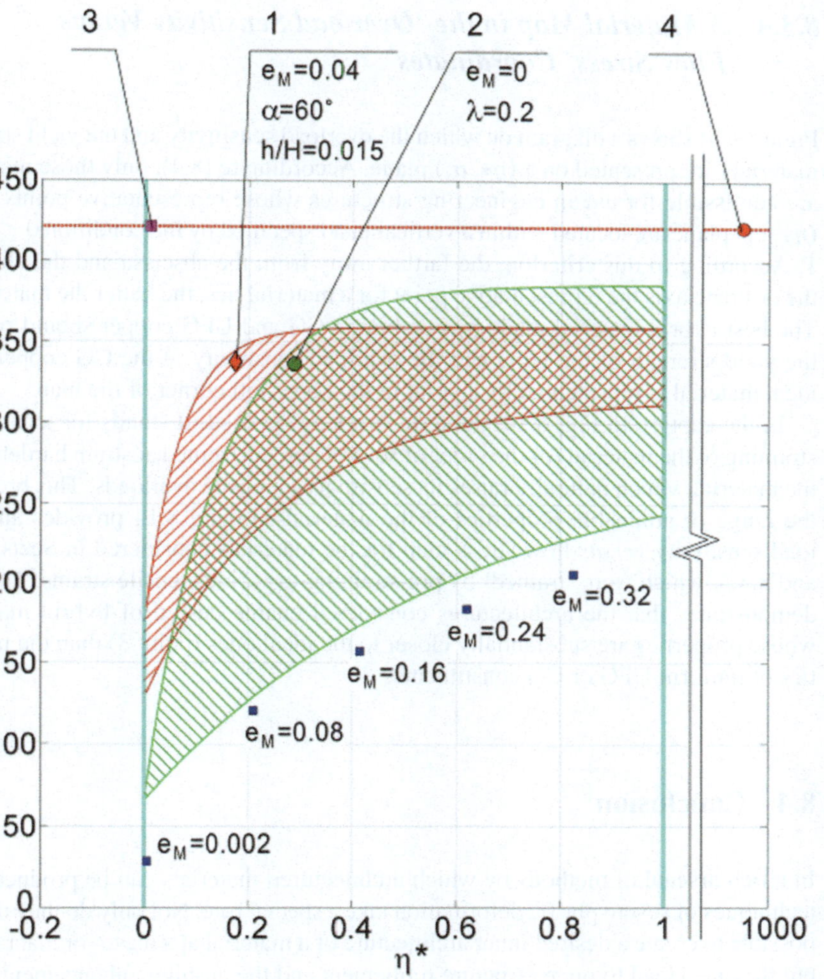


**Fig. 8.14** A material map in the 'overload sensitivity *versus* flow stress' coordinates. The range of parameters admissible for engineering structures corresponds to the vertical band $0 \leq \eta* < 1$. The red-coloured area corresponds to 'trained' artificial crystals with various *h/H* and various layer inclination angles. The best performing material of this kind is represented by a diamond (point 1). The term 'training' implies that the material has been pre-strained before uniaxial tensile testing. The degree of pre-straining $e_M$ is indicated by the blue squares at the lower envelope of the diagram, which corresponds to the stress-strain curve of the CG matrix. The area coloured in green corresponds to the 'trained' soft core/hard sheath type materials with various volume fractions λ of the CG constituent. The best performing soft core/hard sheath material is represented by a circle (point 2). In each region, the calculated point closest to that representing the ideal material (point 3 in the upper left corner of the diagram) is shown. The highest yield strength achieved is indicated by the red horizontal line corresponding to strength of UFG copper (point 4)

### 8.3.4 A Material Map in the 'Overload Sensitivity Versus Flow Stress' Coordinates

Figure 8.14 shows a diagram on which the overload sensitivity and the yield stress of materials are presented on a $(\eta*, \sigma_s)$ plane. According to (8.4), only those materials are admissible for use in engineering structures whose representative points in the $(\eta*, \sigma_s)$ plane are located within a vertical band specified by the condition $0 \leq \eta* < 1$. According to this criterion, the farther away from the abscissa and the closer to the ordinate axis the representative point for a material lies, the better the material is. The best hybrid material made by combining CG and UFG copper should possess the yield strength of the UFG and the overload sensitivity of the CG copper. This ideal material is represented by a point in the upper left corner of the map.

In the caption to Fig. 8.14, the term 'training' is used. It stands for tensile pre-straining of the material (of the order of several percent) aimed at strain hardening of the material, which naturally raises its sensitivity to tensile overloads. This broadens the range of attainable properties of the materials. Figure 8.14 provides an overload sensitivity *versus* flow stress map for the materials considered in Sects. 8.3.2 and 8.3.3, which were 'trained' by pre-straining to various tensile strains. The map demonstrates that the architectures considered enable design of hybrid materials whose properties are substantially closer to the ideal ones (point 3) than the properties of uniform UFG or CG constituents.

## 8.4 Conclusion

In a rich arsenal of methods by which architectured materials can be produced, the techniques of severe plastic deformation take a special role. Not only do they make it possible to create a desired inner architecture of a material at a meso- or macroscale, but they also lead to microstructure refinement and the ensuing enhancement of the mechanical properties. SPD is a mature technology that appears to be at a threshold of achieving breakthroughs in a range of engineering applications. However, it has its limitations and is in search for new fields where its unique potential can best be capitalised on. We believe that fabrication of architectured materials and structures is a very promising field where the strengths of the SPD technology can be exploited in a most efficient way. We hope that through the examples shown in this chapter, we were able to convince the readers that the SPD technology may be entering a new period in its development, with a very promising outlook.

**Acknowledgements** YB would like to express his gratitude to the State Fund for Fundamental Research of the Ukraine for financial support through grant F71/56-2016.
RK acknowledges funding support from the German Research Foundation (DFG) through Grant #IV98/8-1.

## References


1. P.W. Bridgman, Effects of high shearing stress combined with high hydrostatic pressure. Phys. Rev. **48**, 825–847 (1935)
2. V.M. Segal, V.I. Reznikov, A.E. Drobyshevsky, V.I. Kopylov, Plastic treatment of metals by simple shear. Russ. Metall. **1**, 115–123 (1981)
3. R.Z. Valiev, Y. Estrin, Z. Horita, T.G. Langdon, M.J. Zehetbauer, Y.T. Zhu, Producing bulk ultrafine-grained materials by severe plastic deformation: ten years later. JOM **68**, 1216–1226 (2016)
4. Y. Estrin, A. Vinogradov, Extreme grain refinement by severe plastic deformation: a wealth of challenging science. Acta Mater. **61**, 782–817 (2013)
5. A. Bachmaier, R. Pippan, Generation of metallic nanocomposites by severe plastic deformation. Int. Mater. Rev. **58**, 41–62 (2013)
6. N.S. Enikolopian, Superfast polymerization under high pressure and plastic flow. Macromol. Chem. **185**, 1371–1381 (1984)
7. V. Beloshenko, Y. Beygelzimer, Y. Voznyak, *Solid-State Extrusion, Encyclopedia of Polymer Sciences* (Wiley & Sons, New York, 2015). https://doi.org/10.1002/0471440264.pst343.pub2
8. Y. Beygelzimer, R. Kulagin, Y. Estrin, L.S. Toth, H.S. Kim, M.I. Latypov, Twist extrusion as a potent tool for obtaining advanced engineering materials: a review. Adv. Eng. Mater. **19**(8), 1600873 (2017). https://doi.org/10.1002/adem.201600873
9. R. Lapovok, Y. Qi, H.P. Ng, V. Maier, Y. Estrin, Multicomponent materials from machining chips compacted by equal-channel angular pressing. J. Mater. Sci. **49**, 1193–1204 (2014)
10. O. Bouaziz, H.S. Kim, Y. Estrin, Architecturing of metal-based composites with concurrent nanostructuring: a new paradigm of materials design. Adv. Eng. Mater. **15**, 336–340 (2013)
11. M.I. Latypov, Y. Beygelzimer, R. Kulagin, V. Varyukhin, H.S. Kim, Toward architecturing of metal composites by twist extrusion. Mater. Res. Lett. **3**, 161–168 (2015)
12. Y. Beygelzimer, Y. Estrin, R. Kulagin, Synthesis of hybrid materials by severe plastic deformation: a new paradigm of SPD processing. Adv. Eng. Mater. **17**, 1853–1861 (2015)
13. J.Y. Kang, J.G. Kim, H.W. Park, H.S. Kim, Multiscale architectured materials with composition and grain size gradients manufactured using high-pressure torsion. Sci. Rep. **6**, 26590 (2016)
14. X.L. Wu, P. Jiang, L. Chen, F. Yuan, Y.T. Zhu, Extraordinary strain hardening by gradient structure. PNAS **111**, 7197–7201 (2013)
15. T.H. Fang, W.L. Li, N.R. Tao, K. Lu, Revealing extraordinary intrinsic tensile plasticity in gradient nano-grained copper. Science **331**, 1587–1590 (2011)
16. J. Lu, P.S. Mai, C.S. Wen, in *Nanostructured-Lattices Produced by Surface Mechanical Attrition Treatment Method*, Patent US20150033814, 2015
17. L. Ghalandari, M.M. Mahdavian, M. Reihanian, M. Mahmoudiniya, Production of Al/Sn multilayer composite by accumulative roll bonding (ARB): A study of microstructure and mechanical properties. Mater. Sci. Eng., A **661**, 179–186 (2016)
18. K. Lu, J. Lu, Nanostructured surface layer on metallic materials induced by surface mechanical attrition treatment. Mater. Sci. Eng., A **375–377**, 38–45 (2004)
19. O. Bouaziz, Y. Estrin, H.S. Kim, Severe plastic deformation by the cone-cone method: potential for producing ultrafine grained sheet material. Rev. Met. Paris **104**, 318–322 (2007)
20. R. Lapovok, A. Pougis, V. Lemiale, D. Orlov, L.S. Toth, Y. Estrin, Severe plastic deformation processes for thin samples. J. Mater. Sci. **45**, 4554 (2010)
21. R.Z. Valiev, T.G. Langdon, Principles of equal-channel angular pressing as a processing tool for grain refinement. Prog. Mater Sci. **51**, 881–981 (2006)
22. A.P. Zhilyaev, T.G. Langdon, Using high-pressure torsion for metal processing: fundamentals and applications. Prog. Mater Sci. **53**, 893–979 (2008)
23. A. Rosochowski, *Severe Plastic Deformation Technology* (Whittles Publishing, Dunbeath, Scotland, 2017)
24. A. Azushima, R. Kopp, A. Korhonen et al., Severe plastic deformation (SPD) processes for metals. CIRP Ann. Manuf. Technol. **57**, 716–735 (2008)

25. C.P. Wang, F.G. Li, L. Wang et al., Review on modified and novel techniques of severe plastic deformation. Sci. Chin. Technol. Sci. **55**, 2377–2390 (2012)
26. Yu. Ivanisenko, R. Kulagin, V. Fedorov, A. Mazilkin, T. Scherer, B. Baretzky, H. Hahn, High pressure torsion extrusion as a new severe plastic deformation process. Mater. Sci. Eng., A **664**, 247–256 (2016)
27. Y. Beygelzimer, D. Orlov, V. Varyukhin, A new severe plastic deformation method: twist extrusion/ultrafine grained materials II, in *Proceedings of a Symposium held during the 2002 TMS Annual Meeting I*, Seattle, Washington, 2002, pp. 297–304
28. Y. Saito, H. Utsunomiya, N. Tsuji, T. Sakai, Novel ultra-high straining process for bulk materials–development of the accumulative roll-bonding (ARB) process. Acta Mater. **47**, 579–583 (1999)
29. W.L. Li, N.R. Tao, K. Lu, Fabrication of a gradient nano-micro-structured surface layer on bulk copper by means of a surface mechanical grinding treatment. Scripta Mater. **59**, 546–549 (2008)
30. Z. Chen, M.H. Colliander, G. Sundell, R.L. Peng, J. Zhou, S. Johansson, J. Moverare, Nanoscale characterization of white layer in broached Inconel 718. Mater. Sci. Eng., A **684**, 373–384 (2017)
31. Z. Pu, G.L. Song, S. Yang et al., Grain refined and basal textured surface produced by burnishing for improved corrosion performance of AZ31B Mg alloy. Corros. Sci. **57**, 192–201 (2012)
32. Z. Yin, X. Yang, X. Ma et al., Strength and ductility of gradient structured copper obtained by surface mechanical attrition treatment. Mater. Design **105**, 89–95 (2016)
33. Y.L. Wang, A. Molotnikov, M. Diez, R. Lapovok, H.E. Kim, J.T. Wang, Y. Estrin, Gradient structure produced by three roll planetary milling: numerical simulation and microstructural observations. Mater. Sci. Eng. A **639**, 165–172 (2015)
34. M. Diez, H.J. Kim, V.N. Serebryanyi, S.V. Dobatkin, Y. Estrin, Improving the mechanical properties of pure magnesium by three-roll planetary milling. Mater. Sci. Eng. A **612**, 287–292 (2014)
35. O. Prokof'eva, Y. Beygelzimer, R. Kulagin, Y. Estrin, V. Varyukhin, Preparation of UFG composites with large uniform elongation by twist extrusion: mathematical modeling of the process. Russ Metall. **2**, 76–81 (2017)
36. S. Khoddam, Y. Estrin, H.S. Kim et al., Torsional and compressive behaviours of a hybrid material: spiral fibre reinforced metal matrix composite. Mater. Design **85**, 404–411 (2015)
37. O. Bouaziz, Geometrically induced strain hardening. Scripta Mater. **68**, 28–30 (2013)
38. C.W. Passchier, R.A.J. Trouw, *Microtectonics* (Springer, Berlin, Heidelberg, 2005)
39. Y. Beygelzimer, Ruslan Z. Valiev, V. Varyukhin, Simple shear: double-stage deformation. Mater. Sci. Forum **667–669**, 97–102 (2011)
40. R. Kulagin, Y. Beygelzimer, Y. Ivanisenko, A. Mazilkin, B. Straumal, H. Hahn, High pressure torsion: from laminar flow to turbulence, in *2017 IOP Conference Series: Materials Science and Engineering*, vol. 194 (2017), pp. 012045. https://doi.org/10.1088/1757-899x/194/1/012045
41. R. Kulagin, Y. Beygelzimer, Y. Ivanisenko, A. Mazilkin, B. Straumal, H. Hahn, Instabilities of interfaces between dissimilar metals induced by high pressure torsion. Mater. Lett. **222**, 172–175 (2018)
42. M. Pouryazdan, B.J.P. Kaus, A. Rack, A. Ershov, H. Hahn, Mixing instabilities during shearing of metals. Nat. Commun. **8**, 1611 (2017)
43. V. Beloshenko, Y. Beygelzimer, Y. Voznyak, B. Savchenko, V. Dmitrenko, ECAP of polymer billets obtained by the FDM process. Metal. Form. Russ. **44**, 108–114 (2017)
44. S. Timoshenko, *Theory of Elasticity* (McGraw-Hill, New York, 2001)
45. M. Kawasaki, H.J. Lee, J. Jang, T.G. Langdon, Strengthening of metals through severe plastic deformation. Adv. Mater. Sci. **48**, 13–24 (2017)
46. T. Sekiguchi, K. Ono, H. Fujiwara et al., New microstructure design for commercially pure titanium with outstanding mechanical properties by mechanical milling and hot roll sintering. Mater. Trans. **51**, 39–45 (2010)
47. D. Orlov, H. Fujiwara, K. Ameyama, Obtaining copper with harmonic structure for the optimal balance of structure-performance relationship. Mater. Trans. **54**, 1549–1553 (2013)

48. Z. Zhang, S.K. Vajpai, D. Orlov, K. Ameyama, Improvement of mechanical properties in SUS304L steel through the control of bimodal microstructure characteristics. Mater. Sci. Eng., A **598**, 106–113 (2014)
49. S.H. Huang, P. Liu, A. Mokasdar, L. Hou, Additive manufacturing and its societal impact: a literature review. Int. J. Adv. Manuf. Tech. **67**, 1191–1203 (2013)
50. W.E. Frazier, Metal additive manufacturing: a review. J. Mater. Eng. Perform. **23**, 1917–1928 (2014)
51. J.B. Roca, P. Vaishnav, E.R.H. Fuchs, M.G. Morgan, Policy needed for additive manufacturing. Nature Mater. **15**, 815–818 (2016)
52. Y.M. Wang, E. Ma, Three strategies to achieve uniform tensile deformation in a nanostructured metal. Acta Mater. **52**, 1699–1709 (2004)
53. Z.P. Bažant, L. Cedolin, *Stability of Structures: Elastic, Inelastic, Fracture and Damage Theories* (Oxford University Press, New York, 2003)
54. G.M. Ickovich, L.S. Minin, A.I. Vinokurov, *A Guide for Solving Problems on the Resistance of Materials* (High school, Moscow, 1999)
55. E.J. Haug, K.K. Choi, V. Komkov, in *Design Sensitivity Analysis of Structural Systems* (Academic Press, Inc., Harcourt Brace Jovanovich Publishers, New York, 1986)
56. D.C. Drukker, On the postulate of material stability in the mechanics of continua. J Mechanique **3**, 235–249 (1964)
57. Y. Beygelzimer, Vortices and mixing in metals during severe plastic deformation. Mater. Sci. Forum **683**, 213–224 (2011)
58. U.F. Kocks, H. Mecking, Physics and phenomenology of strain hardening: the FCC case. Prog. Mater Sci. **48**, 171–273 (2003)
59. F. Barthelat, Architectured materials in engineering and biology: fabrication, structure, mechanics and performance. Int. Mater. Rev. **60**, 413–430 (2015)
60. R. Hill, *Mathematical Theory of Plasticity* (Oxford University Press, New York, 1950)
61. C. Chen, Y. Beygelzimer, L.S. Toth, Y. Estrin, R. Kulagin, Tensile yield strength of a material pre-processed by simple shear. J. Eng. Mater. Technol. **138**, 031010 (2016)
62. Y. Qi, R. Lapovok, Y. Estrin, Microstructure and electrical conductivity of aluminium/steel bimetallic rods processed by severe plastic deformation. J. Mater. Sci. **51**, 6860–6875 (2016)

# Chapter 9
# Architectured Polymeric Materials Produced by Additive Manufacturing

**Andrey Molotnikov, George P. Simon and Yuri Estrin**

**Abstract** Polymers play an important role in our everyday life. With the advent of additive manufacturing (AM) technologies, the design and manufacture of new polymer-based composite materials has experienced a significant boost. AM enables precise deposition of printable material(s) with micro scale accuracy to build up a desired structure in three dimensions in a layer-by-layer fashion. In this chapter, recent advances in the use of additive manufacturing for the design of architectured polymer-based materials is discussed. A compendium of the existing AM methods is presented, followed by an overview of applications of AM technology to fabrication of polymer-based materials with engineered inner architecture.

## 9.1 Introduction

Design of engineering materials is traditionally focused on attaining a certain functionality of a material by selecting and manipulating its constituents at compositional and/or microstructural levels. Nature teaches us that superior performance of a material can be obtained even if only a small number of constituents are involved. The secret behind this improved performance is the highly optimised architecture at multiple length scales. Among prominent examples studied extensively are bone [1–3], wood [4], nacre [5, 6], and glass sponge spicules [7], just to name a few. These structures possess hierarchical features spanning different length scales [8] and are generally found to have a highly heterogeneous microstructure. Some of these Nature-inspired designs, along with smart engineering designs, motivated the seminal work of Ashby and Bréchet [9] that introduced the notion of *hybrid* materials as "materials which are designed to fulfil multi-objective design requirements and represent a combination of two or more materials". These materials are configured in such a way as to have attributes not offered by any one material alone. This new paradigm, in which shape and mutual arrangement of the constituent materials

A. Molotnikov (✉) · G. P. Simon · Y. Estrin
Department of Materials Science and Engineering, Monash University, Clayton, Australia
e-mail: andrey.molotnikov@monash.edu



Y. Estrin et al. (eds.), *Architectured Materials in Nature and Engineering*, Springer Series in Materials Science 282,
https://doi.org/10.1007/978-3-030-11942-3_9

become additional 'degrees of freedom', thereby extending the design parameter space, and has attracted much interest from researchers [10].

Fabrication of architectured materials remains a challenging task. Combining multiple materials, locally tuning the properties of the resulting material, and building it to the size of a structural member or an entire product go beyond traditional processing routes such as casting. In the last decade, *additive manufacturing* (*AM*) has emerged as a viable tool to overcome the limitations of conventional processing of polymeric materials. Additive manufacturing was conceived in early 70s and was aimed at supporting a variety of industries with rapid prototyping. The method is based on discretising a 3D model into layers and reconstructing the model in a real object by successive deposition and joining of layers. Numerous techniques using synthetic and natural polymers have been developed and commercialised including stereolithography, selective laser sintering (SLS) and Fused Deposition Modelling (FDM) [11]. Each of the AM techniques, often referred to as 3D printing, are linked to a specific class of polymeric materials such as thermoplastics, thermosets or elastomers. The particular deposition method and the details of the solidification process determine the minimum attainable feature size, which can range from a few nanometres to several millimetres. The same factors govern the ability of multiple materials to be integrated in a single printed part. Additive manufacturing is currently transitioning from the traditional rapid prototyping to serial manufacturing of functional

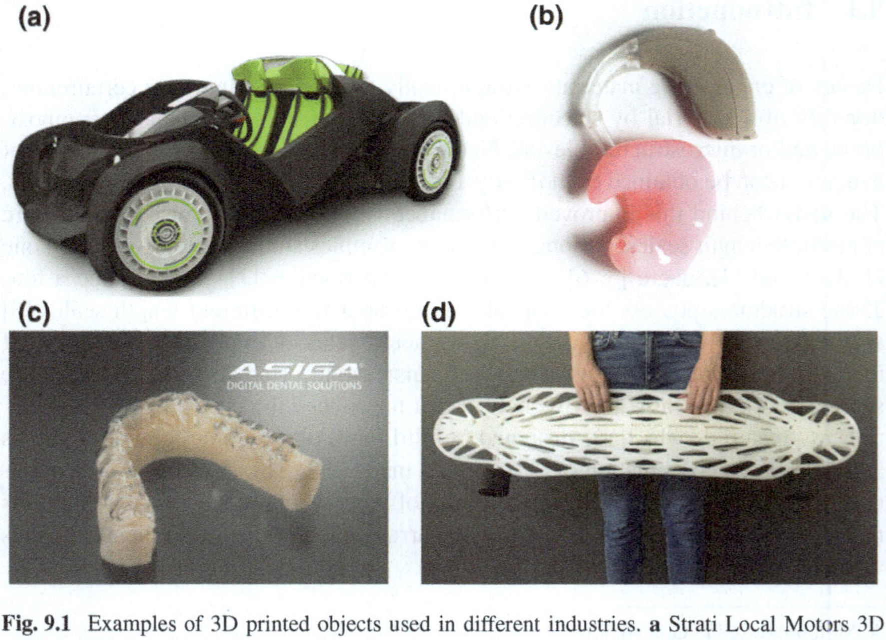


**Fig. 9.1** Examples of 3D printed objects used in different industries. **a** Strati Local Motors 3D printed car produced by FDM, **b** earshell and earmould and **c** orthodontic clear aligner fabricated by SLA (https://www.asiga.com), **d** 3D printed Bionic Longboard (BigRep, https://bigrep.com/sample-prints/)

parts on demand for a wide range of industries including aerospace, automotive, orthopaedics, orthodontics and many others, Fig. 9.1. For instance, the U.S. hearing aid industry converted to full additive manufacturing in less than two years [12] and is now able to produce personalised products such as earshells and earmoulds, see Fig. 9.1b. The uptake of additive manufacturing by industry hinges on the expansion of the palette of the available printable materials, and the broadening of the accessible range of manufacturing speed, precision and overall build volume.

In the context of this book, additive manufacturing is considered as an **emerging and enabling processing method** which allows the control of the deposition of constituent materials in a desired spatial pattern thus enabling the fabrication of architectured materials. While the use of AM for engineering the inner architecture of materials is still in its infancy, the potential it promises is huge. In this chapter, several additive manufacturing techniques are presented and some emerging applications to design and manufacturing of polymeric architectured materials are discussed.

## 9.2 Common Additive Manufacturing Processes for Polymers

The polymeric additive manufacturing processes can be classified depending on the deposition method used to create a layer, and the energy source used to solidify the material upon the deposition. Figure 9.2 presents established additive manufacturing methods including Stereolithography (SLA), selective laser sintering (SLS) of powders, Fused Deposition Modelling (FDM), inkjet printing, and direct ink writing. This section of the chapter will give an overview of the existing AM methods applicable to polymeric materials and outline their advantages and drawbacks in creating polymeric architectured materials, and provide some examples of their application.

### *9.2.1 Stereolithography*

Stereolithography was the first additive manufacturing technology which was commercialised by 3D Systems in 1986 [14]. The process is based on a selective curing of a photosensitive liquid resin using a light-based source such as ultraviolet (UV) laser [15–17], Fig. 9.2a. Typical polymers used in SLA include acrylate, methacrylate, epoxy system and their combinations [11]. A photopolymer solidifies upon reaction due to irradiation from a single-photon source. This source scans across the resin surface and chemically converts a line of monomer to become a polymer to a certain depth and width. The shape of the cured line depends on the resin characteristics and the processing parameters, such as laser energy, laser scan speed, and laser spot size [11]. Figure 9.2a depicts a unit employing a bottom up approach where the build platform is immersed in the resin bath and once the first layer is printed, the platform

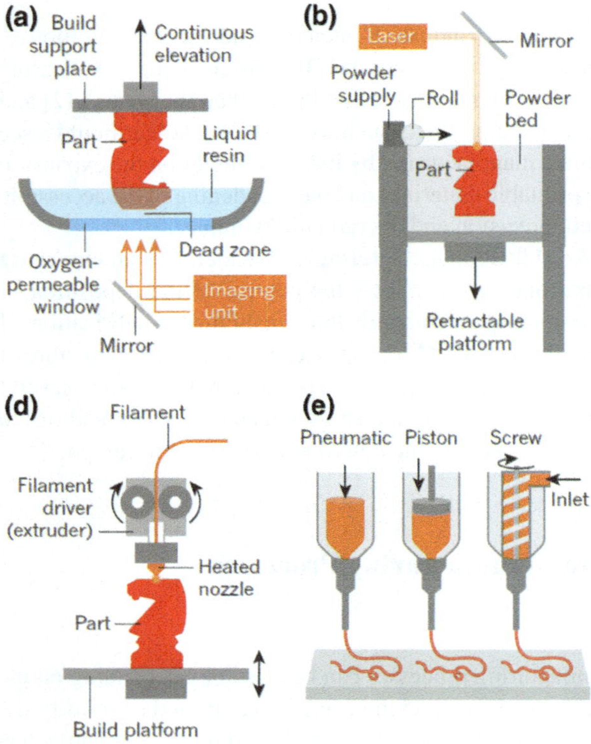


**Fig. 9.2** Common additive manufacturing methods: **a** schematic of modified stereolithography known as continuous liquid interface production, **b** selective laser sintering of thermoplastic powders, **c** inkjet printing of photopolymerisable resin, **d** fused deposition modelling of thermoplastic, and **e** direct ink writing of viscoelastic ink, after [13]

moves up by a defined height, typically in a range of 50–200 μm. The resin bath is then refilled with additional resin and the process repeated for each successive layer, until the part is complete.

Stereolithography enables the fabrication of complex 3D structures ranging from a few microns (Fig. 9.3a) to several metres (Fig. 9.3b) with a high resolution (50–200 μm) and surface finish comparable to traditional manufacturing techniques. Due to the fine resolution of the SLA process, the fabrication of larger components such as that shown in Fig. 9.3b can be time consuming. This shortcoming of the SLA process was addressed by the introduction of digital light projection (DLP) [17] and continuous liquid interface production (CLIP) processes [18]. Instead of using a single point exposure technique, DLP and CLIP project and cure the entire layer using digital mirror devices or dynamic liquid crystal masks.

However, the increase in speed and build volume for these techniques results in a reduction of the printing resolution. Other techniques based on photopolymerisation have also been proposed. For instance, the two-photon [21] and three–photon

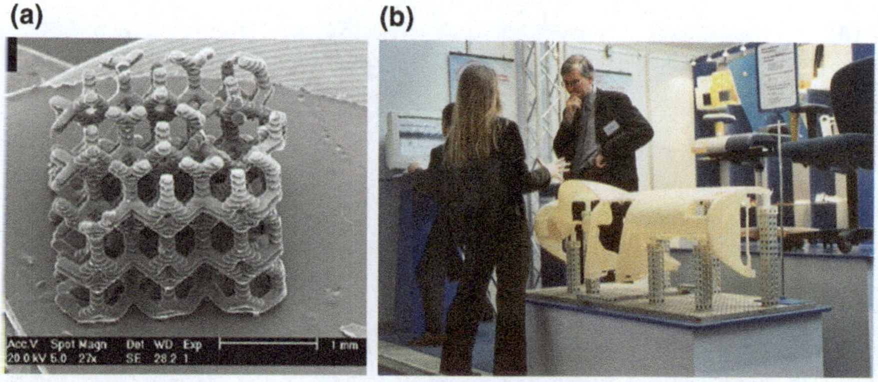


**Fig. 9.3** **a** SEM image of SLA-produced scaffold [19] and **b** a large automotive part presented by Materialise at Euromold 2000 [20]

stereolithography [22] approaches have been demonstrated to produce parts with submicron resolution.

The application of the stereolithography based techniques for design and fabrication of polymeric architectured materials focuses on studying nanocomposites, 3D metamaterials and hierarchical structures such as nanolattices. In particular, the effect of the size of the hierarchically arranged constituents can be assessed and linked to the mechanical properties and damage tolerance of the materials produced. For instance, Meza et al. [23] utilised two-photon stereolithography to fabricate hierarchical nanolattices and perform in situ nanomechanical deformation experiments. Figure 9.4a shows the design principle of a hierarchical nanolattice, which is based on patterning of self-similar unit cells in a fractal-like geometry. In the example shown (Fig. 9.4b), the edges of an octahedron are subdivided into smaller-size octahedra, which, in turn, are subdivided into still smaller octahedra. Using two-photon stereolithography, the authors were able to fabricate samples with the density of less than 1 kg/m$^3$. By studying the different levels of hierarchy, it was shown that strength and the elastic modulus of the hierarchical nanolattices scale with the relative density, approaching the theoretical **linear** relation. Cyclic compression testing, Fig. 9.4c–e, also revealed that hierarchical nanolattices can have improved recoverability and damage tolerance compared to simple periodic lattices.

Bulk manufacturing of such hierarchical structures is desirable for many engineering applications, including energy absorbing panels and biomedical devices. Zheng et al. [24] reported a modification of the traditional two-photon stereolithography and incorporated a spatial light modulator with a coordinated optical scanning system. By combining this method with the electroless nickel deposition technique, bulk metamaterials were produced. These hierarchical metamaterials possess critical features across seven orders of magnitude, over the length scale ranging from 10 nm thick walls of the struts as the smallest building block to the size of a printed bulk sample approaching several centimetres. The specific tensile strength of such samples relative to the density of the lattices was found to be two orders of magnitude higher than

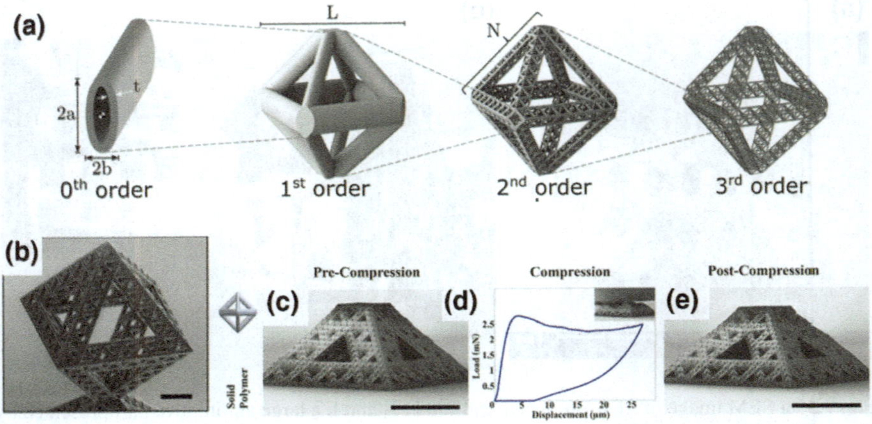


**Fig. 9.4** **a** CAD images of the design principle of third-order hierarchical nanolattice, **b** scanning electron microscopy (SEM) image of a third-order octahedron of octahedra of octahedra (scale bar: 25 μm), and **c**, **d**, **e** images showing half-cell of a third-order octahedron of octahedra of octahedra subjected to cyclic compression to 50% strain (scale bars: 50 μm), after [23]

that of aluminium foams and lattices [24]. It is believed that this result is achieved due to the activation of different deformation mechanisms (viz. bending-stretching-dominated) at different length scales. This development paves the way to mimicking architectural features found at different length scales in biomaterials, such as bone or wood, and can help tailoring and utilising different deformation mechanisms at different length scales.

Another example of the use of the stereolithography technology is the fabrication of polymeric nanocomposites. In a recent study by Kotz et al. [25], a nanocomposite consisting of amorphous silica nanoparticles and hydroxyethylmethacrylate monomer was used to print a polymerised composite. Debinding and sintering of this composite resulted in the formation of silica glass, see Fig. 9.4. This work demonstrates that stereolithography can be used as an enabling tool for processing and fabrication of complex architectures and can widen the materials palette for 3D printing.

Further examples involve the use of nanoparticles such a carbon nanotubes [26], graphene oxides [27] or titanium dioxides [28]. The addition of nanoparticles to polymers enables the creation of functional polymers with enhanced mechanical, electrical or thermal properties. This helps overcome the inherently weak mechanical properties (particularly ductility) of photocurable polymers due to their crosslinked nature and resultant chain mobility. Sandavol et al. [26] reported that the addition of 10 wt% of multi-walled carbon nanotubes to an epoxy resin resulted in an increase in the ultimate tensile strength and fracture stress by 7.5 and 33%, respectively, compared to unfilled printed parts. However, it was found that a uniform dispersion of the multi-walled carbon nanotubes by mechanical mixing or ultrasonic dispersion was difficult to achieve, and the formation of agglomerates remains an issue. Furthermore, it was observed that the improvements in strength were accompanied by a deterioration in elongation to failure and thus a more brittle behaviour. Additionally, different volume fractions of reinforcing particles required new sets of SLA processing parameters because of changes in chemistry, surface tension and viscosity. Lin et al. [27] reported that a small addition of graphene oxide (0.2%) to EnvisionTEC proprietary SLA resin could simultaneously increase the tensile strength and elongation to failure by 62.2 and 12.8%, respectively. These improvements in the mechanical properties were explained by the random alignment of graphene oxide particles and an increase in crystallinity of reinforced polymer [27]. A study by Duan et al. [28] demonstrated that not only does the addition of $TiO_2$ nanoparticles improve the mechanical properties such as tensile and flexural strength, but it also has a positive effect on the thermal stability of the nanocomposite.

It should be noted that none of the stereolithography based methods allow multimaterial printing in a single build process. A manual procedure of swapping resin tank and washing off the unpolymerised resin [29], as well as some automated systems, have been reported [30]. Several examples of the fabricated parts are shown in Fig. 9.5.

Despite promising results offered by multi-material stereolithography, issues associated with laser shadowing, surface tension and interface formation require further research [30] (Fig. 6).

## 9.2.2 Selective Laser Sintering (SLS)

Figure 9.2b shows a schematic of the selective laser sintering process. In this process, fine polymeric powder particles are spread on a build platform and sintered by the application of a high-power $CO_2$ laser beam. Once the printing of a layer is completed, the build platform moves down by about 100 μm and the next layer of powder is spread on top of it using a roller or blade. As distinct from SLA, semi-crystalline thermoplastic polymers are used. The feedstock material for SLS processing are a spherical powders with a particle size distribution of 30–100 μm. The size distribution of the powder controls the minimum thickness of a layer, and a minimum printed feature size of 100 μm has been reported [32]. The regular shape

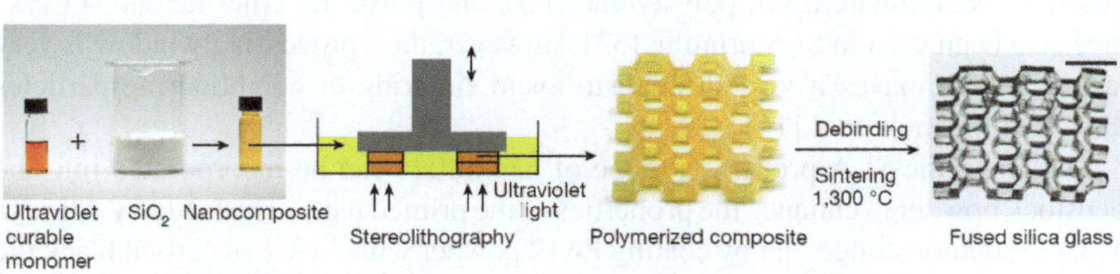


**Fig. 9.5** Stereolithography printing and post processing of photocurable silica nanocomposite, after [25]

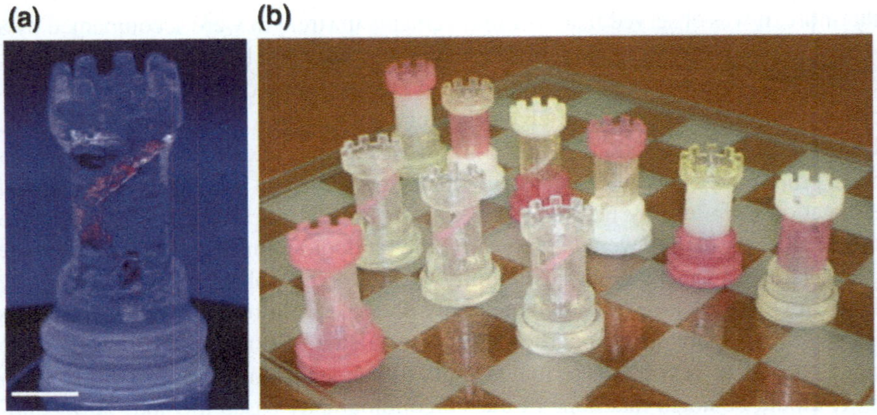


**Fig. 9.6** **a** Multi-material chess rook with embedded fluorescent microspheres in a staircase arrangement fabricated using stereolithography system [31]. The scale bar is 5 mm. **b** Multi-material rooks fabricated using three different resins, after [30]

of the powder particles assists with the flow and spread characteristics, which is important for achieving the highest possible packing density [11]. The SLS process also makes use of preheating of the powder at a temperature close to melting point to facilitate the sintering process and minimises the required energy. Furthermore, preheating reduces accumulation of residual stresses due to thermal gradients and inhibits distortion of a printed part. When the printing is completed, the part is cooled in a controlled manner to reduce the formation of residual stresses and prolong the crystallisation time, thus improving the mechanical properties of the printed part. The unsintered powder which acts as a support structure during the printing process can be removed and recycled for subsequent printing jobs. Fabrication of fully dense parts by SLS processing depends on the careful selection of the thermoplastic polymer and control of the processing parameters, including the laser power, speed and scan pattern, as well as the heating and cooling cycles during the printing process. The most common materials used in SLS are various polyamides (PA12, PA6, PA10). Their printed strengths approach those of injection moulded thermoplastics, but their ductility is significantly lower [33]. An example of a printed functional part is shown in Fig. 9.7a. Other thermoplastic materials such as acrylonitrile–butadiene–styrene (ABS), polycarbonate (PC), polystyrene (PS), and polyether ether ketone (PEEK) have also been used in SLS printing [33]. However, their processing window is very narrow, which makes it very difficult to avoid sintering of neighbouring particles outside of the melt pool [16].

Similar to the SLA process, a range of nanofillers can be incorporated into the feedstock powder to enhance the properties of the printed parts. The study by Athreya et al. [35] demonstrated that by coating PA12 powder with 4 wt% of carbon black the electrical conductivity of the printed nanocomposite was five orders of magnitude higher than that of pure PA12. It was also noted that even after tuning the process

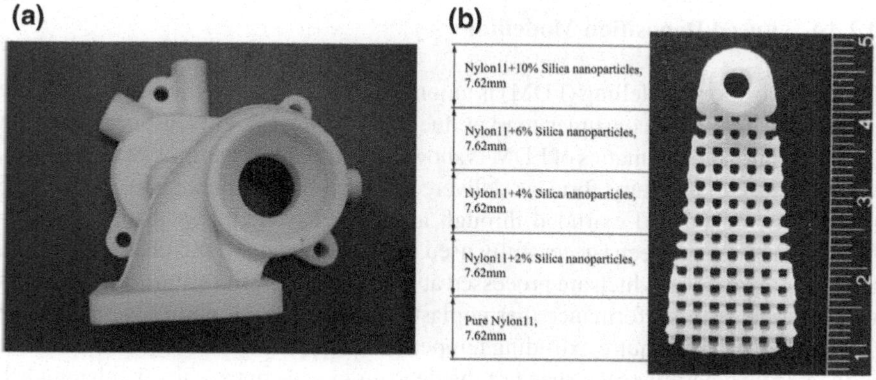


**Fig. 9.7** **a** Example of a functional part printed by SLS using PA12 powder [33]. **b** Example of a functionally graded part printed by SLS using PA11 and silica particles [34]

parameters, a density of around 97 and 96% for pure PA12 and the nanocomposite was attained. The lower density of the nanocomposite was attributed to the formation of weak polymer/filler interfaces and changes in crystallisation behaviour. Other examples of the use of fillers to produce 3D printed nanocomposites include the utilisation of glass beads [36], graphite nanoplatelets [37], $TiO_2$ [37] or $Al_2O_3$ particles [38] and result in improved mechanical properties.

SLS printing was also employed for the fabrication of functionally graded polymer nanocomposites. This is a challenging technical task, since powders with various content of fillers need to be produced and optimisation of the SLS processing parameters is required for each composition. In addition, manual swapping of the powders is required. Despite these difficulties, a proof of concept study by Chung et al. [34] demonstrated the feasibility of this technique. Composites of PA11 with up to 10% volume fraction of silica nanoparticles were produced and components with gradually changing composition were fabricated, see Fig. 9.7b. Due to the limitations of the SLS process, a gradient can only be produced in the vertical building direction. In order to explore the whole potential of the SLS process for manufacture of architectured polymeric materials, further research into alternative powder delivery systems is needed.

### 9.2.3 *Extrusion-Based Systems*

All extrusion-based AM systems have in common that the material required to build an object in layer-by-layer fashion is deposited through a nozzle, Fig. 9.2d. This process is controlled by a computer which prescribes a predefined trajectory of the nozzle that continuously deposits the material in liquid form. The feedstock can either be liquefied in the extruder at a set temperature or be supplied as a viscoelastic fluid that needs to undergo a chemical reaction to trigger solidification.

#### 9.2.3.1 Fused Deposition Modelling

Fused Deposition Modelling (FDM) is among the most popular processes of additive manufacturing in the consumer market due to the relatively low cost of equipment and material. The schematics of FDM is shown in Fig. 9.2d. A thermoplastic material, typically in a filamentary form, is delivered to a vertically mounted extruder. The material is melted and extruded through a nozzle with a chosen diameter. Many thermoplastics have been successfully used in FDM printing. It includes ABS, PLA, PC, and polyamides, which are processed at lower melting temperature (<240 °C), as well as some high performance thermoplastics such as PEEK and Polyetherinmide (ULTEM®), which require extruding temperatures above 350 °C [39]. The diameter of the nozzle determines the speed of the deposition and controls the precision [40]. Typical nozzle diameters range from 100 to 500 μm. Larger nozzle diameters can easily be integrated into FDM provided the gantry system can support the additional weight of the extruder without sacrificing its accuracy of movement. FDM is capable of producing features in the order of 50 μm. The processing parameters affecting the print quality and the mechanical properties, including the layer thickness, fill pattern, printing orientation, and air gap, have been extensively discussed in the literature [41, 42].

One inherent limitation of FDM is the use of feedstock in filament form, although some commercial printers which use pellets have been proposed. The filaments are produced using a twin-screw extruder, which requires a careful control to obtain a filament with the uniform diameter. Dimensional variations in the diameter of the filament would result in pressure drops in the FDM extruder and lead to an interrupted flow of the material and loss of accuracy. Despite this limitation, FDM printing is widely used for fabrication of fibre reinforced composites. Its practical applications include printing of large structures such as automotive panels or even the whole automotive body, see Fig. 9.1a.

Two strategies are commonly investigated with the aim to improve the mechanical properties of fibre reinforced 3D printed composites: the inclusion of short fibres such as glass fibres [43], carbon fibres [44, 45] or the use of co-extrusion technique and the incorporation of continuous fibres [46]. For instance, Tekinalp et al. [44], studied reinforcement of ABS with short (0.2–0.4 mm) carbon fibres by varying their volume fraction from 0 to 40%. The upper limit of 40% is due to clogging of the printer nozzle and disruption of the polymer flow for higher content of fibres. It was shown that 3D printing promotes alignment of fibres in the printing direction, Fig. 9.8a. Increasing fibre content raises the tensile strength and the elastic modulus of 3D-printed samples, see Fig. 9.8b. The highest values were obtained for an ABS carbon fibre composite with 40% fibre loading. It should be noted that despite void (air gap) formation inherent in FDM printing, the properties of the FDM printed composites are comparable with those of the compression moulded ones with the same volume fraction of the fibres, see Fig. 9.8b. This behaviour is attributed to the ability of the FDM process to preferentially align short fibres, which highlights some of the benefits of 3D printing. In addition, reinforcing particles improve the thermal

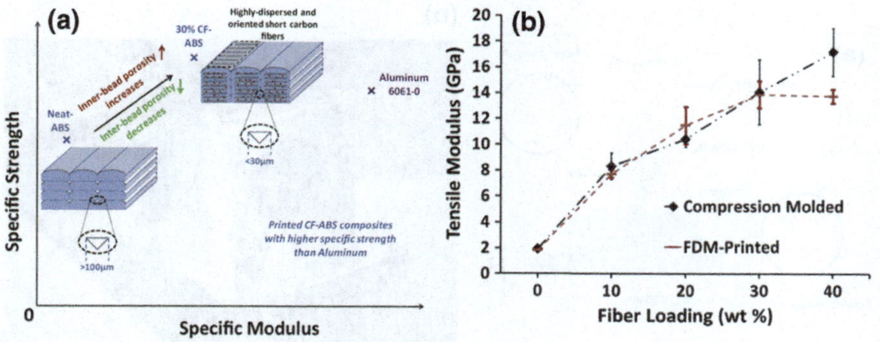


**Fig. 9.8** **a** Schematic representation of the effect of reinforcement of ABS with short carbon fibres on specific strength and modulus of elasticity demonstrating the reduction of the void formation and improved mechanical properties. **b** Effect of the fibre volume fraction on the tensile elastic modulus of samples produced by 3D printing and by conventional compression moulding

stability of the printed materials [47], reduce warping, and allow printing of large scale parts.

Another study conducted by Ning et al. [45] used shorter length carbon fibre reinforcement of ABS (average fibre length of 100 and 150 μm) and varied the fibre volume fraction from 0 to 15 wt%. It was concluded that for specimens with 150 μm long fibres, the best tensile properties were obtained for specimens with carbon fibre content of 5 wt%. Any further increase of the fibre content resulted in a deterioration of the mechanical properties due to the increase of porosity. It was also shown that by increasing the average length of the fibres, the mechanical properties of the 3D printed composite structure improve. Hence, the studies [44, 45] demonstrated the possibility of optimisation of the mechanical properties of fibre reinforced 3D printed composites with regard to the length, volume fraction, and orientation of the fibres.

An extreme case of the fibre reinforced 3D printed composites includes the integration of continuous fibres [46, 48–50]. Several mechanisms for the delivery of continuous fibres were proposed, including co-extrusion, Fig. 9.9a, and deposition through additional nozzles, Fig. 9.9b. In both cases, improvements in the mechanical properties over those of short fibre reinforced composites were reported, see Fig. 9.10. For instance, Tian et al. [46] and Matsuzaki et al. [50] produced a continuous fibre-reinforced PLA composite with 27 and 6.6% fibre volume fraction. The obtained composite samples with 27% fibre volume fraction exhibited a five-fold increase of the flexural modulus—from 6.26 to 30 GPa—and an increase in the tensile elastic modulus and strength to 19.5 GPa and 185.2 MPa, respectively [46].

Similarly, Melenka et al. [49] fabricated a continuous fibre-reinforced nylon composite with 10% fibre volume fraction using a dual extrusion process and also reported an increase in the Young's modulus from 0.35 to 9 GPa. Figure 9.10 shows a comparison of the elastic properties and strength of 3D printed materials obtained by FDM, SLA, and SLS [50]. Continuous carbon fibre-reinforced thermoplastics obtained by FDM exhibit a combination of strength and Young's modulus superior to other 3D

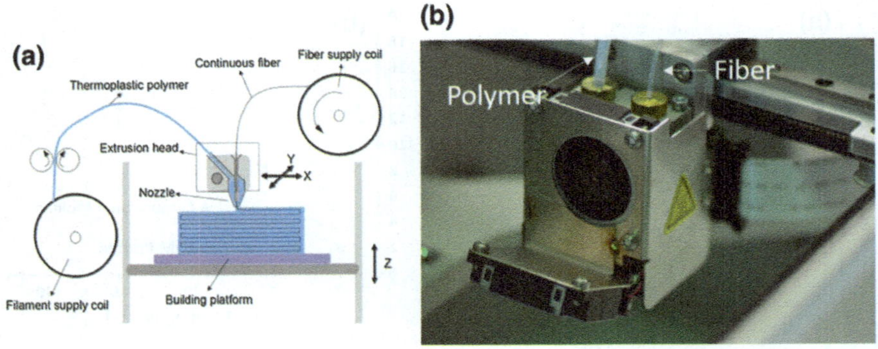


**Fig. 9.9** **a** Schematic representation of the Fused Deposition Modelling with continuous fibre [46] and **b** extruder design for delivery of the polymer and fibre from a commercial desktop 3D printer by MarkForged

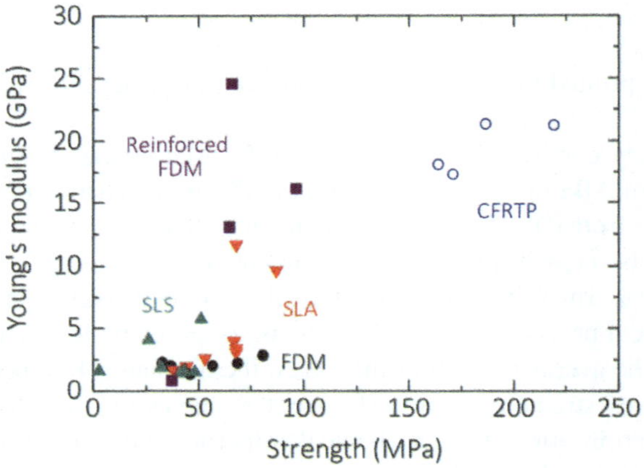


**Fig. 9.10** Comparison of Young's modulus and strength of 3D printed materials obtained by FDM, SLA and SLS, after [50]. Continuous carbon fibre-reinforced thermoplastics (CFRTP) possess a superior combination of strength and Young's modulus compared to other 3D printed samples

printed samples. However, these values are still below those of continuous carbon fibre-reinforced polymers produced by traditional methods, e.g. autoclave processing. These differences are associated with the formation of weaker interfaces between the matrix and the reinforcing agent, as well as the inherent porosity of the FDM samples. Furthermore, due to the discrete layering nature of the FDM printing, the placement of the continuous carbon fibres is limited to the x-y plane for dual extrusion process and discontinues when the printing of the next layer commences.

Further examples of adopting FDM for architectured polymeric materials include integration of metallic particles such as iron or copper [51]. The presence of these metallic particles can improve the thermal properties of the composite materials and permits their use in functional parts, e.g. in injection moulding dies [51].

Owing to the relative ease of use, FDM printing offers a viable manufacturing technique for intricate and complex geometries such as lattice structures [52], auxetic materials, or 3D printed shape memory polymers. For instance, Senatov et al. [53] fabricated PLA based porous scaffolds for bone replacement with the addition of 15 wt% of hydroxyapatite (HA) and studied their mechanical properties and the shape memory effect they exhibit. HA particles were incorporated in the PLA matrix to promote the osseointegration properties, while at the same time enhancing the crack resistance of the material. FDM printing was used as an effective tool to fabricate scaffolds with a designed architecture and control the shape memory effect of the polymer [54]. It was shown that porous scaffolds can tolerate several compression-heating-compression cycles and recover up to 98% of the shape after the deformation cycles. The ability of the shape memory polymers to switch between different shapes upon an application of the external stimuli such as heat opens up the possibility of creating functional materials for a broad range of applications including biomedical devices and soft robotics [55]. Recently, van Manen et al. [56] demonstrated that FDM printing is capable of programming and controlling the shape shifting mechanisms. The principle is based on utilising the printing temperature and layer thickness to control the expansion and shrinkage of the deposited rods, see Fig. 9.11. Printing multiple layers with a predefined pattern can then dictate the shape shifting upon application of an external stimulus. Examples of transition from a flat printed pane to a self-folding dodecahedron and cube are shown in Fig. 9.11b, c, respectively.

FDM printing is also suitable for integrating multi-materials in the same build. Most of the printers have two extrusion nozzles and therefore can deposit two distinct

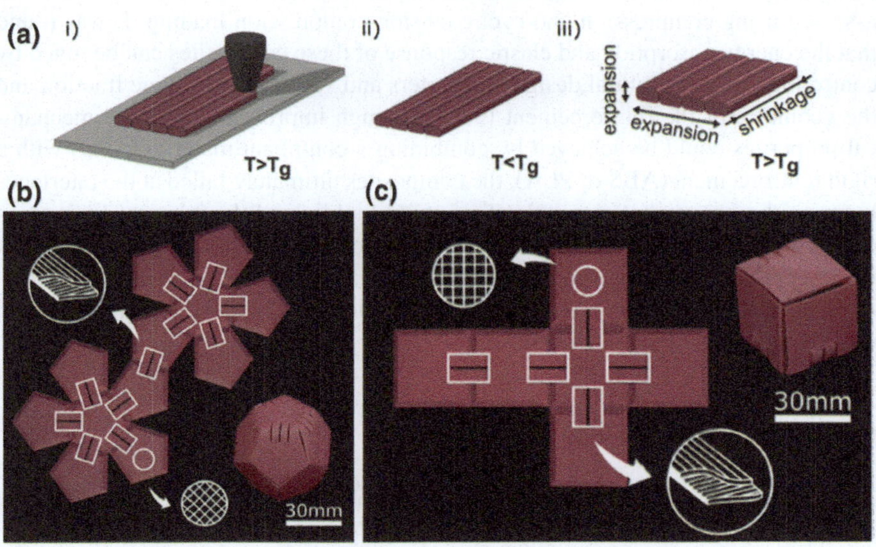


**Fig. 9.11** **a** Deposition process during FDM printing showing the extrusion of the material above its glass transition temperature ($T_g$), solidification and subsequent reheating above $T_g$ to induce the shape memory effect, after [56]. **b** and **c** Self-folding dodecahedron and cube, respectively

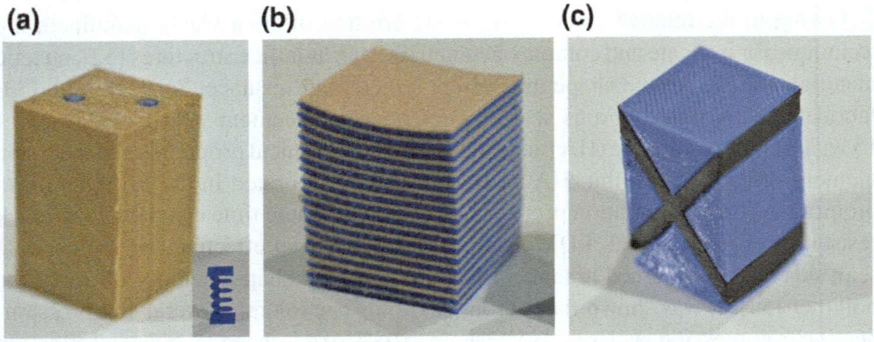


**Fig. 9.12** **a** Thermoplastic polyurethane (TPU—brown colour) block reinforced with two compressive springs made from PLA (blue colour). The inset shows the spring geometry. **b** PLA/TPU layered composite and **c** PLA block with an integrated ABS reinforcement (black colour), after [58]. The sample dimensions are 30 × 30 × 60 mm$^3$

thermoplastic materials [57]. Examples of printed multi-material objects are shown in Fig. 9.12.

The macroscopic composites shown were obtained by printing with combinations of different thermoplastics such as polyurethane (TPU), PLA, and ABS while integrating reinforcements with bespoke geometries. For instance, Fig. 9.12a shows a TPU block with an integrated compressive spring made from PLA, see inset in Fig. 9.12a. A structure of alternating soft TPU (brown colour) and hard PLA (blue colour) layers is shown in Fig. 9.12b. The performance of these composites was assessed using compression and cyclic tension-compression loading. It was found that the energy absorption and elastic response of these composites can be tuned by controlling macrostructural design parameters and varying the volume fraction and the geometry of the reinforcement [58]. Although improvements in the mechanical properties could be achieved by combining a compliant material (TPU) with a rigid reinforcement (ABS or PLA), the composites ultimately failed at the interfaces between the constituent materials. This study highlighted the inherent limitations on the resolution of FDM printing. A further drawback is the formation of weak interfaces due to shrinkage and thermal warping, as well as the occurrence of porosity. These phenomena limit the use of this technique in multi-material printing and design of architectured materials.

#### 9.2.3.2 Direct Ink Writing (DIW)

Another attractive extrusion-based printing technique is Direct Ink Writing (DIW). This process is based on the deposition of *viscoelastic* inks through a nozzle using a precision pump (Fig. 9.2e), which can accommodate a broader range of polymeric materials in comparison to FDM. The extension of the pallet of usable materials is one of the key advantages of this printing technique, which makes it a promising technology for manufacturing architectured materials

These printable inks need to be able to flow upon the application of a shear stress and then rapidly consolidate after the deposition on the build platform or a previous layer. This behaviour is governed by the rheological properties of the inks, which can be influenced in various ways. For instance, using shear-thinning inks, which are characterised by a decrease of viscosity with shear rate, allows the material to be pushed through a syringe nozzle. However, most of the existing polymers exhibiting such behaviour do not possess an elastic shear modulus sufficiently large to retain their shape upon the deposition, especially when the design of the printed structures involves self-supporting features. This shortcoming can be overcome by chemically modifying the polymer (e.g. with methacrylate groups) and the use of photopolymerisation process [11]. Other possibilities include thermal curing, gelation or liquid evaporation, phase transition and incorporation of reinforcement particles [59]. It was shown that such rheological requirements can be fulfilled for a number of natural and synthetic hydrogels [60], epoxy resins, silicon resins [61] and others. It has been found that features down to 1 μm can be achieved using DIW [59]. Figure 9.13 shows several examples of architectured materials fabricated using DIW technique.

Direct Ink Writing opened up a possibility of fabricating structures with intricate features which can exhibit a negative Poisson's ratio. While common engineering materials have a positive Poisson's ratio, materials with a negative Poisson's ratio (or auxetic materials) expand transversally when the material is subjected to tensile loading, see Chaps. 3 and 4. Clausen et al. [62] utilised topology optimisation and created a variety of auxetic materials with Poisson's ratio values varying from −0.8 to 0.8. Samples based on these designs were fabricated using poly(dimethylsiloxane) (PDMS) silicone material, see Fig. 9.13a. They could perform their function over a larger range of strain compared to existing auxetic materials [62].

Another example of computationally-driven design to improve a particular property is the work on architected materials for trapping elastic strain energy, Fig. 9.13b [63]. When elastic beams shown in Fig. 9.13b are deformed under compression, they can buckle and store the deformed energy. Once the applied force is removed the beams can return to their original arrangement. Remarkably, the mechanism of energy absorption only depends on the geometry of the designed beam structures and is independent of the material used [63].

Direct Ink Writing is also an attractive fabrication technique to create lightweight cellular composites. Like the FDM, SLA and SLS techniques, Direct Ink Writing can be applied for 3D printing of fibre reinforced polymers. One of the main benefits of DIW over other 3D printing techniques is its ability to align the reinforced particles or fibres in the printing direction. Compton et al. [65] used epoxy-based inks loaded with nano-clay platelets (1 nm thick, 100 nm in length). The addition of these small platelets assisted in modifying the rheology of the base ink and obtaining the desired shear thinning behaviour. Furthermore, silicon carbide whiskers (0.65 μm in diameter, 12 μm mean length) and carbon fibres (10 μm in diameter, 220 μm mean length) were added to the inks to improve the mechanical properties of the composite, see Fig. 9.14a.

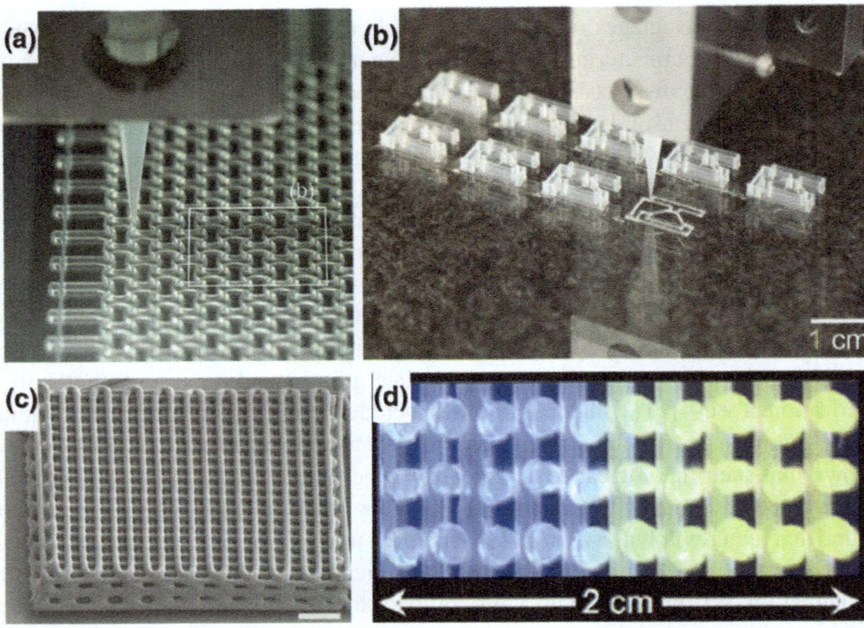


**Fig. 9.13** **a** 3D printing of a structures with programmable Poisson's ratio, after [62]. **b** Elastic beam structures designed for trapping elastic energy (after [63]). **c** An example of a printed periodic lattice structure with 1 μm rod diameter, after [59]. **d** A cross-sectional view of a 3D rectangular lattice structure under UV radiation showing a continuously varying compositional gradient. The structure was printed by mixing two inks, clear und fluorescent, in a specially designed active mixing nozzle, after [64]

When these reinforcement particles are pushed through the nozzle of a printer, their high aspect ratio helps to preferentially arrange them into the printing direction. This approach can be used to obtain synthetic materials with anisotropic mechanical properties, as is often found in natural materials such as balsa wood [65]. Figure 9.14b shows an example of printing of triangular honeycomb composite. A comparison of the mechanical properties of the printed honeycomb composites with other 3D printed materials and balsa wood confirms that preferential alignment of the reinforcement particles is a viable strategy to control the microstructure of the architectured cellular composites. Gladman et al. [60] used the same principle to mimic the dynamic behaviour of plants which are able to respond to environmental changes such as humidity and open or close their leaves. A hydrogel composite ink was loaded with stiff cellulose fibrils and printed using the DIW process resulting in anisotropic material. After subsequent swelling, this material could be programmed to assume a desired shape.

While several studies demonstrated the benefits of DIW in aligning filler particles in the printing direction, the control of placement of such particles in other directions is also desired so that architectures found in some other biological materials could be replicated. One possibility to achieve this goal is to use printing techniques such as 3D

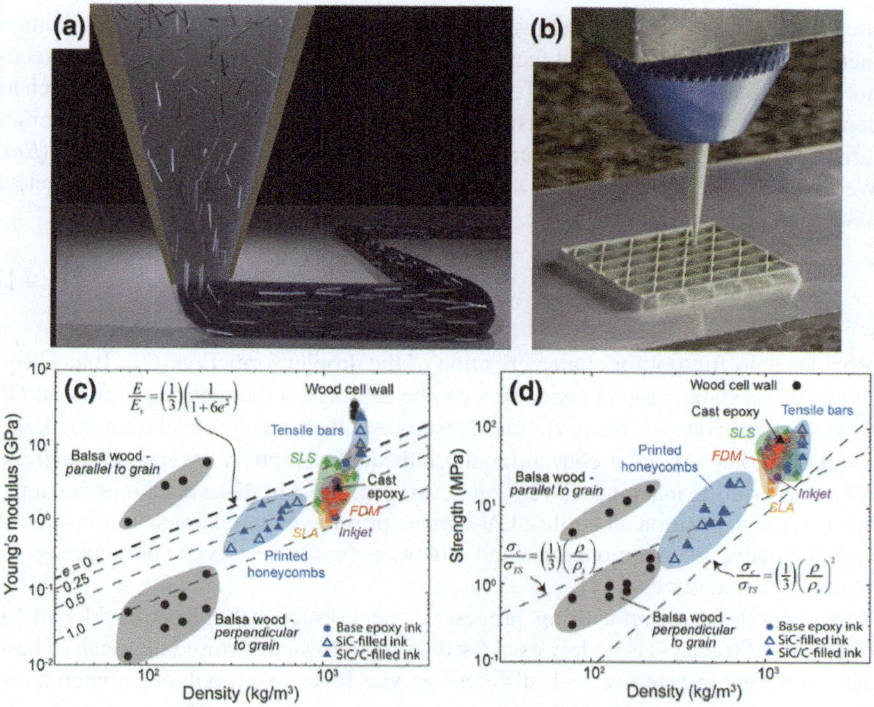


**Fig. 9.14** **a** Illustration of printing of inks with high aspect ratio reinforcement particles and their alignment in the printing direction, after [65]. **b** Example of printed honeycomb composites using DIW. **c** and **d** Comparison of mechanical properties of 3D printed materials and balsa wood

magnetic printing [66] where a magnetic field can be used to control the orientation of anisotropic particles. However, this method also has some limitations, e.g. with regards to the volume fraction of the reinforcement particles, which is restricted to about 15% [67]. Recently, 3D printing with integrated rotational nozzle has been proposed as a way to address these issues [68].

## 9.2.4 Inkjet Printing Technology

Another versatile printing technique which relies on inks is inkjet printing. Conventional two-dimensional inkjet printing has been around since the early 1960s. It has found application in text and graphics printing and is useful in many consumer products, including packaging, food labelling, construction goods, etc. [11, 69]. The main advantage of this technique is the precise placement of small droplets of fluid (typically below 100 picolitres). The printing process is based on the generation of small droplets in the printhead, their deposition on the substrate and subsequent solidification. In recent years, inkjet printing has progressed to a 3D printing fabrication method which allows the manufacture of 3D objects. 3D printing using inkjet technology relies on careful control of the liquid jetted material and its properties such as density $\rho$, viscosity $\eta$, surface tension $\gamma$, velocity of the ejected droplets and characteristic droplet length $a$ [69]. Several physical constants, namely the Reynolds (*Re*), Weber (*We*), and Ohnesorge (Oh) numbers linking the aforementioned parameters have been proposed. In particular, the *Oh* number,

$$Oh = \frac{\eta}{\sqrt{\gamma \rho a}}, \tag{9.1}$$

provides a useful tool for characterisation of the droplet formation [69]. It has been shown that a stable droplet generation can be achieved if the reciprocal value of *Oh* ranges between one and ten [69]. Based on this established theoretical framework and the existing know-how of conventional printhead development, several commercial 3D inkjet printing machines have been developed. Most of these machines decouple the material deposition and utilise UV-curable printing materials such as mixtures of high-molecular weight monomers and oligomers (urethane acrylate or methacrylate resins, urethane waxes) [11].

Among these manufacturing processes, the PolyJet technology developed by Objet (now Stratasys) is widely used for research into architectured materials. These machines enable printing with different acrylic-based photopolymer materials in 16 μm-thick layers with printing heads containing more than 1000 individual nozzles. The deposition technique using small droplets of photocurable resin led to the development of multi-material printing. Figure 9.15 shows an example of Stratasys Connex500 printer which allows printing with 140 different materials. The multi-material capabilities and voxel precision deposition offered by this technology sparked a great interest of the architectured materials community, as it allows mimicking nature-inspired structures/materials (bone, nacre, etc.) and fabrication of compositionally graded and heterogeneous materials.

Examples of the use of this 3D printing technique include the design of damage tolerant structures with a brick-and-mortar architecture adopted from nacre or bone [70–76], morphing or active materials inspired by plants [77–79], control of aerodynamic and hydrodynamic properties of the surfaces [80, 81], and many others. In these examples a soft and a hard phase are combined within a structure. Through controlled local placement of these constituents the mechanical properties (stiffness, strength, and toughness) of the composite structure can be tuned. Furthermore, additional functionalities such as shape memory (morphing) ability [79, 82, 83] can be achieved.

The soft phase can be simulated using the commercially-available, acrylic-based photopolymer TangoBlackPlus shown in black in Fig. 9.15b, while the hard phase can be printed using VeroWhitePlus (shown in white). These polymers possess a stiffness ratio of about 1500 and have been utilised in a number of studies. For instance, Dimas et al. [70] explored the behaviour of hard platelets coated with a thin layer of soft polymer (around 250 μm in thickness) arranged in bone-like, biocalcite-like, and

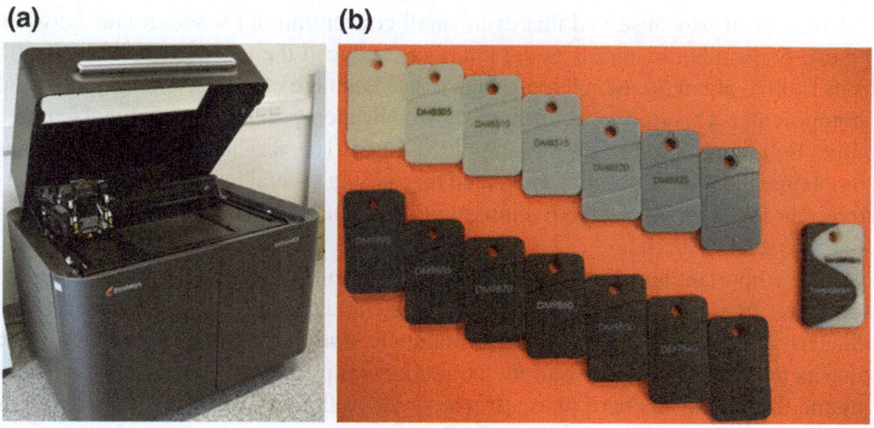

Fig. 9.15 Stratasys Connex500 3D printer at Monash University (**a**) and a set of materials obtained by mixing a soft rubber-like photopolymer (black) with a rigid polymer (white) (**b**)

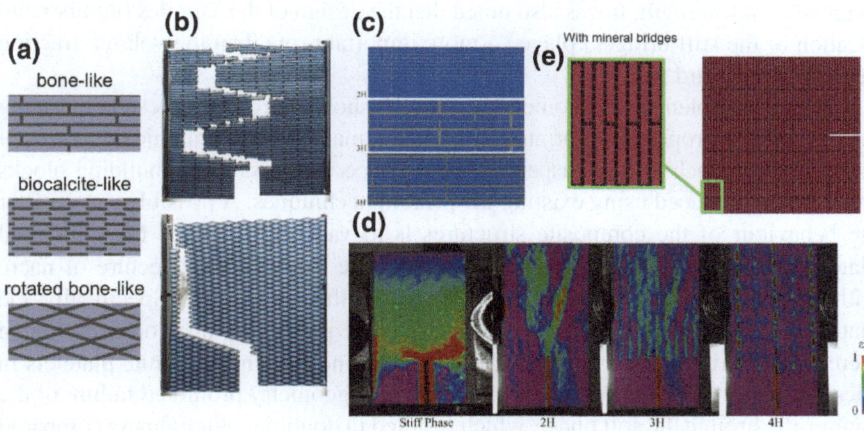


Fig. 9.16 **a** Bio-inspired composite structures obtained using bone-like, biocalcite-like, and rotated bone-like assembly of hard platelets and 250 μm thin layers of soft polymer. **b** Fracture paths in the 3D printed bone-like and rotated bone-like samples, after [70]. **c** Introduction of different levels of structural hierarchy into bone-like composite and the visualisation of the strain field during a mode I tensile test (**d**), after [70]. It is seen that samples with the third and fourth level of hierarchy are capable of delocalising the strain, thus increasing the toughness of the composite structure

rotated bone-like manner, as shown in Fig. 9.16a. These 3D printed assemblies were tested under mode I loading conditions and it was shown that the nature-inspired designs create a favourable fracture propagation path and therefore possess a higher fracture toughness compared to their monolithic counterparts. Notably, the experimentally obtained fracture toughness of these bio-inspired composites exceeded the theoretical predictions based on the rule of mixtures [70].

However, it was observed that even small concentration (3–4%) mixes between the two constituent materials at the interfaces reduced their effective stiffness ratio from 1500 to about 50, which represents a disadvantage of the current acrylic-based photopolymer. This study was further extended by incorporating hierarchical features of the 'building blocks' of the structure [75], Fig. 9.16c, d. The experimental studies complemented with simulations revealed that adding several levels of structural hierarchy can further improve the fracture toughness due to the increased delocalisation of strain, see Fig. 9.16d.

Other important features which contribute to the improved mechanical properties of nature's materials, such as mollusc shells, include mineral bridges connecting the platelets. Gu et al. [84] performed a systematic study of the effects caused by the introduction of stiff bridges into the bone-like composite structure and varying the volume fraction of the hard phase. It was revealed that introduction of stiff bridges alters the stress and strain fields and leads to a change in the load transfer. The stiff bridges carry additional load and change the crack path from a zig-zag one, shown in Fig. 9.16b, to a block-wise pattern. This behaviour results in a further increase of toughness and strength. It was also noted that the design of the samples (number and location of the stiff bridges) played a more important role than the volume fraction of the soft and hard phase.

While the implemented geometrical features showed promising results in tuning the mechanical properties of printed composites, many other hierarchical features of nacre or bone, such as nano-asperities and surface waviness of the building blocks cannot be reproduced using existing 3D printing techniques. A possible way to alter the behaviour of the composite structures is to vary the shape of the individual platelets. Djumas et al. [85] proposed to combine the basic architecture of nacre with the concept of topological interlocking described in Chap. 2. Specifically, the matching concavo-convex contact surfaces of the osteomorphic blocks presented there were used to mimic the natural surface waviness of the aragonite platelets in nacre, see Fig. 9.17. The 3D nature of the platelet geometry promoted failure of the composite through the soft phase, which resulted in doubling of toughness compared to that of the traditional brick geometry of the block.

Bio-inspired composite structures mimicking bone or nacre architecture are also attractive materials for applications requiring high impact resistance [86] and energy dissipation [73]. Gu et al. [86] studied the impact behaviour of printed nacre-like composite consisting of eight plies of stacked up alternating brick and mortar assemblies, see Fig. 9.18a. A drop tower experiment combined with finite element modelling revealed that impacting the stiff bulk material with a projectile travelling at a speed of 2.3 m/s leads to a catastrophic failure and projectile penetration. By contrast, the bio-inspired composite structure was able to defuse the deformation and stopped the projectile, Fig. 9.18b. The observed superior impact performance is attributed to the complex crack path which propagates through the stiff and soft materials. The same group of researchers [87] also investigated the impact performance of conch shell inspired composite, Fig. 9.18c. Conch shells have a ten times higher toughness compared to nacre due to their sophisticated hierarchical cross-lamellar architecture [88]. Mimicking this lamellar architecture and fabricating specimens with two orders

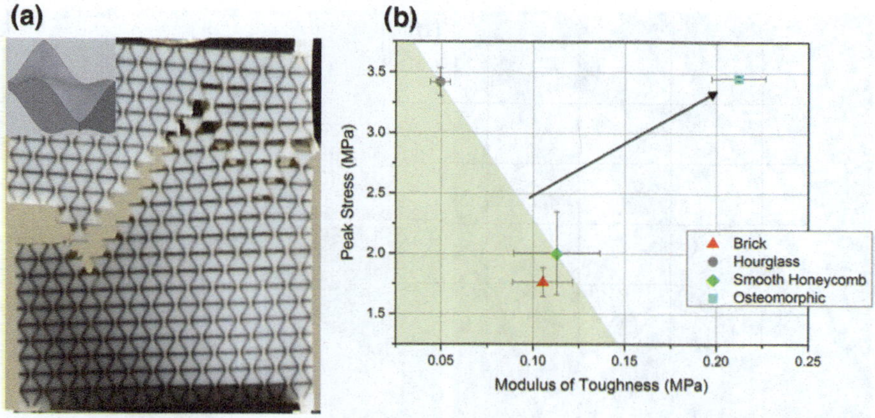


**Fig. 9.17** **a** Fracture path in a bio-inspired composite produced using topologically interlocking platelets (see inset), after [85]. **b** The plot of toughness versus peak stress for four different geometries of the platelets shows that the osteomorphic design results in simultaneous doubling of peak stress and toughness compared to the conventional brick geometry. Sample dimensions are 64.75 × 58.5 × 3.125 $mm^3$. The designation of the block geometries in the diagram on the right-hand-side is defined in [85]

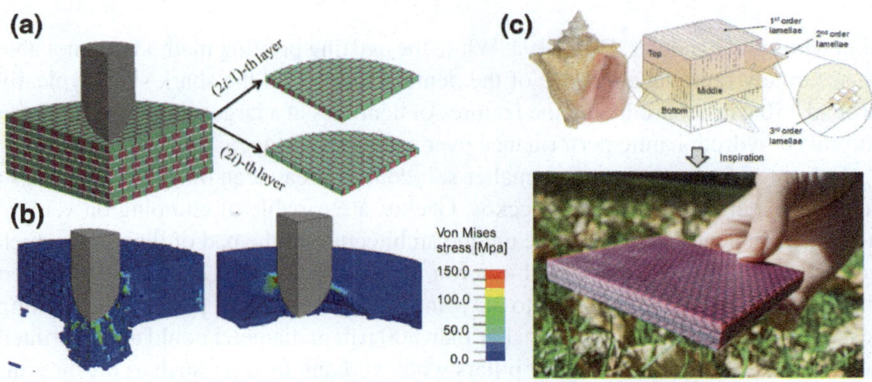


**Fig. 9.18** **a** Bio-inspired composite structures mimicking nacre architecture showing alternating layers. The stiff and soft phases are shown in magenta and green, after [86]. **b** Failure mode and stress distribution in the composite structure (left) and bulk sample (right). **c** Hierarchical features of a conch shell (top) and additive manufactured specimen for impact testing, after [87]

of hierarchy, it was found that the impact performance could be improved up to 85% compared to the stiff constituent.

Another major benefit of the PolyJet technology is the fine resolution of deposited layers, down to 12 μm. This opens up the possibility to mimic some of the functional surface features found on the skin of the animals, thus allowing tuning of hydrodynamic [80], aerodynamic [81] and frictional behaviour of the architectured structure [89]. For instance, Wen et al. [80] fabricated a flexible biomimetic shark

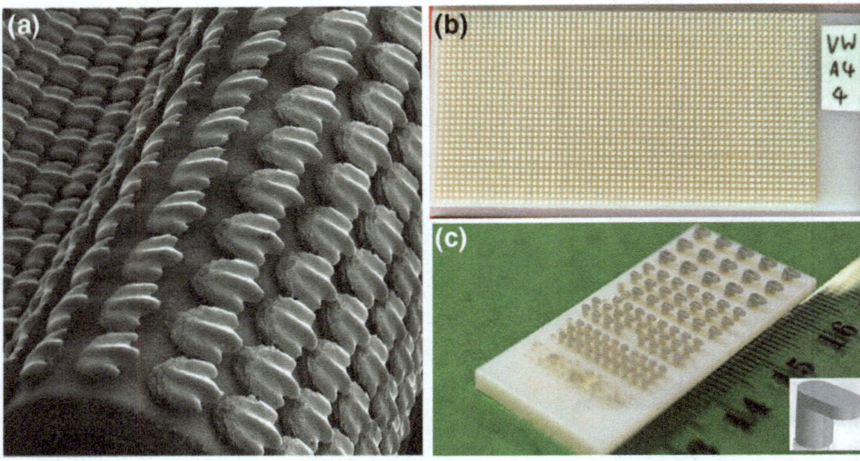


**Fig. 9.19** **a** Biomimetic flexible shark skin with integrated rigid denticles each measuring 1.5 mm in length, after [80]. **b** Example of an array of pillars mimicking a fine structure of a gecko's toe with a diameter and height of pillars equal to 0.5 and 2 mm, respectively. **c** 3D printed spatula tips with diameters decreasing from 2 to 0.5 mm

skin with rigid denticles, Fig. 9.19a. While the existing printing methods are not able to accurately reproduce the size of the denticles found on the shark skin (typically around 150 μm), reproducing the features of denticles at a larger scale resulted in an improved hydrodynamic performance over a smooth surface.

Splitting of a large area into smaller sub-contact areas seen on shark skin is also found in other animals such as geckos. Geckos are capable of climbing on vertical and horizontal surfaces due to the unique architecture of the pad of their toes which consists of a large number of sub-micron fibrils [40, 90, 91]. Zheng [89], studied the ability of PolyJet technology to 3D print such geometries of pillars. The printing test demonstrated that features smaller than 500 μm in diameter could not be printed reliably and it was also noted that pillars with overhang features such as mushroom-shaped tip or a spatula tip, shown in Fig. 9.19c, require support material whose removal is difficult and can damage the intricate features of the printed structure. Zheng [89], studied the effect of pillar size on the frictional properties of the structured samples shown in Fig. 9.19b. The total contact was kept constant and the diameter of the pillars was decreased from 2 mm (128 pillars) to 0.5 mm (2048 pillars). It was shown that the static coefficient of friction increases with a decrease in the pillar size. Overall, the limitations of PolyJet technology in producing features in the sub-micron range were highlighted [89]. The study also established that the requirement for placement of support material for overhang features limits the applicability of the PolyJet technology to replicate many of the features found in natural composites.

Another application of multi-material 3D printing using PolyJet technology is in morphing or responsive materials inspired by plants [77–79] or 4D printing [79, 82, 83]. Guiducci et al. [77–79] exploited the difference in swellability of two base materials shown in Fig. 9.15b. The rigid and stiff material based on polyurethane (VeroWhite) is not-swellable, while the soft and rubbery material (TangoBlack) can swell. The printer allows the user to mix them in different proportions and obtain digital materials with tunable properties, notably the degree of swellabiity. Inspired by unfolding action of the ice plant seeds, Guiducci et al. [77] fabricated swelling-honeycomb actuators, Fig. 9.20a, which are able to expand in a specific direction upon swelling and produce different types of movements. Studying the design parameters, such as wall thickness of the honeycombs and placement of the soft inclusions, by analytical and finite element models allowed the researchers [77] to demonstrate that various aspects of actuated movement can be tailored through the geometry of an assembly. The swelling of a multi-material printed part has also been used to create origami-like shape changes, for instance a self-assembly of various 3D shapes including cubes [83], pyramids [92], truncated octahedra [82] and so on.

Other examples of applications of PolyJet technology include the use of shape memory polymers (SMP) in the form of fibres incorporated into an elastomeric matrix [93], see Fig. 9.20b. The precise placement of SMP fibres, control of their size and orientation, and subsequent thermomechanical training permit printing of composite structures able to follow complex folding patterns, see Fig. 9.20c. For example, Ge et al. [92] developed a theoretical model which takes the aforementioned parameters into account, and provides design guidelines for fabrication of active composite materials. These composite materials target applications in soft robotics where such tasks as actuation or locomotion can be performed using the elastomeric materials.

Multi-material 3D printing has also been used to create compliant mechanisms [94–96] and auxetic materials [97]. For example, Bafekrpour et al. [96] designed machine-augmented composites (MACs, [97]) able to convert compression displacement into shear displacement. Guided by FEM simulations, the shape of the struts of MACs resembling the letter "J" was proposed. The performance of these MACs was linked to the selection of the materials (with suitable elastic moduli) of the embedded 'machines' and the matrix. Applications of such MACs in impact energy absorbers, mechanical sensors and actuators are envisaged. An interesting spin-off of this research is the development of devices exhibiting strongly direction-dependent friction was also reported [95].

Overall, multi-material ink-jet technology has many advantages for fabrication of architectured materials. It enables precise jetting of small droplets of a liquid polymer which allows local control of the composition of the material. However, the required rheology of the material dictated by the design of the printheads limits the use of reinforcing particles (which have to be smaller than 100 nm in size) or other functional additions to the ink. Furthermore, the existing commercial printers utilise proprietary inks, which hinders the development of the next generation of materials with improved mechanical and functional properties.

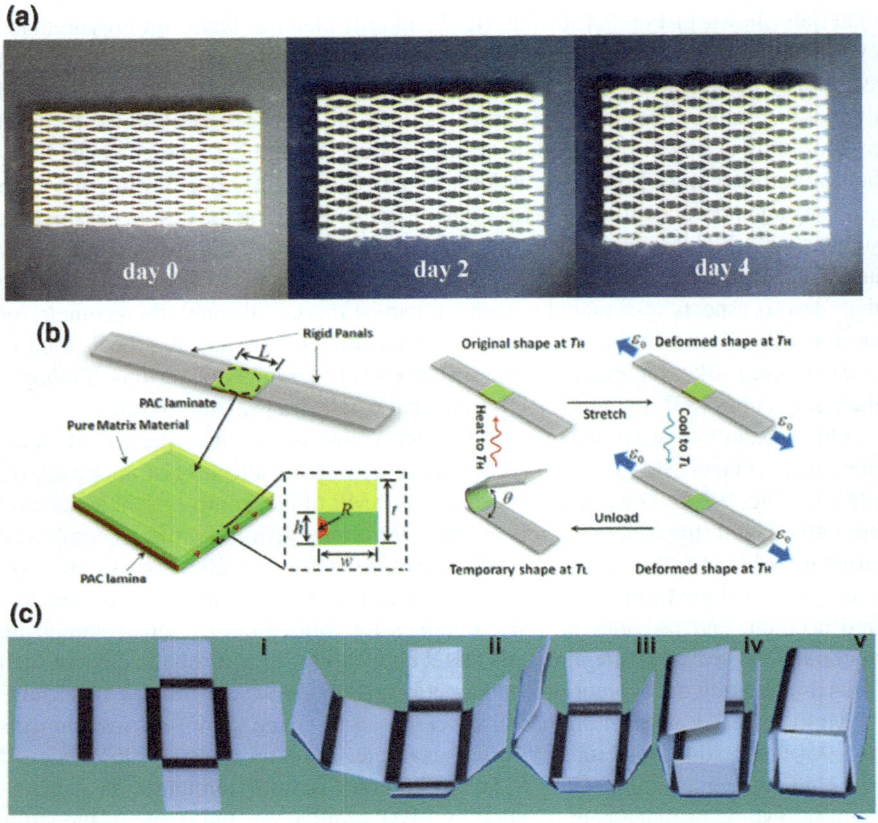


**Fig. 9.20** **a** Unfolding of a honeycomb structure after four days due to swelling of the soft phase placed in the centre of the honeycomb, after [77]. **b** Design of printed active composites with elastomeric matrix and rigid fibres exhibiting a shape-memory effect and a schematic of the thermomechanical training, after [92]. **c** 3D printed flat box with active hinges exhibiting folding upon heating, after [93]

## 9.3 Conclusions and Future Outlook

In the last decade, additive manufacturing emerged as an enabling fabrication method to design, study and produce the next generation of polymeric architectured materials. AM allows the control of the deposition of constituent materials in a desired spatial pattern and can combine multiple materials and locally tune the properties of the resulting material. These attributes have led to an increased use of AM for engineering the architecture of materials. In this chapter, recent advances in the use of additive manufacturing for the design of architectured materials have been presented.

A variety of the existing additive manufacturing techniques, including fused deposition modelling, selective laser sintering, stereolithography, inkjet printing and some emerging techniques such as 3D magnetic printing [66] or direct ink writing with integrated rotational nozzle [68] have been successfully used to produce architectured materials. Among those, functionally graded polymer nanocomposites, multi-material compliant mechanisms, auxetic materials and bio-inspired composite structures mimicking nacre architecture are of much current interest. A small selection of these research successes has been reviewed in this chapter. Each of the 3D printing techniques presented has its own advantages and drawbacks and to take a full advantage of these techniques, the existing hardware, software and materials need to be improved. Most of the commercially-available printers have a limited closed-loop feedback control and their resolution is below a critical level of 100 nm. This resolution is required to tap into the plethora of structures found in Nature where the hierarchical features span many length scales—from nano to macroscale—and control of 3D printing over this range is needed [98–102]. Techniques for translating a CAD file to a 3D printing file format also require additional research efforts; some promising concepts of voxel-based fabrication have recently been reported [103].

The use of multiple materials in the design of architectured polymeric structures is desired to produce heterogeneous structures replicating some of the sophisticated inner architectures and hierarchical features found in natural materials. However, most of the existing 3D printing platforms can only accommodate a limited number of proprietary materials, which limits their application. Further improvement of the mechanical and functional properties of printable materials and development of more flexible 3D printing hardware are necessary for the progress in this area.

Multi-materials design also poses a challenge for computational modelling which is needed to guide and inform future development of archimats. Specifically, current optimisation techniques are only capable of optimising functional properties of a material by using only two or three input materials [104, 105], while voxel-based fabrication will produce truly heterogeneous structures. New methods of topology optimisation emphasising the role of interfaces in the design of heterogeneous archimats are discussed in Chap. 6. Nascent methods of machine learning [106] offer a potent tool for the search for architectured materials engineered over several length scales and employing multiple input materials.

**Acknowledgements** The authors would like to thank their colleagues, collaborators and former students, especially Ehsan Bafekrpour, Lee Djumas, Yunhe Zheng, and Nathan Way who contributed to some of the content shown.

## References

1. S. Weiner, H.D. Wagner, The material bone: structure-mechanical function relations. Annu. Rev. Mater. Sci. **28**(1), 271–298 (1998)
2. P. Fratzl et al., Structure and mechanical quality of the collagen-mineral nano-composite in bone. J. Mater. Chem. **14**(14), 2115–2123 (2004)
3. J.D. Currey, *Bones: Structure and Mechanics* (Princeton University Press, Princeton, 2002)
4. P. Fratzl, R. Weinkamer, Nature's hierarchical materials. Prog. Mater Sci. **52**(8), 1263–1334 (2007)

5. B. Pokroy, V. Demensky, E. Zolotoyabko, Nacre in mollusk shells as a multilayered structure with strain gradient. Adv. Funct. Mater. **19**(7), 1054–1059 (2009)
6. M.A. Meyers et al., Mechanical strength of abalone nacre: role of the soft organic layer. J. Mech. Behav. Biomed. Mater. **1**(1), 76–85 (2008)
7. J. Aizenberg et al., Biological glass fibers: correlation between optical and structural properties. Proc. Natl. Acad. Sci. U.S.A. **101**(10), 3358 (2004)
8. R.O. Ritchie, The conflicts between strength and toughness. Nat. Mater. **10**(11), 817–822 (2011)
9. M.F. Ashby, Y.J.M. Bréchet, Designing hybrid materials. Acta Mater. **51**(19), 5801–5821 (2003)
10. Y.J.M. Brechet, Chapter 1 Architectured materials: an alternative to microstructure control for structural materials design? A possible playground for bio-inspiration? in *Materials Design Inspired by Nature: Function Through Inner Architecture* (The Royal Society of Chemistry, 2013), pp. 1–16
11. I. Gibson, D. Rosen, B. Stucker, *Additive Manufacturing Technologies* (Springer, New York, 2015)
12. R. D'Aveni, The 3-D printing revolution. Harvard Bus. Rev. **93**(5), 40–48 (2015)
13. R.L. Truby, J.A. Lewis, Printing soft matter in three dimensions. Nature **540**(7633), 371–378 (2016)
14. C.W. Hull, *Apparatus for Production of Three-Dimensional Objects by Stereolithography* (USA, 1986)
15. R. Raman, R. Bashir, Chapter 6—Stereolithographic 3D bioprinting for biomedical applications, in *Essentials of 3D Biofabrication and Translation*, ed. by A. Atala, J.J. Yoo (Academic Press, Boston, 2015), pp. 89–121
16. J.W. Stansbury, M.J. Idacavage, 3D printing with polymers: challenges among expanding options and opportunities. Dent. Mater. **32**(1), 54–64 (2016)
17. F.P.W. Melchels, J. Feijen, D.W. Grijpma, A review on stereolithography and its applications in biomedical engineering. Biomaterials **31**(24), 6121–6130 (2010)
18. J.R. Tumbleston et al., Continuous liquid interface production of 3D objects. Science **347**(6228), 1349 (2015)
19. C. Heller et al., Vinyl esters: low cytotoxicity monomers for the fabrication of biocompatible 3D scaffolds by lithography based additive manufacturing. J. Polym. Sci. Part A Polym. Chem. **47**(24), 6941–6954 (2009)
20. Materialise, *Materialise's Mammoth Stereolithography—3D Printing on a Grand Scale!* February 28, 2014 [cited 2018 2.07.2018]. Available from https://www.materialise.com/en/blog/materialises-mammoth-stereolithography-3d-printing-on-a-grand-scale
21. B.H. Cumpston et al., Two-photon polymerization initiators for three-dimensional optical data storage and microfabrication. Nature **398**, 51 (1999)
22. M. Farsari et al., Three-dimensional biomolecule patterning. Appl. Surf. Sci. **253**(19), 8115–8118 (2007)
23. L.R. Meza et al., Resilient 3D hierarchical architected metamaterials. Proc. Natl. Acad. Sci. **112**, 11502–11507 (2015)
24. X. Zheng et al., Multiscale metallic metamaterials. Nat. Mater. **15**(10), 1100–1106 (2016)
25. F. Kotz et al., Three-dimensional printing of transparent fused silica glass. Nature **544**(7650), 337–339 (2017)
26. J.H. Sandoval, B.W. Ryan, Functionalizing stereolithography resins: effects of dispersed multi-walled carbon nanotubes on physical properties. Rapid Prototyping J. **12**(5), 292–303 (2006)
27. L. Dong et al., 3D stereolithography printing of graphene oxide reinforced complex architectures. Nanotechnology **26**(43), 434003 (2015)
28. Y. Duan et al., Nano-$TiO_2$-modified photosensitive resin for RP. Rapid Prototyping J. **17**(4), 247–252 (2011)
29. J.-W. Choi, E. MacDonald, R. Wicker, Multi-material microstereolithography. Int. J. Adv. Manuf. Technol. **49**(5), 543–551 (2010)

30. J.-W. Choi, H.-C. Kim, R. Wicker, Multi-material stereolithography. J. Mater. Process. Technol. **211**(3), 318–328 (2011)
31. K. Arcaute, B. Mann, R. Wicker, Stereolithography of spatially controlled multi-material bioactive poly(ethylene glycol) scaffolds. Acta Biomater. **6**(3), 1047–1054 (2010)
32. R.D. Goodridge, C.J. Tuck, R.J.M. Hague, Laser sintering of polyamides and other polymers. Prog. Mater Sci. **57**(2), 229–267 (2012)
33. J.P. Kruth et al., Consolidation phenomena in laser and powder-bed based layered manufacturing. CIRP Ann. **56**(2), 730–759 (2007)
34. H. Chung, S. Das, Functionally graded Nylon-11/silica nanocomposites produced by selective laser sintering. Mater. Sci. Eng. A **487**(1), 251–257 (2008)
35. S.R. Athreya, K. Kalaitzidou, S. Das, Processing and characterization of a carbon black-filled electrically conductive Nylon-12 nanocomposite produced by selective laser sintering. Mater. Sci. Eng. A **527**(10), 2637–2642 (2010)
36. H. Chung, S. Das, Processing and properties of glass bead particulate-filled functionally graded Nylon-11 composites produced by selective laser sintering. Mater. Sci. Eng. A **437**(2), 226–234 (2006)
37. H.C. Kim, H.T. Hahn, Y.S. Yang, Synthesis of PA12/functionalized GNP nanocomposite powders for the selective laser sintering process. J. Compos. Mater. **47**(4), 501–509 (2012)
38. H. Zheng et al., Effect of core–shell composite particles on the sintering behavior and properties of nano-Al2O3/polystyrene composite prepared by SLS. Mater. Lett. **60**(9), 1219–1223 (2006)
39. X. Wang et al., 3D printing of polymer matrix composites: a review and prospective. Compos. B Eng. **110**, 442–458 (2017)
40. E. Kroner, E. Arzt, Gecko Adhesion, in *Encyclopedia of Nanotechnology*, ed. by B. Bhushan (Springer Netherlands, 2012), pp. 934–943
41. B.N. Turner, S.A. Gold, A review of melt extrusion additive manufacturing processes: II. Materials, dimensional accuracy, and surface roughness. Rapid Prototyping J. **21**(3), 250–261 (2015)
42. B.N. Turner, R. Strong, S.A. Gold, A review of melt extrusion additive manufacturing processes: I. Process design and modeling. Rapid Prototyping J. **20**(3), 192–204 (2014)
43. W. Zhong et al., Short fiber reinforced composites for fused deposition modeling. Mater. Sci. Eng. A **301**(2), 125–130 (2001)
44. H.L. Tekinalp et al., Highly oriented carbon fiber–polymer composites via additive manufacturing. Compos. Sci. Technol. **105**, 144–150 (2014)
45. F. Ning et al., Additive manufacturing of carbon fiber reinforced thermoplastic composites using fused deposition modeling. Compos. B Eng. **80**, 369–378 (2015)
46. X. Tian et al., Interface and performance of 3D printed continuous carbon fiber reinforced PLA composites. Compos. Part A Appl. Sci. Manuf. **88**, 198–205 (2016)
47. X. Wei et al., 3D printable graphene composite. Sci. Rep. **5**, 11181 (2015)
48. Y. Chuncheng et al., 3D printing for continuous fiber reinforced thermoplastic composites: mechanism and performance. Rapid Prototyping J. **23**(1), 209–215 (2017)
49. G.W. Melenka et al., Evaluation and prediction of the tensile properties of continuous fiber-reinforced 3D printed structures. Compos. Struct. **153**, 866–875 (2016)
50. R. Matsuzaki et al., Three-dimensional printing of continuous-fiber composites by in-nozzle impregnation. Sci. Rep. **6**, 23058 (2016)
51. M. Nikzad, S.H. Masood, I. Sbarski, Thermo-mechanical properties of a highly filled polymeric composites for fused deposition modeling. Mater. Des. **32**(6), 3448–3456 (2011)
52. A.A. Zadpoor, J. Malda, Additive manufacturing of biomaterials, tissues, and organs. Ann. Biomed. Eng. **45**(1), 1–11 (2017)
53. F.S. Senatov et al., Mechanical properties and shape memory effect of 3D-printed PLA-based porous scaffolds. J. Mech. Behav. Biomed. Mater. **57**, 139–148 (2016)
54. Q. Zhang et al., Pattern transformation of heat-shrinkable polymer by three-dimensional (3D) printing technique. Sci. Rep. **5**, 8936 (2015)

55. Z.X. Khoo et al., 3D printing of smart materials: a review on recent progresses in 4D printing. Virtual Phys. Prototyping **10**(3), 103–122 (2015)
56. T. van Manen, S. Janbaz, A.A. Zadpoor, Programming 2D/3D shape-shifting with hobbyist 3D printers. Mater. Horiz. **4**(6), 1064–1069 (2017)
57. E. David et al., Multi-material, multi-technology FDM: exploring build process variations. Rapid Prototyping J. **20**(3), 236–244 (2014)
58. N. Way, *Additive Manufacturing of Multi-Material Composite Flexible Structures in Department of Materials Science and Enginering* (Monash University, Clayton, 2017)
59. J.A. Lewis, Direct ink writing of 3D functional materials. Adv. Funct. Mater. **16**(17), 2193–2204 (2006)
60. A. Sydney Gladman et al., Biomimetic 4D printing. Nat. Mater. **15**, 413 (2016)
61. Y. Kim et al., Printing ferromagnetic domains for untethered fast-transforming soft materials. Nature **558**(7709), 274–279 (2018)
62. A. Clausen et al., Topology optimized architectures with programmable Poisson's ratio over large deformations. Adv. Mater. **27**(37), 5523–5527 (2015)
63. S. Shan et al., Multistable architected materials for trapping elastic strain energy. Adv. Mater. **27**(29), 4296–4301 (2015)
64. T.J. Ober, D. Foresti, J.A. Lewis, Active mixing of complex fluids at the microscale. Proc. Natl. Acad. Sci. **112**(40), 12293–12298 (2015)
65. B.G. Compton, J.A. Lewis, 3D-printing of lightweight cellular composites. Adv. Mater. **26**(34), 5930–5935 (2014)
66. J.J. Martin, B.E. Fiore, R.M. Erb, Designing bioinspired composite reinforcement architectures via 3D magnetic printing. Nat. Commun. **6**, 8641 (2015)
67. A.R. Studart, Additive manufacturing of biologically-inspired materials. Chem. Soc. Rev. **45**(2), 359–376 (2016)
68. J.R. Raney et al., Rotational 3D printing of damage-tolerant composites with programmable mechanics. Proc. Natl. Acad. Sci. (2018)
69. B. Derby, Inkjet printing of functional and structural materials: fluid property requirements, feature stability, and resolution. Annu. Rev. Mater. Res. **40**(1), 395–414 (2010)
70. L.S. Dimas et al., Tough composites inspired by mineralized natural materials: computation, 3d printing, and testing. Adv. Funct. Mater. **23**(36), 4629–4638 (2013)
71. L.S. Dimas, M.J. Buehler, Modeling and additive manufacturing of bio-inspired composites with tunable fracture mechanical properties. Soft Matter **10**(25), 4436–4442 (2014)
72. E. Lin et al., 3D printed, bio-inspired prototypes and analytical models for structured suture interfaces with geometrically-tuned deformation and failure behavior. J. Mech. Phys. Solids **73**, 166–182 (2014)
73. P. Zhang, M.A. Heyne, A.C. To, Biomimetic staggered composites with highly enhanced energy dissipation: modeling, 3D printing, and testing. J. Mech. Phys. Solids **83**, 285–300 (2015)
74. V. Slesarenko, N. Kazarinov, S. Rudykh, Distinct failure modes in bio-inspired 3D-printed staggered composites under non-aligned loadings. Smart Mater. Struct. **26**(3), 035053 (2017)
75. R. Mirzaeifar et al., Defect-tolerant bioinspired hierarchical composites: simulation and experiment. ACS Biomater. Sci. Eng. **1**(5), 295–304 (2015)
76. F. Libonati et al., Bone-inspired materials by design: toughness amplification observed using 3D printing and testing. Adv. Eng. Mater. **18**(8), 1354–1363 (2016)
77. L. Guiducci et al., Honeycomb actuators inspired by the unfolding of ice plant seed capsules. PLoS ONE **11**(11), e0163506 (2016)
78. L. Guiducci et al., The geometric design and fabrication of actuating cellular structures. Adv. Mater. Interfaces **2**(11), 1–6 (2015)
79. D. Raviv et al., Active printed materials for complex self-evolving deformations. Sci. Rep. **4**, 7422 (2014)
80. L. Wen, J.C. Weaver, G.V. Lauder, Biomimetic shark skin: design, fabrication and hydrodynamic function. J. Exp. Biol. **217**(10), 1656–1666 (2014)

81. A.G. Domel et al., Shark skin-inspired designs that improve aerodynamic performance. J. R. Soc. Interface **15**(139) (2018)
82. T. Skylar, C. Kenny, Programmable materials for architectural assembly and automation. Assembly Autom. **32**(3), 216–225 (2012)
83. S. Tibbits, Design to self-assembly. Architectural Des. **82**(2), 68–73 (2012)
84. G.X. Gu et al., Printing nature: unraveling the role of nacre's mineral bridges. J. Mech. Behav. Biomed. Mater. **76**, 135–144 (2017)
85. L. Djumas et al., Enhanced mechanical performance of bio-inspired hybrid structures utilising topological interlocking geometry. Sci. Rep. **6**, 26706 (2016)
86. G.X. Gu et al., Biomimetic additive manufactured polymer composites for improved impact resistance. Extreme Mech. Lett. **9**, 317–323 (2016)
87. G.X. Gu, M. Takaffoli, M.J. Buehler, Hierarchically enhanced impact resistance of bioinspired composites. Adv. Mater. **29**(28), 1700060 (2017)
88. Y. Jingjie et al., Materials-by-design: computation, synthesis, and characterization from atoms to structures. Phys. Scr. **93**(5), 053003 (2018)
89. Y. Zheng, Design of hybrid materials using multi-material 3D printer, in *Department of Materials Science and Engineering* (Monash University, Clayton, Australia, 2015)
90. M. Kamperman et al., Functional adhesive surfaces with "Gecko" effect: the concept of contact splitting. Adv. Eng. Mater. **12**(5), 335–348 (2010)
91. M. Micciché, E. Arzt, E. Kroner, Single macroscopic pillars as model system for bioinspired adhesives: influence of tip dimension, aspect ratio, and tilt angle. ACS Appl. Mater. Interfaces **6**(10), 7076–7083 (2014)
92. G. Qi et al., Active origami by 4D printing. Smart Mater. Struct. **23**(9), 094007 (2014)
93. Q. Ge, H.J. Qi, M.L. Dunn, Active materials by four-dimension printing. Appl. Phys. Lett. **103**(13), 131901 (2013)
94. A.T. Gaynor et al., Multiple-material topology optimization of compliant mechanisms created via PolyJet three-dimensional printing. J. Manuf. Sci. Eng. **136**(6), 061015 (2014)
95. E. Bafekrpour et al., Internally architectured materials with directionally asymmetric friction. Sci. Rep. **5**, 10732 (2015)
96. E. Bafekrpour et al., Responsive materials: a novel design for enhanced machine-augmented composites. Sci. Rep. **4**, 3783 (2014)
97. G.F. Hawkins, Augmenting the mechanical properties of materials by embedding simple machines. J. Adv. Mater. **34**, 16–20 (2002)
98. F. Javid et al., Dimpled elastic sheets: a new class of non-porous negative Poisson's ratio materials. Sci. Rep. **5**, 18373 (2015)
99. Z. Liu et al., Functional gradients and heterogeneities in biological materials: design principles, functions, and bioinspired applications. Prog. Mater. Sci. **88**, 467–498 (2017)
100. P.-Y. Chen, J. McKittrick, M.A. Meyers, Biological materials: functional adaptations and bioinspired designs. Prog. Mater. Sci. **57**(8), 1492–1704 (2012)
101. M.A. Meyers et al., Biological materials: a materials science approach. J. Mech. Behav. Biomed. Mater. **4**(5), 626–657 (2011)
102. I.H. Chen et al., Armadillo armor: mechanical testing and micro-structural evaluation. J. Mech. Behav. Biomed. Mater. **4**(5), 713–722 (2011)
103. E.L. Doubrovski et al., Voxel-based fabrication through material property mapping: a design method for bitmap printing. Comput. Aided Des. **60**, 3–13 (2015)
104. M. Osanov, J.K. Guest, Topology optimization for architected materials design, in *Annual Review of Materials Research*, ed. by D.R. Clarke, vol 46 (Annual Reviews, Palo Alto, 2016), pp. 211–233
105. P. Zhang et al., Efficient design-optimization of variable-density hexagonal cellular structure by additive manufacturing: theory and validation. J. Manuf. Sci. Eng. **137**(2), 021004 (2015)
106. G.X. Gu, C.-T. Chen, M.J. Buehler, De novo composite design based on machine learning algorithm. Extreme Mech. Lett. **18**, 19–28 (2018)

# Chapter 10
# Mechanics of Arthropod Cuticle-Versatility by Structural and Compositional Variation

**Yael Politi, Benny Bar-On and Helge-Otto Fabritius**

**Abstract** The arthropod cuticle may be seen as a multifunctional material displaying a wide range of physical properties. The materials properties of the cuticle are brought about by compositional and structural gradients at multiple hierarchical levels. In the following chapter we first discuss the main components of the cuticle, namely, chitin, proteins, water, mineral and tanning agents and their relevance in determining the mechanical properties of the cuticle. We then describe the hierarchical organization of the cuticle and how it contributes to tuning the mechanical properties of the material. Finally we show several examples of cuticular structures with increasing structural complexity to exemplify the discussed principles.

## 10.1 Arthropod Cuticle as a Material: An Introduction

The cuticle of the Arthropoda is an intrinsically multifunctional tissue serving both as an integument and a skeleton. It must provide a barrier to the environment and a shell that supports the body and a diverse set of "tools" in the form of functional appendages for locomotion, foraging and defense such as legs, wings, mandibles, and claws. In addition, the cuticle plays an important role in environmental perception and inter- and intraspecific communication with a variety of embedded sensors and by generating color and sound [1–3]. The diverse material properties required to fulfill all of these various functions call for special adaptations and compromises

Y. Politi (✉)
Department of Biomaterials, Max Planck Institute of Colloids and Interfaces, Am Mühlenberg 1, 14424 Potsdam, Germany
e-mail: yael.politi@mpikg.mpg.de

B. Bar-On
Department of Mechanical Engineering, Ben-Gurion University of the Negev, 84105 Beer Sheva, Israel

H.-O. Fabritius
Department Microstructure Physics and Alloy Design, Max-Planck-Institut für Eisenforschung GmbH, Max-Planck-Straße 1, 40237 Düsseldorf, Germany



Y. Estrin et al. (eds.), *Architectured Materials in Nature and Engineering*, Springer Series in Materials Science 282,
https://doi.org/10.1007/978-3-030-11942-3_10

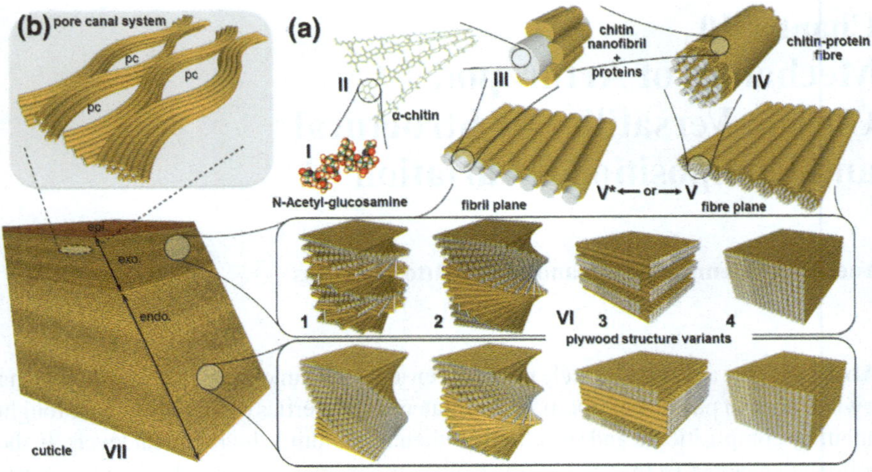


**Fig. 10.1** **a** Schematic depiction of the structural organization of the organic phase in arthropod cuticle. N-Acetylglucosamine molecules (I) polymerize to form anti-parallel chains of α-chitin (II). Chitin crystallites coated with a protein layer form nanofibrils (III) that organize in parallel to form single fibril layers (V*). Alternatively, the nanofibrils can assemble to larger chitin-protein fibres (IV) that form horizontal fibre planes. (V). The fibril or fibre planes are horizontally stacked and form lamellae (VI) that can either be helicoidally stacked forming twisted plywood structures (VI 1, 2, 3) or unidirectionally stacked (VI, 4). Different variants of level (VI) constitute the procuticle layers. Together with the outermost epicuticle, the exo- and endocuticle form the basic material of the exoskeleton (VII). **b** The pore canal system is generated by deposition of fibrils/fibres around cellular protrusions during cuticle formation and it pervades the procuticle in normal direction

making the cuticle one of the most versatile biological materials. The adaptability of the cuticle's material is indeed thought central to the evolution of the more than a million species that render the Arthropoda the largest and probably most diverse, animal group [1]. It is also what makes the cuticle a particularly attractive archetype for bio-inspired multifunctional materials design.

From a materials perspective, the arthropod cuticle is a hierarchical structure composite in which α-chitin crystallites draped by proteins form nanofibrils that organize into higher micro-structural motifs [4] (Fig. 10.1). The physical properties of the cuticle material are consequently determined by the chitin content and the composition of the associated proteins as well as the higher-level structural arrangement. Post-deposition hardening by dehydration and chemical cross-linking or mineralization renders the cuticle its final qualities.

In its role as integument, the cuticle functions as a barrier to the external environment, protecting against desiccation, osmotic pressures and infection by fungi and bacteria. It forms the interface with the external world and many sensors such as mechano-sensors, chemo-sensors and heat (Infrared, IR) sensors are embedded within it. The cuticle's external layer, the epicuticle, also carries contact-pheromones for intra-species communication. As an exoskeleton, the main task of the cuticle is to shape the body and provide muscle attachment sites. Being segmented animals, arthropods have a varying number of body tagmata whose originally uniform morphologies have diversified into a large number of specialized body plans over the course of evolution [5–7]. Consequently, the cuticle forms rigid elements (sclerites) connected by joint structures and flexible arthrodial membranes that enable locomotion. The sclerites hold the muscle attachment sites. Besides providing protection and stability, stiff tubular sclerites are also advantageous for flight as they can sustain high loads at a largely reduced weight. Arthrodial membranes can also support locomotion by storing elastic energy.

Not only the outer surface of Arthropoda is covered by the cuticle. Since the epidermal cells that secrete the cuticle are of ectodermal origin all the internal organs that are ontogenetically descendent of the ectoderm are lined by the cuticle. This includes parts of the digestive system and specialized ectodermal invaginations such as the respiratory system of the insects, the tracheae. Where it covers the lining of the gut, gland tracts, and respiratory structures the cuticle is involved in controlling the diffusion, uptake and secretion of material to and from the environment.

The consequence of having an exoskeleton is that it hinders continuous growth in size. The cuticle has therefore to be shed and replaced by a larger one several times in the animal's lifetime. In this process, called ecdysis, part of the old cuticle is digested for reuse and the rest of the cuticle is shed as an exuvia. Before molting, the new cuticle is deposited by the underlying epidermal cells as an initially undifferentiated layer. It is accepted that the apical end of the epidermal cells forms microvilli at the top of which chitin is secreted into the extracellular space where the interaction with cuticular proteins, secreted from the grooves between microvilli, takes place (for more details see latest review by Moussian [8, 9]). The process of chitin secretion, crystallization and deposition is, however, still poorly understood. It is also still unknown how fiber orientation is controlled, although self-assembly of the fibers into helicoids resembling cholesteric liquid crystals guided by the physical conditions of the extra-cellular space [10] or by specific chitin-binding proteins has been proposed [11, 12]. The newly formed cuticle is highly hydrated and compliant to facilitate leaving the old exoskeleton and the size increase after ecdysis. It will only harden and dehydrate a few hours after molting during the sclerotization and/or mineralization processes. All specialized cuticular "tools" and, importantly, sensors are also replaced and newly formed during molting.

The cuticle that forms the exoskeleton is built of two main anatomical layers: the chitin-devoid epicuticle, and the fibrous procuticle, made primarily of chitin and proteins. The epicuticle is protein-rich and usually a few micrometers thick. It is deposited first by the secretion of cuticulin, a lipo-protein complex that dominates the epicuticle composition in most arthropod groups [13, 14]. This layer is covered by a waxy layer, the envelope, containing multiple fatty acids and long chain hydrocarbons. In some cases it becomes additionally covered by a so-called cement layer of tanned proteins. The envelope and cement layers are deposited after ecdysis, meaning that their components are transported through the already formed exo- and epicuticle by pore- and wax canals, respectively [15].

The chitin-rich procuticle exhibits two and in some cases three anatomical layers: the exo-, the meso- (occurs in arachnids and some parts of insects cuticles), and the endocuticle. The main biochemical difference between the three is the degree of sclerotization, which is the process of cuticle hardening by the enzymatic addition of a wide range of catechol derivatives. Sclerotization also occurs in crustaceans [16]. However, their main mechanism of cuticle hardening is mineralization, most commonly by calcium carbonate and phosphate mineral phases that can occur in either crystalline or amorphous states. Both processes, sclerotization and mineralization, are accompanied by changes in the cuticle water content. Structurally, the procuticle consists of stacked planar layers of protein coated chitin fibers, where a large variability in terms of fiber orientation has been described, as detailed (Chap. 3). The exocuticle is most often built exclusively from helicoidally arranged fiber planes, the so-called twisted plywood or Bouligand structure [17]. The fiber organization in the endocuticle may vary significantly from stacks of parallel fibers to twisted plywood with varying rotation angles between layers.

In the following we outline important aspects of the cuticle biochemical composition (Sect. 10.2) and describe key features in its structural hierarchy and organization (Sect. 10.3). Finally we summarize the current understanding on how composition and structure govern the basic mechanical properties of the cuticle (Sect. 10.4). We finalize with the description of several systems where the relationships between structure, mechanical properties and function are well exemplified (Sect. 10.5).

## 10.2 Cuticle Composition

### *10.2.1 Chitin*

The main component of arthropod cuticle is crystalline chitin, a water insoluble polysaccharide composed of β-1,4-linked N-acetyl-D-glucosamine (GlcNAc) residues. Chitin crystallizes mainly into one of two polymorphs: α and β. They differ in the polymer chain orientation within the crystal. This is parallel in β-chitin, and antiparallel in α-chitin [18, 19]. The polymer chain orientation dictates the density of available inter-chain hydrogen bonds, which in turn determines the stability of the crystal. For this reason α-chitin is the most stable form with multiple inter-chain hydrogen bonding in both the *a* and *b* crystallographic directions [20]. In β-chitin, on the other hand, there is no hydrogen bonding between chains along its *b* axis leading to decreased stability and significant swelling of the crystal in aqueous (and other) solutions [21]. Both polymorphs are found in the animal kingdom, but α-chitin is the most abundant.

Chitin in arthropods is partly deacetylated. In crab-shell, one every 6 amino sugars has been found to be deacetylated [1], and deacetylation degree ranges of 5–25% are found in different species [22]. Deacetylation leaves the polymer chain charged and increases its solubility. This is likely important for the interaction of chitin with the matrix proteins (see below). Chitin is synthesized and translocated across the cell membrane by the trans-membrane enzyme complex chitin synthase. However, the mechanistic details of this process are still unknown [22, 23]. Chitin crystallization and the attachment of chitin-binding proteins occur in the extracellular space, leading to the assembly of the cuticle. However, not much is known about these processes and their control. A recent study has shown that chitin deposition in *Drosophila melanogaster* not only requires chitin synthase but also the activity of two other genes [expansion (exp), and rebuf (reb)]. The products of these genes are localized to the apical cell membrane of the epithelium and are essential for fibril formation in yet an unknown mechanism. Interestingly, the expression of these genes in addition to chitin synthase is sufficient to promote chitin deposition in tissues that normally do not form chitin [24]. Whatever the mechanisms involved, within the cuticle, the chitin crystallites in most arthropods are about 3 nm thick and at least 300 nm in length. Draped by chitin-binding proteins together they form the main structural unit of the cuticle—the nanofibrils.

## 10.2.2 Cuticular Proteins

The versatility and adaptability of the cuticle's properties originate to a large extent from the variability in cuticular protein (CP) composition [25]. However, the molecular relations between the proteins' amino-acid composition and the cuticle's mechanical properties are not yet fully understood [9]. Tens of different proteins may be found in each cuticle type. Some are structural, others are enzymatic and some have both structural and enzymatic domains. Enzymatic functions include deacetylation [12, 26], post-translation modification such as glycosylation [27, 28], oxidation, isomerization and cross-linking of sclerotization agents [29–31]. These enzymes are most likely active during cuticle formation and sclerotization.

Structural CPs were shown to play various roles in cuticle formation and in defining its properties. For example, CPs have been shown to be important for cuticle morphology and for chitin organization into the rotated plywood [11, 26, 27, 32, 33], or vertical fiber arrangement [34] in several model organisms. The disruption of chitin organization then leads to deformed body shape, as the resulting random fiber orientation is unable to provide the anisotropic mechanical properties of the cuticle required for morphogenesis. Such mutations are often lethal, emphasizing the importance of cuticle organization for animal survival. The molecular mechanisms leading to the structural organization of the cuticle by CPs are however yet unclear.

### 10.2.2.1 Chitin Binding Proteins

A few hundreds of arthropods CPs have been identified and classified into about a dozen families based on their amino-acid sequences [35]. Some of these families are found throughout the Arthropoda and others are restricted to particular taxonomic groups. Most of these protein families seem to be restricted to arthropods [35]. It seems to date that all the structural CP contain one of two types of chitin-binding domains (CBD), the R&R domain or the more universal CBD type 2 [9]. The latter is found in enzymatic proteins across many phyla and was also identified in structural proteins from the squid beak [36]. Chitin-binding CPs exhibit at least one CBD, but most families have two or three CBDs [35]. This could, in principle, allow them to bind several chitin crystals.

The most prominent arthropod-specific chitin-binding protein family is the CPR family, named for the presence of the R&R consensus sequence discovered by Rebers and Riddiford in 1988 [37]. The R&R motif has been shown to bind chitin in vitro and to be sufficient to render chitin-binding activity to proteins otherwise unable to bind to the polysaccharide [38]. Several modifications of the consensus sequence have been identified, some of which display a preferential distribution within soft and extensible cuticle (RR-1) whereas others (RR-2) are more frequently found in hard, rigid cuticles [11, 39]. It is suggested that the type of chitin-biding interactions are permanent in RR-2 domains and reversible in RR-1 domains, allowing the proteins to slide along the chitin fiber when the cuticle is extended [40]. The division of R&R motifs into two subgroups and their association with soft and hard cuticles is however not valid for all arthropod cuticles studied [41] and a comprehensive understanding of the origin and significance of the observed variability in CBDs is still lacking.

One of the most intensively studied CPR variant is resilin, a highly elastic protein found in specialized cuticular structures of some insects such as locusts, fleas and beetles [42, 43]. These animals use temporary deformations of specific exoskeletal parts to store elastic energy for sudden release, allowing for very fast movements, such as jumping or flapping of wings [42, 44, 45].

Despite their abundance, it is still unclear how these proteins bind chitin and what are the intermolecular interactions involved. As mentioned above, it is postulated that deacetylation facilitates the interaction of chitin with the proteins, possibly allowing covalent linkage [46]. However, recent proteomic analysis argues that chitin-binding by covalent linkage is unlikely [41]. Other possibilities have also been suggested including hydrophobic interactions and hydrogen bonding [25, 47]. Although the nature of the molecular binding between chitin and the CBP is still unknown, it seems that many CBD's adopt beta sheet conformations that form the interface with the chitin crystal [11, 48, 49]. This was earlier suggested by Fraenkel and Rudall [50] and by Atkins [51], who demonstrated the stereo-chemical complementarity between the chitin crystal structure and repeats within the beta strand structure. Evidence for the abundance of beta-sheet motifs in cuticular protein matrices are numerous (for example [48]), including co-ordering of a beta-sheet motif along the (010) chitin surface and its alignment with respect to the fiber's long axis [52].

Not all cuticular proteins bind chitin, and many are introduced into the cuticle sometime after the chitin-protein fibers are deposited. These proteins are even less well characterized. However, it is postulated that unstructured domains within them are associated with inter-protein interactions [40].

#### 10.2.2.2 Composition of CPs and Protein-Protein Interactions

Characteristic amino acids of CPs are mainly glycine and alanine but also proline, valine and tyrosine [25, 29, 40, 53]. It has been frequently suggested that the balance between hydrophobic and hydrophilic groups in the proteins should have important consequences for the mechanical properties of the cuticle since a high degree of hydrophobicity increases protein-protein interactions and decreases hydration level while the opposite is true for increased hydrophilicity [29, 54]. Indeed there are several indications supporting that soft cuticles tend to be richer in polar and charged amino acid residues as compared to stiff cuticles, which are enriched with hydrophobic amino acids [55]. This view might be however too simplified and ignores the possible increase in hydrophobicity caused by sclerotization [47, 56] and the increased hydrophilicity caused by glycosylation [29].

The molecular interactions within the cuticle and the resulting mechanical properties may be manipulated by hormonally controlled pH-changes. Such an example is the abdomen cuticle of the fifth instar nymphs of the blood-sucking bug *Rhodnius prolixus* that changes its properties from stiff to highly stretchable after a blood-meal in a fully reversible process [57]. Cuticle plasticization in this case was originally related to increased water content, however, it was later suggested that the rapid change in properties results from the weakening of the bonds between proteins following the change in pH [58]. Prior to acidification, the cuticle's pH is maintained close to the isoelectric point of the major proteins present, leading to the strongest inter-molecular binding. Thus, a pH change is a fast and reversible way to alter the strength of inter-protein interactions and thereby to easily manipulate the cuticle's mechanical properties.

A special case of cuticular amino-acid composition is found in the fangs of the spider *Cuppiennius salei*. Although no single protein has yet been sequenced from the fangs, the amino-acid composition was shown to vary gradually from tip to base. While glycine is dominant along the whole length of the fang, the relative proportion of alanine increases from tip to base. Histidine shows the opposite trend and reaches 25% of the total amino acid composition at the fang tip [59]. The rise in His levels is accompanied by increased levels of Zn and Cl. Spectroscopic data shows that Zn coordinates with His residues to cross link the protein matrix [60]. In accordance, increased hardness and stiffness were measured in the Zn rich tip of the fang. The mechanical properties of metal-ion enriched cuticle are comparable to those of mineralized skeletal tissues, such as the lobster carapace, but have lower material density [61]. Other arthropods are also known to incorporate Zn and other metal ions in specialized cuticular tools, such as claws, ovipositors, stingers etc. [54, 61–67]. The most frequent metal ions present in the cuticle are Ca, Mn and Zn, and the concomitant increase in mechanical properties (in the order of 1.5 to threefold in stiffness and hardness compared to similar, but metal-devoid structures) was often demonstrated. How these metal ions are bound within the protein matrix and to which amino acids, is however still unknown.

In grasshoppers, the presence of Zn ions in the mandibles was demonstrated to protect against wear [68]. In this context it is interesting that metal incorporation is often localized to the distal layers (epi- and exocuticle) of the cuticular tool [69, 70]. Metal ion incorporation is often accompanied by halogens such as Br and Cl, however it is still unclear if and how these incorporations are related.

Metal-ion coordination cross-link bonds are stronger relative to hydrogen bonding, hydrophobic or electrostatic interactions. They are weaker though than covalent bonds, but have the advantage that they can easily reform after perturbation [71]. Stable covalent bonding between proteins in the matrix could have a large effect on the matrix properties. Disulfide bonds are likely not common as cysteine is rare in arthropod cuticle [29]. Direct covalent bonds occur in resilin rich cuticles, which form di- and tertyrosine bridges [43, 45, 56]. Other covalent bonding occurs via sclerotization agents, usually via Tyr or Hys residues [72, 73]. The long standing hypothesis of covalent cross-linking of the protein matrix by sclerotization agents has recently been substantiated by Mun et al. [39], who identified a specific CP in the rigid cuticle of the red flour beetle *Tribolium castaneum*. This particular CP lacks CBDs and is cross-linked via catechol derivatives to two CPRs (CPs containing the R&R CBD). Thus, although direct covalent bonds between proteins are not common in the cuticle, the protein matrix can be highly cross-linked via sclerotization (see below).

#### 10.2.2.3 Structure of CPs

CBD have been proposed to fold into various beta-sheets motifs, and large proportion of beta-sheet has been demonstrated in cuticles by FTIR and XRD [48, 52]. Beta sheet motifs are known in many structural proteins with mechanical functionality, the most common example is spider silk in which the presence of beta sheet crystalline domains endow the material with exceptional strength [74]. Nonetheless, many CPs contain various sizes of low-complexity sequence regions predicted as intrinsically disordered and suggested to have functional importance [40]. In contrast to beta sheets, which are highly stabilized by intra-molecular hydrogen bonding of the peptide backbone, disordered regions are highly flexible and allow extensibility and deformability of the proteins enabling the cuticle to store elastic energy. Andersen suggested that under load, stiff sclerotized cuticles would store elastic energy as small deformations of the individual protein molecules, as the protein molecules cannot move relative to each other due to cross-linking. In contrast, proteins in soft, flexible cuticles can be both deformed and they can move relative to each other. The extreme case in which disorder in cuticular proteins is of high importance is the rubber-like protein resilin. Resilin can be extended to 300% of its resting length due to extension of disordered domains. Here, minimal intramolecular interactions occur, but the protein chains are connected by stable di- and trityrosine covalent bonds [43].

### 10.2.3 Sclerotization

In the process of sclerotization, low-molecular-weight compounds impregnate the cuticle drastically changing its properties. This process leads to covalent cross-linking of proteins and chitin [72, 73, 75, 76] via a wide variety of adduct products as well as a significant reduction in the cuticle's water content [77]. Sclerotization usually occurs shortly after ecdysis, although specialized cuticle parts can be sclerotized before ecdysis in a process called pre-ecdysal sclerotization. Sclerotization is often, but not always, accompanied by color change and darkening, hence often referred to as tanning.

Sclerotization is most extensively studied in insects and not much is known about the process in other arthropods. Several reviews on sclerotization describe the variability in sclerotization agents and their formation pathways [31, 78, 79]. In short, the process involves enzymatic addition of quinones produced by the oxidation of N-acylcatechols. The cathecholic precursors Dihydroxyphenylalanine (DOPA), dopamine and a variety of dopamine derivatives are synthesized within the epithelial cells and released to the cuticle at the time of sclerotization. The amino acids that often participate in sclerotization are histidine, lysine and tyrosine linked via their functional groups and glycine and alanine linked by their amino groups. A wide variety of quinones and catechol adducts have been extracted from cuticles of various groups. The type of adducts and their relative abundance vary significantly between cuticular regions within one species, different insect species and between insects and other arthropod taxa. In melanization, a process related to sclerotization, melanin is formed from dopamine precursors resulting in black cuticles. Melanin may be contained in granules or distributed diffusively [80–82]. Melanin polymerization may also result in hardening of the cuticle.

The molecular basis for cuticle hardening by sclerotization is thought to stem from the covalent cross-links between the macromolecular building blocks. It is however well acknowledged that sclerotization is accompanied by a high degree of dehydration, which by itself can account for the observed increase in mechanical properties [83]. It was therefore suggested that the incorporation of phenols into the cuticle drives the dehydration process, allowing the formation of a large number of weak interactions between the proteins [83]. The two mechanisms are, nonetheless, not mutually exclusive.

### 10.2.4 Halogenation

Halogens, mainly Br and Cl, are sometimes found in arthropod cuticle. Andersen has shown that various types of sclerotized cuticle contain halogen-substituted amino acids [29] and in some cases halogens are co-localized with metal ions [59, 69, 70]. The function of halogenation is yet unclear. The participation of halogens in metal ions coordination has been demonstrated in Polychaeta worm cuticles [84], but not yet in arthropods. It has also been suggested that halogenated amino acids may be functionally important during the sclerotization process [30, 78]. Alternatively, Schofield et al. [85] suggested that halogenated cuticles are more resistant to fracture by impact due to reduction in molecular resonant frequencies. They have also shown that unmineralized brominated cuticle in crustaceans has a higher elastic modulus and hardness compared to unmineralized cuticle that does not contain bromine.

### *10.2.5 Mineralization*

Within the Arthropoda, the Crustacea and some groups within the polyphyletic Myriapoda harden the rigid parts of their exoskeletons by incorporating biominerals into their cuticle. In Crustacea, a large variety of mineral phases have been described. The type and phase of minerals found in the cuticle differ not only between species, but also between and within the different procuticle layers. Thus, organic-inorganic composite materials can be formed with properties tailored for specific functions of different exoskeleton parts. The most widespread and commonly found mineral is calcium carbonate. Mostly, it occurs in the form of Mg-calcite and amorphous calcium carbonate (ACC). Calcium phosphates are less abundant and occur mostly in functionally specialized skeletal elements, but also co-localized with calcium carbonates. In crustacean cuticles, amorphous calcium phosphate (ACP), hydroxyapatite and even fluoroapatite have been identified so far [86–91]. In general, Crustacea not only precisely control the type of mineral deposited, but also the localization and amount of mineral incorporated into their skeletal elements. This control is exerted not only on the level of entire skeletal elements or specific layers of the cuticle, but can be very finely used to tailor properties within specialized skeletal elements. Here, different mineral phases and/or morphologies can be co-localized in the same cuticle layer. Also, gradual changes in mineral type and amount are possible both vertically, i.e. within and across the layers as well as horizontally, i.e. at transitions between mineralized and unmineralized parts of cuticle such as articulations between segments [92]. The mineralization process generally starts after ecdysis. Mineral ions are then transported over the epithelium and the pore canal system to the distal regions of the cuticle, where mineralization is initiated. Over the course of time the cuticle mineralizes from the outside to the inside until the process is complete [93]. In many cases, the final mineral phase is formed by a precursor phase, usually ACC [93]. How the animals control the localization and phase of the mineral they form is largely unknown. Morphologically, the mineral in cuticle is either present as crystals of varying sizes or in the form of small spherical particles which are aligned along the organic fibres. Mineral structure thus mostly follows the hierarchical organization of the organic matrix [94, 95]. Mineral morphology can greatly vary between layers and/or species and seems rather determined by the function of the skeletal element. So far, four prominent mineral morphologies have been described that occur on different levels of the structural hierarchy: small, 20–50 nm thick spherical particles loosely distributed around the nanofibrils (Fig. 10.2a) [94] or solid tubes enclosing

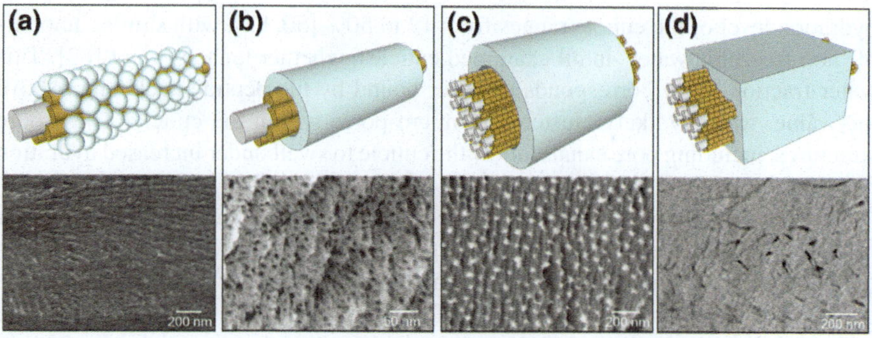


**Fig. 10.2** Schemes and STEM micrographs illustrating different morphologies of the mineral phase observed in various Crustacea species and skeletal elements. **a** Spherical ACC particles (dark) decorating individual nanofibrils (bright) in the endocuticle of the lobster *Homarus americanus*. **b** Individual nanofibrils (dark) enclosed by mineral tubes (bright) as found in the endocuticle of the isopod *Tylos europaeus* (altered from [97]). **c** Larger chitin-protein fibers (bright) surrounded by fused mineral tubes (dark) from the endocuticle of the crab *Cancer pagurus*. **d** Irregularly shaped chitin-protein fiber bundles (dark) embedded in the mineral phase (bright) forming the distal exocuticle of *T. europaeus* (altered from [92])

the nanofibrils (Fig. 10.2b) [96, 97] on the nanofibril level; solid tubes enclosing larger fibres (Fig. 10.2c) [95, 98] or large mineral blocks surrounding irregularly shaped clusters of nanofibrils (Fig. 10.2d) [96, 97] on the chitin-protein fiber level. In the latter two cases, the mineral is fused to a compact block after the completion of the mineralization process. Interestingly, Crustacea are able not only to control the crystalline phase, but also local crystallographic properties of their minerals, such as the orientation and texture. This has been shown for calcite in the exocuticle of decapods and isopods [87, 97, 99]. In all cases the preferred orientation of the mineral is such that the mechanical properties of the layer are enhanced for a specific function, such as increased mechanical strength. Overall, in Crustacea cuticle the mineral phase is governing the final mechanical properties of the cuticle, as has been shown in a number of examples. However, there is an elaborate balance between the contributions of minerals and organic constituents as the chitin-protein-fibers play an important role in rendering the cuticle mechanically anisotropic and providing interfaces for inhibition of crack propagation as discussed below.

### *10.2.6 Hydration*

Water content is a key factor determining the cuticle's mechanical properties. In fact, Vincent and coworkers pointed out that dehydration alone can account for the observed increased stiffening of arthropod cuticles during sclerotization without the need to form covalent cross links between the matrix components [83]. They suggested the view of sclerotization as a process of controlled dehydration. The hydration level of the cuticle ranges from 12 to 50% [50, 83], with similar amounts of ‘non freezable water’ in all examined cuticles, whether tanned or not [83]. This water fraction likely corresponds to water bound by the peptide backbone. Above this value, water is likely clustered in micro-pores within the cuticle. These pore structures, including pore canals, allow the cuticle to swell under increased hydration conditions.

The mechanical properties of fresh, dried and rehydrated cuticles from various taxa have been so far investigated, showing large modulations of properties as a function of hydration level. For example after drying, the endocuticle of locust sternal cuticle hardened by a factor of up to 9 and stiffened by a factor of up to 7.4 [100]. The increase in mechanical properties was less significant in the exocuticle, resulting in the elimination of the gradual decrease in hardness and Young’s modulus from the exo- to the endocuticle, which is characteristic of native cuticle [100]. Dirks and Taylor [101] analyzed the fracture toughness of locust tibiae. They showed that the stiffness and strength increase of cuticle upon water loss compromises its fracture toughness, making the legs more brittle with reduced strain to fracture [101, 102]. In addition, the cuticle’s defect tolerance was highly sensitive to water content; the critical length of a crack was significantly larger in fresh vs. dried leg cuticle. Similar effects have been observed in mineralized cuticle from a number of crustaceans, notably the lobster *Homarus americanus*. Here, water also acts as a plasticizer that provides the cuticle the ability to plastically deform to delay brittle fracture as it occurs in dried material [103–105] and this phenomenon is even more pronounced in unmineralized arthrodial membranes [98].

In summary, variations in the cuticle composition may lead to a large variation in the materials local mechanical properties, for example around six orders of magnitude in Young’s modulus [47]. Modulation of the composition across the cuticle forming an anatomical structure is a path to tailor and fine-tune the mechanical performance of that part. However, the overall properties of the functional element are further governed by its local fiber arrangement and its morphology as discussed next.

## 10.3 Structural Organization

### *10.3.1 Structural Hierarchy*

After discussing the different constituents present in the cuticle, this section shall illustrate how they structurally organize to form the complex architectures observed in nature. Acetylglucosamine and its polymer chitin represent the two lowest levels of the structural hierarchy in arthropod cuticle on the molecular scale (Fig. 10.1, I, II). The first real composite structure appears with the protein-coated chitin nanofibrils (Fig. 10.1 III). They consist of 18–25 chitin molecules arranged in antiparallel α-conformation resulting polygonal section contour with a typical diameter of 2–5 nm at lengths of about 300 nm [106]. Molecular dynamics simulations of chitin fibrils with different chain numbers have shown that the thickness observed in arthropod chitin nanofibrils corresponds to the lowest molecule numbers required to form stable α-chitin [107]. Within their assembly zone, an approximately 1.5 nm thick sheath of protein forms around the presumably self-organizing chitin fibrils. The chitin-protein nanofibrils are the smallest building blocks in the arthropod cuticle and all higher order structural motifs found in the organic phase, with the exception of the epicuticle, are based on them regardless of group or species, skeletal element or function. In the majority of cases, the nanofibrils form the next level of structural hierarchy, the fibril planes (Fig. 10.1, V*). However, in some species the nanofibrils cluster to form larger, clearly defined fibers (Fig. 10.1, V) with diameters between 50 and 250 nm. These have mostly been observed in mineralized crustacean cuticle such as various crab species [93, 95]. How the number of fibrils and thus the size of the larger fibers are determined is not yet understood. Where present, the chitin-protein fibers organize analogous to the fibrils and form planar arrays where the long axes of the fibers are parallel (Fig. 10.1, V).

The lamellar appearance of the procuticle layers is formed by stacking of fibril or fiber layers (Fig. 10.1, VI). Thereby, a large number of configurations are possible, the most common and characteristic one being the twisted plywood or Bouligand structure [7, 17, 108]. It is characterized by the fact that the long axes of the fibrils gradually rotate around the normal axis of the cuticle from one sheet to the next. The distance in which the superimposed fiber sheets complete a 180° rotation is defined as the stacking height or the lamellar thickness of the twisted plywood. The helicity defined by the interplanar rotation angle of the plywood structure can vary anywhere between 90° down to a few degrees (Fig. 10.1, VI 1–3). More rarely, the fibril layers become stacked with the long axes of their fibrils in parallel, forming unidirectional arrangements. Such a cuticular architecture is mostly found in the endocuticle of structures that have to sustain unidirectional loading such as mouth parts [59, 99] leg joints [96], tendons [52] and occasionally also in parts of the body shell [109]. In exocuticle, unidirectionally arranged fibres have so far only been found in the tip of the incisive edge of mandibles from the isopod *Porcellio scaber* [99]. In some cases, combinations of unidirectional and rotating architectures have been observed such as in the cuticle forming the elytra of many beetle species. Here, several unidirectionally organized stacks alternate at an angle of approximately 90° (Fig. 10.1, VI, 3).

The second-to last level of structural organization is the bulk cuticle (Fig. 10.1, VII) that forms the material of which the arthropod exoskeleton in its enormous variety of body plans consists [6]. Classically, the cuticle is subdivided into three main layers from the outside to the inside: epicuticle, exocuticle, and endocuticle [7]. However, more subdivisions have been introduced based on specific morphologies found in certain arthropod groups such as the mesocuticle in chelicerates or the membraneous layer in crustaceans. A concise definition, e.g. based on ontogenetic sequences, is still lacking. It is safe to assume that the usually very thin epicuticle does not contribute significantly to the mechanical properties of the bulk cuticle, but it has many other vital functions for the animals as detailed above. Twisted plywood and unidirectional lamellae can occur in both the exo- and endocuticle. Generally, plywood found in the exocuticle has a smaller stacking height than that forming the endocuticle, but this is merely a rule of thumb and exceptions have been observed [110]. Cuticle is usually horizontally layered, meaning the lamellae are oriented parallel to the cuticle surface, following the curvatures present in skeletal elements. However, there are exceptions from this rule, such as in the pigmented dorsal carapace of the crab *Cancer pagurus*, where the exocuticle is partially substituted by parts of the endocuticle bulging out in distal direction [95]. In these bulges, the twisted plywood lamellae change their orientation from horizontal to vertical.

### 10.3.2 Pore Canals

The already complex architecture of the cuticle's organic matrix is further altered by the pore canal system (Fig. 10.1b). Pore canals are tubular ducts that penetrate the cuticle in vertical direction and form a connection to the underlying epithelium. They function as a transport system through which the epithelial cells can provide components for post-ecdysial hardening such as tanning and mineralization and enable the animals to control the hydration state of the tissue. They are further important for the secretion of wax layers [46] and are thought to play a role for cuticular repair mechanisms [111]. The pore canals originate from cellular extensions present during secretion of the cuticle. During successive deposition of chitin layers, the fibrils are deposited around these extensions, leaving canals with an ellipsoid cross section and a thin inner lining of organic material. If the deposited cuticle has a twisted plywood structure, the pore canals consequently have the shape of twisted ribbons [46]. In different cuticles and even in different layers of the same cuticle the canals can vary in number, spatial distribution patterns, dimensions and shape. In Crustacea, the pore canal system is often particularly well developed, because large amounts of mineral precursors have to be transported to the cuticle during mineralization. Species like the lobster *Homarus americanus* have pore canals that take up almost 40% of the entire cuticle volume in the endocuticle, generating a honeycomb-like structure [112]. This design leads to remarkable mechanical properties and an anisotropic deformation behavior [94, 105]. It is of interest that despite having a higher volume fraction of mineral than *H. americanus* [113], the pore canal system of the crab *Cancer pagurus* takes up less space than that of the lobster. It is therefore possible that in lobster the pore canals also help to reduce weight and thus energy consumption for the animals. In *C. pagurus*, it has been found that the pore canal margins are reinforced by internal mineralized fibrils [95], which can also be interpreted as an adaptation serving mechanical purposes. In some cases, pore canals can also be completely closed by minerals when the post-ecdysial hardening process is completed [97].

### 10.3.3 Structural Colors

In addition to its structural functions, the cuticle also plays an important role for the ecophysiology of the organisms by generating colors used for signaling, camouflage or warning. Arthropods do this by two general principles, the first being incorporation of pigments into the cuticle and the second being modifications of the cuticular architecture on the same length scale as the wavelength range of visible light, resulting in photonic crystal structures [114]. Common pigments found in arthropod cuticle are carotenoids and melanins, which are generating most dark and reddish colours observed. The brilliant green, blue and often also vividly iridescent coloration displayed by many insects are structural colors generated by constructive interference of light reflected by cuticular photonic structures. These are either modifications in structure and chemical composition of the outermost layers of the cuticle, mostly the epi- and exocuticle, or specialized structural elements like scales formed on the cuticle's surface. The colors of many beetle species originate from thin multilayer structures formed by the epicuticle that act like a simple Bragg reflector. Other beetle families use their plywood lamellae as pseudo-Bragg reflectors, exploiting the optical anisotropy of the chitin fibres and their rotation within the tissue to generate a periodic refractive index contrast leading to selective wavelength reflection. By controlling the thickness of the helical lamellae in their outer exocuticle, they can control the reflected color. Uniform thickness of the lamellae leads to pure colors, while variations in thickness, so called "chirped layers", lead to mixed reflection resulting in metallic silver or gold tones [115]. The by far most elaborate photonic crystals occur in cuticular scales such as those present on the exoskeletons of many curculionid beetles and the wings of butterflies. Butterfly scales often combine diffraction gratings with three-dimensional photonic crystals of the gyroid type [116, 117], while the beetles form diamond-type photonic crystals within their scales [116, 118]. The processes involved in the formation of these highly complex nanostructures are still not fully investigated [119].

## 10.4 Consequences of Architecture and Composition for the Mechanical Properties of Cuticle

As introduced above, the cuticle is a complex, hierarchically built bio-composite structure with a number of different architectural motifs spanning from the nanometer to the millimeter length scale. This section draws some of the basic relationships between structural and mechanical characteristics of the cuticle, so as to gain fundamental physical insights to its functionalities from a mechanical point of view.

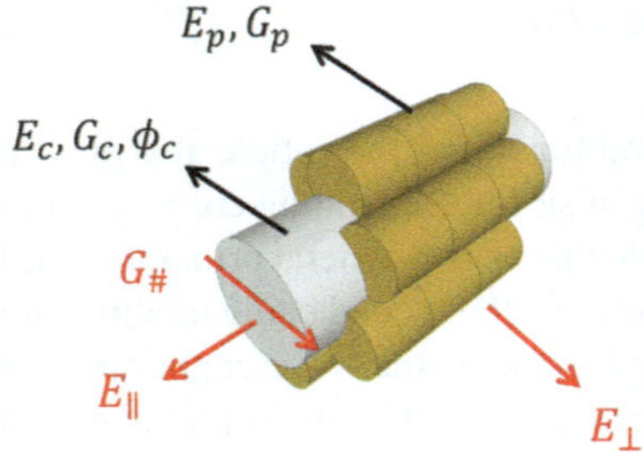


**Fig. 10.3** Schematic illustration of the basic mechanical characteristics of the chitin crystals and the protein matrix that comprise an individual nano-composite chitin–protein fibril. Indicated are the Young's and shear moduli and the volumetric fraction of the chitin crystals ($E_c$, $G_c$, $\phi_c$), those of the protein matrix ($E_p$, $G_p$, $\phi_p$), and the anisotropic axial and shear moduli of the nano-composite ($E_{\parallel}$, $E_{\perp}$, and $G_{\#}$)

### *10.4.1 Resistance to Deformation*

#### 10.4.1.1 The Nano-Composite Level: An Individual Chitin–Protein Nanofibril

The basic structural unit of the cuticle is the individual nanofibril, which is a chitin–protein nano-composite (level III, Fig. 10.1). Thus, by approximation, the elastic behavior of the cuticle at this material level is essentially dominated by the stiffness of the individual phases and their relative content in the composite (Fig. 10.3), namely (a) the resistance to axial tension, i.e., the Young's moduli of the chitin crystallite and the protein matrix ($E_c$ and $E_p$, respectively); (b) their resistance to shape change, i.e., their shear moduli, $G_c$ and $G_p$, respectively; and (c) the volumetric fraction of each phase, $\phi_c$ and $\phi_p$, respectively, such that, $\phi_c + \phi_p = 1$. The alignment of the chitin crystals within the nano-composite gives rise to a clear structural—and thus mechanical—anisotropy, exhibited by substantial differences in stiffness along, and perpendicular to the direction of the chitin fiber long axis. Following the fundamental micro-mechanical models of composite materials (Voigt and Reuss models [120]), the axial moduli of the nano-composite fibril along, and perpendicular to, the chitin fiber direction ($E_{\parallel}$ and $E_{\perp}$) and its shear modulus ($G_{\#}$), as illustrated in Fig. 10.3, can be evaluated with the following direct and inverse rules of mixtures:

$$E_{\parallel} = \phi_c \cdot E_c + (1 - \phi_c) \cdot E_p \quad (10.1)$$

$$E_{\perp} = \left[\frac{\phi_c}{E_c} + \frac{1 - \phi_c}{E_p}\right]^{-1} \quad (10.2)$$

$$G_{\#} = \left[\frac{\phi_c}{G_c} + \frac{1 - \phi_c}{E_p}\right]^{-1} \quad (10.3)$$

The Young's modulus of the crystalline α-chitin is commonly evaluated by means of *ab initio* calculations and molecular dynamics simulations, and it can be estimated to be $E_c \approx 100$ GPa (range: $E_c \approx 88$–120 GPa [121–123]). In contrast to the chitin, there is no typical value for the Young's modulus of the proteinaceous matrix, and its magnitude may span the range of $E_p \approx 10$ MPa–1 GPa [47]. This high variability is attributed to several bio-chemical and bio-mechanical factors, including: (a) variations in the composition and organization of the protein at different parts of the cuticle [47, 124]; (b) the unknown effects of chitin–protein interactions on the protein itself; (c) the local presence of mineral particles in the protein (e.g., in crustaceans such as lobsters and crabs), which effectively makes the proteinaceous matrix a nano-composite material in itself [109, 125, 126]; (d) the local presence of ion metals (e.g., in mandibles, claws and fangs), which promotes additional cross-linking between the protein chains [127]; and (e) the effect of conditions such as the hydration state or pH level which change both the protein–protein and the protein–chitin molecular bonds [52, 128]. Moreover, even for a specific bio-composite system in well-controlled environmental conditions, determining the exact $E_p$ value through in situ experiments is often impossible. While advanced testing procedures (such as AFM or nanoDMA [129–131]) do enable measuring the nano-mechanical properties of materials, probing only the matrix phases within the bio-composite complex is an extremely challenging task, due to their extreme softness and sub-micron dimensions [132]. The most difficult challenge that has to be overcome is to preserve the natural state of the samples during the experiments in order to obtain experimental results that reflect as much as possible the in vivo conditions. Cutting-edge experimental techniques, based on nanoDMA of micro-beams, promote new capabilities to resolve some of the technical limitations in evaluating the matrix phases of composites [133, 134], but they have yet to be applied to the chitin-based bio-composites of cuticles in biologically relevant conditions. In fact, the suggested range of $E_p \approx 10$ MPa–1 GPa was identified by indirect methods, such as mechanical testing of chitin-free tissues (e.g., the epicuticle [135]), and by theoretical fittings of bio-composite experiments (e.g., [126, 136]). Clearly, due to the various bio-chemical and mechanical factors described above and the lack of direct experimental evidence, any attempt to select a "representative" modulus value from the $E_p$ range mentioned above must be taken with a pinch of salt. Despite the immense uncertainty in the absolute modulus value of the proteinaceous matrix, it can be said with confidence that $E_c \gg E_p$. Note that while the chitin fibril is in fact anisotropic in its elastic properties, $E_c$ in all crystallographic directions likely exhibit $E_c \gg E_p$. As detailed below, this key insight provides a significant relaxation to most mechanical treatments of the chitin–protein bio-composite complex; in fact, the specific magnitude of $E_p$ is not essential for understanding the essence of the cuticle mechanics.

In order to complete the mechanical characterization of an individual chitin–protein fibril, the shear modulus of the chitin and of the protein ($G_c$ and $G_p$, respectively), and the volume fraction of chitin in the nano-composite ($\phi_c$) need to be characterized. By employing the relation $G = E/[2(1 + \nu)]$ with typical Poisson's ratios of 0.3–0.45 for bio-polymers, the shear moduli of the chitin and protein can be estimated as $G_c \approx E_c/2.8$ and $G_p \approx E_p/2.8$; similarly, the relation $G_c \gg G_p$ also

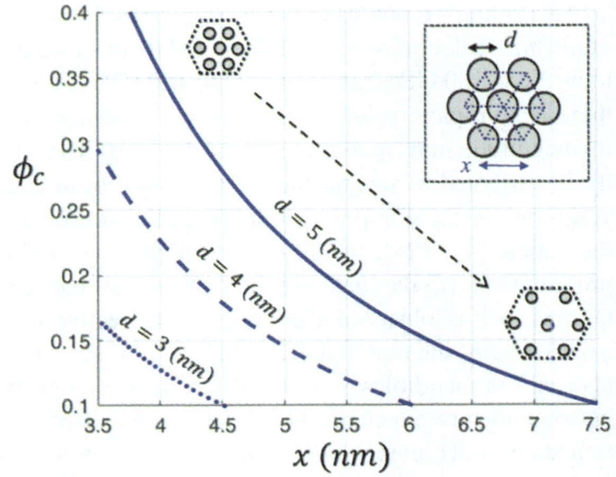


**Fig. 10.4** Correlation between the volume fraction of the chitin crystals ($\phi_c$) in the nano-composite to the diameter ($d$) and spacing ($x$) of the chitin crystals. A decrease in the spacing between adjacent chitin crystals and an increase in their diameter monotonically increase $\phi_c$

applies. For the chitin content, a range of values of $\phi_c \approx 0.1-0.4$ is commonly considered (e.g. [47, 126]). As evident from (10.1–10.3), and as further discussed below, $\phi_c$ plays a significant role in determining the mechanical properties of the cuticle. Thus, the variability range mentioned above should be narrowed down by using designated experiments. By employing X-ray diffraction measurements, the diameter of the chitin ($d$) and the spacing between adjacent chitin crystals ($x$) of a specific cuticle material of interest can be characterized. Then, by considering hexagonal packing of the nano-composite [61] and using elementary trigonometry, the chitin content $\phi_c$ can be calculated by:

$$\phi_c = \frac{\pi}{8\sqrt{3}}\left(\frac{d}{x}\right)^2 \tag{10.4}$$

As shown in Fig. 10.4, an increase in fiber spacing monotonically decreases their volume fraction in the bio-composite. Considering typical values of $d = 3$–$5$ nm for the chitin diameter, the customary range of $\phi_c \approx 0.1-0.4$ emerges for a chitin spacing of $x = 3.5$–$7.5$ nm. This corresponds well to fibril dimensions observed by transmission electron microscopy (TEM) (e.g. [137]). As is also shown in (10.4) and in Fig. 10.4, the diameter ($d$) and spacing ($x$) of the chitin crystals exert a complementary effect, such that $\phi_c$ varies only when the ratio $d/x$ changes.

#### 10.4.1.2 The Micro-Composite Level: Lamellar Arrays

On the micro-structural level, the cuticle is composed of several distinct layers, each of which is structured as a stack of sheets of parallel nano-composite fibrils. As discussed above, the layup configuration may vary from one cuticle layer to another [138, 139], and we focus here on two common types of architecture: the twisted plywood architecture and the unidirectional architecture (level VI 3 and 4, Fig. 10.1). As detailed above, the twisted plywood architecture is perhaps the most common architectural motif in arthropod cuticles [137]. In this configuration, the in-plane orientation of the nano-composite layout gradually varies between two successive sheets, resulting in a helical periodic architecture (level VI, 1 and 2, Fig. 10.1). The somewhat less common unidirectional architecture may appear either as a part of the exocuticle (the proximal exocuticle) or a part of the endocuticle [59, 92]. In the unidirectional architecture, all nano-composite sheets are aligned in the same direction, which is often closely related to the specific loading conditions the respective cuticle part has to face.

After obtaining the fundamental characteristics of an individual chitin-protein fibril, i.e., $E_c$, $E_p$, $G_c$, $G_p$, and $\phi_c$, we turn to estimate the mechanical properties of the micro-composite architectures of the cuticle. As introduced above, the anisotropic moduli of an individual nano-composite fibril are evaluated by the direct and inverse rule-of-mixtures in (10.1–10.3). By further considering the typical moduli characteristics of the chitin and protein phases, $E_c \gg E_p$ and $G_c \gg G_p$, and the typical chitin content in the cuticle, $\phi_c \approx 0.1-0.4$, the moduli of an individual chitin-protein fibril (Fig. 10.3) can be approximated via:

$$E_{\parallel} \approx \phi_c \cdot E_c, \tag{10.5a}$$

$$E_{\perp} \approx \frac{E_p}{1-\phi_c}, \tag{10.5b}$$

$$G_{\#} \approx \frac{G_p}{1-\phi_c} \quad \left(E_f, G_f \gg E_m, G_m\right) \tag{10.5c}$$

From these simplified relations, it is evident that the chitin crystals and the proteinaceous matrix dominate different mechanical properties of the fibril; the chitin dominates the axial stiffness of the fibril along the chitin direction, while the protein dominates the perpendicular and shear stiffness of the fibril. Next, the expressions in (10.5a, b, c) are used to evaluate the moduli of the twisted plywood and unidirectional architectures.

Clearly, the twisted plywood architecture is structurally—and, thus, mechanically—in plane isotropic, and its resistance to axial and shear deformations is characterized by Young's and shear moduli, $E_{ply}$ and $G_{ply}$, respectively (Fig. 10.5). Following benchmark composite mechanical tools, $E_{ply}$ and $G_{ply}$ are obtained by averaging the moduli of the nano-composite over the entire in-plane rotation angles, thus providing the following analytical expressions [120, 140]:

$$E_{ply} = \frac{3}{8} \cdot E_{\parallel} + \frac{5}{8} \cdot E_{\perp} \approx \frac{3}{8} \cdot \phi_c \cdot E_c \tag{10.6a}$$

$$G_{ply} = \frac{1}{4} \cdot E_{\parallel} + \frac{1}{8} \cdot E_{\perp} \approx \frac{1}{4} \cdot \phi_c \cdot E_c \tag{10.6b}$$

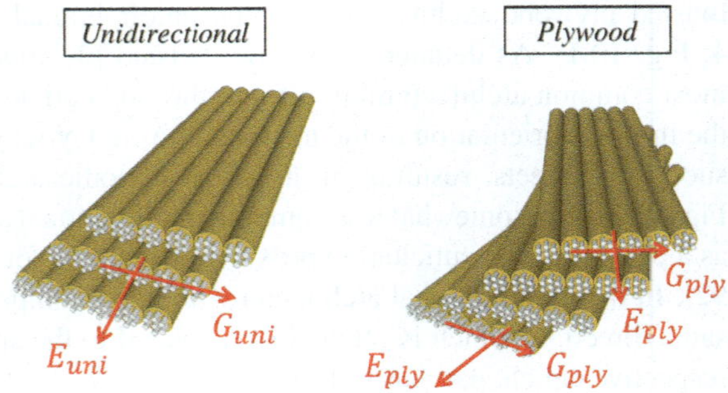


**Fig. 10.5** Schematic illustration of the axial and shear moduli of unidirectional ($E_{uni}$, $G_{uni}$) and twisted plywood ($E_{ply}$, $G_{ply}$) micro-composites. Note that the twisted plywood architecture is mechanically isotropic, i.e., it has the same $E_{ply}$ and $G_{ply}$ in all directions

The unidirectional architecture can be considered as a direct up-scaling of the individual fibril, and it is characterized by a Young's modulus in the alignment direction $E_{uni}$ and its shear modulus $G_{uni}$ (Fig. 10.5):

$$E_{uni} = E_{\parallel} \approx \phi_c \cdot E_c \tag{10.7a}$$

$$G_{uni} = G_{\#} \approx \frac{G_p}{1 - \phi_c} \tag{10.7b}$$

As can be seen from (10.6a, 10.6b and 10.7a, 10.7b), the characteristics of the chitin ($E_c$ and $\phi_c$) dominate both the twisted plywood moduli ($E_{ply}$ and $G_{ply}$) and the unidirectional axial modulus ($E_{uni}$), where the mechanical effects of the proteinaceous matrix are negligible—and thus disregarded. The shear modulus of the unidirectional architecture ($G_{uni}$), on the other hand, is dominated by the characteristics of the protein ($E_p$ and $\phi_p = 1 - \phi_c$), and the mechanical significance of the crystalline chitin is negligible. Figure 10.6a, b plots the Young's modulus and shear modulus of the twisted plywood and unidirectional architectures of the micro-composite, as calculated from (10.6a, 10.6b and 10.7a, 10.7b), employing typical chitin and protein characteristics. Representative experimental ranges for the Young's modulus of the twisted plywood [100, 126, 141] and unidirectional [59, 61, 136] architectures of the micro-composite are indicated in Fig. 10.6. The experimental range for the twisted plywood architecture incorporates the results of exocuticular layers (thinner lamellae with higher densities and higher moduli) and endocuticular layers (thicker lamellae with lower densities and lower moduli), which are indicated separately. The axial modulus of the unidirectional architecture is approximately twice that of the twisted plywood architecture ($E_{uni} \sim 2 \cdot E_{ply}$), while its shear modulus is much smaller than that of the twisted plywood architecture ($G_{uni} \ll G_{ply}$). These simple observations suggest that the twisted plywood and unidirectional architectures are complementary mechanical configurations; their selective integration into the layered structure of the cuticle gives rise to possible functional adaptations of the cuticle for addressing various bio-mechanical states as exemplified in Sect. 10.5.

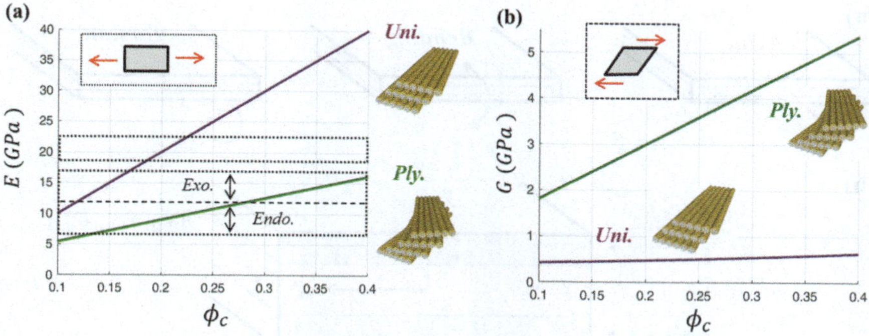


**Fig. 10.6** The elastic moduli of the twisted plywood and unidirectional lamellar architectures: **a** Young modulus, and **b** shear modulus, calculated for $E_c = 100$ GPa. The shear modulus of the unidirectional configurations is calculated for $E_p = 1$ GPa, and thus represents the upper boundary of the possible range of $G_{uni}$ values. The typical range of experimental results for $E_{ply}$ (exo- and endocuticle) and for $E_{uni}$ are indicated by the dashed rectangles

#### 10.4.1.3 The Macro-Structural Level: The Layered Cuticle

From a bio-mechanical viewpoint, the load-bearing parts of the cuticle are required to sustain various loading conditions, such as axial tension and compression, bending, and torsion (Fig. 10.7a, b). These functionalities are essentially provided by the macro-mechanical characteristics of the cuticle, including its axial stiffness ($K_{axial}$, force vs. displacement), bending stiffness ($K_{bend}$, moment vs. flexural curvature), and twisting stiffness ($K_{twist}$, torsion vs. twist angle gradient) [120]. These stiffness parameters are all tightly related to the internal architecture of the cuticle. From a simplified structural perspective, the cuticle can be viewed as a layered (laminate) material, which comprises a varying number of lamellar subunits. When the cuticle includes both twisted plywood and unidirectional micro-composite layers, the axial, bending, and twisting stiffness of the cuticle ($K_{axial}$, $K_{bend}$, and $K_{twist}$, respectively) are obtained by:

$$K_{axial} = E_{ply} \cdot A_{ply} + E_{uni} \cdot A_{uni} \quad (10.8a)$$

$$K_{bend} = E_{ply} \cdot I_{ply} + E_{uni} \cdot I_{uni} \quad (10.8b)$$

$$K_{twist} = G_{ply} \cdot J_{ply} + G_{uni} \cdot J_{uni} \quad (10.8c)$$

where the axial stiffness is provided by the axial moduli of the twisted plywood and unidirectional layers ($E_{ply}$ and $E_{uni}$, respectively), multiplied by the corresponding cross-sectional area of each layer ($A_{ply}$ and $A_{uni}$). The bending stiffness is also provided by the axial moduli $E_{ply}$ and $E_{uni}$, multiplied by the corresponding moments of inertia of the layers ($I_{ply}$ and $I_{uni}$). The twisting stiffness is provided by the shear moduli of the plywood and unidirectional layers, $G_{ply}$ and $G_{uni}$, multiplied the

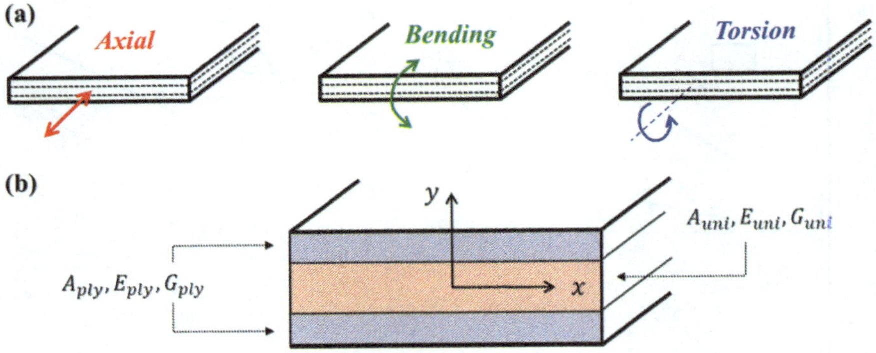


**Fig. 10.7** **a** Schematic illustration of possible mechanical loads on the cuticle: axial tension/compression (left), bending (middle), and twisting (right). **b** Schematic illustration of the moduli and cross section geometry that determine the axial, bending and twisting stiffness of the cuticle

corresponding polar moment of each layer ($J_{ply}$ and $J_{uni}$). Explicitly, the moments of inertia are calculated via $I_{ply} = \int_{A_{ply}} y^2 \cdot dA$ and $I_{uni} = \int_{A_{uni}} y^2 \cdot dA$, and the polar moments via $J_{ply} = \int_{A_{ply}} \sqrt{x^2 + y^2} \cdot dA$ and $J_{uni} = \int_{A_{uni}} \sqrt{x^2 + y^2} \cdot dA$. Note that while $K_{axial}$ incorporates merely the area of the cuticular layers (regardless of their location), $K_{bend}$ and $K_{twist}$ incorporate both the area and the square distance of the layers from the cuticle centerline. Thus, external layers will provide a greater effect on the bending stiffness and twisting stiffness of the cuticle than internal layers of the same area. In a slightly more elegant manner, the expressions for $K_{axial}$, $K_{bend}$, and $K_{twist}$ can be uniformly rephrased as a weighted average of the twisted plywood and unidirectional moduli, as follows:

$$K_{axial} = A \cdot \left[E_{ply} \cdot x_A + E_{uni} \cdot (1 - x_A)\right] \tag{10.9a}$$

$$K_{bend} = I \cdot \left[E_{ply} \cdot x_I + E_{uni} \cdot (1 - x_I)\right] \tag{10.9b}$$

$$K_{twist} = J \cdot \left[G_{ply} \cdot x_J + G_{uni} \cdot (1 - x_J)\right] \tag{10.9c}$$

where $A$, $I$, and $J$ indicate the total cross-sectional area, the moment of inertia, and the polar moment of the cuticle as a whole, respectively. The weighting parameters $x_A = A_{ply}/A$, $x_I = I_{ply}/I$, and $x_J = J_{ply}/J$ represent the relative contribution of the twisted plywood layers to the laminate architecture of the cuticle. Note that in cases where the cuticle possesses only the twisted plywood architecture and is devoid of the unidirectional architecture $x_A = 1$, $x_I = 1$, and $x_J = 1$, and the stiffness parameters simplify into:

$$K_{axial} = A \cdot E_{ply} \tag{10.10a}$$

$$K_{bend} = I \cdot E_{ply} \tag{10.10b}$$

$$K_{twist} = J \cdot E_{ply} \tag{10.10c}$$

To gain a deeper insight in the stiffness of the cuticle, the moduli of the twisted plywood and unidirectional micro-composites (10.6a, b and 10.7a, b) can be incorporated into (10.9a–c); the results, normalized by the maximal stiffness, are plotted in Fig. 10.8. As evident from Figs. 10.6 and 10.8, the unidirectional architecture, which includes a high axial modulus, is well suited to sustain axial and bending loadings along the fibers direction; however, its low Young's modulus and low shear modulus perpendicular to the fibers create poor-performance for sustaining bi-axial (tension and bending) and twisting loads. The twisted plywood architecture, by contrast, has a lower axial modulus than the unidirectional architecture, and it is, therefore, of poor-performance for sustaining unidirectional axial and bending loads—but it can provide both bi-axial and twisting resistance, which are absent in the unidirectional architecture. While the individual twisted plywood and unidirectional layers have certain limitations to specific loading conditions, the integration of both architectures in the cuticle appears to be an adequate evolutionary outcome toward an efficient bio-mechanical element. In fact, as shown in Fig. 10.8, the cuticle is, in principle, a highly versatile structure, and changes in its laminate architecture (i.e., in $x_A$, $x_I$, and $x_J$) may significantly change its mechanical characteristics. For example, a proper combination of the twisted plywood and unidirectional layers in the cuticle may produce $K_{bend} \approx K_{twist}$, which would make the cuticle mechanically robust, i.e., equally stiff under different all states of flexural loading. In a broader perspective, the content of the twisted plywood and unidirectional layers may, in fact, vary between different locations of the cuticle, enabling local architectural and mechanical adaptations to address various bio-mechanical functionalities. The fang of the spider is a good example for such graded cuticle architectures, as further discussed below.

## *10.4.2 The Resistance to Damage*

#### 10.4.2.1 Resistance to Penetration Damage: Hardness

The hardness ($H$) of a material represents its resistance to a permanent penetration damage (impression), which is generated by a penetration force ($F$) applied to the surface by an indenting element (Fig. 10.9) [142–144]. After force unloading, the residual penetration depth ($h_r$) at the surface is related to the material hardness via:

$$h_r \propto \sqrt{F/H}, \tag{10.11}$$

where the proportion of the penetration depth is a function of the geometry of the indenter. Clearly, a harder material will experience a shallower residual penetration

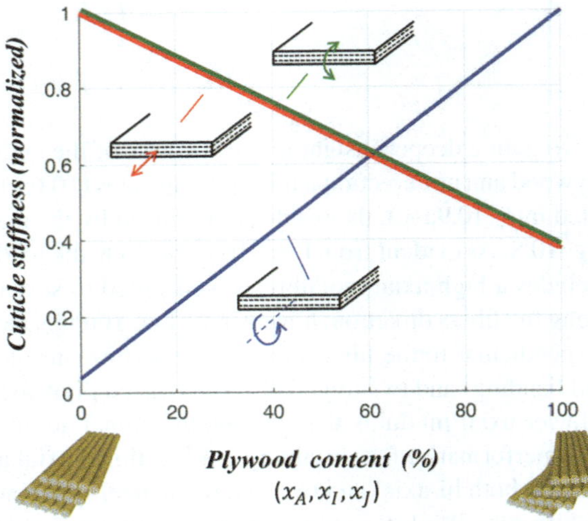


**Fig. 10.8** Variations in the axial, bending and twisting stiffness of the cuticle (red, green, and blue lines, respectively), normalized by their maximal value, shown as a function of the corresponding geometric portion of the plywood in the cuticle ($x_A$, $x_I$, and $x_J$, respectively)

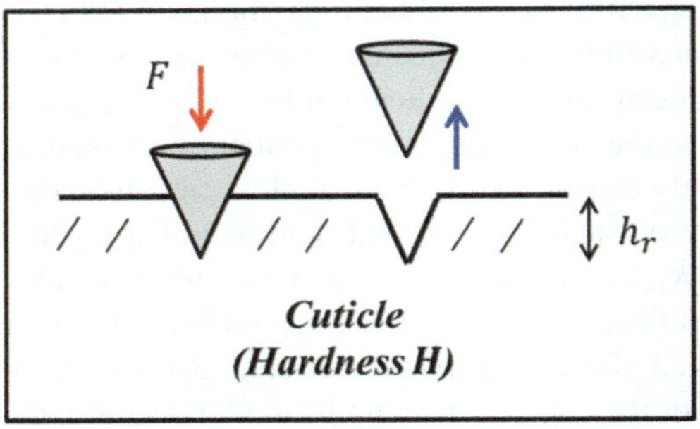


**Fig. 10.9** Schematic illustration of the penetration damage of a cuticle surface of hardness $H$. An indenting element is pressed into the surface by a force $F$ (left), producing a residual impression of depth $h_r$ to the surface after unloading and pulling out the indenter from the surface (right)

and, therefore, lesser surface damage than a softer material under the same load. In classical engineering and metallurgy, the penetration damage arises from irreversible plastic deformations and the hardness is thus related to the yield stress of the material ($\sigma_y$) via $H \approx \sigma_y/1.5 - \sigma_y/3$ [143, 145, 146]. In brittle materials, the penetration damage arises from material crushing [147] and crack propagation [143, 148], and the hardness is therefore related to the modulus of rupture ($\sigma_r$) and to the work-of-fracture (toughness; $G_c$) of the material. In biological materials, which are mostly composites, the mode of the penetration damage (plasticity or brittle) cannot be identified a priori, and it may vary with the mechanical characteristics of the bio-composite, the content and microstructure of the individual phases. In many biological materials, and in cuticles, in particular, plastic damage occurs at much lower forces than brittle damage, and the hardness can mostly be treated in the classical manner of surface yielding [141, 149].

In a simplistic micro-mechanical perspective, upon cuticle penetration, the stiff and hard chitin crystals sustain part of the load via reversible elastic deformations, while the softer proteinaceous phase bears the remaining load and experiences irreversible plastic deformations. Qualitatively, the hardness of the cuticle will typically increase with its chitin content ($\phi_c$) and with the hardness of its proteins ($H_p$). In addition, the hardness of the cuticle will typically change with the local orientation of the chitin in the tissue ($\theta_c$), due to variations in the relative load-sharing between the chitin and the proteins:

$$H = f(\phi_c, \theta_c, H_p) \quad (10.12)$$

The effect of the protein hardness ($H_p$) can be observed by manipulating the hydration state of the cuticle, which changes the protein interactions and effectively makes the proteinaceous phase harder (and stiffer) upon dehydration. This effect cascades across the multi-scale mechanical behavior of cuticles; consequently, at the nano-, micro-, and macro-structural levels, the hardness (and elastic moduli) of dry cuticles are higher than those of wet cuticles [68, 100, 150], including mineralized crustacean cuticles [103–105]. The effect of chitin orientation ($\theta_c$) is nicely elucidated by local nano-indentation measurements across the lamellar structure of exo- and endocuticular layers. In these experiments, gradual and periodic variations in the local hardness (and modulus) of the cuticle are observed, and are well synchronized with the periodicity of the twisted plywood architecture of these cuticular layers [126, 131].

Interestingly, the resistance to penetration damage of biological materials—and of cuticles in particular—can be further improved by employing hardness and modulus modification to the near-surface region of the material, which may significantly affect the resulting damage profile [151, 152]. In non-mineralized cuticles, these near-surface mechanical adaptations are mostly generated by local compositional modifications, mainly in the form of sclerotization or metal ion based cross-linking of the proteinaceous matrix (e.g., in the spider fangs [60]), as is further discussed, in the context of contact resistance, below. In mineralized cuticles, local differences in the mineral phases incorporated into the exocuticle as observed in a number of Crustacea may serve the same purpose [95, 153]. This can go as far as to local changes in the arrangement of the plywood layers from horizontal to vertical as has been shown for the carapace of the large crab *Cancer pagurus* [95].

#### 10.4.2.2 Resistance to Contact Damage

The resistance of the cuticle to penetration protects the animal against surface damage, such as against a predator's attack; however, in other cases, the cuticle needs to function as a bio-mechanical tool for penetrating other materials—such as when mandibles cut food or when a stinger or a fang inject venom. In such cases, the tip of the cuticular penetrating tool is subjected to extreme, localized contact loads, which

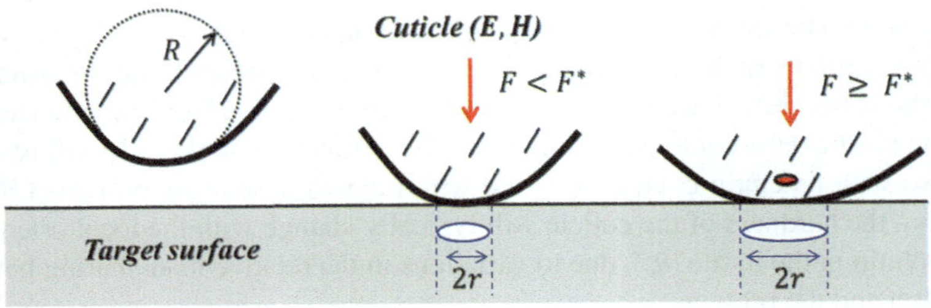


**Fig. 10.10** Schematic illustration of the contact loading of a cuticle tip with a Young's modulus $E$, a hardness $H$, and a tip curvature $\kappa = 1/R$ (where $R$ is the radius of the tip). For low-contact forces below the critical force for onset of tip yielding ($F < F^*$, middle), the tip deforms elastically and increases the tip-surface contact area of radius $r$ (in blue). For higher contact forces ($F \geq F^*$, right), plastic damage is initiated and propagates throughout the tip (red circle)

may damage the tip (e.g., tip blunting or even breakage). The contact resistance of the tip is the key feature that prevents such damages, and it is, therefore, critical to the survival of the organism.

To illustrate the idea of contact resistance, consider a convex element with curvature $\kappa$ ($\kappa = 1/R$, where $R$ is the radius of the tip), Young's modulus $E$, and hardness $H$, which represents the tip of a cuticular penetrating tool. This convex element approaches the target surface with an increasing contact force ($F$), as shown in Fig. 10.10. At the initial state of a low-force contact, the tip of the cuticle undergoes reversible elastic deformations, which generate a certain tip-surface contact area that progressively increases with the contact force. According to the Hertz theory [154, 155], the radius of the contact area scales with the contact force and the Young's modulus of the tip, as $r \propto \sqrt[3]{F/(\kappa \cdot E)}$. Thus, sharper and stiffer tips (i.e., tips with higher moduli) will produce smaller tip-surface contact areas for a given contact force. When the applied contact force reaches a critical value ($F^*$), plastic damage is initiated at a depth of $\sim r/2$ below the contact area and propagates throughout the tip upon increasing force, $F > F^*$. The resistance of the penetrating tip to contact damage is defined, in a somewhat conservative manner, as the force that initiates the damage to the tip ($F^*$). Note that this threshold force merely represents the onset of the contact damage—and should not be referred to as the "abrasion resistance" of the material (as is often done), since it cannot grasp wear effects that involve material removal and the propagation of the damaged zone with time. Following the Hertz model, the threshold force ($F^*$) scales with the Young's modulus, hardness, and curvature of the tip ($E$, $H$, and $\kappa$, respectively) as [144, 155]:

$$F^* \propto \frac{1}{\kappa^2} \cdot \frac{H^3}{E^2} \tag{10.13}$$

As seen from (10.13), sharper tips (i.e., tips with a higher $\kappa$) will reach contact damage at lower forces than blunter tips, which may explain why ultra-sharp cuticle tips are scarce. In addition, and quite intuitively, harder tips (higher $H$) can resist contact damage better than softer tips. However, somewhat less intuitive is the notion that more compliant tips (lower $E$) resist contact damage better than rigid tips; this effect is attributed to the increase in the radius of the contact area ($r$) with increasing compliance, which effectively decreases the contact pressure (force per unit area) that generates the contact damage. In conclusion, to better resist contact damage, the cutting-edge of penetrating cuticle tools must not only be of greater hardness ($H$), but they also need to possess a high hardness-to-modulus ratio ($H/E$).

From a bio-mechanical point of view, it is interesting to note that in many insects and chelicerates, the near-tip region of penetrating cuticular tools is typically enriched with metal ions, such as Ca, Zn and Mn; these local material adaptations increase the $H^3/E^2$ ratio of the tip material, which in turn increases the critical force $F^*$ and thus the resistance of the tip to contact damage. The tips of the jaws of the annelid *Nereis* sp., for example, are enriched with zinc, which increases their hardness by a factor of 2 and the modulus by only $\sim 5/4$, and thus increases the $H^3/E^2$ ratio by a factor of almost 5 [84, 156]. A similar increase in the hardness and modulus of the cuticle is also evident in the fangs of spiders [59] and in the jaws of the annelid *Glycera* sp. [150].

#### 10.4.2.3 Resistance to Fracture: The Twisted Plywood Architecture

Cuticles are highly resistant to fracture damage, both under static and impact-loading conditions [47, 90]. As cuticles are multi-scale bio-composites, their fracture mechanics is rather intricate and incorporates various toughening mechanisms across the nano-, micro-, and macro-structural scales [157] including the many interfaces inherent to the hierarchical structural organization such as fibril/fibril, fiber layer/fiber layer, exo-/endocuticle but also between different chemical components such as chitin/protein, organic/inorganic and also additional structures such as pore canals that help in stopping or deviating cracks. These complexities make the fracture analysis of the cuticle especially challenging, and the understanding of many of its mechanical characteristics is still vague. The twisted plywood architecture of the cuticle is considered to play a major role in the fracture resistance capabilities of the cuticle. Nevertheless, even today, little is known about the specific relationship between the structural and the mechanical characteristics of the twisted plywood architecture and how it promotes fracture resistance, many parameters of this relationship such as the lamellar thickness (stacking height), twisting angle, and material composition, are currently under extensive investigation.

Figure 10.11 provides several examples of fractured surfaces of the twisted plywood in exo- and endocuticles. Qualitatively, the twisted plywood fracture appears to be dominated by two competing mechanisms promoting the propagating crack normal to the surface in two possible paths—direct and spiral—as shown in Fig. 10.12. In the direct path mode (Fig. 10.12a), the crack orientation is preserved upon propagation, breaking successive layers with gradual variations in the orientation of the in-plane fibers orientations, accompanied by changes in the effective modulus at the tip of the crack [158]. During a direct path crack growth, the strain energy at the tip of

**Fig. 10.11** SEM micrographs of a cross-fractured cuticle, showing the fracture mechanisms of different mineralized and unmineralized twisted plywood structures. **a** The mineralized endocuticle of the lobster *Homarus americanus*, in which the fibrils that are mineralized with ACC (modified from [94]). **b** The mineralized exocuticle of the crab *Cancer pagurus*, in which the organic fibres are embedded in a solid calcite matrix (modified from [95]). **c** The non-mineralized exocuticle of the chelicerate *Limulus polyphemus*, in which the rotation angle between superimposed planes is large, resulting in thin "lamellae" (Fabritius, unpublished)

the crack varies with the direction of the local fibers, resulting in periodic variations in the fracture driving-force, which are correlated with the twisted plywood lamellar architecture. In the spiral-path mode (Fig. 10.12b), the crack propagation occurs by breaking the interface between adjacent fibers (i.e., fiber-splitting), but with no fiber breakage, resulting in alteration of the tip of the crack propagation direction. Such spiral-like crack propagation significantly increases the crack path and changes its opening plane compared to the external loading – and by that potentially increases the energy required for fracture [159–161]. According to the fundamentals of fracture mechanics, the propagation mode for an incremental crack growth will be on the path of the maximal crack-tip deriving force ($\propto$ modulus) and minimal resisting force ($\propto$ toughness) [162]. As shown in Fig. 10.11, both crack paths are essentially present in twisted plywood fractured surfaces; the spiral path emerges where the local lamellar orientation is approximately aligned with the crack direction, and the direct path emerges where the crack path is approximately perpendicular to the lamellar orientation. For intermediate cases, in which the local twisted plywood direction at the tip of the crack is neither aligned nor perpendicular to the crack plane, no conclusive statement can be made, and the crack propagation path may be either direct or spiral, depending on the specific characteristics of the twisted plywood structure.

## 10.5 Structural–Mechanical Aspects of Selected Bio-mechanical Cuticular Elements

### *10.5.1 Uni-Axial Loads—Tendons and Ovipositors*

Tendons or apodemes of spiders [52], insects (e.g. locusts) [163] and crustaceans [20] are made of parallel arrangement of chitin-protein fibers. Indeed, these structures are only loaded uniaxially by the muscles. Similarly, the ovipositors of several insects

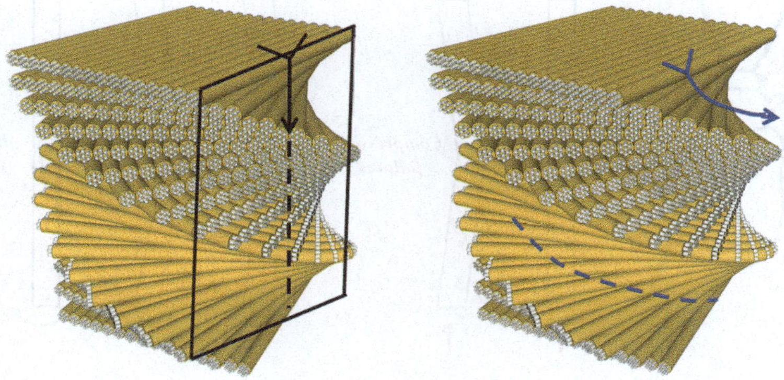


**Fig. 10.12** Schematic illustrations of possible crack propagation paths in the twisted plywood architecture of the cuticle: **a** a direct crack path; **b** a spiral crack path

have been studied by either TEM or x-ray diffraction and showed high alignment of fiber orientation with no structural heterogeneity [4, 164]. These structures are used by female insects (flies, wasps and others) for boring to lay eggs within different substrates. The substrate maybe soft like fruit flesh, caterpillars etc. or stiff like wood, in the latter case metal ions are often found hardening the tip of the structure [64]. These elements are structurally simple consistent with their biomechanical function and their predominantly uniaxial loading manner.

### *10.5.2 Moving on Ground—Leg Elements*

Specialized segmented appendages made of the cuticle form legs that enable locomotion in arthropods [165]. From a mechanical viewpoint, the leg segments are elements that are mostly loaded by reaction forces that arise from the walking surface during muscle operation, which may range between low-frequency and low-magnitude forces during walking, through high-frequency and medium forces during running and up to temporary but extreme forces during jumping. The cuticular leg elements typically possess high stiffness, strength, toughness, and resistance to fatigue damage, which enable the legs to sustain the various loading modes [102, 166]. From a structural–mechanical perspective, the leg elements are viewed as long, thin-walled tubes, which are effectively characterized by anisotropic elastic moduli ($E_L$ and $E_T$) and an anisotropic stress-to-failure ($\sigma_L$ and $\sigma_T$) in their longitudinal and transverse (circumferential) directions, respectively. During locomotion, compressive forces and bending moments are generated within the cross-section of the leg, which may lead to its mechanical failure in a number of possible ways: axial failures via (1) Euler's buckling ($\propto E_L$), and (2) local buckling ($\propto E_L$), and bending failures via

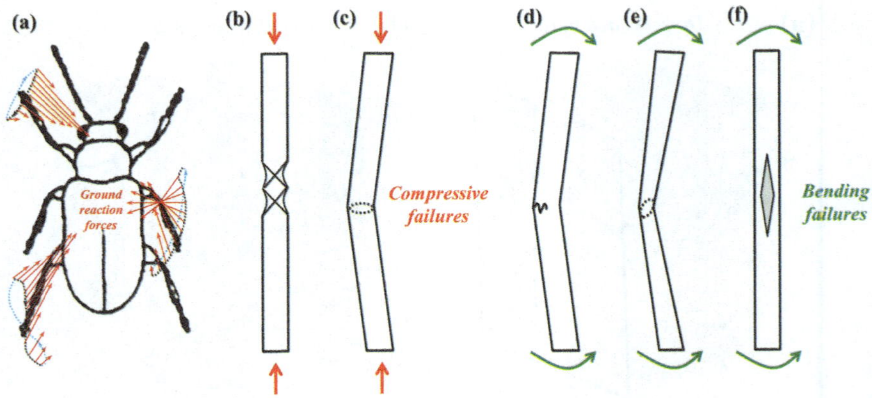


**Fig. 10.13** **a** Schematic illustration of the reaction forces (red arrows) on the cuticular leg elements (blue arrows) during arthropod locomotion (adapted from [165]). **b**, **c** Schematic illustrations of compression failures via local buckling (**b**) and Euler's bucking (**c**) modes. **d–f** Schematic illustrations of bending failures via axial fracture (**d**), ovalization instability (**e**), and transverse splitting (**f**) (adapted from [101])

(3) axial fracture ($\propto \sigma_L$), (4) cross-sectional ovalization instability ($\propto \sqrt{E_L \cdot E_T}$), and (5) transverse (circumferential) splitting ($\propto \sqrt{E_L \cdot \sigma_T}$) (Fig. 10.13, adapted from [101]). Clearly, no single "design" solution can address all these failure modes, and, therefore, proper mechanical adaptations in the anisotropic elastic moduli and strength of the cuticle are required to achieve "optimal" efficiency for each specific geometrical configuration under a specific loading state. Arthropods typically achieve this by selectively increasing or decreasing the thickness of the cuticle in defined areas of the tubular segments and by adopting the local shape of the elements to withstand the load distributions they have to face. The locust tibia, for example, is loaded almost exclusively by bending moments and its structural–mechanical characteristics appear to provide a close-to-optimal bending stiffness and a high resistance against ovalization failure. The structural–mechanical characteristics of the crab merus, on the other hand, appear to give rise to both compressive and bending stiffness, where the bending stiffness is, again, close to its optimal value [101].

### *10.5.3 Moving Through Air—Wing Elements*

Wing elements formed by the cuticle efficiently exploit relatively weak thrust forces, such as high-lift forces, thus providing specialized maneuvering capabilities and unusual aerodynamic functionalities. To promote their aeronautical capabilities, wings must be stiff, strong and tough, but they also need to be light-weight. The internal wing architecture typically includes thin veins embedded in a membrane

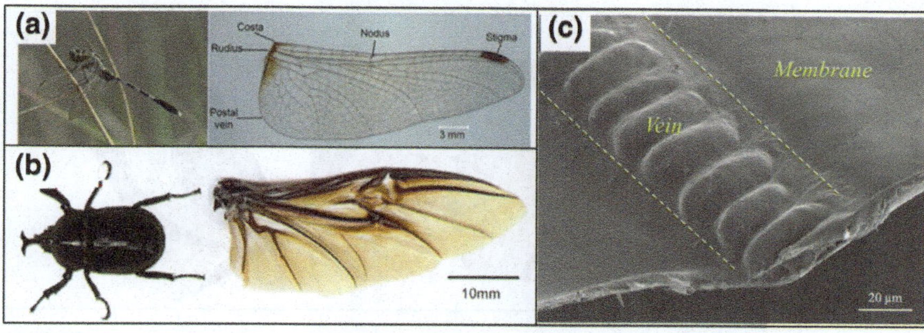


**Fig. 10.14** The architecture of various insect wings. **a** Pictures of a dragonfly (left) and the architecture of its hind wing (right) (adapted from [168]). **b** Picture of the *Allomyrina dichotoma* beetle (left) and the architecture of its hind wing (adapted from [169]). **c** Picture of the fractured hind wing of *Schistocerca gregaria*, indicating the vein and the membrane (adapted from [167])

material (see, for example, Fig. 10.14) [164–169], resembling the structure of kites. The veins function as reinforcing elements that provide stiffness and stability to the wing, while the inter-vein membrane provides the aerodynamic functionality. On the material level, insect wings are multi-scale chitin-based composites, such as the rest of the cuticle, made of stiff chitin nanofibrils and a much softer proteinaceous matrix material. The ratio of chitin to proteins varies significantly between the stiff veins, which are chitin rich, and the more compliant inter-vein membrane, which in some cases may be devoid of chitin [170]. The highly anisotropic and inhomogeneous multi-scale architecture of the cuticular wing elements provides irregular mechanical functionalities to the wing, such as coupled flexural–torsional deformations, localized stiffening effects, and improved damage resilience [167, 168–173], which play an important role in the aeronautical efficiency the wing [171–178].

### 10.5.4 Venom-Injection—Fangs and Stingers

Many arachnids and some insects are equipped with designated cuticular elements that are used to penetrate the integument of their target upon biting or stinging, i.e. to inject venom or suck blood. Bees and mosquitos, for example, possess a syringe needle-like injection structure with a specialized near-tip architecture (e.g., multi-tip or jigsaw), which promotes their smooth penetration into relatively soft target surfaces (e.g., the skin of vertebrates) [176–182]. Other examples include the fangs of the spider and the stinger of the scorpion, which are functionally more challenging because they must puncture prey with an exoskeleton, that can be as hard and rigid as their own injection elements [183–186]. To fulfill their bio-mechanical function, these elements are usually architectured in a graded manner, both at the macro-structural level and at the material level, which makes them mechanically superior compared to the target surface [59, 187, 188]. The spider fang (Fig. 10.15),

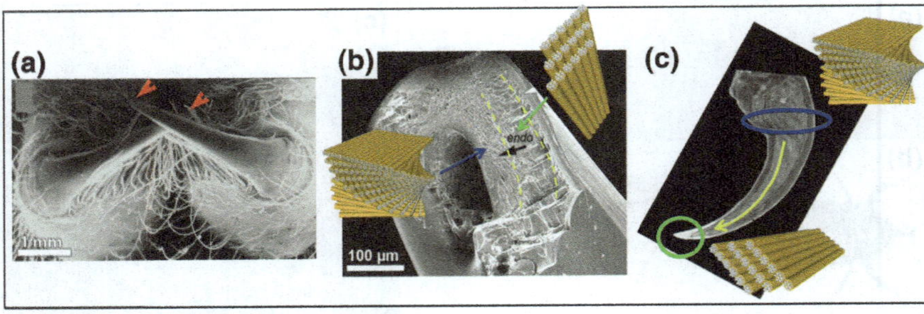


**Fig. 10.15** The architecture of the fang of the *Cupiennius salei* spider (adapted from [59, 136]). **a** Macroscopic view of the two fangs in the chelicera (red arrows). **b** Cross-sectional view of a single fang, indicating the unidirectional and plywood architectures of the endocuticle. **c** Sagittal view of the fang, showing its curved conical structure; the gradients in the architecture of the endocuticle (unidirectional close to the tip and twisted plywood close to the base) are schematically illustrated

for example, is curved and conical in shape, tapered from the sharp tip toward the wider base of the fang—a structure that gives rise to high (close-to-optimal) mechanical rigidity per unit mass [136]. At the material level, the spider fang has a hybrid endo(meso)-cuticular architecture that combines both unidirectional and twisted plywood structures. The relative fractions of both architectures vary gradually—from plywood at the base of the fang resisting multi-directional loadings, to unidirectional at the tip, supporting the localized penetration forces. The near-tip region of the fang is enriched with metal ions, which give rise to a material that is almost twice as hard and stiff as that of the exoskeleton of its typical target.

### *10.5.5 Mechano-Sensory—Setae*

Mechanosensors in arthropods are extremely widespread. Insects, crustaceans, and arachnids sense touch, medium flow, and substrate vibration to detect the presence of prey or predators, as well as for intra-species communication [182–192]. For instance, the spider's cuticle is densely covered by cuticular hairs (setae) (around 400 hairs, or setae, per $mm^2$, Fig. 10.16a, b). These hairs include trichobothria, which are air-flow sensors (comparable to filiform hairs in insects), and tactile hairs, which react to mechanical touch stimuli [192, 193]. The hair shaft varies in length and in structure and may consist of various micro-trichs with differing shapes. The hair is connected to the cuticle by a socket delineated by a compliant membrane. A conspicuous difference between the trichobothria and the tactile hairs in spiders is that the former do not bend during deflection, while the latter are both deflected and bent by the applied force, which is usually in the order of a few micro-newtons. This distinction stems mainly from the differences in the spring stiffness values of the hair articulation, which is four orders of magnitude higher in tactile hairs than in trichobothria. At the micro-structural level, many spider hairs are feather-like or

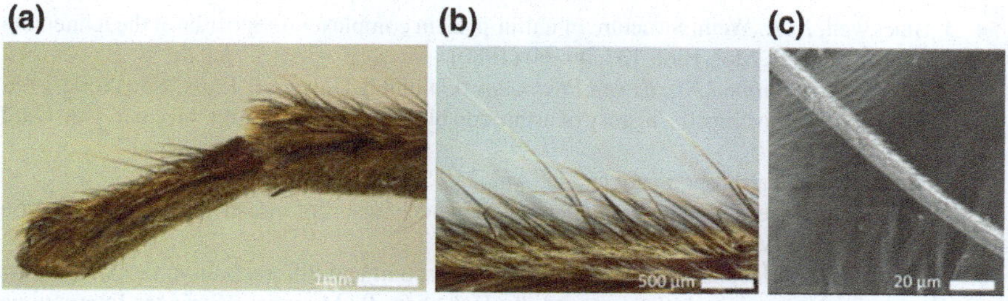


**Fig. 10.16** Spider mechanosensors. **a** Light microscopy of the last two segments of the leg of the spider *Cupiennius salei*, showing the multitude of sensory setae. **b** Higher magnification of the metatarsus hairs including air-flow and touch receptors. **c** Scanning electron microscopy of one air-flow sensor showing the rough surface with multiple micro-trichs. Images courtesy of Maryam Tadayon, Max Planck Institute of Colloids and Interfaces, Germany

contain micro-trichs. It has been suggested that these two structures serve different functions: In air-flow, they largely reduce the hair mass, thus lowering its inertia, but still increase the hair friction with air. Micro-trichs in tactile hairs that are used for propriosensing have been shown to slightly interlock with micro-trichs of adjacent hairs, thus coupling the deflection of both types of hair (Fig. 10.16d).

## 10.6 Conclusion

The enormous diversity of body plans and evolutionary adaptations of Arthropoda and the wide variety of functions the cuticle has to fulfil in its role as exoskeleton make it without doubt one of the most versatile natural materials. The wide range of mechanical properties covered by different cuticle types are brought about by the high structural and compositional complexity that is generated in a large number of mostly not fully understood metabolic processes controlled by the epithelial cells forming the integument. Thus, the cuticle is equally interesting for scientists from biology and materials science as a rewarding subject both for basic research and as inspiration for the development of new bioinspired synthetic materials that incorporate its fascinating design principles.

## References

1. A.C. Neville, *Biology of the Arthropod Cuticle* (Springer, Berlin, 1975)
2. S. Manton, *The Arthropoda. Habits, Functional Morphology, and Evolution* (Clarendon Press, Oxford, 1977)
3. H.R. Hepburn, *The Insect Integument* (Elsevier Scientific Publishing Company, Amsterdam, 1976)

4. J. Blackwell, M.A. Weih, Structure of chitin-protein complexes: ovipositor of the ichneumon fly Megarhyssa. J. Mol. Biol. **137**, 49–60 (1980)
5. N. Eldredge, *Arthropod Fossils and Phylogeny* (Columbia University Press, New York, 1998)
6. R.C. Brusca, Unraveling the history of arthropod biodiversification. Ann. Missouri Bot. Gard. **87**, 13–25 (2000)
7. N.F. Hadley, The arthropod cuticle. Sci. Am. **255**, 98–106 (1986)
8. B. Moussian, Recent advances in understanding mechanisms of insect cuticle differentiation. Insect Biochem. Mol. Biol. **40**, 363–375 (2010)
9. B. Moussian, Molecular model of skeletal organization and differentiation, in *Extracellular Composite Matrices in Arthropods*, ed. by E. Cohen, B. Moussian (Springer International Publishing, Berlin, 2016), pp. 67–87
10. S.B. Murray, A.C. Neville, The role of pH, temperature and nucleation in the formation of cholesteric liquid crystal spherulites from chitin and chitosan. Int. J. Biol. Macromol. **22**, 137–144 (1998)
11. S.J. Hamodrakas, J.H. Willis, V.A. Iconomidou, A structural model of the chitin-binding domain of cuticle proteins. Insect Biochem. Mol. Biol. **32**, 1577–1583 (2002)
12. R. Yu et al., Helicoidal organization of chitin in the cuticle of the migratory locust requires the function of the chitin deacetylase 2 enzyme (LmCDA2). J. Biol. Chem. **291**, 24352–24363 (2016)
13. M. Locke, The structure and formation of the cuticulin layer in the epicuticle of an insect, *Calpodes ethlius* (Lepidoptera, Hesperiidae). J. Morphol. **118**, 461–494 (1966)
14. R. Zuber et al., The ABC transporter Snu drives formation of the lipid-based inward and outward barrier in the skin of *Drosophila* (2017)
15. M. Locke, D. Ph, Pore canals and related structures in insect cuticle. J. Biophys. Biochem. Cytol. **10**, 589–618 (1961)
16. R.D. Roer, R.M. Dillaman, Molt-related change in integumental structure and function, in *The Crustacean Integument Morphology and Biochemistry*, vol. 240, ed. by M.N. Horst, J. Freeman (CRC Press, Boca Raton, FL, 1993)
17. Y. Bouligand, Twisted fibrous arrangements in biological materials and cholesteric mesophases. Tissue Cell **4**, 189–217 (1972)
18. M. Rinaudo, Chitin and chitosan: properties and applications. Prog. Polym. Sci. **31**, 603–632 (2006)
19. B. Focher, A. Naggi, G. Torri, A. Cosani, M. Terbojevich, Structural differences between chitin polymorphs and their precipitates from solutions—evidence from CP-MAS $^{13}$C-NMR, FT-IR and FT-Raman spectroscopy. Carbohydr. Polym. **17**, 97–102 (1992)
20. P. Sikorski, R. Hori, M. Wada, Revisit of alpha-chitin crystal structure using high resolution X-ray diffraction data. Biomacromol **10**, 1100–1105 (2009)
21. K. Kobayashi, S. Kimura, E. Togawa, M. Wada, Crystal transition between hydrate and anhydrous β-chitin monitored by synchrotron X-ray fiber diffraction. Carbohydr. Polym. **79**, 882–889 (2010)
22. S. Muthukrishnan, H. Merzendorfer, Y. Arakane, Q. Yang, Chitin metabolic pathways in insects and their regulation, in *Extracellular Composite Matrices in Arthropods*, ed. by E. Cohen, B. Moussian (Springer International Publishing, Berlin, 2016), pp. 31–65
23. H. Merzendorfer, Insect chitin synthases: a review. J. Comp. Physiol. B. **176**, 1–15 (2006)
24. B. Moussian et al, Deciphering the genetic programme triggering timely and spatially-regulated chitin deposition. PLOS Genet. 1–24 (2015)
25. J.H. Willis, Cuticular Proteins in Insects and Crustaceans. Integr. Comp. Biol. **39**, 600–609 (1999)
26. S. Luschnig, T. Bätz, K. Armbruster, M.A. Krasnow, Serpentine and vermiform encode matrix proteins with chitin binding and deacetylation domains that limit tracheal tube length in *Drosophila*. Curr. Biol. **16**, 186–194 (2006)
27. X. Zhao et al., Identification and expression of cuticular protein genes based on *Locusta migratoria* transcriptome. Sci. Rep. **7**, 45462 (2017)

28. D.L. Cox, J.H. Willis, Post-translational modifications of the cuticular proteins of *Hyalophora cecropia* from different anatomical regions and metamorphic stages. Insect Biochem. **17**, 469–484 (1987)
29. S.O. Andersen, Biochemistry of insect cuticle. Ann. Rev. Entomol. **24**, 29–61 (1979)
30. M.G.M. Pryor, On the hardening of the ootheca of Blatta orientalis. Proc. R. Soc. B Biol. Sci. **128**, 378–393 (1940)
31. S. Andersen, M. Peter, P. Roepstorff, Cuticular sclerotization in insects. Comp. Biochem. Physiol. Part B Biochem. Mol. Biol. **113**, 689–705 (1996)
32. S.S. Chaudhari et al., Knickkopf protein protects and organizes chitin in the newly synthesized insect exoskeleton 1–6 (2011)
33. R. Tajiri, N. Ogawa, H. Fujiwara, T. Kojima, Mechanical control of whole body shape by a single cuticular protein Obstructor-E in *Drosophila melanogaster*. PLoS Genet. **13**, e1006548 (2017)
34. S. Abehsera et al., MARS: a protein family involved in the formation of vertical skeletal elements. J. Struct. Biol. **198**, 92–102 (2017)
35. J.H. Willis, Structural cuticular proteins from arthropods: annotation, nomenclature, and sequence characteristics in the genomics era. Insect Biochem. Mol. Biol. **40**, 189–204 (2010)
36. Y. Tan et al., Infiltration of chitin by protein coacervates defines the squid beak mechanical gradient. Nat. Chem. Biol. **11**, 488–495 (2015)
37. J.E. Rebers, L.M. Riddiford, Structure and expression of a *Manduca sexta* larval cuticle gene homologous to *Drosophila* cuticle genes. J. Mol. Biol. **203**, 411–423 (1988)
38. J.E. Rebers, J.H. Willis, A conserved domain in arthropod cuticular proteins binds chitin. Insect Biochem. Mol. Biol. **31**, 1083–1093 (2001)
39. S. Mun et al., Cuticular protein with a low complexity sequence becomes cross-linked during insect cuticle sclerotization and is required for the adult molt. Sci. Rep. **5**, 10484 (2015)
40. S.O. Andersen, Are structural proteins in insect cuticles dominated by intrinsically disordered regions? Insect Biochem. Mol. Biol. **41**, 620–627 (2011)
41. Y. Zhou et al., Distribution of cuticular proteins in different structures of adult *Anopheles gambiae*. Insect Biochem. Mol. Biol. **75**, 45–57 (2016)
42. T. Weis-Fogh, A rubber-like protein in insect cuticle. J. Exp. Biol. **37**, 889–907 (1960)
43. S.O. Andersen, Characterization of a new type of cross-linkage in resilin, a rubber-like protein. Biochim. Biophys. Acta **69**, 249–262 (1963)
44. H.C. Bennet-Clark, E.C. Lucey, The jump of the flea: a study of the energetics and a model of the mechanism. J. Exp. Biol. **47**, 59–67 (1967)
45. J. Michels, E. Appel, S.N. Gorb, Resilin—the pliant protein, in *Extracellular Composite Matrices in Arthropods*, ed. by E. Cohen, B. Moussian (Springer International Publishing, Berlin, 2016), pp. 89–136
46. A.C. Neville, *Biology of Fibrous Composites: Development Beyond the Cell Membrane* (Cambridge University Press, New York, 1993)
47. J.F.V. Vincent, U.G.K. Wegst, Design and mechanical properties of insect cuticle. Arthropod Struct. Dev. **33**, 187–199 (2004)
48. V.A. Iconomidou, G.D. Chryssikos, V. Gionis, J.H. Willis, S.J. Hamodrakas, 'Soft'-cuticle protein secondary structure as revealed by FT-Raman, ATR FT-IR and CD spectroscopy. Insect Biochem. Mol. Biol. **31**, 877–885 (2001)
49. V.A. Iconomidou, J.H. Willis, S.J. Hamodrakas, Unique features of the structural model of 'hard' cuticle proteins: implications for chitin-protein interactions and cross-linking in cuticle. Insect Biochem. Mol. Biol. **35**, 553–560 (2005)
50. G. Fraenkel, K.M. Rudall, The structure of insect cuticles. Proc. R. Soc. London. Ser. B **134**, 111–143 (1947)
51. E. Atkins, Conformations in polysaccharides and complex carbohydrates. J. Biosci. **8**, 375–387 (1985)
52. C. Valverde Serrano et al., Ordering of protein and water molecules at their interfaces with chitin nano-crystals. J. Struct. Biol. **193**, 124–131 (2016)

53. R.S. Cornman, The distribution of GYR- and YLP-like motifs in *Drosophila* suggests a general role in cuticle assembly and other protein-protein interactions. PLoS ONE **5**, 1–10 (2010)
54. J.E. Hillerton, J.F. Vincent, The specific location of zinc in insect mandibles. J. Exp. Biol. **101**, 333–336 (1982)
55. J.E. Hillerton, J.F.V. Vincent, Consideration of the importance of hydrophobic interactions in stabilizing insect cuticle. Int. J. Biol. Macromol. **5**, 163–166 (1983)
56. J.F.V. Vincent, J.E. Hillerton, The tanning of insect cuticle—a critical review and a revised mechanism. J. Insect Physiol. **25**, 653–658 (1979)
57. S.E. Reynolds, The mechanism of plasticization of the abdominal cuticle in *Rhodnius*. J. Exp. Biol. **62**, 81–98 (1975)
58. R.H. Hackman, M. Goldberg, Comparative study of some expanding arthropod cuticles: the relation between composition, structure and function. J. Insect Physiol. **33**, 39–50 (1987)
59. Y. Politi et al., A spider's fang: how to design an injection needle using chitin-based composite material. Adv. Funct. Mater. **22**, 2519–2528 (2012)
60. Y. Politi et al., Nano-channels in the spider fang for the transport of Zn ions to cross-link His-rich proteins pre-deposited in the cuticle matrix. Arthropod Struct. Dev. **46**, 30–38 (2016)
61. M. Erko et al., Structural and mechanical properties of the arthropod cuticle: comparison between the fang of the spider *Cupiennius salei* and the carapace of American lobster *Homarus americanus*. J. Struct. Biol. **183**, 172–179 (2013)
62. J. Eric Hillerton, B. Robertson, J.F.V. Vincent, The presence of zinc or manganese as the predominant metal in the mandibles of adult, stored-product beetles. J. Stored Prod. Res. **20**, 133–137 (1984)
63. R. Schofield, H. Lefevre, Short communication: high concentrations of zinc in the fangs and manganese in the teeth of spiders. J. Exp. Biol. **144**, 577 (1989)
64. D.L.J. Quicke, P. Wyeth, J.D. Fawke, H.H. Basibuyuk, J.F.V. Vincent, Manganese and zinc in the ovipositors and mandibles of hymenopterous insects. Zool. J. Linn. Soc. **124**, 387–396 (1998)
65. R.M.S. Schofield, H.W. Lefevre, PIXE-STIM microtomography: zinc and manganese concentrations in a scorpion stinger. Nucl. Instruments Methods Phys. Res. Sect. B Beam Interact. with Mater. Atoms **72**, 104–110 (1992)
66. R.M.S. Schofield, Zinc is incorporated into cuticular 'tools' after ecdysis: the time course of the zinc distribution in 'tools' and whole bodies of an ant and a scorpion. J. Insect Physiol. **49**, 31–44 (2003)
67. T.D. Morgan, P. Baker, K.J. Kramer, H.H. Basibuyuk, D.L.J. Quicke, Metals in mandibles of stored product insects: do zinc and manganese enhance the ability of larvae to infest seeds? J. Stored Prod. Res. **39**, 65–75 (2003)
68. T. Schöberl, I.L. Jäger, Wet or dry—hardness, stiffness and wear resistance of biological materials on the micron scale. Adv. Eng. Mater. **8**, 1164–1169 (2006)
69. B.W. Cribb et al., Insect mandibles—comparative mechanical properties and links with metal incorporation. Naturwissenschaften **95**, 17–23 (2008)
70. B.W. Cribb et al., Structure, composition and properties of naturally occurring non-calcified crustacean cuticle. Arthropod Struct. Dev. **38**, 173–178 (2009)
71. E. Degtyar, M.J. Harrington, Y. Politi, P. Fratzl, The mechanical role of metal ions in biogenic protein-based materials. Angewandte. 12026–12044 (2014)
72. J. Schaefer et al., Aromatic cross-links in insect cuticle: detection by solid-state 13C and 15N NMR. Science **235**, 1200–1204 (1987)
73. R. Xu, X.I.N. Huang, T.L. Hopkins, K.J. Kramer, Rapid communication catecholamine and histidyl protein cross-linked structures in sclerotized insect cuticle. **9651748** (1997)
74. S. Zhang, I.-M. Tso, Spider silk: factors affecting mechanical properties and biomimetic applications, in *Extracellular Composite Matrices in Arthropods*, ed. by E. Cohen, B. Moussian (Springer International Publishing, Berlin, 2016), pp. 489–513
75. K.J. Kramer, T.L. Hopkins, J. Schaefer, Applications of solids NMR to the analysis of insect sclerotized structures. Insect Biochem. Mol. Biol. **25**, 1067–1080 (1995)

76. J.L. Kerwin et al., Mass spectrometric analysis of catechol-histidine adducts from insect cuticle. Anal. Biochem. **268**, 229–237 (1999)
77. P. Wappner, L.A. Quesada-Allué, Water loss during cuticle sclerotization in the medfly *Ceratitis capitata* is independent of catecholamines. J. Insect Physiol. **42**, 705–709 (1996)
78. S.O. Andersen, Insect cuticular sclerotization: a review. Insect Biochem. Mol. Biol. **40**, 166–178 (2010)
79. T.L. Hopkins, K.J. Kramer, Insect cuticle sclerotization. Annu. Rev. Entomol. **37**, 273–302 (1992)
80. K. Hiruma, L.M. Riddiford, Granular phenoloxidase involved in cuticular melanization in the tobacco hornworm: regulation of its synthesis in the epidermis by juvenile hormone. Dev. Biol. **130**, 87–97 (1988)
81. H. Nakagawa, Y. Hori, S. Sato, T.B. Fitzpatrick, R.L. Martuza, The nature and origin of the melanin macroglobule. J Invest Dermatol **83**, 134–139 (1984)
82. H. Kayser, Pigments, in *Comprehensive Insect Physiology, Biochemistry and Pharmacology*, ed. by G.A. Kerkut, L.I. Gilbert (Pergamon Press, New York, 1985)
83. J.F.V. Vincent, S. Ablett, Hydration and tanning in insect cuticle. J. Insect Physiol. **33**, 973–979 (1987)
84. H.C. Lichtenegger et al., Zinc and mechanical prowess in the jaws of Nereis, a marine worm. Proc. Natl. Acad. Sci. **100**, 9144–9149 (2003)
85. R.M.S. Schofield et al., Br-rich tips of calcified crab claws are less hard but more fracture resistant: a comparison of biomineralized and heavy-element biomaterials. J. Struct. Biol. **166**, 272–287 (2009)
86. A. Becker, M. Epple, The mineral phase in the cuticle of two species of Crustacea consists of magnesium calcite, amorphous calcium carbonate and amorphous calcium phosphate. Dalt. Trans. 1814–1820 (2005)
87. A. Al-Sawalmih et al., Microtexture and chitin/calcite orientation relationship in the mineralized exoskeleton of the American lobster. Adv. Funct. Mater. **18**, 3307–3314 (2008)
88. H.A. Lowenstam, S. Weiner, *On Biomineralization* (Oxford University Press, Oxford, 1989)
89. S. Bentov et al., Enamel-like apatite crown covering amorphous mineral in a crayfish mandible. Nat. Commun. **3**, 839 (2012)
90. J.C. Weaver et al., The stomatopod dactyl club: a formidable damage-tolerant biological hammer. Science **336**, 1275–1280 (2012)
91. F. Neues, A. Ziegler, M. Epple, The composition of the mineralized cuticle in marine and terrestrial isopods: a comparative study. CrystEngComm **9**, 1245–1251 (2007)
92. H.O. Fabritius et al., Functional adaptation of crustacean exoskeletal elements through structural and compositional diversity: a combined experimental and theoretical study. Bioinspiration Biomimetics **11**, 55006 (2016)
93. R. Dillaman, S. Hequembourg, M. Gay, Early pattern of calcification in the dorsal carapace of the blue crab, *Callinectes sapidus*. J. Morphol. **263**, 356–374 (2005)
94. H.O. Fabritius, C. Sachs, P.R. Triguero, D. Raabe, Influence of structural principles on the mechanics of a biological fiber-based composite material with hierarchical organization: the exoskeleton of the lobster *Homarus americanus*. Adv. Mater. **21**, 391–400 (2009)
95. H.-O. Fabritius et al., Correlation of structure, composition and local mechanical properties in the dorsal carapace of the edible crab Cancer pagurus. Z. Krist. **227**, 766–776 (2012)
96. S. Ruangchai, C. Reisecker, S. Hild, A. Ziegler, The architecture of the joint head cuticle and its transition to the arthrodial membrane in the terrestrial crustacean *Porcellio scaber*. J. Struct. Biol. **182**, 22–35 (2013)
97. B. Seidl et al., Ultrastructure and mineral distribution in the tergite cuticle of the beach isopod *Tylos europaeus* Arcangeli, 1938. J. Struct. Biol. **174**, 512–526 (2011)
98. H. Fabritius, C. Sachs, D. Raabe, S. Nikolov, M. Friák, J. Neugebauer, Chitin in the exoskeletons of arthropoda: from ancient design to novel materials science, in *Chitin: Formation and Diagenesis*, ed. by N.S. Gupta (Springer, Berlin, 2011), pp. 35–60
99. J. Huber, H.-O. Fabritius, E. Griesshaber, A. Ziegler, Function-related adaptations of ultrastructure, mineral phase distribution and mechanical properties in the incisive cuticle of mandibles of *Porcellio scaber* Latreille, 1804. J. Struct. Biol. **188**, 1–15 (2014)

100. D. Klocke, H. Schmitz, Water as a major modulator of the mechanical properties of insect cuticle. Acta Biomater. **7**, 2935–2942 (2011)
101. D. Taylor, J.-H.J.-H. Dirks, Shape optimization in exoskeletons and endoskeletons: a biomechanics analysis. J. R. Soc. Interface **9**, 3480–3489 (2012)
102. J.-H. Dirks, E. Parle, D. Taylor, Fatigue of insect cuticle. J. Exp. Biol. **216**, 1924–1927 (2013)
103. C. Sachs, H. Fabritius, D. Raabe, Experimental investigation of the elastic-plastic deformation of mineralized lobster cuticle by digital image correlation. J. Struct. Biol. **155**, 409–425 (2006)
104. C. Sachs, H. Fabritius, D. Raabe, Hardness and elastic properties of dehydrated cuticle from the lobster *Homarus americanus* obtained by nanoindentation. J. Mater. Res. **21**, 1987–1995 (2006)
105. C. Sachs, H. Fabritius, D. Raabe, Influence of microstructure on deformation anisotropy of mineralized cuticle from the lobster *Homarus americanus*. J. Struct. Biol. **161**, 120–132 (2008)
106. A.C. Neville, D.A. Parry, J. Woodhead-Galloway, The chitin crystallite in arthropod cuticle. J. Cell Sci. **21**, 73–82 (1976)
107. Z. Střelcová, P. Kulhánek, M. Friák, H.-O. Fabritius, M. Petrov, J. Neugebauer, J. Koča, The structure and dynamics of chitin nanofibrils in an aqueous environment revealed by molecular dynamics simulations. RSC Adv. **6**, 30710–30721 (2016)
108. M.M. Giraud-Guille, Fine structure of the chitin-protein system in the crab cuticle. Tissue Cell **16**, 75–92 (1984)
109. P.Y. Chen, A.Y.M. Lin, J. McKittrick, M.A. Meyers, Structure and mechanical properties of crab exoskeletons. Acta Biomater. **4**, 587–596 (2008)
110. M. Erko et al., Micro- and nano-structural details of a spider' s filter for substrate vibrations: relevance for low-frequency signal transmission. R. Soc. Interface **12**, 2014111 (2015)
111. E. Parle, J.H. Dirks, D. Taylor, Damage, repair and regeneration in insect cuticle: the story so far, and possibilities for the future. Arthropod Struct. Dev. **46**, 49–55 (2017)
112. D. Raabe, C. Sachs, P. Romano, The crustacean exoskeleton as an example of a structurally and mechanically graded biological nanocomposite material. Acta Mater. **53**, 4281–4292 (2005)
113. F. Boßelmann, P. Romano, H. Fabritius, D. Raabe, M. Epple, The composition of the exoskeleton of two crustacea: the American lobster *Homarus americanus* and the edible crab *Cancer pagurus*. Thermochim. Acta **463**, 65–68 (2007)
114. S. Kinoshita, *Structural Colors in the Realm of Nature* (Word Scientific Publishing, Singapore, 2008)
115. P.R. Andrew, D. McKenzie, M. Large, Multilayer reflectors in animals using green and gold beetles as contrasting examples. J. Exp. Biol. **201**, 1307–1313 (1998)
116. B.D. Wilts, K. Michielsen, H. De Raedt, D.G. Stavenga, Hemispherical Brillouin zone imaging of a diamond-type biological photonic crystal. J. R. Soc. Interface **9**, 1609–1614 (2012)
117. V. Saranathan et al., Structure, function, and self-assembly of single network gyroid (I4132) photonic crystals in butterfly wing scales. Proc. Natl. Acad. Sci. **107**, 11676–11681 (2010)
118. X. Wu, A. Erbe, D. Raabe, H.O. Fabritius, Extreme optical properties tuned through phase substitution in a structurally optimized biological photonic polycrystal. Adv. Funct. Mater. **23**, 3615–3620 (2013)
119. H. Ghiradella, Insect cuticular surface modifications: scales and other structural formations. Adv. In Insect Phys. **38**, 135–180 (2010)
120. D. Hull, T.W. Clyne, *An Introduction to Composite Materials* (Cambridge University Press, Cambridge, 1996)
121. K. Jin, X. Feng, Z. Xu, Mechanical properties of chitin-protein interfaces: a molecular dynamics study. Bionanoscience **3**, 312–320 (2013)
122. Z. Yu, D. Lau, Molecular dynamics study on stiffness and ductility in chitin–protein composite. J. Mater. Sci. **50**, 7149–7157 (2015)
123. S. Nikolov et al., Revealing the design principles of high-performance biological composites using ab initio and multiscale simulations: the example of lobster cuticle. Adv. Mater. **22**, 519–526 (2010)

124. F.G. Barth, *A Spider's World: Senses and Behavior* (Springer Science & Business Media, Berlin, 2013)
125. J. Lian, J. Wang, Microstructure and mechanical anisotropy of crab *Cancer magister* exoskeletons. Exp. Mech. **54**, 229–239 (2014)
126. H. Fabritius, C. Sachs, D. Raabe, S. Nikolov, M. Friák, J. Neugebauer, Chitin in the exoskeletons of Arthropoda: From ancient design to novel materials science, Chitin: formation and diagenesis, ed. By S.N. Gupta (Springer, Germany, 2011)
127. E. Degtyar, M.J. Harrington, Y. Politi, P. Fratzl, The mechanical role of metal ions in biogenic protein-based materials. Angew. Chemie Int. Ed. **53**, 12026–12044 (2014)
128. J.F.V. Vincent, Arthropod cuticle: a natural composite shell system. Compos. Part A Appl. Sci. Manuf. **33**, 1311–1315 (2002)
129. S.R. Cohen, E. Kalfon-Cohen, Dynamic nanoindentation by instrumented nanoindentation and force microscopy: a comparative review. Beilstein J. Nanotechnol. **4**, 815–833 (2013)
130. D.M. Ebenstein, L.A. Pruitt, Nanoindentation of biological materials. Nano Today **1**, 26–33 (2006)
131. I. Zlotnikov, E. Zolotoyabko, P. Fratzl, Nano-scale modulus mapping of biological composite materials: theory and practice. Prog. Mater Sci. **87**, 292–320 (2017)
132. I. Zlotnikov et al., In situ elastic modulus measurements of ultrathin protein-rich organic layers in biosilica: towards deeper understanding of superior resistance to fracture of biocomposites. RSC Adv. **3**, 5798 (2013)
133. B. Bayerlein et al., Inherent role of water in damage tolerance of the prismatic mineral-organic biocomposite in the shell of *Pinna nobilis*. Adv. Funct. Mater. **26**, 3663–3669 (2016)
134. B. Bar-On, B. Bayerlein, H. Blumtritt, I. Zlotnikov, Dynamic response of a single interface in a biocomposite structure. Phys. Rev. Lett. **115**, 1–5 (2015)
135. S.L. Young et al., A spider's biological vibration filter: micromechanical characteristics of a biomaterial surface. Acta Biomater. **10**, 4832–4842 (2014)
136. B. Bar-On, F.G. Barth, P. Fratzl, Y. Politi, Multiscale structural gradients enhance the biomechanical functionality of the spider fang. Nat. Commun. **5**, 3894 (2014)
137. M.-M. Giraud-Guille, Plywood structures in nature. Biomaterials **3**, 221–227 (1998)
138. F.G. Barth, Microfiber reinforcement of an arthropod cuticle. Laminated composite material in biology. Z. Zellforsch. Mikrosk. Anat. **144**, 409–433 (1973)
139. S.A. Wainwright, *Mechanical Design in Organisms* (Princeton University Press, Princeton, 1982)
140. B. Bar-On, H.D. Wagner, Structural motifs and elastic properties of hierarchical biological tissues—a review. J. Struct. Biol. **183**, 149–164 (2013)
141. D. Labonte, A.K. Lenz, M.L. Oyen, On the relationship between indentation hardness and modulus, and the damage resistance of biological materials. Acta Biomater. **57**, 373–383 (2017)
142. D. Tabor, The hardness of solids. Rev. Phys. Technol. **1**, 145 (1970)
143. A.C. Fischer-Cripps, *Introduction to Contact Mechanics*. Springer Science+Business Media, LLC (Springer US) (2007)
144. B. Bhushan, *Principles and Applications of Tribology* (Wiley, New York, 2013)
145. A.K. Bhattacharya, W.D. Nix, Finite element simulation of indentation experiments. Int. J. Solids Struct. **24**, 881–891 (1988)
146. A.K. Bhattacharya, W.D. Nix, Finite element analysis of cone indentation. Int. J. Solids Struct. **27**, 1047–1058 (1991)
147. M.F. Ashby, L.J. Gibson, *Cellular Solids: Structure and Properties* (Press Syndicate of the University of Cambridge, Cambridge, 1997)
148. F.W. Zok, A. Miserez, Property maps for abrasion resistance of materials. Acta Mater. **55**, 6365–6371 (2007)
149. S. Amini, A. Miserez, Wear and abrasion resistance selection maps of biological materials. Acta Biomater. **9**, 7895–7907 (2013)
150. D.N. Moses, M.G. Pontin, J.H. Waite, F.W. Zok, Effects of hydration on mechanical properties of a highly sclerotized tissue. Biophys. J. **94**, 3266–3272 (2008)

151. Y. Shelef, B. Bar-On, Surface protection in bio-shields via a functional soft skin layer: lessons from the turtle shell. J. Mech. Behav. Biomed. Mater. 1 (2017)
152. S. Suresh, Graded materials for resistance to contact deformation and damage. Science **292**, 2447–2451 (2001)
153. J.G. Kunkel, W. Nagel, M.J. Jercinovic, Mineral fine structure of the American lobster cuticle. J. Shellfish Res. **31**, 512–526 (2012)
154. H. Hertz, Üeber die berührung fester elastischer Körper. J. fur die reine und Angew. Math. **92**, 156–171 (1882)
155. K.L. Johnson et al., *Contact Mechanics* (Cambridge University Press, Cambridge, 1985)
156. C.C. Broomell, F.W. Zok, J.H. Waite, Role of transition metals in sclerotization of biological tissue. Acta Biomater. **4**, 2045–2051 (2008)
157. R.O. Ritchie, The conflicts between strength and toughness. Nat. Mater. **10**, 817–822 (2011)
158. F.D. Fischer, O. Kolednik, J. Predan, H. Razi, P. Fratzl, Crack driving force in twisted plywood structures. Acta Biomater. **55**, 349–359 (2017)
159. N. Suksangpanya, N.A. Yaraghi, D. Kisailus, P. Zavattieri, Twisting cracks in Bouligand structures. J. Mech. Behav. Biomed. Mater. **76**, 38–57 (2017)
160. S.M. Chen, H.L. Gao, Y.B. Zhu, H.B. Yao, L.B. Mao, Q.Y. Song, S.H. Yu, Biomimetic twisted plywood structural materials. Nat. Sci. Rev. **5**(5), 703–714 (2018)
161. N. Suksangpanya, N.A. Yaraghi, R.B. Pipes, D. Kisailus, P. Zavattieri, Crack twisting and toughening strategies in Bouligand architectures. Int J Solids Struct. **150**, 83–106 (2018)
162. A.G. Atkins, Y.-W. Mai, *Elastic and Plastic Fracture: Metals, Polymers, Ceramics, Composites, Biological Materials* (Ellis Horwood; Halsted Press, 1985)
163. R.H. Hackman, Chitin and the fine structure of cuticles, in *Chitin and Benzoylphenyl Ureas*, pp. 1–32 (1987)
164. M.M. Giraud-Guille, H. Chanzy, R. Voung, Chitin crystals in arthropod cuticles revealed by diffraction contrast transmission electron microscopy. J. Struct. Biol. **103**, 232–240 (1990)
165. M.H. Dickinson et al., How animals move: an integrative view. Source Sci. New Ser. **288**, 100–106 (2000)
166. J.-H.J.-H. Dirks, D. Taylor, Fracture toughness of locust cuticle. J. Exp. Biol. **215**, 1502–1508 (2012)
167. J.H. Dirks, D. Taylor, Veins improve fracture toughness of insect wings. PLoS One **7**, e43411 (2012)
168. J. Sun, B. Bhushan, The structure and mechanical properties of dragonfly wings and their role on flyability. Comptes Rendus—Mec. **340**, 3–17 (2012)
169. N.S. Ha, T.L. Jin, N.S. Goo, H.C. Park, Anisotropy and non-homogeneity of an *Allomyrina dichotoma* beetle hind wing membrane. Bioinspiration Biomimetics **6**, 46003 (2011)
170. M. Sun et al., Influence of cuticle nanostructuring on the wetting behaviour/states on cicada wings. PLoS One **7**, e35056 (2012)
171. S.A. Combes, T.L. Daniel, Flexural stiffness in insect wings I. Scaling and the influence of wing venation. J. Exp. Biol. **206**, 2979–2987 (2003)
172. S.A. Combes, Flexural stiffness in insect wings II. Spatial distribution and dynamic wing bending. J. Exp. Biol. **206**, 2989–2997 (2003)
173. S.A. Combes, T.L. Daniel, Into thin air: contributions of aerodynamic and inertial-elastic forces to wing bending in the hawkmoth *Manduca sexta*. J. Exp. Biol. **206**, 2999–3006 (2003)
174. A.M. Mountcastle, S.A. Combes, Wing flexibility enhances load-lifting capacity in bumblebees. Proc. Biol. Sci. **280**, 20130531 (2013)
175. F.O. Lehmann, The mechanisms of lift enhancement in insect flight. Naturwissenschaften **91**, 101–122 (2004)
176. S.P. Sane, The aerodynamics of insect flight. J. Exp. Biol. **206**, 4191–4208 (2003)
177. L. Zhao, Q. Huang, X. Deng, S.P. Sane, Aerodynamic effects of flexibility in flapping wings. J. R. Soc. Interface **7**, 485–497 (2010)
178. T. Nakata, H. Liu, Aerodynamic performance of a hovering hawkmoth with flexible wings: a computational approach. Proc. Biol. Sci. **279**, 722–731 (2012)

179. H. Izumi, M. Suzuki, S. Aoyagi, T. Kanzaki, Realistic imitation of mosquito's proboscis: electrochemically etched sharp and jagged needles and their cooperative inserting motion. Sensors Actuators A Phys. **165**, 115–123 (2011)
180. M.K. Ramasubramanian, O.M. Barham, V. Swaminathan, Mechanics of a mosquito bite with applications to microneedle design. Bioinspiration Biomimetics **3**, 1–10 (2008)
181. Z.-L. Zhao et al., Structures, properties, and functions of the stings of honey bees and paper wasps: a comparative study. Biol. Open **4**, 921–928 (2015)
182. M.A. Meyers, A.Y.M. Lin, Y.S. Lin, E.A. Olevsky, S. Georgalis, The cutting edge: sharp biological materials. JOM **60**, 19–24 (2008)
183. A. van der Meijden, T. Kleinteich, A biomechanical view on stinger diversity in scorpions. J. Anat. **230**(4), 497–509 (2017)
184. C. Broeckhoven, A. du Plessis, Has snake fang evolution lost its bite? New insights from a structural mechanics viewpoint. Biol. Lett. **13**(8), 20170293 (2017)
185. B. Bhushan, Insects Locomotion, Piercing, *Sucking and Stinging Mechanisms, in Biomimetics. Springer Series in Materials Science*, vol. 279 (Springer, Cham, 2018)
186. B. Bar-On, On the form and bio-mechanics of venom-injection elements. Acta Biomater. **85**, 263–271 (2019)
187. Z. Liu et al., Enhanced protective role in materials with gradient structural orientations: lessons from nature. Acta Biomater. **44**, 31–40 (2016)
188. A. van der Meijden, T. Kleinteich, A biomechanical view on stinger diversity in scorpions. J. Anat. **230**, 497–509 (2017)
189. W. Gnatzy, J.J. Tautz, Ultrastructure and mechanical properties of an insect mechanoreceptor: stimulus-transmitting structures and sensory apparatus of the cereal filiform hairs of *Gryllus*. Cell Tissue Res. **213**, 441–463 (1980)
190. P. Fratzl, F.G. Barth, Biomaterial systems for mechanosensing and actuation. Nature **462**, 442–448 (2009)
191. M.E. McConney et al., Surface force spectroscopic point load measurements and viscoelastic modelling of the micromechanical properties of air flow sensitive hairs of a spider (*Cupiennius salei*). J. R. Soc. Interface **6**, 681–694 (2009)
192. F.G. Barth, Spider mechanoreceptors. Curr. Opin. Neurobiol. **14**, 415–422 (2004)
193. H.-E. Dechant, F.G. Rammerstorfer, F.G. Barth, Arthropod touch reception: Stimulus transformation and finite element model of spider tactile hairs. J. Comp. Physiol.—A Sensory, Neural Behav. Physiol. **187**, 313–322 (2001)

# Chapter 11
# The Multiscale Architectures of Fish Bone and Tessellated Cartilage and Their Relation to Function

**Ronald Seidel, Aravind K. Jayasankar, Ron Shahar and Mason N. Dean**

**Abstract** When describing the architecture and ultrastructure of animal skeletons, introductory biology, anatomy and histology textbooks typically focus on the few bone and cartilage types prevalent in humans. In reality, cartilage and bone are far more diverse in the animal kingdom, particularly within fishes, where cartilage and bone types exist that are characterized by features that are anomalous or even pathological in human skeletons. Here, we discuss the curious and complex architectures of fish bone and shark and ray cartilage, highlighting similarities and differences with their mammalian skeletal tissue counterparts. By synthesizing older anatomical literature with recent high-resolution structural and materials characterization work, we frame emerging pictures of form-function relationships in these tissues and of the evolution and true diversity of cartilage and bone.

## 11.1 Introduction

A major innovation in vertebrate evolution was the emergence of calcium phosphate-based mineralized tissues, which enabled the evolution of a diversity of protective armors, teeth and an internal skeleton [1–3]. Mineralized skeletons first appeared in

R. Seidel · A. K. Jayasankar · M. N. Dean (✉)
Department of Biomaterials, Max Planck Institute of Colloids and Interfaces, 14424 Potsdam, Germany
e-mail: Mason.Dean@mpikg.mpg.de

R. Seidel
e-mail: Ronald.Seidel@mpikg.mpg.de

A. K. Jayasankar
e-mail: Aravind.Jayasankar@mpikg.mpg.de

R. Shahar
The Robert H. Smith Faculty of Agriculture, Food and Environmental Sciences, Koret School of Veterinary Medicine, The Hebrew University of Jerusalem, PO Box 12, 76100 Rehovot, Israel
e-mail: Ron.Shahar1@mail.huji.ac.il



Y. Estrin et al. (eds.), *Architectured Materials in Nature and Engineering*, Springer Series in Materials Science 282,
https://doi.org/10.1007/978-3-030-11942-3_11

fishes, some 500 million years ago, as dermal exoskeletons covering the bodies of basal jawless fishes. This outer bony armor was lost in evolution and later supplanted by the internal skeletons we know from modern species that, for the most part, are either entirely bony (as in the vast majority of vertebrates) or consist of a cartilaginous core stiffened by a mineralized collar (as in sharks and rays, the "cartilaginous fishes") [1, 4, 5].

Regardless of whether the primary skeletal tissue type is mineralized cartilage or bone, all vertebrate mineralized tissues are comprised of water, non-collagenous proteins, and collagen fibrils mineralized with carbonated apatite (e.g. [6]). The range of material properties observed in vertebrate skeletal tissues are therefore not determined by changes in chemistry (e.g. by introducing heavy metals, as in some invertebrate tissues) (e.g. [7]), but rather through variation in the proportions of the building blocks—proteins, mineral and water—and in the architectures of the tissues [6, 8]. The bulk mineral content of mammalian bone, for example, ranges from ~50% to nearly 100% ash content (from antler to beaked whale rostrum, respectively), resulting in large variations in stiffness and toughness [9].

The natural diversity of mammalian bone has made it a particularly useful system for studying the effects of composition and structure on mineralized tissue material properties. There are bone tissues with extremely high or low bulk mineral densities (e.g. otic vs. antler bone); bone tissue with similar bulk mineral density, but drastically different ultrastructure or mesostructure (e.g. osteonal vs. fibrolamellar and cortical vs. cancellous bone, respectively); and bone tissue with local variation of both composition and structure (e.g. cortical bone with fiber-lamellar and osteonal regions) (see review in [10]). Studies of these tissues have informed our understanding of structural and compositional complexity in bony skeletons, however, have focused predominantly on a small fraction of mammalian species, thereby largely ignoring the wide range of habitat, bone, and skeletal tissue types of the rest of vertebrates.

To understand the true scope of variation of vertebrate skeletal tissues, with regard to structure, composition, and mechanics, it is important to look beyond the limited number of the often-used model species and sample vertebrates as broadly as possible. By examining evolutionary alternatives to mammalian bone, we can expand and reframe our understanding of form-function relationships in vertebrate skeletal tissues and the factors that drove their evolution. Here, we discuss the two major mineralized tissue types of fishes: the bone of bony fishes and the tessellated cartilage of sharks and rays (Fig. 11.1). These tissues are comprised of the same basic components as the bone in mammalian skeletons, but exhibit distinct 'mesostructural' tissue arrangements that result in distinctive mechanical properties. Furthermore, as fishes occupy a huge range of habitats and ecological niches, exhibit extensive morphological diversity, and represent at least 55% of all vertebrate species, they

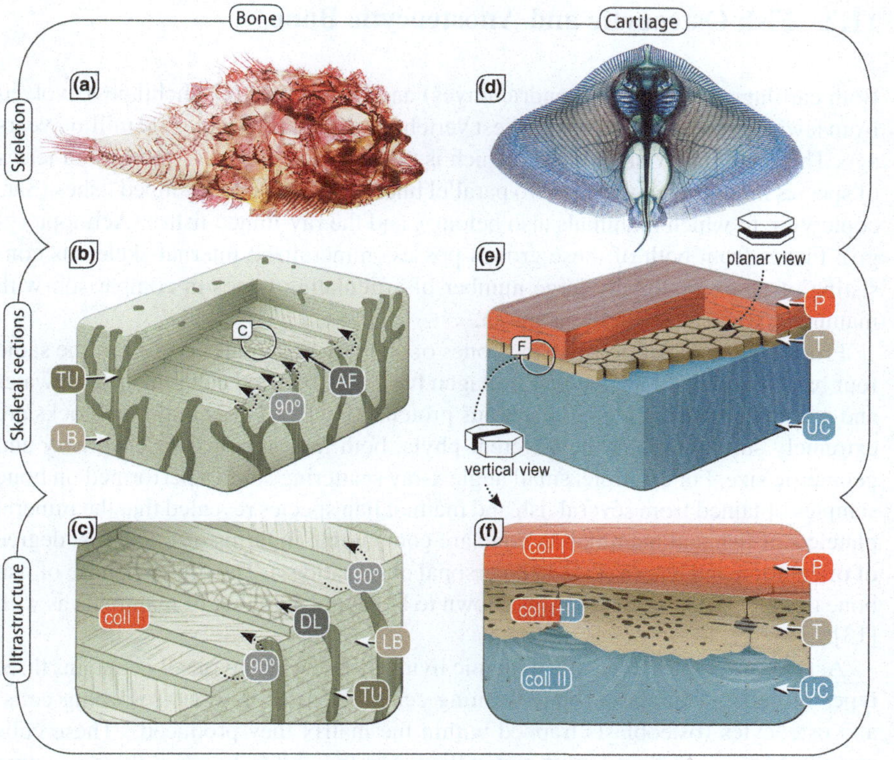


**Fig. 11.1 General organization of the two primary mineralized skeletal tissues in fish.** (**a–c** Bony fish bone; **d–f** Shark and ray tessellated cartilage) Detailed descriptions of the two tissue types are provided later in text. **a** Cleared and stained bony fish skeleton (lumpsucker, *Eumicrotremus orbis*). **b–c** Fish bone, depending on the species, is either osteocytic or anosteocytic, with perforating tubules, and exhibiting a laminated tissue organization. Each layer is patterned on fibers, smoothly and gradationally spiraled from layer to layer (plywood-like structure) until a turn of 180° is reached, and a layer of anisotropically oriented fibers (AF) is incorporated. **d** Cleared and stained butterfly ray skeleton (*Gymnura* sp.). Specimen is young and not yet fully mineralized: blue color shows the cartilage of the skeleton, which will form a mineralized layer later in life. **e, f** Structure of tessellated cartilage of sharks and rays, comprising mineralized tiles (tesserae) covering the skeletal cartilage. Abbr.: DL = disordered layer; coll = collagen; LB = layered bone; P = Perichondrium; T = Tesserae; TU = Tubules; UC = uncalcified cartilage. Images in **a** and **d** courtesy of A. Summers

offer unparalleled opportunity for defining and understanding the parameter space of form-function relationships in vertebrate skeletal tissues. The study of structure-function relationships in fish skeletal tissues is still incipient: this chapter aims to describe the current understanding of fish skeletal tissue architecture, drawing on recent high-resolution ultrastructural data, and discussing potential links between these features and available mechanics data.

## 11.2 Fish Osteocytic and Anosteocytic Bone

Both cartilaginous fishes (Chondrichthyes) and bony fishes (Osteichthyes) evolved from jawless fish ancestors, the earliest vertebrates appearing some 500 million years ago. The superclass Osteichthyes, which is the largest group of vertebrates in terms of species numbers, evolved in two parallel lines, namely the lobe-finned fishes (Sarcopterygii, to which mammals also belong), and the ray-finned fishes (Actinopterygii). Fishes from both of these groups possess mineralized internal skeletons consisting of an exceptionally large number of articulating bones in comparison with mammals, particularly in the cranium.

From a material perspective, the bones of all bony vertebrates consist of the same four basic building blocks: type-I collagen fibrils, carbonated calcium apatite, water and a small amount of non-collagenous proteins (NCPs). These building blocks are extremely similar among the different phyla, both in terms of their chemistry and geometric size. For example, small angle x-ray scattering studies performed on bone samples obtained from several fish and mammalian species revealed that the mineral platelets of fish and mammalian bone are comparable in terms of thickness, degree of orientation and length ([11, 12]; personal observations). The organic phase of fish bone (mostly type-I collagen) was shown to be identical to that of mammals as well [13].

Across vertebrates, bone is a dynamic living tissue, which typically contains three types of cells: osteoblasts (bone forming cells), osteoclasts (bone resorbing cells) and osteocytes (osteoblasts trapped within the matrix they produced). These cells provide bones with their active metabolism and the ability to adapt their structure and composition to changing loading circumstances (a process called modeling). In addition, bone cells allow for continual renewal of the tissue, by replacing packets of old bone with new and flawless material (remodeling). Modeling and remodeling are carried out by the coordinated activity of osteoclasts and osteoblasts, and are believed to be regulated by the osteocytes, perhaps by responding to changes in fluid flow in the lacuna-canalicular network and/or local bone strains [10, 14, 15].

Although the cellular and matrix components of the bones of the fish skeleton are largely similar to those of other vertebrates (Fig. 11.2a–f), two remarkable observations, made already 150 years ago, delineated drastic differences. Firstly, it was found that the bones of a substantial number of fish species—more than half of the several tens of thousands of fish species [16]—do not contain osteocytes (i.e. are anosteocytic), although they do contain osteoblasts and osteoclasts (Fig. 11.2g–i). Secondly, it was observed that anosteocytic bone occurs predominantly in the advanced teleosts, whereas the bones of the basal teleosts are typically osteocytic [17]. This observation suggests that osteocytes were effectively lost during the evolution of the bony fish lineage and therefore that the anosteocytic state represents a functional advantage of some sort to the fish skeleton, although the nature of this advantage is still unclear.

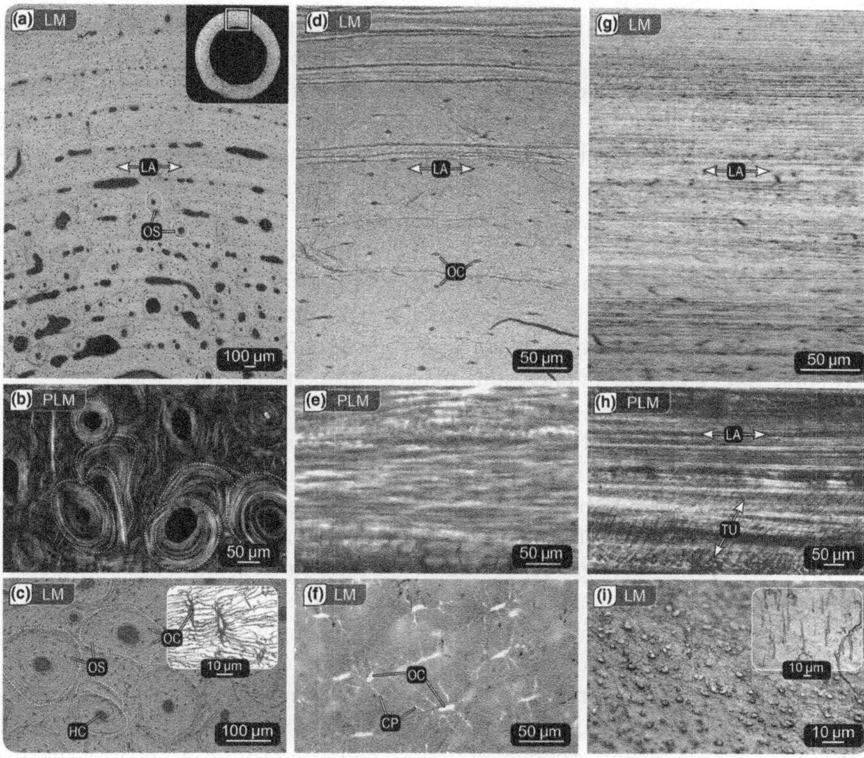


**Fig. 11.2 Ultrastructural differences of vertebrate bone types**. (**a–c** mammalian bone; **d–f** fish osteocytic bone; **g–i**: fish anosteocytic bone). **a** Cross-section of a mammalian tibia exhibiting a lamellar appearance with multiple primary osteons. **b** Secondary osteons under polarized light, their overlapping indicating bone remodeling. **c** Light microscopy image of secondary osteons, showing their central Haversian canals, which contain blood vessels and nerves, and are surrounded by concentric lamellae. Osteocytic lacunae can be seen within the lamellae, with small canaliculi connecting each lacuna to its neighbors, together forming the lacuna-canalicular network. **d–f** Osteocytic bone from carp (*Cyprinus carpio*) and **g–i** anosteocytic bone from tilapia (*Oreochromis aureus*) have a similar lamellar appearance, but the latter bone type lacks the osteocytes present in the osteocytic carp bone (shown in **f**). Tubules (**h**) perforate the lamellae of both fish bone types. Abbr.: CP = cell processes; HC = Haversian canal; LA = Laminae/Layers; LS = Lacunar spaces; OC = Osteocytes; OS = Osteon; TU = Tubules. Image in **c** courtesy of A. Roschger

Of particular interest is how advanced fish can maintain the health of their bones and avoid failure without having osteocytes. The belief that osteocytes are vital to bone health and mutability is a central paradigm in bone biology [10, 18] and argues that bone without osteocytes will be unable to model and remodel, and will therefore frequently fail disastrously. This is supported by studies showing that the removal of osteocytes has a detrimental effect on bone health and material properties (reviewed in [10]). And yet, the anosteocytic bone of neoteleost fish is not only an extremely common bone type—there are likely 16,000 species of anosteocytic fishes—but also the skeletons of neoteleost fishes perform their mechanical function for very long times without failure (some fish have life expectancies of more than 100 years [19]).

The predominant lack of osteocytes among fishes can have several possible explanations. One possibility is that osteocytes do not perform the functions we have assigned to them, at least not exclusively, and the regulation of modeling and remodeling can be accomplished without the need for osteocytes. Another (even more intriguing) possibility is that the structural features of anosteocytic bone material are such that modeling and remodeling are not required for the tissue's health. This could mean, for example, that anosteocytic bone material is inherently highly resistant to crack initiation and/or crack propagation. We discuss below several recently characterized structural features that appear to be unique to fish bone and that may play some role in mediating interesting material properties of the tissue.

Although many fish species lack a fundamental cellular component of bone (osteocytes), the bone matrices of both anosteocytic and osteocytic fish species typically exhibit a lamellar structure, a basic meso-scale structural motif shared with the bone of all other vertebrates. In both mammals and fishes, lamellae are typically <7 μm thick, consist of mineralized collagen fibrils (~100 nm in diameter) and ground substance (water and non-collagenous proteins), and form strata in the bone tissue that are visible in various forms of microscopy. In cross sections of mammalian long bones, for example, lamellae may form concentric rings surrounding the entire interior marrow cavity (Fig. 11.2a), or form smaller structural units (osteons) surrounding blood vessels [10]. Osteons are structural results of remodeling of the bone matrix (renovation of small regions) and can be packed in high densities (20–70 osteons/mm$^2$, [20]), especially in heavily remodeled tissues, and as a result can impart a striking structural patterning to mammalian bone (Fig. 11.2b, c) [10]. In contrast, fish bone was long described as being largely featureless in appearance, due to its often simple layered structure and comparative rarity of vasculature (Fig. 11.2d, g) [21]. The layered appearance of most fish bone (both osteocytic and anosteocytic) results from its formation process, where new lamellae are deposited upon older ones throughout the fish's life [21].

The recent use of modern structural and material analysis techniques, however, has shown that fish bone is in fact characterized by a wide array of rather complex architectural features, some of which are not seen in mammalian bone. These include hyperostoses (non-pathological, spherical bone growths attached to fin elements), acellular osteon-like structures, and porous feather- or honeycomb-like bone tissue morphologies (Fig. 11.3) (reviewed in [10, 21]). Modern characterization advances have also revealed nanostructural aspects distinct to fish bone lamellae. A recent study used serial surface view, focused ion beam milling—SSV FIB-SEM, a technique combining ablation of thin layers (5–15 nm thickness) and serial, high resolution imaging—to characterize the 3D arrangement of collagen fibrils at the nanometer scale in the anosteocytic opercular (gill cover) bone of tilapia (*Oreochromis aureus*), and compare it to the collagenous architecture of several types of mammalian cortical bone [21]. In mammals, bone lamellae are comprised of sub-lamellar layers that vary in their spatial arrangement, degree of order (co-alignment of fibers) and fiber directionality according to species and taxon [22]. The analysis of the anosteocytic fish bone lamellae revealed some similarity to mammalian bone in being comprised of planar sub-lamellar layers of mineralized collagen fibrils [21]. However, whereas mammalian bone seems to involve complicated interweavings of ordered and disordered sub-lamellar layers (the latter containing the cells and their processes), the lamellae of the anosteocytic fish bone exhibited a conceptually simpler arrangement. Fibrils in the sub-lamellar layers smoothly and gradationally spiraled from layer to layer, resulting in a layered plywood-like structure (Figs. 11.1a–c and 11.4a). When the sequence of fibril orientations completed a rotation of 180°, a thin layer of poorly-oriented fibrils intervened before the next spiral of sub-lamellar layers. As the bones of only a tiny fraction of existing fish species have been examined at this level of detail to date, it remains unclear how universal this 'lamellated bone' motif is among fishes and whether it is linked to the distinct mechanical properties of fish bone (see below).

In addition to the basic layered structure of fish bone, over the last 150 years several authors have noted the presence of micron-sized tubules within the bulk of the bone, in some cases present in very high densities [17, 23–28]. The nature of these tubules (e.g. their content and function) were not clear, and led to several contradictory speculations, positing that the tubules either contained unmineralized collagen fibrils, or sensory extensions of surface osteoblasts, or even osteocytes (i.e. contradicting the axiom that advanced teleost bone lacks osteocytes). The FIB-SEM study by Atkins and colleagues [21] confirmed the presence of these tubules, but also enabled the characterization of their contents and structure at high resolution and in three dimensions. The tubules were shown to form an intricate 3D array, with each 1–4 μm diameter tubule containing a bundle of tightly-packed hypo-mineralized collagen fibrils. The tubules were oriented mostly orthogonal to the lamellar plane (Fig. 11.4b–d), suggesting that they, and the internal fibers, are linking bone lamellae, in a manner analogous to Coptic bookbinding where pages are sewn together with thread. It is reasonable to assume that they serve a mechanical purpose, such as pinning together the bone layers and resisting shear failure, although this remains to be shown. They may also perhaps be associated with a sensory function, given that the tubules extend so deeply into the bone tissue and could somehow monitor the strains experienced within the bulk of the bone material.

The unique structural features of fish bone and the possibility that these affect an inherent failure resistance offer a strong motivation for the study of fish bone material, in particular its mechanical properties and behavior. Nevertheless, the mechanical properties of fish bone (and particularly anosteocytic bone) have received far less attention than those of the bones of tetrapods, especially mammals. Limited available data suggest that fish bone is much more compliant than tetrapod bone, having an elastic modulus of 5–8 GPa [17]. This is less than half of typical mammalian cortical bone stiffness in the principal direction (axial orientation of the long bones ~15–25 GPa [8]). In contrast, in post-yield mechanical testing, fish bone was observed to be considerably tougher than mammalian bone (can absorb much more energy before failure), with strains at failure often exceeding 5%, approximately two times the failure strains commonly seen in mammalian bone (Fig. 11.5a). Since fish bone consists of the same basic building blocks as tetrapod bone (mineralized

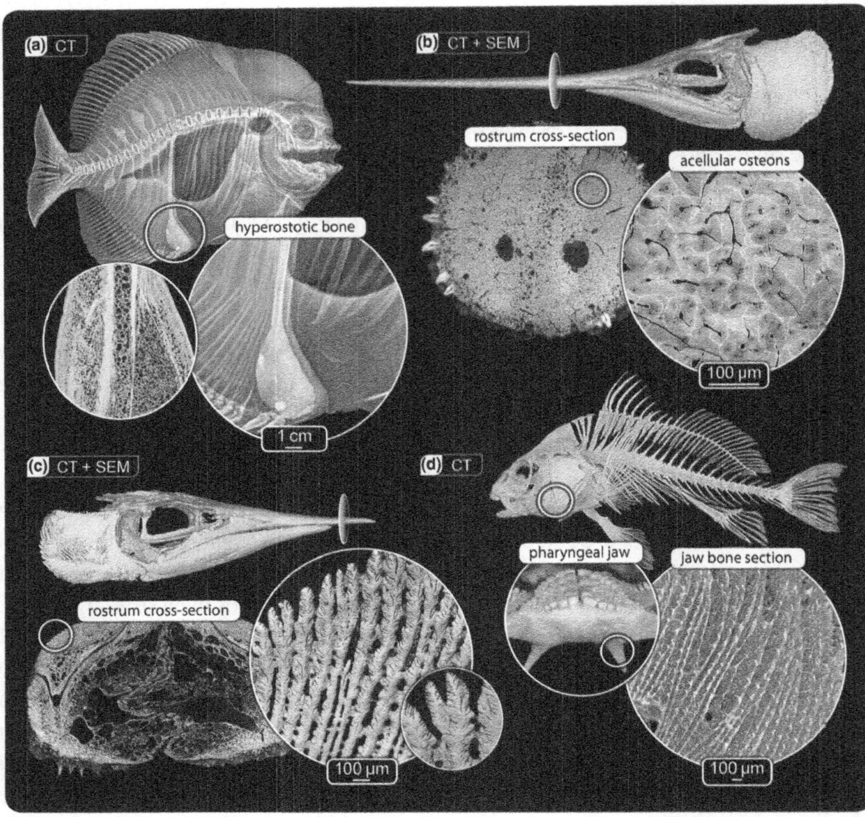


**Fig. 11.3 Variation in fish bone ultrastructure.** CT scan, light microscopy and backscatter SEM images of non-lamellar fish bone: **a** hyperostotic bone in longfin spadefish (*Platax teira*), **b** acellular osteons in the rostrum of blue marlin (*Makaira nigricans*), **c** feather-like bone in the rostrum of shortbill spearfish (*Tetrapturus angustirostris*) and **d** honeycomb-like organization of bone in the pharyngeal jaw of black drum (*Pogonias cromis*). Modified with permission from [64], copyright John Wiley Sons, Inc.

collagen fibrils and water), the mechanical data suggest that some features of the 3D structural arrangement of fish bone and its mineral density are responsible for its unusual mechanical behavior. We hypothesize that the mineral density of anosteocytic fish bone and its unique tubules-linking-lamellae ('bookbinding') structure are responsible for these unusual mechanical properties (e.g. that post-yield deformation is controlled and fracture is arrested by the hypo-mineralized tubular fiber bundles). Deeper investigation of the mechanical properties of fish bone, particularly post-yield behaviors and fracture toughness according to different load axes, will help to establish the structural roots of this tissue's impressive resistance to failure.

An additional and particularly intriguing observation about fish bones concerns the effect of wetting dry samples, or *vice versa*. All biological tissues are affected

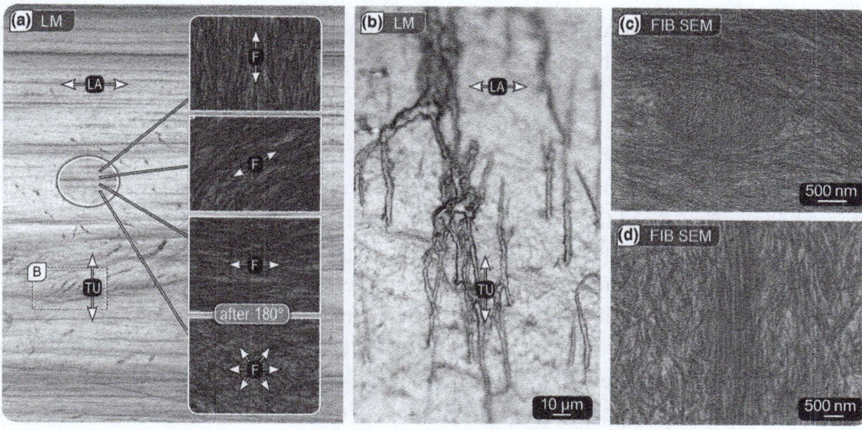


**Fig. 11.4 Fibrous organization and tubules in acellular fish bone (*Oreochromis aureus* operculum).** **a** In successive tissue lamellae of fish bone, the predominant fiber orientation is rotated, with a layer of random fiber orientation intervening after the fiber orientation has rotated 180°. **b–d** Tubules run perpendicular to and perforate the lamellae; in FIB-SEM images hypomineralized fiber bundles are visible within the tubules (**c** Tubule cross-section; **d** Tubule longitudinal section). Abbr.: F = Collagenous fibers; LA = Laminae/Layers; TU = Tubules

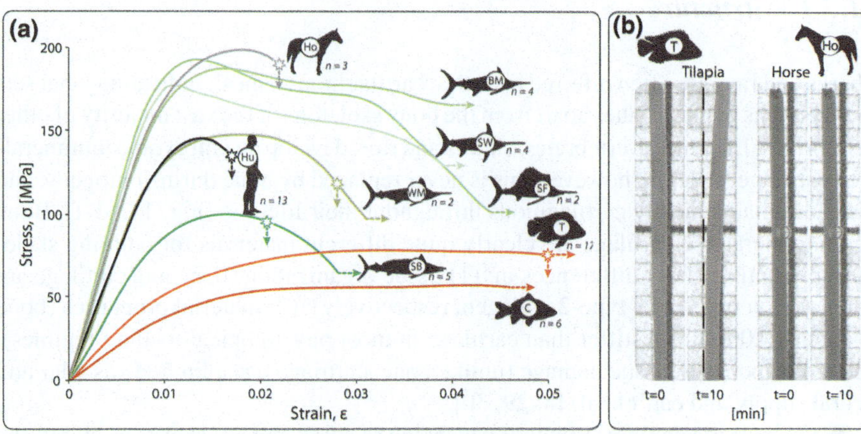


**Fig. 11.5 Variation in structure-function relationships in vertebrate bones.** **a** Stress-strain curves of bone beams from mammal and fish species tested in three-point bending; the initial slope is the stiffness, regions after the initial peak represent post-yield deformation, and stars indicate the average fracture strain for those samples fractured during testing. Lines lacking stars bent without failure until the displacement limit of the testing machine was reached. Note the high toughness (large post-yield region) of fish bone and combination of high stiffness and high toughness of billfish bones, relative to human and horse bone. Modified from [18], copyright 2014 National Academy of Sciences. **b** Comparison of the behavior of hydrated tilapia and horse bone beams (t = 0) following dehydration (t = 10 min); the large deflection of tilapia bone is reversible with rehydration

structurally and mechanically when they change from a wet to a dry state. This change is usually quite subtle in mineralized materials, involving only a slight deformation of tissue samples. However, we have observed, using samples from a variety of species, that this change can be quite dramatic in fish bone. For example, results of a test comparing beams of identical dimensions made of equine cortical bone and *Oreochromis aureus* (tilapia) opercular bone illustrate this difference: when hydrated equine beams were dried they deformed only slightly, whereas when hydrated beams of anosteocytic fish bone were dried they exhibited pronounced bending and some torsion (Fig. 11.5b). This transformation was completely reversible upon re-hydration in beams from both species. We postulate that the cause for the pronounced effect of wetting and drying on fish bone samples results from their unique 3D architecture and structural arrangements of the tubular hypo-mineralized and tightly-packed collagen bundles linking adjacent bone laminae (Fig. 11.4b–d), and we are currently investigating the phenomenon and its mechanics further.

## 11.3 Shark and Ray Tessellated Cartilage

### *11.3.1 Structure*

Sharks and rays are often referred to as the 'cartilaginous fishes', indicating what sets the skeletons of these fishes apart from the bony skeletons of the vast majority of other vertebrates. Like most vertebrates, sharks and rays develop an embryonic unmineralized cartilage skeleton; however, this is never replaced by bone during ontogeny, and instead remains mostly cartilaginous throughout their lifetime (Fig. 11.1d–f). Bone and unmineralized cartilage are clearly quite different materials for building skeletons, exhibiting major differences in: (1) tissue organization (bone and cartilage are patterned on type-I and type-2 collagen, respectively); (2) material properties (bone is about 10,000 times stiffer than cartilage in most physiological loading regimes); and (3) response to tissue damage (unlike bone, cartilage has a limited vascular and neural supply and can't heal) [8, 29, 30].

We believe the distinct structural patterning of shark and ray skeletons allows the cartilage to perform many of the same mechanical roles that bone performs in the other 98% of vertebrates: each piece of the cartilaginous endoskeleton is covered in a thin layer of thousands of mineralized, polygonal tiles called tesserae, typically hundreds of microns wide and deep [4, 31, 32] (Fig. 11.6). This tessellated crust of mineralized tissue is sandwiched between the unmineralized cartilage core of the skeleton and an outer fibrous perichondrium wrapping each skeletal element, resulting in a layered fibro-mineral composite. This unique endoskeletal tiling typically occupies only 30% or less of each skeletal element by volume [33], yet appears to be an important evolutionary innovation of elasmobranch fishes. Tesserae have characterized elasmobranch skeletons for more than 400 million years [34] and are vital to shark and ray skeletal biology. Tesserae not only permit interstitial growth

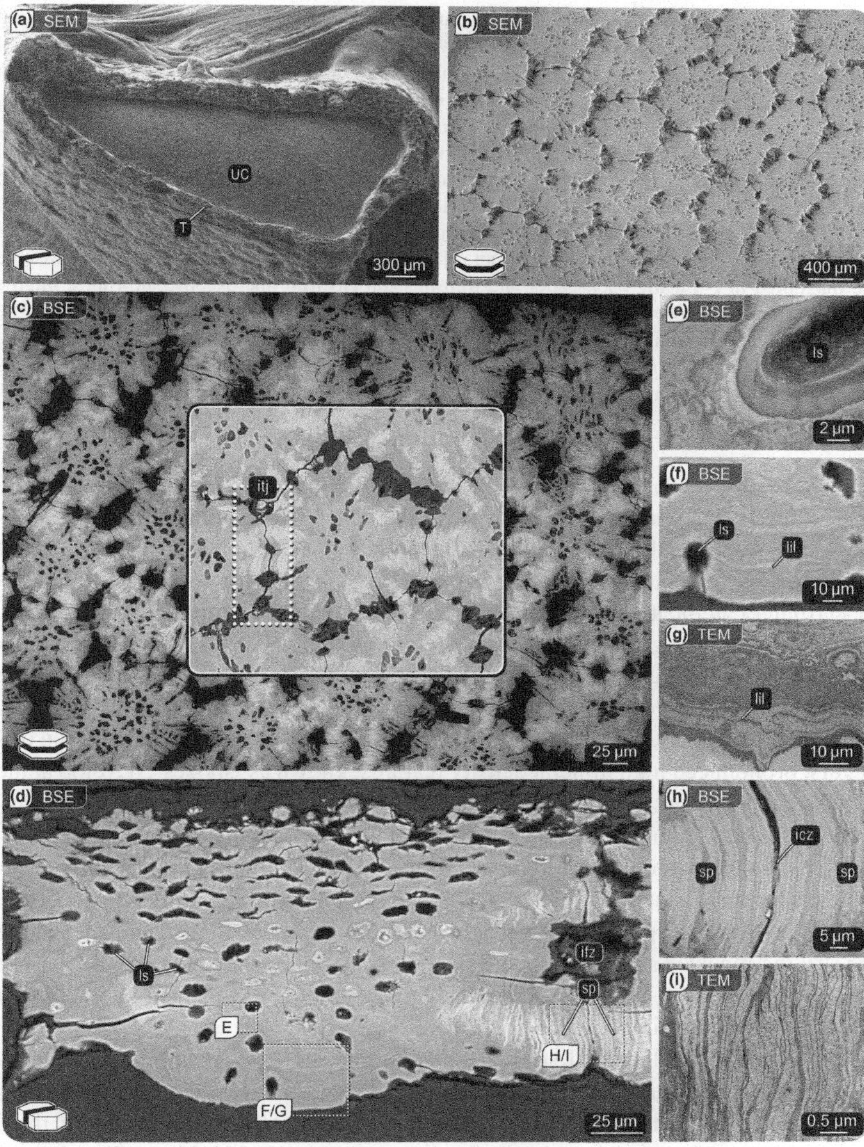


of the mineralized layer—via deposition of new mineral at the margins of tesserae [4, 31, 35], a growth mechanism impossible with a continuously mineralized crust that cannot remodel—but also afford stiffness to skeletal elements (see Sect. 11.3.2 below).

◂**Fig. 11.6 Tessellated cartilage of elasmobranch fish (sharks and rays).** Note icons showing section orientation, introduced in Fig. 11.1. **a** Cryo-electron microscopy image of an elasmobranch skeletal piece in cross-section showing the unmineralized cartilage core (*UC*) sheathed in a thin layer of mineralized tiles, called tesserae (*T*). **b, c** Planar and **d** vertical views of the tesseral layer, showing abutting tesserae from an adult specimen (**b** SEM; **c, d** backscatter SEM imaging, showing only mineralized tissue), revealing the variation in the shapes of tesserae and their intertesseral joints (comprised of regions of direct contact and gaps of fibrous connection between tesserae; see Fig. 11.7). Note the regional variation in cell lacunae shape and in gray value within tesserae in **b**, reflecting local differences in mineral density and showing regions of high mineralization associated with zones of intertesseral contact. The most prominent, diagnostic features of tesserae are magnified in **e–i** using backscatter and transmission electron microscopy, showing **e** filigree mineralization pattern surrounding a lacuna, **f, g** Liesegang lines of accretive growth and **h, i** the laminated, hypermineralized 'spokes' reinforcing intertesseral contact zones. All samples from the round stingray, *Urobatis halleri*. Abbr.: ICZ = intertesseral contact zone; IFZ = intertesseral fibrous zones; ITJ = intertesseral joint; LIL = Liesegang lines; LS = lacunar space; SP = spokes

The dogfish *Scyliorhynus canicula* and round stingray *Urobatis halleri* are the best-studied elasmobranch species as far as the general development and ultrastructure of the tessellated endoskeleton are concerned (e.g. [4, 31, 33, 36–38]). During development, tesserae first appear in the embryonic skeleton as isolated platelets of cartilage calcification, embedded in and separated by unmineralized cartilage. The individual tiles grow by mineral accretion on their existing surfaces, a process reflected in periodic, concentric layers of varying mineral density (Liesegang lines) in the mineralized tissue [4, 32, 33] (Fig. 11.6d–g). This accretionary growth eventually brings young tesserae into contact with one another at their lateral edges. Once this occurs, pronounced high mineral density features, known as 'spokes', develop at the regions of direct contact of two adjacent tesserae (Fig. 11.6d, h, i). Spokes are laminated structures, comprised of densely packed layers of oscillating mineral density, stacked parallel to intertesseral contact surfaces. As the animal and its skeleton grow, tesserae continue to enlarge by accretion, spokes lengthen, radiating outward from tesseral centers like spokes on a wheel (Fig. 11.6c). The growth mechanisms behind the striking repeated structural pattern in spokes are unknown, but the association of spokes with zones of intertesseral contact argues that the mechanical interaction of tesserae may be a guiding factor.

Both Liesegang lines and spokes illustrate that tesserae are more than simply homogeneous mineralized blocks, instead having local mineral density variation as heterogeneous as that seen in bone [4, 10]. Unlike mammalian bone, however, there is no evidence of remodeling or repair in tesserae [4, 29, 33]. If tessellated cartilage—which apparently performs many of the same functional roles as bone—truly is a deposition-only tissue with no healing capacity, it may also possess in-built strategies for avoiding catastrophic damage, similar to the neoteleost bone described in the previous section. This is an enticing suggestion of the potential for tessellated cartilage as an inspiration for manmade design, since the mechanical performance of the tissue might be reproducible by mimicking structural and material properties, rather than biological action (e.g. cellular involvement, tissue remodeling).

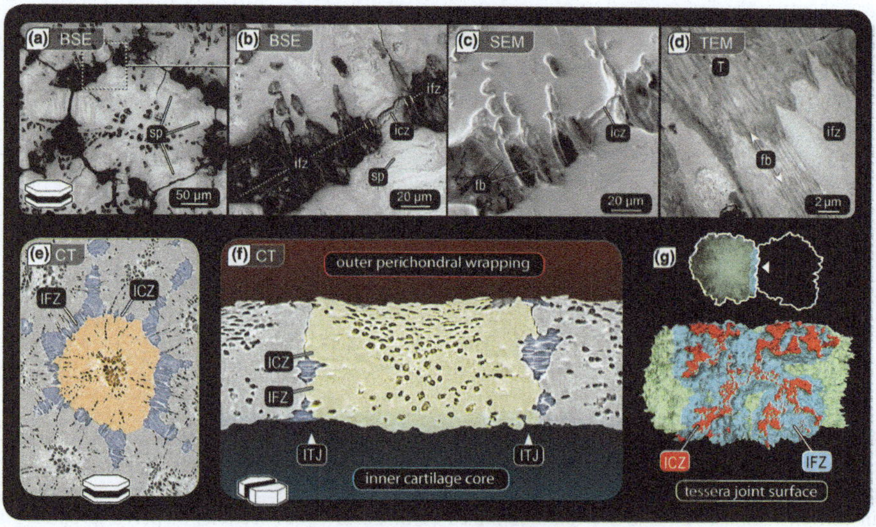


**Fig. 11.7 Flexible linkage of tesserae—collagen fibers at the intertesseral joints. a–d** Backscatter SEM, environmental SEM and TEM images showing the structural complexity of a joint of two abutting tesserae in planar section: **a**, **b** joints are comprised of regions where tesserae are in direct contact (*ICZ*) and gaps (*IFZ*) filled with cells and densely aligned fiber bundles (*fb*) linking adjacent tesserae. **e–g** High-resolution synchrotron µCT scans show no macroscopic tesseral overlapping or interdigitation, but fibrous and contact zones interact in complex ways as illustrated in **g** viewed from the perspective of the neighboring tessera. See abbreviations in previous figure

Our understanding of the features driving tessellated cartilage mechanical properties are in their infancy, but the tissue's performance appears to hinge to a large degree on the interactions and spatial arrangements of softer and harder materials (Fig. 11.7). There is some evidence, for example, that the serial laminae in spokes, by possessing differing mineral densities, introduce interfaces to redirect cracks and dissipate their energy [4]. At a larger scale, the interactions between tesserae are also characterized by material heterogeneities: the sides of tesserae exhibit smooth patches where they come in direct contact with their neighbors, but these are surrounded by regions of densely aligned collagenous fibers, tethering the tesserae together [4, 33] (Fig. 11.7a–d). The bipartite nature of intertesseral joints is thought to impart an interesting mechanical anisotropy to the skeleton as a whole, providing stiffness or flexibility to the tessellated composite layer, depending on the loading conditions. However, these interactions have never been expressly visualized, largely due to technical challenges of visualizing movements of small features in 3D, at adequate resolution and in hydrated conditions. Recent high-resolution synchrotron micro-CT (Fig. 11.7e–g) and modeling studies of tessellated cartilage (see below) are making headway overcoming these difficulties.

The general structural features and tissue arrangements of tessellated cartilage described above appear to be largely universal for sharks and rays [4]. Our high-resolution electron microscopy and synchrotron tomography data, however, indicate

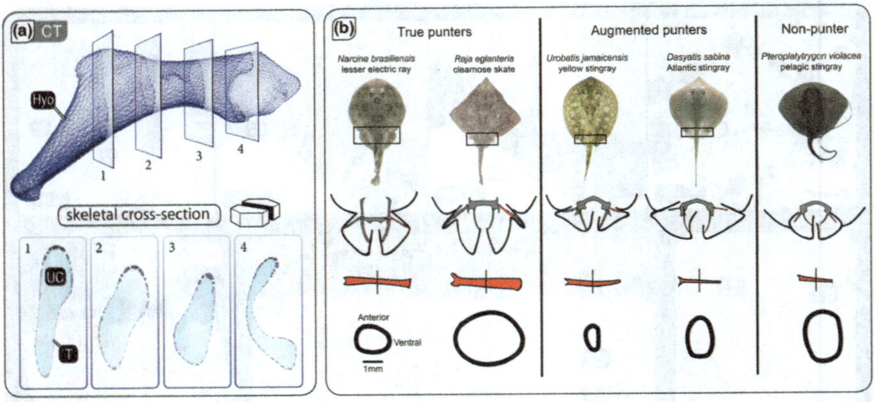


**Fig. 11.8 Local variation in tesserae form and suggested relationship to function. a** Micro-computed tomography of a stingray hyomandibula (*Hyo*; skeletal piece connecting the jaws with the skull), virtually sectioned in 4 different anatomical positions, showing thicker tesserae (T) in convex regions. Abbr: UC = unmineralized cartilage. **b** Cross-sections of pelvic propterygia (skeletal piece supporting the pelvic fins) from different species, showing cross-sectional shape variation according to the species' reliance on 'punting' behavior (use of pelvic fins to move along the sea floor). Modified with permission from [35], copyright Company of Biologists Ltd.

that the shape and structure of tesserae can vary, both within individuals (e.g. between different regions of the skeleton) and among species, in ways that further suggest that the interactions between tesserae are functionally important. For example, the tesserae of different shark and ray species have been shown to vary enormously in size (from <100 μm to nearly 1 mm in width and thickness) and shape, ranging from disc-like plates to cuboid blocks [4, 34, 39]. The shapes of tesserae also appear to vary according to their location on a skeletal element (e.g. depending on the skeleton's local surface curvature; [4, 40]) and according to the skeletal element they cover (e.g. consistently cuboid in the rostral cartilages of some sharks and specific regions of the jaws of some rays; [34, 39]) (Fig. 11.8a). These observations are strongly suggestive of a form-function relationship between the shape of tesserae, their joints and the effective mechanical behavior of the tesseral mat and whole skeletal elements, but these links are only beginning to be established.

### 11.3.2 Mechanics

In general, the study of tessellated cartilage mechanics lags far behind the study of skeletal anatomy and tesseral ultrastructure. However, the results of several works, taken together, begin to paint a picture of how the structural and mechanical properties of elasmobranch cartilage interrelate and how tesserae play an important role in tailoring skeletal properties to specific ecological roles and high load-bearing activities.

An understanding of the global mechanical properties of tessellated cartilage—a composite with relatively discrete material phases—demands characterization of the properties of the primary tissue constituents: uncalcified cartilage, unmineralized joint fibers and mineralized tissue. Capturing *in vivo* properties, however, is complicated by the fine scale 3D structural arrangements of tessellated cartilage and the need for testing conditions that mimic physiological conditions (e.g. hydration and load rates). Available evidence indicates that elasmobranch uncalcified cartilage has a proteoglycan and collagen content similar to mammalian hyaline cartilage and suggests that it can be at least as stiff, if not several orders of magnitude stiffer for similar loading rates (mammalian: 0.45–19 MPa vs. elasmobranch: 2–775 MPa—[41–43]. These properties depend apparently on the species and the skeletal element tested; more rigorous studies are required to understand the relationships between mechanical properties and composition, loading rate and phylogeny.

Whereas covering a cartilage-like gel with a hard, continuous shell is expected to increase the stiffness but decrease the flexibility of a composite, there is some indication—from tessellated cartilage, but also fabricated arrays (e.g. [44])—that a tessellated shell with interacting tiles can be a 'best of both worlds' configuration, maximizing desirable properties of both tissue phases. Fahle and Thomason [45] showed that compared with embryonic (non-tessellated) small-spotted catsharks (*S. canicula*), adult individuals have jaw cartilage that has a higher ability to damp mechanical energy, but is also stiffer. A large portion of the stiffness is surely due to the tessellated layer in adult animals [4, 31, 37, 46]. From the biological perspective, this change in properties permits adults to consume harder prey than newborns [45], but is also particularly intriguing for engineering considerations since stiffness and damping are typically negatively correlated in manmade materials.

The arrangement of the tessellated layer relative to the direction of loading plays a considerable role in elasmobranch skeletal tissue mechanics. Tessellated cartilage cubes from blue sharks (*Prionace glauca*) loaded normal to the tesseral mat (in stress relaxation experiments) behaved similarly to non-tessellated cubes, being ~45 times softer than tessellated cubes with the load applied in-plane with the tesseral mat [42] (Fig. 11.9). These results are supported by indentation experiments performed on hydrated jaw samples from two large sharks (*Carcharodon carcharias*, *Carcharias leucas*) [47]. Nanoindentation experiments typically involve pushing a very small, hard tip (e.g. with a tip radius of hundreds of nanometers) into a material to examine hardness and elastic modulus at very small scales. However, as the indenter used by Ferrara et al. was very large (100 μm) and approached the dimensions of some tesserae [4, 31, 48], we believe their data are more representative of the properties of the composite material (e.g. tesserae and their surrounding soft tissues), in that they report values considerably softer than either the tesserae themselves [42, 49] or whole skeletal elements [35]. Deeper investigation into the relationship between local properties and emergent skeletal properties is required to untangle the contradictions in available data.

Variations around the generalized tessellated cartilage anatomy described above, when interpreted in the context of animal ecology, also provide perspectives on *in vivo* skeletal performance, as well as the functional limits of the tissues. For

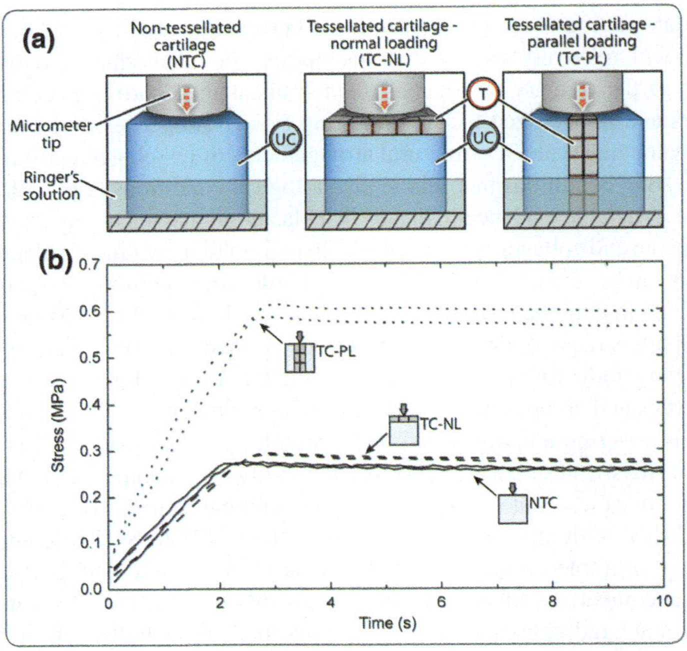


**Fig. 11.9 Contribution of the tesseral layer and its orientation to mechanical properties**. Stress relaxation behavior of blue shark (*Prionace glauca*) cartilage was tested either with tesserae (T) oriented normal (*TC-NL*) or parallel (*TC-PL*) to the loading direction, or without tesserae (*NTC*). Stress relaxation behavior of non-tessellated cartilage (*NTC*) and tessellated cartilage under normal loading (*TC-NL*) were similar, but the behavior of tessellated cartilage samples under parallel loading (*TC-PL*) was far stiffer, indicating that the performance of tessellated cartilage is strongly dependent on the orientation of loads relative to the tesseral layer. Modified with permission from [42], copyright 2014 Elsevier

example, in addition to the outer tessellated layer, the jaws of many batoid fishes (rays and relatives) contain hollow tessellated struts (trabeculae), typically hundreds of microns in diameter, spanning the uncalcified cartilage-filled lumen of the jaws [50–52] (Fig. 11.10a). These appear to be arranged along lines of principal loading, often in narrow regions of the jaws or jaw joints (e.g. [51, 53]), and are therefore structurally and functionally convergent with the trabecular bone found in tetrapods. The importance of trabeculae to the reinforcement of tessellated cartilage is underlined by the high density of trabeculae in the jaws of species that experience high skeletal loads during feeding, such as the lesser electric ray which uses explosive jaw protrusion to retrieve buried prey [50] or myliobatid stingrays which employ high bite forces to crush hard shelled mollusks ([51, 52, 54]; see also Fig. 1 in [55]).

Additional mechanisms described for reinforcing tessellated cartilage against bending involve either thickening of the skeleton's hard, outer cortex (i.e. the tesseral layer) or modifications to the cross-sectional shape of skeletal elements. In the former, the tesseral layer is often locally thickened (e.g. the tesserae in skeletal cross-sections

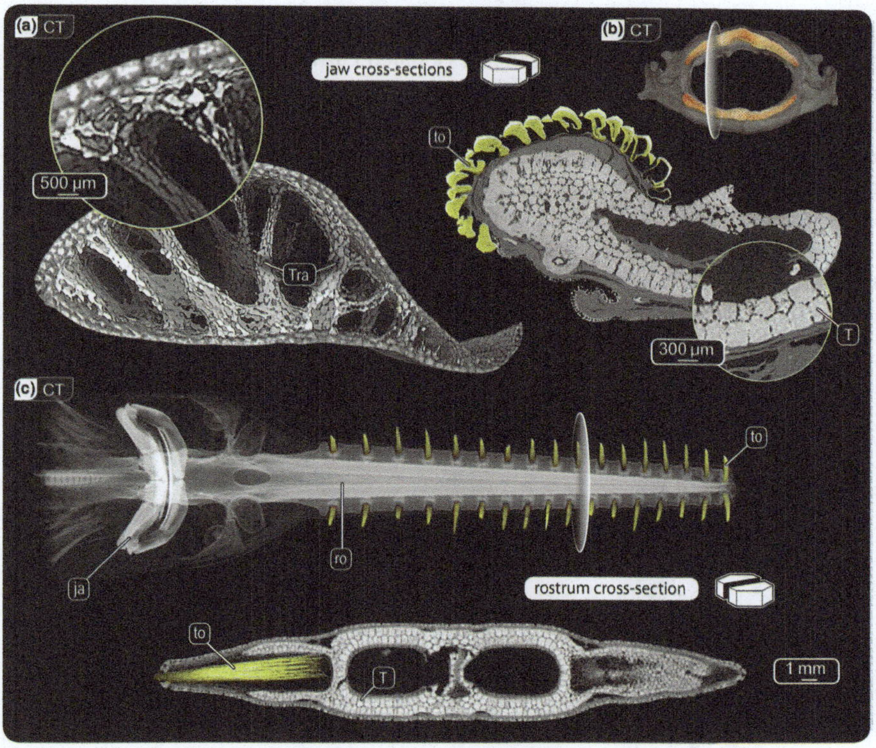


**Fig. 11.10 Methods of reinforcement of tessellated cartilage.** All images are volume renderings or virtual cross-sections of micro-CT scans. **a** Cross-section of the jaws of the cownose ray (*Rhinoptera bonasus*), a hard prey specialist, showing mineralized trabeculae (struts) running through the jaw, reinforcing the primary biting direction. Alternatively, tessellated cartilage can be reinforced by multiple tesseral layers, as in **a** the outer layers of the cownose ray jaw, **b** the jaws of a wedgefish (*Rhynchobatus* sp.) and in **c** the rostral 'saw' of a sawfish (*Pristis microdon*). Abbr: JA = jaws; RO = rostrum; T = tesserae; Tra = Trabeculae; To = Tooth

in Fig. 2 of [4] vary up to 4× in thickness), particularly in regions in line with principal loading or in areas of high curvature [4, 40, 53, 56–58]. The cortex can also be thickened via introduction of additional tesseral layers, particularly in the jaws of large carnivorous sharks [59], the "saws" of sawfish [33, 51], or the jaws of species with diets containing large proportions of hard shelled prey (Fig. 11.10b, c) [52, 55]. Up to 10 supernumerary tesseral layers have been recorded in the jaws of some extant species [57]. High flexural stiffness in whole skeletal elements appears to result from either high mineral content (a proxy for the proportion of tesserae in an element or cross-section), skeletal cross-sectional shapes with high second moment of area or combinations of the two [35, 56, 58, 60]. Skeletal element flexural stiffness has also been shown to be correlated with differences in ecology—for example, supporting specific locomotion [35, 60] or feeding modes [56]—but mineral content and cross-sectional shape/size do not always vary predictably (e.g. animals with large skeletal cross-sections can have either very high or very low mineral content [35, 56, 58]) (Fig. 11.8b). This indicates that tessellated cartilage has evolved to be 'modular', where functionally and ecologically relevant skeletal mechanical properties can be achieved through a variety of structural mechanisms (e.g. cortical thickening, increased second moment of area, introduction of trabeculae), rather than compositional alterations to the apatite mineral in the tessellated layer (e.g. the introduction of heavier elements as can occur in invertebrate tissues [7].

The mechanical influence of the *tessellation* itself—the tiling of the cortex and the varying patterns formed by tesserae—has yet to be investigated in situ, likely due to the experimental difficulties posed by the size of tesserae and their complex

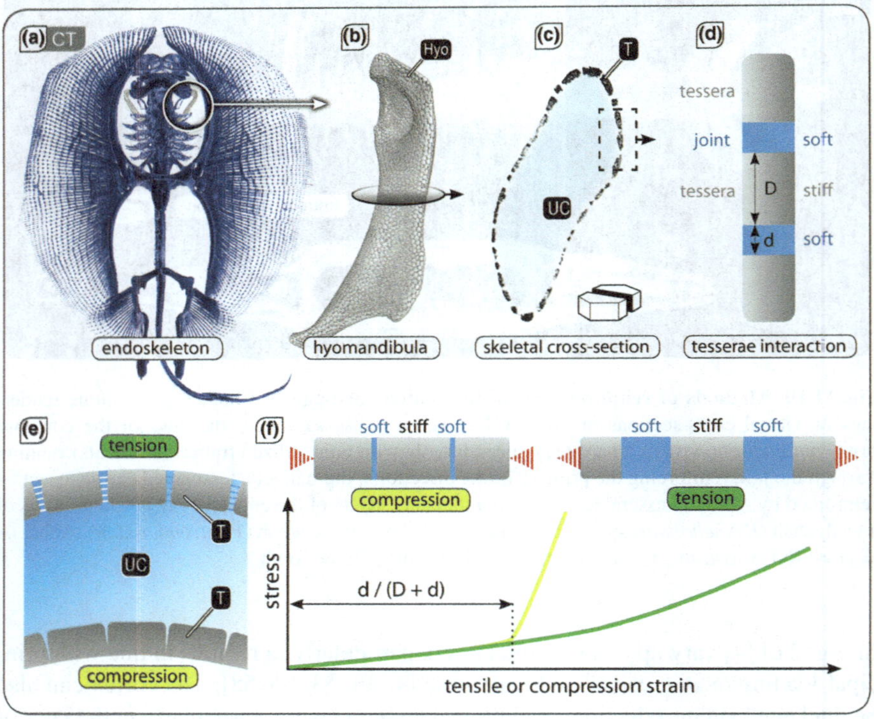


**Fig. 11.11 Proposed compression-tension asymmetry in tessellated cartilage. a–c** Micro-CT scans of the skeleton and hyomandibula (Hyo) of a stingray (*Urobatis halleri*) illustrating the composite nature of tessellated cartilage, formed by stiff mineralized tiles (T) separated by softer unmineralized joints. Abbr: UC = unmineralized cartilage. **d** Schematic of alternating soft and hard constituents in a tessellated system. **e** Schematic section of tessellated skeletal element in bending, with the top and bottom tessellated layers experiencing tension and compression, respectively. **f** Hypothetical stress-strain curves illustrating the proposed tension-compression asymmetry of a tessellation. In tension, the deformation will be dominated by the stretching of the soft joint interlayer (yellow line), whereas in compression, the behavior is stiffer (green line), dominated by the stiff tiles colliding once the thickness of the soft interlayer is exhausted [at a compressive strain of approximately d/(D + d)]. Modified from [61]

tissue connections. Several studies, however, have incorporated existing materials and/or structural data from the biological tissue in computational simulations and mathematical models of tessellated cartilage, helping to shape informed and testable hypotheses relating to tissue growth and mechanics. Empirical models of tessellated cartilage cross sections, for example, based on biological material and ultrastructural data, suggest that the tessellation plays a role in controlling the stress distribution within the skeletal tissue during bending [61, 62]. This is a function of the narrow joints between tesserae, the structure of which is hypothesized to result in strikingly different properties in tension than compression (Fig. 11.11). In a hypothetical laminated tessellated cartilage beam—a monolithic core of unmineralized cartilage sandwiched between two thin tessellated layers—subjected to bending, the tesserae on the side of the beam loaded in tension should pull apart from one another, whereas tesserae on the compressive side of the beam should readily collide. Such a compression-tension asymmetry would impart a constrained flexibility in tessellated cartilage that could also play a role in the tissue's ability to resist damage. In general, the combination of a stiff outer cortex and a soft inner core will tend to ensure that higher stresses are concentrated more safely in the stiffer cortical/mineralized tissue rather than the softer core/unmineralized tissue [41, 61, 62]. This is true, even if the cortex is continuous rather than tessellated, as was simulated in Finite Element models of the jaws of two large shark species subjected to biologically-relevant bite forces [41].

Tessellating the cortex, however, serves additional functions, such as protecting the mineralized tissue from fracture on the tension side of the skeleton by localizing tensile stresses in the intertesseral joint fibers rather than in tesserae. At the same time, the compressive stiffness of tesserae should shift the skeleton's neutral axis of bending closer to the compressive side of the skeleton, concentrating potentially damaging compressive stresses in the tessellated layer rather than the unmineralized cartilage [61, 62]. In this way, the tessellation can manage bending loads and increase resistance to damage by distributing the highest stresses to the tissues and loading regimes best able to bear them. Stresses may also be mitigated by the properties of the unmineralized cartilage itself, which Wroe et al. [49] showed would tend to result in considerably lower stresses and higher strains than simulated shark jaws made of bone and subjected to the same bite forces (Fig. 11.12). The hypothesized stress-management behavior of tessellated cartilage may therefore serve a protective function, in a skeletal tissue that apparently cannot heal [4, 29, 38, 57], although it is surely cyclically loaded a massive number of times over an animal's lifetime during swimming and feeding behaviors.

Most models of tessellated cartilage have focused largely on the mechanical result of combining soft and hard tissues; the geometry of the tessellation—the distribution of shapes and sizes of tesserae—surely also plays an important role in the mechanics of the skeleton, yet this has hardly been investigated. Tesserae are typically polygonal in shape and apparently predominantly hexagonal, but have been observed to vary in the regularity of their form and range from squares to twelve sided polygons ([4, 39, 40, 63]; Dean, pers. obs.). Analytical models of 2D tiled arrays with joint and tile geometries and material properties varied in a range around those of tesserae, indicate

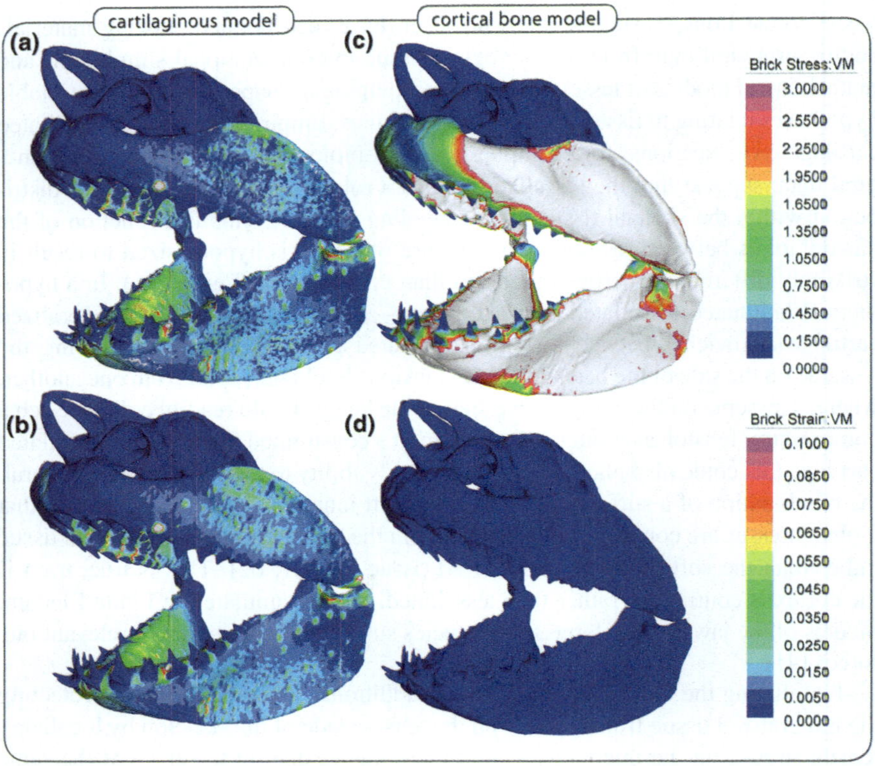


**Fig. 11.12 Simulation of feeding mechanics in white shark jaws with varying material properties.** **a**, **c** Von Mises stress and **b**, **d** strain distributions for maximal bilateral bites in Finite Element simulations of jaws with tessellated cartilage properties (**a**, **b**) and cortical bone properties (**c**, **d**). Note that stress is much lower, but strain is much higher in the cartilaginous model. Modified with permission from [49], copyright John Wiley Sons, Inc.

that the stiffness of the tessellation will increase the narrower joint spaces are and the stiffer tiles are relative to their joint fibers [39] (Fig. 11.13: modified from [39]). This holds true for triangular, square and hexagonal tiles, with hexagons being most sensitive to changes in model parameters and therefore allowing for a tessellation that can be more subtly 'tuned' to specific material properties. These results demonstrate that the shape of tesserae should affect the properties of the tissue composite, yet the performance of more complicated, bio-realistic tessellations (e.g. three-dimensional and/or less uniform tilings) remains to be investigated.

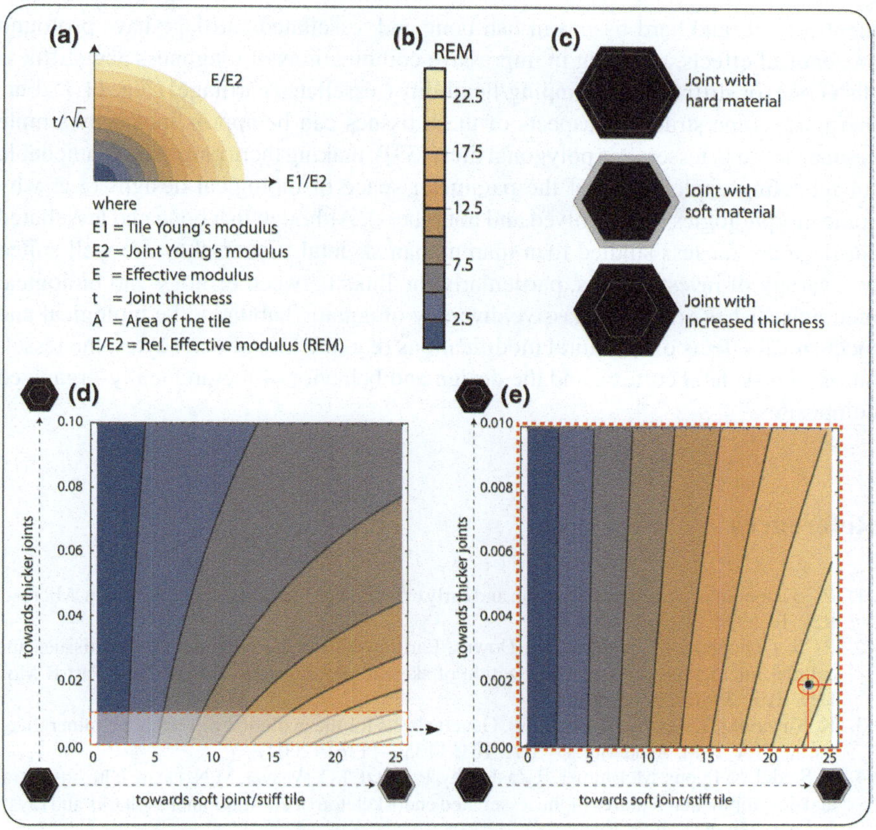


**Fig. 11.13 The relative effective modulus (REM) of tessellated cartilage-inspired hexagonal tilings as a function of morphology and material properties. a–c** Variables varied in analytical simulations of hexagonal tessellations with intervening joint material under in-plane loading. **d, e** Simulations show that a decrease in joint thickness (decreasing y-axis value) and an increase in the stiffness of tiles relative to joints (increasing x-axis value) results in a stiffer composite material. **e** Zoomed in view of the range of biologically relevant x- and y-axis values from **d**; structural and material properties for round stingray (*U. halleri*) tesserae are marked with red lines in the bottom right. Modified with permission from [39], copyright 2017 Elsevier

## 11.4 Conclusions

Although mammalian bone is commonly considered the paradigmatic vertebrate skeletal tissue, there are 'many other fish in the sea': anosteocytic and osteocytic fish bone and elasmobranch tessellated cartilage are two of Nature's skeletal design alternatives, offering valuable fodder for wider investigations on the interplay of form and function in mineralized biological materials. These tissues also present largely untapped opportunities for exploring the mechanical effects of alternate structuring in tissues with otherwise similar composition. Clearly, the distinct geometric arrangements of soft and hard tissues in fish bone and tessellated cartilage have profound mechanical effects, resulting in impressive combinations of toughness and stiffness (fish bone) or stiffness and damping/flexibility (tessellated cartilage) (Fig. 11.1). Furthermore, some structural aspects of these tissues can be approximated by simple geometries (e.g. tesserae as polygonal tiles; [39]), making them particularly amenable for modeling explorations of the parameter space of biological designs (e.g. why some morphologies have evolved and not others). Although fish bone and tessellated cartilage are far less studied than mammalian skeletal tissues, they are well suited to a variety of investigations, particularly the links between ecology and biological materials (fishes occupy a massive diversity of aquatic habitats), the biological and mechanical effects of structural modifications (e.g. the loss of osteocytes, the tessellation of a skeletal cortex), and the design and behavior of hierarchically-organized composites.

## References

1. P.C.J. Donoghue, I.J. Sansom, Origin and early evolution of vertebrate skeletonization. Microsc. Res. Tech. **59**, 352–372 (2002)
2. P.C.J. Donoghue, I.J. Sansom, J.P. Downs, Early evolution of vertebrate skeletal tissues and cellular interactions, and the canalization of skeletal development. J. Exp. Zool. Part B Mol. Dev. Evol. **306B**, 278–294 (2006)
3. K. Kawasaki, T. Suzuki, K.M. Weiss, Genetic basis for the evolution of vertebrate mineralized tissue. Proc. Natl. Acad. Sci. U. S. A. **101**, 11356–11361 (2004)
4. R. Seidel, K. Lyons, M. Blumer, P. Zaslansky, P. Fratzl, J.C. Weaver, M.N. Dean, Ultrastructural and developmental features of the tessellated endoskeleton of elasmobranchs (sharks and rays). J. Anat. **229**, 681–702 (2016)
5. J.-Y. Sire, A. Huysseune, Formation of dermal skeletal and dental tissues in fish: a comparative and evolutionary approach. Biol. Rev. Camb. Philos. Soc. **78**, 219–249 (2003)
6. J.D. Currey, Collagen and the mechanical properties of bone and calcified cartilage, in *Collagen: Structure and Mechanics, An Introduction*, ed. by P. Fratzl (Springer US, Boston, MA, 2008), pp. 397–420
7. E. Degtyar, M.J. Harrington, Y. Politi, P. Fratzl, The mechanical role of metal ions in biogenic protein-based materials. Angew. Chem. Int. Ed. **53**, 12026–12044 (2014)
8. J.D. Currey, The design of mineralised hard tissues for their mechanical functions. J. Exp. Biol. **202**, 3285–3294 (1999)
9. J.D. Currey, The structure and mechanics of bone. J. Mater. Sci. **47**, 41–54 (2011)
10. J.D. Currey, M.N. Dean, R. Shahar, Revisiting the links between bone remodelling and osteocytes: insights from across phyla. Biol. Rev. Camb. Philos. Soc. **92**, 1702–1719 (2017)
11. M.L. Moss, A.S. Posner, X-ray diffraction study of acellular teleost bone. Nature **188**, 1037–1038 (1960)
12. M.R. Urist, Calcium and phosphorus in the blood and skeleton of the Elasmobranchii, 1–24 (1961)
13. J.W. Smith, Collagen fibre patterns in mammalian bone. J. Anat. **94**, 329–344 (1960)
14. L.F. Bonewald, The amazing osteocyte. J. Bone Miner. Res. **26**, 229–238 (2011)
15. S.L. Dallas, M. Prideaux, L.F. Bonewald, The osteocyte: an endocrine cell … and more. Endocr. Rev. **34**, 658–690 (2013)
16. P.E. Witten, A. Huysseune, T. Franz-Odendaal, T. Fedak, M. Vickaryous, A. Cole, B.K. Hall, Acellular teleost bone: primitive or derived, dead or alive? Palaeontol. Assoc. Newsl. **55**, 37–41 (2004)

17. L. Cohen, M. Dean, A. Shipov, A. Atkins, E. Monsonego-Ornan, R. Shahar, Comparison of structural, architectural and mechanical aspects of cellular and acellular bone in two teleost fish. J. Exp. Biol. **215**, 1983–1993 (2012)
18. A. Atkins, M.N. Dean, M.L. Habegger, P.J. Motta, L. Ofer, F. Repp, A. Shipov, S. Weiner, J.D. Currey, R. Shahar, Remodeling in bone without osteocytes: billfish challenge bone structure-function paradigms. Proc. Natl. Acad. Sci. U. S. A. **111**, 16047–16052 (2014)
19. G.S. Helfman, B.B. Collette, D.E. Facey, B.W. Bowen, *The Diversity of Fishes: Biology* (Wiley-Blackwell, 2009)
20. R. Shahar, C. Lukas, S. Papo, J.W.C. Dunlop, R. Weinkamer, Characterization of the spatial arrangement of secondary osteons in the diaphysis of equine and canine long bones. Anat. Rec. **294**, 1093–1102 (2011)
21. A. Atkins, N. Reznikov, L. Ofer, A. Masic, S. Weiner, R. Shahar, The three-dimensional structure of anosteocytic lamellated bone of fish. Acta Biomater. **13**, 311–323 (2015)
22. N. Reznikov, R. Shahar, S. Weiner, Bone hierarchical structure in three dimensions. Acta Biomater. **10**, 3815–3826 (2014)
23. L.B. Halstead, Calcified tissues in the earliest vertebrates. Calcif. Tissue Int. **3**, 107–124 (1969)
24. D.R. Hughes, J.R. Bassett, L.A. Moffat, Histological identification of osteocytes in the allegedly acellular bone of the sea breams *Acanthopagrus australis*, *Pagrus auratus* and *Rhabdosargus sarba* (Sparidae, Perciformes, Teleostei). Anat. Embryol. **190**, 163–179 (1994)
25. A. Kölliker, On the different types in the microscopic structure of the skeleton of osseous fishes (1857)
26. M.L. Moss, Studies of the acellular bone of teleost fish. Cells Tissues Organs **46**, 343–362 (1961)
27. J.Y. Sire, F.J. Meunier, The canaliculi of Williamson in holostean bone (Osteichthyes, Actinopterygii): a structural and ultrastructural study. Acta Zool. **75**, 235–247 (1994)
28. T. Ørvig, Histologic studies of placoderms and fossil elasmobranchs, 1–152 (1950)
29. D.E. Ashhurst, The cartilaginous skeleton of an elasmobranch fish does not heal. Matrix Biol. **23**, 15–22 (2004)
30. B.K. Hall, *Bones and Cartilage*, 2nd edn. (Academic Press, 2014)
31. M.N. Dean, C.G. Mull, S.N. Gorb, A.P. Summers, Ontogeny of the tessellated skeleton: insight from the skeletal growth of the round stingray *Urobatis halleri*. J. Anat. **215**, 227–239 (2009)
32. N.E. Kemp, S.K. Westin, Ultrastructure of calcified cartilage in the endoskeletal tesserae of sharks. J. Morphol. **160**, 75–101 (1979)
33. R. Seidel, M.J.F. Blumer, E.J. Pechriggl, K. Lyons, B.K. Hall, P. Fratzl, J.C. Weaver, M.N. Dean, Calcified cartilage or bone? Collagens in the tessellated endoskeletons of cartilaginous fish (sharks and rays). J. Struct. Biol. (2017)
34. J.G. Maisey, The diversity of tessellated calcification in modern and extinct chondrichthyans. Rev. Paléobiol. Genève **32**, 355–371 (2013)
35. L.J. Macesic, A.P. Summers, Flexural stiffness and composition of the batoid propterygium as predictors of punting ability. J. Exp. Biol. **215**, 2003–2012 (2012)
36. J.G. Clement, Re-examination of the fine structure of endoskeletal mineralization in chondrichthyans: implications for growth, ageing and calcium homeostasis. Aust. J. Mar. Freshw. Res. **43**, 157–181 (1992)
37. S. Enault, D.N. Muñoz, W.T.A.F. Silva, V. Borday-Birraux, M. Bonade, S. Oulion, S. Ventéo, S. Marcellini, M. Debiais-Thibaud, Molecular footprinting of skeletal tissues in the catshark *Scyliorhinus canicula* and the clawed frog *Xenopus tropicalis* identifies conserved and derived features of vertebrate calcification. Front. Genet. **6**, 3133–3144 (2015)
38. R. Seidel, M. Blumer, P. Zaslansky, D. Knötel, D.R. Huber, J.C. Weaver, P. Fratzl, S. Omelon, L. Bertinetti, M.N. Dean, Ultrastructural, material and crystallographic description of endophytic masses; a possible damage response in shark and ray tessellated calcified cartilage. J. Struct. Biol. **198**, 5–18 (2017)
39. A.K. Jayasankar, R. Seidel, J. Naumann, L. Guiducci, A. Hosny, P. Fratzl, J.C. Weaver, J.W.C. Dunlop, M.N. Dean, Mechanical behavior of idealized, stingray-skeleton-inspired tiled composites as a function of geometry and material properties. J. Mech. Behav. Biomed. Mater. **73**, 1–35 (2017)

40. M.N. Dean, R. Seidel, D. Knoetel, K. Lyons, D. Baum, J.C. Weaver, P. Fratzl, To build a shark-3D tiling laws of tessellated cartilage. Integr. Comp. Biol. **56**, E50 (2016)
41. T.L. Ferrara, P. Clausen, D.R. Huber, C.R. McHenry, V. Peddemors, S. Wroe, Mechanics of biting in great white and sandtiger sharks. J. Biomech. **44**, 430–435 (2011)
42. X. Liu, M.N. Dean, H. Youssefpour, A.P. Summers, J.C. Earthman, Stress relaxation behavior of tessellated cartilage from the jaws of blue sharks. J. Mech. Behav. Biomed. Mater. **29**, 68–80 (2014)
43. M.E. Porter, J.L. Beltran, S.M. Kajiura, T.J. Koob, A.P. Summers, Stiffness without mineral: material properties and biochemical components of jaws and chondrocrania in the Elasmobranchii (sharks, skates, and rays). PeerJ **1**, e47v1 (2013)
44. R. Martini, Y. Balit, F. Barthelat, A comparative study of bio-inspired protective scales using 3D printing and mechanical testing. Acta Biomater. **55**, 360–372 (2017)
45. S.R. Fahle, J.C. Thomason, Measurement of jaw viscoelasticity in newborn and adult lesser spotted dogfish *Scyliorhinus canicula* (L., 1758). J. Fish Biol. **72**, 1553–1557 (2008)
46. M. Egerbacher, M. Helmreich, E. Mayrhofer, P. Böck, Mineralisation of the hyaline cartilage in the small-spotted dogfish *Scyliorhinus canicula* L. *Scripta Medica (BRNO)* (2006)
47. T.L. Ferrara, P. Boughton, E. Slavich, S. Wroe, A novel method for single sample multi-axial nanoindentation of hydrated heterogeneous tissues based on testing great white shark jaws. PLoS ONE **8**, e81196 (2013)
48. S. Applegate, A survey of shark hard parts. In *Sharks, Skates and Rays* (1967), pp. 37–57
49. S. Wroe, D.R. Huber, M. Lowry, C. McHenry, K. Moreno, P. Clausen, T.L. Ferrara, E. Cunningham, M.N. Dean, A.P. Summers, Three-dimensional computer analysis of white shark jaw mechanics: how hard can a great white bite? J. Zool. **276**, 336–342 (2008)
50. M.N. Dean, A.P. Summers, Mineralized cartilage in the skeleton of chondrichthyan fishes. Zoology **109**, 164–168 (2006)
51. A.P. Summers, Stiffening the stingray skeleton—an investigation of durophagy in myliobatid stingrays (Chondrichthyes, Batoidea, Myliobatidae). J. Morphol. **243**, 113–126 (2000)
52. A.P. Summers, R.A. Ketcham, T. Rowe, Structure and function of the horn shark (*Heterodontus francisci*) cranium through ontogeny: development of a hard prey specialist. J. Morphol. **260**, 1–12 (2004)
53. M.N. Dean, D.R. Huber, H.A. Nance, Functional morphology of jaw trabeculation in the lesser electric ray *Narcine brasiliensis*, with comments on the evolution of structural support in the Batoidea. J. Morphol. **267**, 1137–1146 (2006)
54. M.A. Kolmann, S.B. Crofts, M.N. Dean, A.P. Summers, N.R. Lovejoy, Morphology does not predict performance: jaw curvature and prey crushing in durophagous stingrays. J. Exp. Biol. **218**, 3941–3949 (2015)
55. R. Seidel, M. Blumer, E.-J. Pechriggl, K. Lyons, B.K. Hall, P. Fratzl, J.C. Weaver, M.N. Dean, Calcified cartilage or bone? Collagens in the tessellated endoskeletons of cartilaginous fish (sharks and rays). J. Struct. Biol. **200**, 1–35 (2017)
56. J.P. Balaban, A.P. Summers, C.A. Wilga, Mechanical properties of the hyomandibula in four shark species. J. Exp. Zool. Part A Ecol. Genet. Physiol. **323**, 1–9 (2014)
57. M.N. Dean, J.J. Bizzarro, B. Clark, C.J. Underwood, Z. Johanson, Large batoid fishes frequently consume stingrays despite skeletal damage. R. Soc. Open Sci. **4**, 170674–11 (2017)
58. C.A.D. Wilga, S.E. Diniz, P.R. Steele, J. Sudario-Cook, E.R. Dumont, L.A. Ferry, Ontogeny of feeding mechanics in smoothhound sharks: morphology and cartilage stiffness. Integr. Comp. Biol. **56**, 442–448 (2016)
59. G. Dingerkus, B. Seret, Multiple prismatic calcium phosphate layers in the jaws of present-day sharks (Chondrichthyes; Selachii). Experientia **47**, 38–40 (1991)
60. W. Huang, W. Hongjamrassilp, J.-Y. Jung, P.A. Hastings, V.A. Lubarda, J. McKittrick, Structure and mechanical implications of the pectoral fin skeleton in the Longnose Skate (Chondrichthyes, Batoidea). Acta Biomater. **51**, 393–407 (2017)
61. P. Fratzl, O. Kolednik, F.D. Fischer, M.N. Dean, The mechanics of tessellations—bioinspired strategies for fracture resistance. Chemical Society Reviews **45**, 252–267 (2016)

62. X. Liu, M.N. Dean, A.P. Summers, J.C. Earthman, Composite model of the shark's skeleton in bending: a novel architecture for biomimetic design of functional compression bias. Mater. Sci. Eng. C **30**, 1077–1084 (2010)
63. M.N. Dean, J.T. Schaefer, Patterns of growth and mineralization in elasmobranch cartilage. Faseb J. **19**, A247–A247 (2005)
64. L.A. Jawad, Hyperostosis in three fish species collected from the Sea of Oman. Anat. Rec. **296**, 1145–1147 (2013)

# Chapter 12
# Insect-Inspired Distributed Flow-Sensing: Fluid-Mediated Coupling Between Sensors

**Gijs J. M. Krijnen, Thomas Steinmann, Ram K. Jaganatharaja and Jérôme Casas**

**Abstract** Crickets and other arthropods are evolved with numerous flow-sensitive hairs on their body. These sensory hairs have garnered interest among scientists resulting in the development of bio-inspired artificial hair-shaped flow sensors. Flow-sensitive hairs are arranged in dense arrays, both in natural and bio-inspired cases. Do the hair-sensors which occur in closely-packed settings affect each other's performance by so-called viscous coupling? Answering this question is key to the optimal arrangement of hair-sensors for future applications. In this work viscous coupling is investigated from two angles. First, what does the existence of many hairs at close mutual distance mean for the flow profiles? How is the air-flow around a hair changed by it's neighbours proximity? Secondly, in what way do the incurred differences in air-flow profile alter the drag-torque on the hairs and their subsequent rotations? The first question is attacked both from a theoretical approach as well as by experimental investigations using particle image velocimetry to observe air flow profiles around regular arrays of millimeter sized micro-machined pillar structures. Both approaches confirm significant reductions in flow-velocity for high density hair arrays in dependence of air-flow frequency. For the second set of questions we used dedicated micro-fabricated chips consisting of artificial hair-sensors to controllably and reliably investigate viscous coupling effects between hair-sensors. The experimental results confirm the presence of coupling effects (including secondary) between hair-sensors when placed at inter-hair distances of less than 10 hair diameters ($d$). Moreover, these results give a thorough insight into viscous coupling effects. Insight which can be used equally well to further our understanding of the biological

Parts of this chapter related to the interaction between 2 and 3 hairs have been adapted from the Ph.D. thesis of one the authors, Ram K. Jaganatharaja, see [1].

G. J. M. Krijnen (✉) · R. K. Jaganatharaja
TechMed Centre, University of Twente, Enschede, The Netherlands
e-mail: gijs.krijnen@utwente.nl

T. Steinmann · J. Casas
Institute de Recherche en Biologie de l'Insecte, Université de Tours, Tours, France

 
Y. Estrin et al. (eds.), *Architectured Materials in Nature and Engineering*, Springer Series in Materials Science 282,
https://doi.org/10.1007/978-3-030-11942-3_12

implications of high density arrays as well as have a better base for the design of biomimetic artificial hair-sensor arrays where spatial resolution needs to be balanced by sufficiently mutually decoupled hair-sensor responses.

## 12.1 Introduction

Insects are fabulous templates for bio-inspired design of miniature sensors because they harbour a multiplicity of sensors, often placed at the outer surface of their relatively hard-shell bodies to increase the sensing of environmental stimuli. This makes insects particularly amenable to comparative architectural and hierarchical studies and bio-inspiration. The neural and physiological mechanisms of single sensors or entire sensor systems have been well studied over a century [2]. The study of and inspiration from the integration of sensors is also well advanced regarding identical sensory units, epitomized by the production of insect-inspired artificial eyes [3]. These considerations are also partially valid for one of the best studied systems of mechanosensors in insects, the hair flow sensing escape system of crickets and their bio-inspired micro-mechanical systems. The recent decade has indeed seen a major increase in our understanding of the neural and biomechanical functioning of single hairs and implementation as single micro-electromechanical systems (MEMS) based hairs [4].

These filiform hairs, present on a pair of special sensory appendages called cerci on the abdomen of crickets, are considered to be one of the most sensitive (down to $30\,\mu m/s$ oscillating flows) and energy efficient sensors found in nature, operating at energy levels comparable to the thermal-mechanical noise floor [4–18]. Recently, these filiform hairs have inspired scientists to successfully realize artificial hair-sensor arrays using advanced MEMS technologies (Fig. 12.1) [4, 19, 20]. Arrays of high-density artificial hair-sensors are required for high spatial resolution in flow signature detection applications. Despite successful sensor-array developments, several characteristics of sensory hairs in arrays (both in natural and bio-inspired cases) still remain unknown or less understood. One unexplained and interesting aspect of cricket's cerci is related to the occurrence of its sensory hairs in high-densities [5–7] and the associated effects on single hair-sensor performance [5, 6]. The impact of one hair-sensor on its neighbors is mediated by viscous coupling [5]. This effect has been debated in literature, both from theoretical and experimental viewpoints [5, 8–10, 12, 13], but poses many practical difficulties for experimental studies on living animals [6]. For example, hairs have preferential planes of movement [17], and it is nearly impossible to find a cricket hair arrangement that ensures a common plane of movement for two hairs [8]. Misalignments of these preferential planes, therefore, lead to biased estimates.

The presence of viscous coupling impacts the sensitivity of the artificial hair-sensors. Hence, investigation of viscous coupling between artificial hair-sensors can make a significant contribution to the design of optimally arranged sensor arrays. Simultaneously it helps biologists to further their understanding of the functioning

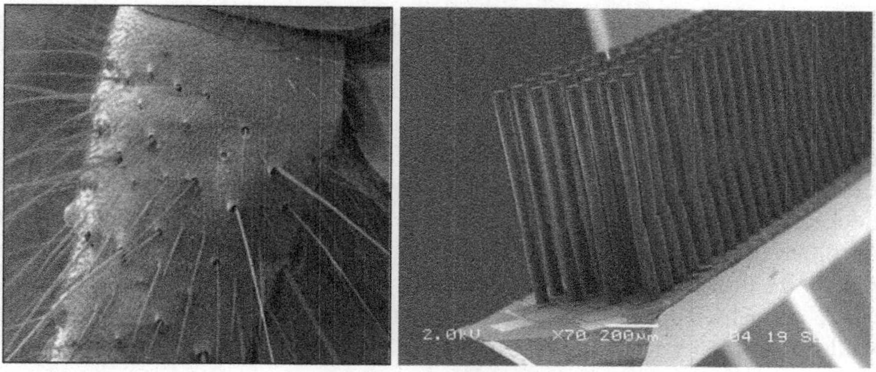


**Fig. 12.1** Filiform hairs on a cercus of a cricket (left) (SEM image courtesy: G. Jeronimides, University of Reading, UK) and an example of a bio-inspired artificial hair-sensor array (right)

and bauplan of such high density sensory arrays as found on crickets, appreciating the difficulties to investigate viscous coupling effects on the actual cerci. With the major advantage of exerting control over the operation and geometry of the artificial sensory hairs, a wide range of experimental characterisations is now possible.

### *12.1.1 Hair-Sensor Arrays: Natural and Bio-inspired Cases*

Flow-sensitive hairs on the cerci of crickets occur in varying length and diameters and are arranged at certain specific spatial positions on the circumferential surface of the cercus [14–16]. The length of the hairs on the cerci is observed to have a bimodal distribution, with a group of short hairs and a group of long hairs [6, 7]. Moreover, short hairs are abundant (more than 75%) and long hairs are relatively scarce on a cercal surface [7]. The hair diameters, as measured near the socket, range between $\approx 1$ and 9 $\mu$m, depending allometrically on their length, which ranges between $\approx 100$ to 1500 $\mu$m [5–7]. Based on careful experimental observations of the crickets of various growth stages, hair densities of $\approx 400$ (first instar) to $\approx 100$ (adults) hairs (of different lengths) per mm$^2$ were found at average nearest neighbour distances between 30 and 60 $\mu$m [7]. Also, it is to be noted that the adjacent hairs are of different orientation of mechanical movements and thereby sensitive to flows of specific directions [5–7]. It has also been studied that there exists an 'identifiable' pattern with respect to a hair's preferred directionality and its occurrence on the cercal surface, observed on individuals of the same species [17]. Miller et al. [21] have quantitatively characterised the hair patterns on cercus of different crickets Acheta domesticus, determining their lengths, positions and orientations, recalling that the inter-individual variation of the patterns of structural organisation was very low. Heys et al. [11] have also recently proposed a predictive model of distribution of the preferential orientations of the hairs. These authors also quantified the spatial distribution of hairs for various sizes of circular search windows. They determined that for a small search area less than 750 μm, the pattern of hairs corresponded to what would be expected for a random Poisson process of positioning.

On the other hand, our artificial hair-sensor arrays feature, geometrically speaking, thicker hairs ($\approx$50 to 60 μm) and have a uniform length ($\approx$700 to 800 μm). Comparing the density of the artificial hair-sensor arrays to the actual situations in cricket hair-sensor arrays, the latter are relatively denser, but upon taking into account the differences in hair geometries, the normalised spacing between the hairs, i.e. the ratio of the distance $s$ over hair-diameter $d$, are more or less in a comparable range. These normalised spacings range approximately between 5 and $\approx$100 for crickets while in our artificial case, it ranges between 5 and 68.

### *12.1.2 Viscosity-Mediated Coupling Effects*

When two hairs (of length $L$ and diameter $d$) are present close to each other (inter-hair distance, $s$) in the direction of flow, it is intuitive to consider that each hair could cast a hydrodynamical shadow on the other. The coupling effect depends on several factors related to the hairs and the character of the flow. While such coupling could negatively affect the individual hair performances, the paradoxical occurrence of hairs in highly dense arrangements raises an unexplained and less understood natural phenomenon. The possible existence of viscosity-mediated coupling between the filiform hairs was first suggested by Humphrey et al. [18] in their studies involving the mechano-sensory hairs found in crickets and arachnids. This term, viscosity-mediated coupling, suggestively denotes the dominance of viscous forces in low-Reynolds' number flow regimes. The coupling is denoted negative when the hairs are submitted to flows of lower velocities due to the presence of nearby hairs, and positive in the reverse case.

Bathellier et al. made an extensive investigation, both theoretically and experimentally, on negative viscous coupling between hairs found in the trichobothria of Cupiennes salei spiders [9]. In this work, they defined a dimensionless coefficient to define the viscous coupling and conducted several experiments on spider hairs. For the experimental tests, pairs of hairs with varying normalised distances were chosen. The angular deflection of a reference hair was optically measured, in presence of a perturbing hair and measured again in the absence of the perturbing hair (after removal). The role of the perturbing hair was investigated in three different scenarios: (i) free moving, (ii) immobilised at its base and (iii) mechanically actuated. They measured the coupling effect for the second and third cases, and determined the effect was negligible for the first case. Bathellier et al. also proposed a theoretical model as a framework for their experimental results. Significant insights provided by their model can be summarized as: (i) the coupling coefficient is dependent on the flow frequency, (ii) the coupling effect drops with increasing distance between the hairs, (iii) a fixed or mechanically driven hair perturbs the flow more than a free-moving hair,[1] and (iv) the coupling effect depends on the length of the hairs (longer hairs have more prominent effects on the reference hair than shorter ones). This paper turned out to be a seminal study for later experimental and computational works.

These results were in part substantiated by our PIV studies on MEMS hairs [8]. Indeed, working with cricket hairs is very difficult. The motivation to use MEMS hairs for studying the viscosity-mediated coupling has its prominent advantage in providing us a far more reliable control over the hair geometry and hair arrangements. Any approach to perform similar measurements on the actual cricket hairs poses many practical difficulties starting from finding the appropriate hairs for investigation. With MEMS technologies, it is relatively simple to place hairs in specific arrangements, at different distances and also, to some extent, alter the hair lengths and hair diameter. Our work on MEMS hairs thus filled in a, up to then, remaining void in this research by actually measuring the flow field around the hairs, instead of focusing on the hair response itself. Flow field perturbations were measured using Particle Image Velocimetry (PIV) between the fixed hairs, arranged in wide range of inter-hair distances ($9 < s/d < 56$). Significant coupling and larger phase shifts between the far-field flow and inter-hair flow, at smaller inter-hair distance and at low flow frequencies, were reported.

Computational studies of viscous coupling also complemented analytical and experimental ones, starting with Cummins et al. [10] and Heys et al. [11]. Cummins et al. provided results that suggested significant coupling effects to be present for a wide range of normalised inter-hair distances. Lewin et al. [13] developed a finite element model to study the viscosity-mediated coupling between filiform hairs and extended it to a pair of MEMS-based hair-sensors, providing an approximate theoretical prediction for the presented work. Their models used the actual values of the MEMS hair-sensors and suggested significant coupling when $s/d < 10$. Their work also considered the perpendicular configuration, i.e. hairs placed next to each other, both facing the flow at the same time (contrary to the cases discussed until now). For such configuration, they observed positive coupling, suggesting an increased sensitivity of the reference hair to the flow.

## 12.2 Theoretical Backgrounds

### *12.2.1 General*

Viscosity-mediated coupling effects between mechano-receptive hairs in arthropod have been studied experimentally [9] and modeled in literature [9–11, 13]. The work of Bathellier et al. [9] is dedicated both to experimental observation of viscous coupling as well as to deriving analytical expressions for the effects in the approximation of a flat surface, making for a coherent analysis of the effects between two hairs.

[1] With respect to a free-moving reference hair (of same length). The reason for this negligible coupling between free-moving hairs is due to the cancellation of driving and damping torques [9].

The work of Cummins et al. [10] is based on a cylindrically approximation of the conical cercal geometry with discretisation of the conical hairs in many slices along the length of the hair, each exposed to the subsequent flow at the discretised distances above the substrate. Due to the discretised nature the method is highly numerical, making it both time-consuming as well as powerful. E.g. with the method arbitrary collections of hairs with varying hair-length and inter-hair distances may be simulated. The method has produced very dependable results for single hairs. It also largely reproduces the results of Bathellier et al. for the case of two hairs, both for the situation of a free moving and fixed hairs, only deviating at large normalised spacing.

Since a comparable size-match exists between the actual cricket hairs and our artificial hairs and hair-arrays, these models on viscosity-mediated coupling can be applied to our experimental systems using the actual parameters.

Our approach unfolds in two steps. First, we develop a generic model dedicated to the flow in complex geometries of a finite array of hairs where the air-flow direction is arbitrarily chosen relative to the array orientation and where the hairs are equidistantly spaced along both $x$ and $y$ directions of the array. With this general model we set out to predict the flow profiles in given arrays of hairs which we then test with PIV. We then apply in a second step the model to a specific arrangement of three hairs sitting next to each other in one row, parallel to the air-flow direction. We expand the model to also address the torque acting on the hairs and their subsequent rotation. Again, we test our predictions with dedicated experiments.

### *12.2.2 Harmonic Flow in Arrays of Stationary Hairs*

To interpret our PIV measurements on harmonic air-flow profiles in arrays of immovable hairs we develop the theory by extending the model earlier developed by Bathellier et al. [9]. We extend this model to an array of $N \times M$ hairs with arbitrary angle between flow direction and array orientation. Then in the experimental Sect. 12.3.1 we will show that the total perturbation produced by two hairs in tandem can be validated for the transversal flow, and that the prediction of an increase of the velocity component perpendicular to a hair ($V_x$) is validated by measurements. Subsequently we compare the results of our model with PIV measurements on larger arrays. These measurements confirm that the increase of spacing in the array translates into an increase of expected and measured velocities between the hairs. Further we use the model in the discussion Sect. 12.4.1 to model the arrangement of $3 \times 3$ hairs in arrays of different spacings and show that the velocity in the center of this array is increased compared to the one expected in a simple row of axially arranged hairs. By varying the length we prove that we may expect very strong perturbation in small spacing arrays of 250 μm and very low perturbation in large spacing arrays of 1700 μm. We finally use the $N \times M$ model to predict the perturbations caused by arrays of hairs of various spacing at various frequencies.

#### 12.2.2.1 Perturbation by a Single, Immobile Hair

When a harmonic flow $V_{//}$ with far field amplitude $V_\infty$ impinges on an infinite immobile cylinder which is perpendicular to the flow it will cause the flow to be perturbed in a certain area around the cylinder. At the position of the cylinder the flow will be zero and forced to move around it, thereby locally also causing flow in the direction perpendicular ($V_\perp$) to the initial flow. The distance at which the presence of the cylinder is noticeable depends on various parameters involved: the diameter of the cylinder $d$, the dynamic viscosity $\mu$ and mass-density $\rho$, combined in the kinematic viscosity $\nu = \mu/\rho$, of the medium.

Obviously, when the cylinder is not infinite but rather a hair of finite length sitting on a substrate things become far more complicated and complex to solve analytically. Bathellier et al. [9] have given a first analysis of this situation. In their seminal paper they start from the Navier-Stokes equations, but to come to a practical description they assume the boundary layers on the substrate and around the hairs to be separately solvable, the complete viscous effects becoming a superposition of the two effects. Hence, in this approach the effects of viscosity mediated effects in hair-sensor arrays can be largely appreciated by studying the effects in the plane parallel to the substrate and perpendicular to the hairs. In this work we will not re-establish the conditions nor iterate the assumptions that Bathellier have made. In fact we do not develop new theory but rather extend their results to include air flow from an arbitrary chosen direction and impinging on an array of regularly spaced hairs. To do so we will first look at the perturbation caused by one hair, then look at the boundary conditions for multiple hairs and subsequently derive the expressions for the flow velocity components parallel and perpendicular to the velocity far away from the perturbing array.

Following Bathellier et al. [9], the perturbation to a harmonic flow at a point $P(r, \psi)$ relative to an immobile hair, is given by:

$$
\begin{aligned}
D_{//}(r,\psi) &= 1 - \frac{V_{//}(r,\psi,z)}{V_\infty(z)} \\
&= -\Re\left\{\left(\frac{K_2(\lambda)}{K_0(\lambda)}\frac{1}{4\eta^2} + \frac{2K_1'(2\lambda\eta)}{K_0(\lambda)}\right)\sin^2\psi + \left(\frac{K_2(\lambda)}{K_0(\lambda)}\frac{1}{4\eta^2} - \frac{K_1(2\lambda\eta)}{\lambda\eta K_0(\lambda)}\right)\cos^2\psi\right\} \\
D_\perp(r,\psi) &= \frac{V_\perp(r,\psi,z)}{V_\infty(z)} \\
&= -\Re\left\{\frac{K_2(\lambda)}{K_0(\lambda)}\left(\frac{1}{4\eta^2} - \frac{K_2(2\lambda\eta)}{K_2(\lambda)}\right)\sin(2\psi)\right\}
\end{aligned}
\tag{12.1}
$$

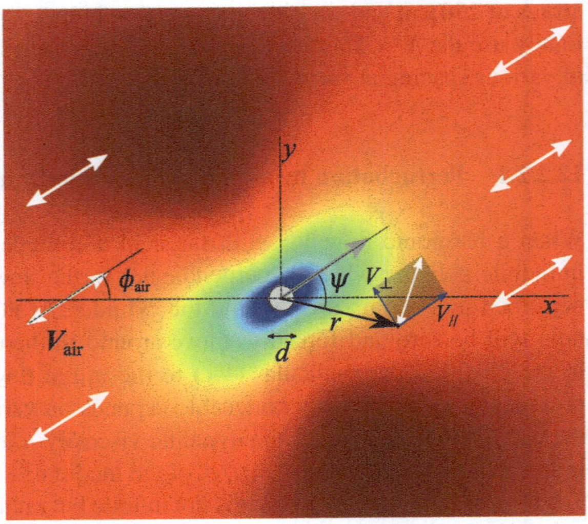


**Fig. 12.2** Single hair perturbing a harmonic airflow directed under an angle $\phi_{air}$ relative to the $x$-direction. In a point at $(r, \psi)$ the hair will cause the flow to be changed in magnitude along the parallel direction ($V_{//}$) and cause some flow in perpendicular direction ($V_{\perp}$). The background image was derived using (12.1)

where

$$\eta = \frac{r}{d}$$
$$\lambda = (1+i)\frac{d}{2}\sqrt{2\zeta}$$
$$\zeta = \sqrt{\frac{\nu}{2\pi f}} \tag{12.2}$$

$i = \sqrt{-1}$, $V_{\infty}(z)$ is the far-field flow amplitude at height $z$ and $\Re$ denotes the real part of the expression in brackets. The $K_i$ are the modified Bessel functions of the second kind. The various geometrical definitions are illustrated in Fig. 12.2. Note that all quantities in the above expressions are amplitudes where the harmonic time-dependence has been suppressed.

#### 12.2.2.2 Arrays of Immobile Hairs

In order to compare the theoretical flow velocity and perturbation in arrays of hairs to those measured in the hair arrays we extend the previous model using superposition of perturbations to a $N \times M$ hair array model. We assume the array to be rectangular with $N$ columns and $M$ rows of hairs, spaced in $x$-direction by $s_x$ and in the $y$-direction by $s_y$. The point of interest is $P = (x_p, y_p)$, see Fig. 12.3.

In the same spirit as Bathellier et al. have analysed the situation for the perturbation of two hairs in tandem we analyse the situation for multiple hairs. Equation (12.1) describes the perturbation of only a single hair. Inspection of $D_{//}$ shows that it is 1 at the position of the hair whereas $D_{\perp}$ is 0. So for a single hair the no-slip

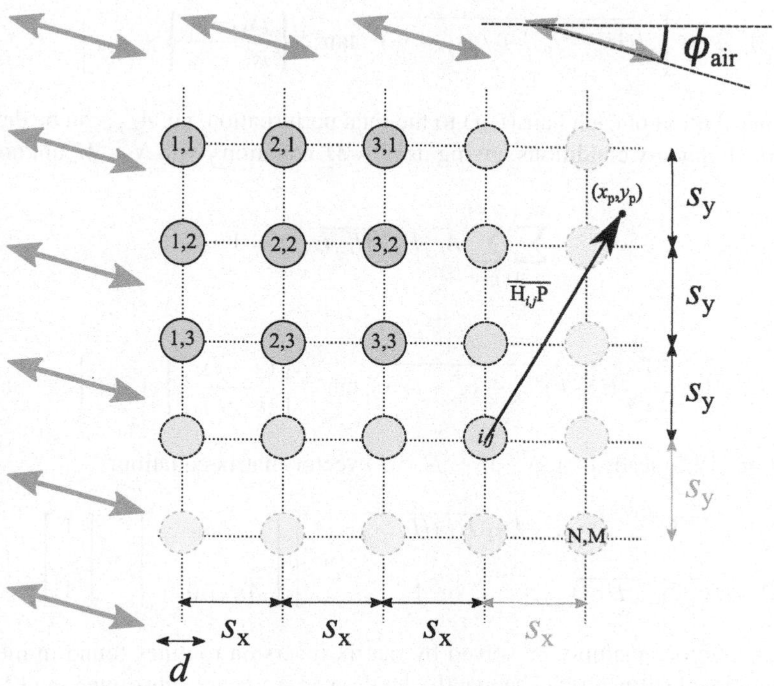


**Fig. 12.3** Illustration of the geometry and definitions used for the analysis of the combined perturbation of multiple hairs. The origin (0, 0) is chosen at the position of hair (1, 1)

boundary condition of zero radial flow at the hair perimeter is fulfilled. However, when multiple hairs are perturbing the flow the combined perturbation still needs to obey the boundary condition of zero radial flow on all the hair perimeters. Hence the task is to determine a linear combination of the perturbation of all the hairs that obeys zero radial velocity at the hair sites. For an arbitrary point in the plane of analysis we can write the linear combination of perturbations as:

$$
\begin{aligned}
D_{//,\mathrm{tot}}(x_\mathrm{p}, y_\mathrm{p}) &= \sum_{i=0}^{N-1}\sum_{j=0}^{M-1} A_{i,j} D_{//}(\overline{H_{i,j}P}) \\
D_{\perp,\mathrm{tot}}(x_\mathrm{p}, y_\mathrm{p}) &= \sum_{i=0}^{N-1}\sum_{j=0}^{M-1} A_{i,j} D_{\perp}(\overline{H_{i,j}P})
\end{aligned}
\tag{12.3}
$$

where $(i, j)$ are the column and row numbers respectively, $A_{i,j}$ determines the strength of the perturbation by hair $(i, j)$ and where the vectors $\overline{H_{i,j}P}$ are expressed in cylindrical coordinates by:

$$\overline{H_{i,j}P} = \left[\sqrt{(is_x - x_p)^2 + (js_y - y_p)^2}, \tan^{-1}\left\{\frac{js_y - y_p}{is_x - x_p}\right\} + \phi_{air}\right] \quad (12.4)$$

The contribution of each hair $(i, j)$ to the total perturbation, i.e. $A_{i,j}$, can be derived from the boundary conditions, giving us $N \times M$ equations with $N \times M$ unknowns:

$$\sum_{i=0}^{N-1}\sum_{j=0}^{M-1} A_{i,j} D_{//}(\overline{H_{i,j}H_{k,l}}) = 1 \quad (12.5)$$

$$\overline{H_{i,j}H_{kl}} = \left[\sqrt{(i-k)^2 s_x^2 + (j-l)^2 s_y^2}, \tan^{-1}\left\{\frac{(j-l)s_y}{(i-k)s_x}\right\} + \phi_{air}\right] \quad (12.6)$$

Equation (12.5) leads to a $(N \cdot M \times N \cdot M)$ vector-matrix equation:

$$\begin{bmatrix} 1 & \ldots & D_{//}(\overline{H_{1,1}H_{N-1,M-1}}) \\ \ldots & \ldots & \ldots \\ D_{//}(\overline{H_{N-1,M-1}H_{1,1}}) & \ldots & 1 \end{bmatrix} \begin{bmatrix} A_{0,0} \\ .. \\ A_{N-1,M-1} \end{bmatrix} = \begin{bmatrix} 1 \\ .. \\ 1 \end{bmatrix} \quad (12.7)$$

This equation can simply be solved by matrix inversion routines found in modern computational software. Subsequently the derived $A_{i,j}$'s are substituted in (12.3) to calculate the flow profile for arbitrary points $P = (x_p, y_p)$, both inside and outside the array.

It should be remarked that the method outlined above is not necessarily limited to regular arrays of fixed diameter hairs. It is relatively straight forward to have arbitrary positions of the hairs, thereby changing (12.4) and (12.6) requiring slight changes in the numerical code without changing the actual method.

Computationally the method is practically doable: for an array of 20 × 20 (400!) hairs the coefficients $A_{i,j}$ can be calculated in about 3.4 s on a machine with 2 quad core Xeon processors at 2.3 GHz. Calculation of the entire flow field for the same array in a window of 500 × 500 points [requiring at least $50 \times 10^6$ evaluations of (12.1)] takes about 4′32″ on the same computer.

#### 12.2.2.3 3D Prediction of Harmonic Flows in Hair Arrays

In order to compare our models with experimental findings we need to extend our 2D perturbation model to 3D since our measurements are in the $(x, z)$ plane. It is important to note that the effect of the finite length of the hair is not described in the model though it will influence the flow velocity near the tip of the hair. To extend our 2D model we use the expressions for the boundary effect in the $(x, z)$ plane [6] and add those to the 2D perturbation in the $(x, y)$ plane. Figure 12.4 illustrates the two different planes as used in the measurements and the model. The comparison

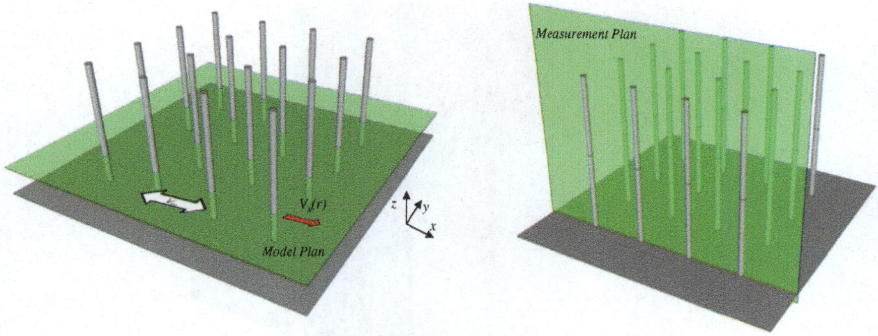


**Fig. 12.4** Geometry as used in the model and measurements. The oscillatory velocity in the far field is directed along $x$

between both measurement and model will only be done for the amplitude of the $V_x$ component of the flow.

The change of the flow velocity with respect to the distance from the substrate can be determined using the theory of Stokes's second problem describing the flow produced by an oscillatory flat and infinitely large plate [22]. This boundary layer is only dependant on the distance $z$ and its influence on the instantaneous velocity $V_\infty$ is expressed as follows:

$$V_\infty(z) = V_0\sqrt{\left\{1 - e^{-\beta z}\cos(\beta z)\right\}^2 + \left\{e^{-\beta z}\sin(\beta z)\right\}^2} \tag{12.8}$$

where $\beta = \sqrt{2\pi f/2\nu}$, $\nu$ is the kinematic viscosity (m/s$^2$) and $V_0$ is the velocity amplitude far from the substrate. The velocity at a point $P(x_p, y_p, z_p)$ at a distance $z$ from the substrate is determined as:

$$\begin{aligned} V_{//}(r, \psi, z) &= V_\infty(z)\left(1 - D_{//,\mathrm{tot}}(x_p, y_p)\right) \\ V_\perp(r, \psi, z) &= V_\infty(z) D_{\perp,\mathrm{tot}}(x_p, y_p) \end{aligned} \tag{12.9}$$

with $D_{//,\mathrm{tot}}(x_p, y_p)$ and $D_{\perp,\mathrm{tot}}(x_p, y_p)$ as in (12.3).

In the comparisons between model and measurements, we will only constrain our analysis to the first 750 µm over the substrate, where the length of the hairs is approximately 850 µm. Figure 12.5 illustrates the effect of the boundary layer over the substrate and the expected velocities between the hairs. As expected, the flow in the vicinity of the hairs is further perturbed by the boundary layer effect over the substrate and largely decreased at the bottom of the hair as represented in Fig. 12.5.

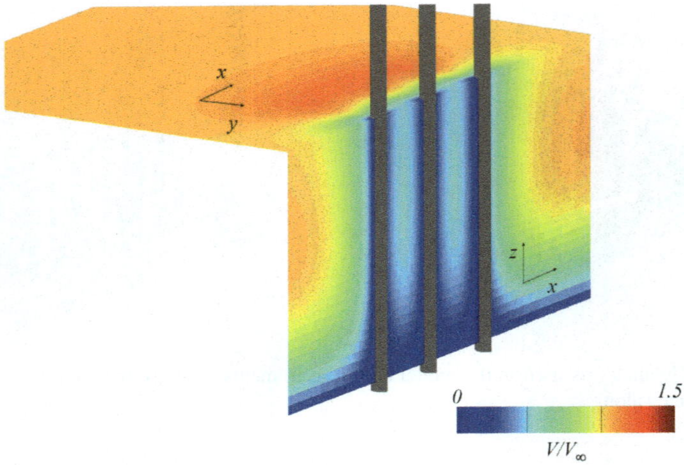


**Fig. 12.5** Expected velocities between and around three hairs in line ($s$ = 500 μm) in a 60 Hz oscillatory flow depicted in the ($x$, $z$) and ($x$, $y$) planes

## 12.2.3 Viscous Coupling Acting on Sets of Two and Three Hairs

In the previous section we have analysed how the harmonic flow-fields in arrays of immovable hairs are perturbed by viscous effects. However, in the end, the importance lies in the hair-rotations evoked by these flows which ultimately transmit the flow information to the hair-sensor array. Here we develop a model to calculate the response of a single hair (reference), for given oscillatory flow conditions and then re-calculate the response of the same hair, in presence of one or two flow perturbing hairs. A coefficient for the coupling, based on unperturbed and perturbed responses of the reference hair, is defined. An existing model was used for the case of a pair of tandem hairs (one reference and one perturbing hairs) [9] and is reiterated in Appendix 1 for completeness. As a special case, for the interest of our experiments, we extend this theoretical model to study the viscosity-mediated coupling between three hairs in tandem.

### 12.2.3.1 Viscosity-Mediated Coupling Between Three Hairs in Tandem

We will consider the case where three hairs are arranged in line parallel to the direction of a harmonic oscillating flow (see Fig. 12.6). The hair in the middle is treated as our reference, which is perturbed by the presence of two hairs, one on either side. We chose the direction of the flow in the $x$-direction (i.e. $D_{//} = D_x$ and $D_{\perp} = D_y$).

From the symmetry of the flow perturbing function (12.1), visualised in Fig. 12.2 we can easily see that the mutual perturbation by each of the two identical hairs

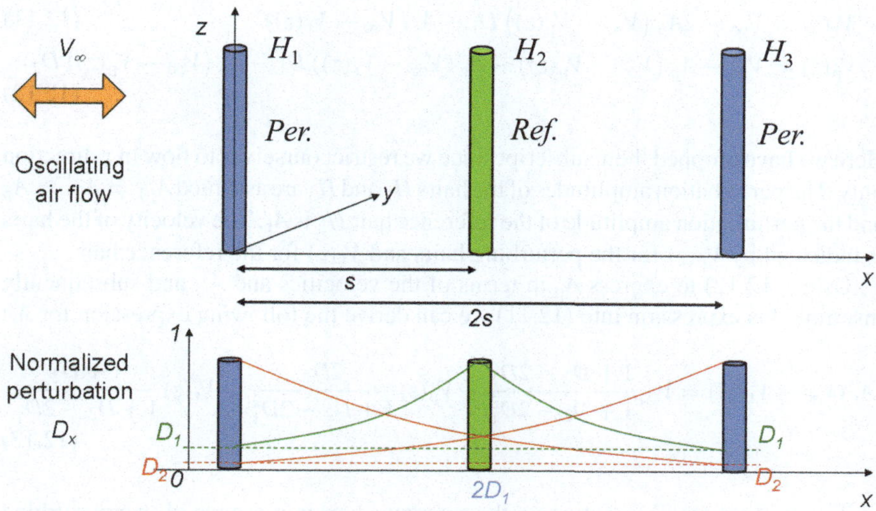


**Fig. 12.6** Simple model for viscous coupling between three hairs showing: (i) $D_1$, between a reference hair and two flow-perturbing hairs (of same length), one on both side and (ii) $D_2$, between two flow-perturbing hairs

on the other hair is equal. For hairs separated by a distance $s$ along the $x$-axis this perturbation is given by $D_1 = D_x(s)$.

Since the hairs are not immobile in this analysis the perturbation is not only a function of $V_\infty(z)$ but depends on the flow velocity difference relative to the hair velocity. This means that we have to adjust (12.1) to become:

$$D_x(s) = \frac{V_\infty(z) - V_x(s,z)}{V_\infty(z) - V^*_{x,pr}(z)}$$

$$D_y(s) = \frac{V_y(s,z)}{V_\infty(z) - V^*_{x,pr}(z)} \tag{12.10}$$

where $V^*_{x,pr}(z)$ is the velocity of the reference ($r$) or one of the perturbing ($p$) hairs in $x$-direction at height $z$ above the substrate.

The analysis of two hairs is readily extended to the case of three hairs, see Fig. 12.6. In this case the reference hair is subjected to the flow perturbations from both hairs $H_1$ and $H_2$ on either side, i.e. a perturbation of strength $2D_1$ in total. Furthermore, both perturbing hairs mutually influence each other, at a distance $2s$ leading to a total perturbation of $D_x(s) + D_x(2s) = D_1 + D_2$. Applying the no-slip condition on the surface of the hairs leads to the modified form of (12.3) for the $x$-component of the velocity and at the positions of the hair only:

$$V_r(z) = V_\infty - 2A_p\left(V_\infty - V_p(z)\right)D_1 - A_r\left(V_\infty - V_r(z)\right) \tag{12.11}$$

$$V_p(z) = V_\infty - A_p\left(V_\infty - V_p(z)\right) - A_r\left(V_\infty - V_r(z)\right)D_1 - A_p\left(V_\infty - V_p(z)\right)D_2 \tag{12.12}$$

Here we have dropped the $x$ subscript since we restrict ourselves to flow in $x$ direction only. The perturbation amplitudes of the hairs $H_1$ and $H_3$ are assumed $A_{p1} = A_{p3} = A_p$ and the perturbation amplitude of the reference hair $H_2$ is $A_r$. The velocity of the hairs is indicated by $V_p(z)$ for the perturbing hairs and $V_r(z)$ for the reference hair.

Using (12.12) to express $A_p$ in terms of the velocities and $A_r$, and subsequently inserting this expression into (12.11) we can derive the following expression for $A_r$:

$$A_r\left(V_\infty - V_r(z)\right) = V_\infty\frac{1+D_2-2D_1}{1+D_2-2D_1^2} + V_p(z)\frac{2D_1}{1+D_2-2D_1^2} - V_r(z)\frac{1+D_2}{1+D_2-2D_1^2} \tag{12.13}$$

The drag-torque $T_{d,2P}$ acting on the reference hair in presence of two perturbing hairs can then be determined as:

$$\begin{aligned} T_{d,2P} &= \alpha\int_0^L A_r z\,dz \\ &= \alpha\int_0^L\left\{V_\infty\frac{1+D_2-2D_1}{1+D_2-2D_1^2} + V_p(z)\frac{2D_1}{1+D_2-2D_1^2} - V_r(z)\frac{1+D_2}{1+D_2-2D_1^2}\right\}z\,dz \end{aligned} \tag{12.14}$$

where $\alpha$ is a constant depending on flow - hair interaction and is given following an analysis based on Stokes' theory (see [9] and Appendix 1). This drag torque will cause the reference hair to rotate. Conservation of angular momentum requires:

$$J\frac{\partial^2\theta}{\partial t^2} + R\frac{\partial\theta}{\partial t} + S\theta = T_{d,2P} \tag{12.15}$$

with $J$, $R$ and $S$ parameters of the hair representing the moment of inertia, torsional damping constant and torsional stiffness respectively.

Subsequently, the resulting complex rotational angle of the doubly-perturbed reference hair (in response to induced flow perturbations by two perturbing hairs) is given as:

$$\theta_{2P}(\omega) = \frac{V_\infty R_\mu}{2LJ}\frac{3\varepsilon_4 + 2\varepsilon_6\gamma_p}{(\omega_0^2 - \omega^2) + i\left(\frac{\omega}{J}\right)\left(R + \varepsilon_5 R_\mu\right)} \tag{12.16}$$

where $\gamma_p$ is a complex quantity of which the magnitude represents the tip velocity of the perturbing hair, normalised by the far-field velocity $V_\infty$. It is to be noted that

**Table 12.1** Viscous coupling terms for the three-hair model

| $\varepsilon_4$ | $\varepsilon_5$ | $\varepsilon_6$ |
|---|---|---|
| $\frac{1+D_2-2D_1}{1+D_2-2D_1^2}$ | $\frac{1+D_2}{1+D_2-2D_1^2}$ | $\frac{2D_1}{1+D_2-2D_1^2}$ |

this ratio is not known a priori. However, our artificial hairs have relatively large rotational stiffness and $\gamma_p$ is rather small; consequently $\gamma_p$ is taken in the limit to zero. The various $\varepsilon$'s stem from the terms following the velocity terms in (12.14). Note that at long distance between the hairs both $D_1$ and $D_2$ will become very small and consequently (12.14) reduces to the drag-torque expression for a single hair (12.32). Table 12.1 shows the values of $\varepsilon_4$, $\varepsilon_5$ and $\varepsilon_6$, for the case that all the hairs (reference and perturbing) are of equal length.

From (12.33) and (12.35) (see Appendix 1) and (12.16), we have rotation responses of a hair for three different cases respectively as: (i) $\theta_R$ corresponding to the unperturbed response, (ii) $\theta_{1P}$ corresponding to the response while being perturbed by one of the hairs and (iii) $\theta_{2P}$ corresponding to the response while being doubly perturbed by the hairs on either side. From these three responses, three coupling coefficients can be defined for the case of three hairs: The coefficient for the viscosity-mediated coupling of only one perturbing hair on the reference hair is defined as [same as (12.36)]:

$$\kappa_1 = \frac{\theta_R - \theta_P}{\theta_R} \tag{12.17}$$

The coefficient for the viscosity-mediated coupling of a second perturbing hair on the reference hair (when already one perturbing hair is present) is defined as:

$$\kappa_2 = \frac{\theta_{1P} - \theta_{2P}}{\theta_{1P}} \tag{12.18}$$

On the other hand, the coefficient for viscosity-mediated coupling of two perturbing hairs, relative to an unperturbed hair, is defined as:

$$\kappa_3 = \frac{\theta_R - \theta_{2P}}{\theta_R} \tag{12.19}$$

The presented analysis of the viscous interaction of three hairs (and that of the two-hair model presented in Appendix 1) provides interesting insights which can be applied to our hair-sensors. The novelty in the model is the presence of the secondary coupling effects between three hairs. The effect of viscosity-mediated coupling between the hair-sensors can be experimentally determined and the results can be compared with this model.

This work features specially modified hair-sensors (cylindrical hairs, shorter membranes, varying hair lengths etc.) and measurements of the coupling effect based on the hair response itself. References are made to the paper of Lewin et al. [13] for comparisons.

## 12.3 Experimental

After the introduction of the theory in the previous section, in this section we describe the experiments and compare the results with the predictions of the models. This experimental section is divided in two parts, one related to flow-field observations in arrays of hairs and one to drag-torques experienced by a set of three hairs in line.

### *12.3.1 Flow Field Observations in Arrays of Hairs*

#### 12.3.1.1 Methods and Experimental

MEMS hairs were made of SU-8, an epoxy that can be structured by photolithography. To obtain sufficiently long hairs, two deposition/exposure cycles of about 450 μm each are used. Design values for the hair diameters of the first and second layers were 50 and 25 μm, respectively. Owing to technical limitations in alignment (because of the large thickness of the layers) as well as curing-induced deformations, the second part of the hair can be slightly eccentric over the first part. The diameter of the second part of the hairs was tapered from top to bottom. The total length of the MEMS hairs was ≈825 μm. Only fixed hairs were used in this study. Details of the fabrication of fully functional hair sensors are described in [23]. The hair-to-hair distance varied between 450 and 2800 μm, corresponding to ≈9 to 56 hair diameters[2] ($s/d$).

MEMS hairs, fixed on plates of dimensions 10 mm × 10 mm, were placed in a glass container (dimensions: 10 cm × 10 cm × 10 cm). A loudspeaker (40 W) was connected at one side of the glass container and driven by a sinusoidal signal generator. The air inside the sealed glass box was seeded with 0.2 μm oil particles (diethylhexyl-sebacat, DEHS) using an aerosol generator. A pulsed laser (NewWave Research Solo PIV 2, 532 nm, 30 mJ, Nd:YAG, dual pulsed; Dantec Dynamics A/S) illuminated the flow through the glass from the top and produced a light sheet (width =17 mm, thickness at focus point = 50 μm). A target area was imaged onto a CCD array of a digital camera (696 × 512 pixels) using a stereomicroscope. The field of view was 2700 μm × 2000 μm. Particle velocities were extracted from the images by cross correlation. Owing to the information contained in the grey-level values

[2]Due to fabrication tolerances the diameter of the hairs may deviate from the design values, thus also impacting the ratio of $s/d$.

as captured by CCD as well as the relative large number of particles used in the cross-correlation calculations, a particle displacement precision of 0.1 pixels can be obtained. Given the entire set-up this translates into a measurable displacement of 0.4 µm. We set the time between two pairs of images at 500 µs to provide a sufficient dynamic range of velocities, giving a velocity precision of 0.8 mm/s. We chose a precision level larger than 1% (1% of the velocity amplitude), setting the far-field flow amplitude to between 50 and 80 mm/s. We used the stroboscopic principle to measure different phases of sinusoidal flow with a PIV system limited to 20 Hz. As explained in [24], it consists of sampling a signal of high frequency at a frequency slightly lower than a sub-multiple of the signal frequency. This technique allows us to estimate the flow phase by interference. In the following figures, we only represent the amplitude of the oscillatory flows.

#### 12.3.1.2 Results

Figure 12.7 illustrates the comparison between theory and measurements of the velocity amplitude[3] in the cross section plane of a tandem of hairs with 50 µm diameter, separated by 1200 µm, in an oscillatory flow of 80 Hz. As expected and already shown in [8], the result obtained by the addition of the perturbation of two tandem hairs is correctly estimated by (12.3) for the axial velocity profile in between the two hairs. The novel result is that (12.3) describes as well and with good accuracy the velocity perturbation caused by tandem hairs at a transversal distance from the hairs. Figure 12.7 also clearly illustrates that $V_{//}$, which is the amplitude of the longitudinal component of the velocity, is not only decreased in the axial direction of the flow but also increased at certain distance away from the hair in the transversal ($\perp$) direction. The essential point is that we may expect that this increase of velocity translates into a *positive viscous coupling* for hairs spaced transversally and therefore will play a role in 2D arrays of hairs.

*Measurements in arrays at* $s = 250\,\mu\text{m}$ $(s/d \approx 5)$. We have also performed measurements on an array of 7 × 7 hairs with a spacing of 250 µm and applied the model to this geometry as well. Figure 12.8 shows the result of model and measurements. In the PIV measurements we estimate that the velocity between the hairs was insignificant, whereas in the model we determine that the average normalised velocity $(V_x/V_\infty)$ inside the array was equal to 0.12.

*Measurements in arrays at* $s = 500\,\mu\text{m}$ $(s/d \approx 10)$. PIV measurements were performed in the $(x, z)$ plane between hairs in a line with a spacing of $s = 500\,\mu\text{m}$ (Fig. 12.9). Figure 12.9a (left) is the result of the modeled velocities in the $(x, y)$ plane and in the $(x, z)$ plane. The present model is unable to correctly predict the flow near the tip of the hairs as hairs are modeled as infinitely long cylinders. However, if we limit the prediction of the models to distances over the substrate smaller than three quarters of the length of the hairs, the comparison of modeled velocities

[3] In the figures where velocity amplitudes are displayed these will be the values normalised to the far field velocity amplitudes $V_\infty$.

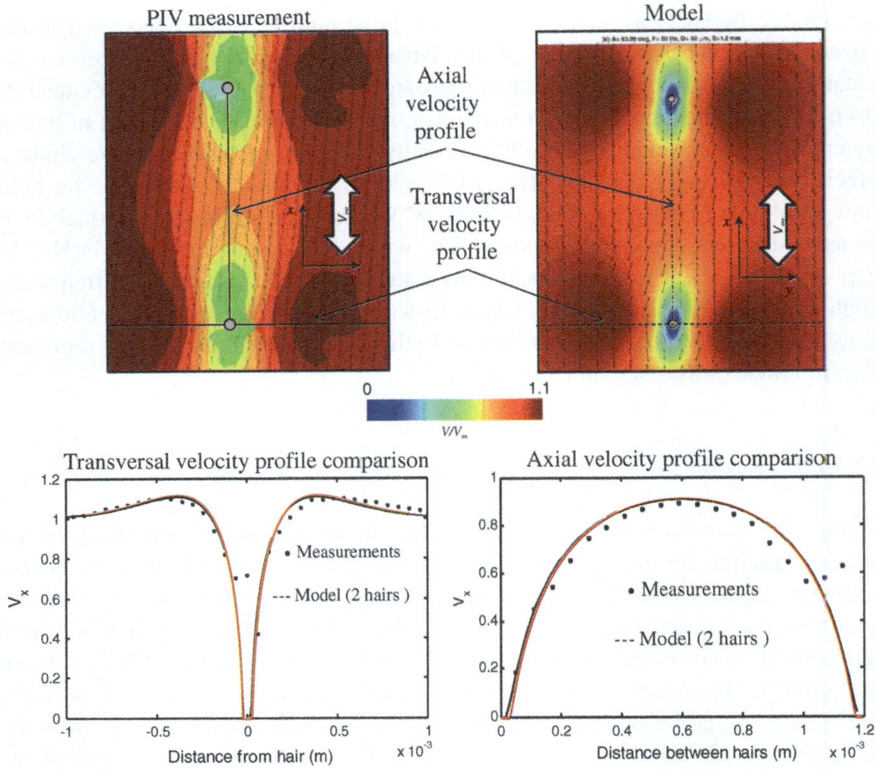


**Fig. 12.7** Comparison of PIV measurement and model for the normalised transversal flow velocity ($V_x/V_\infty$) perturbation caused by a tandem of hairs ($s$ = 1200 μm) on a 80 Hz oscillatory flow

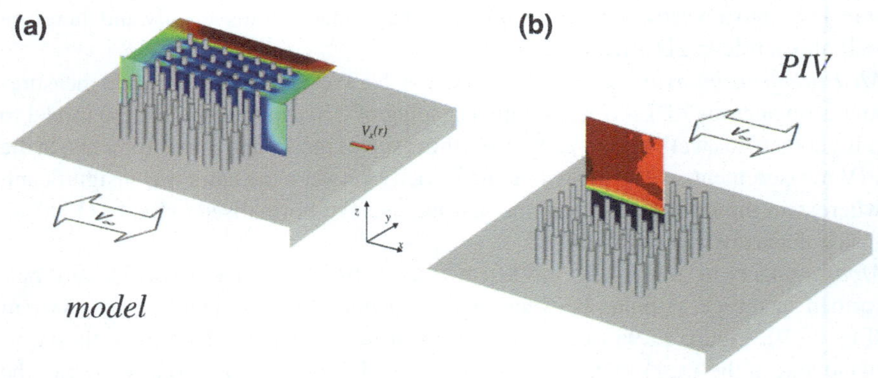


**Fig. 12.8** **a** Results of the model of the $V_x$ component of the 60 Hz oscillatory flow velocity field in the transversal horizontal $(x, y)$ plane and in the vertical plane $(x, z)$ of an array of $7 \times 7$ hairs with a spacing of 250 μm. **b** PIV measurement of the $V_x$ component in the vertical $(x, z)$ plane of the same array. The PIV Measurement is performed in the axial orientation of the first line of hairs. The far field velocity $V_\infty$ is oriented along $x$

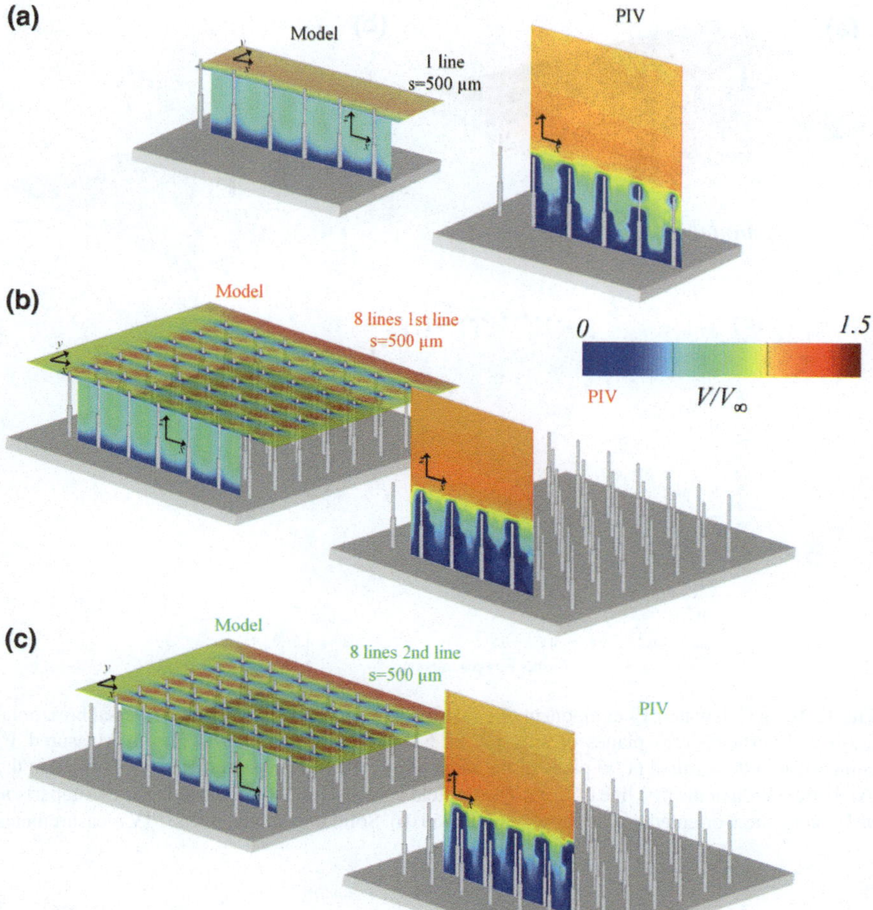


**Fig. 12.9** **a** Left: Calculated $V_x$ component of the 60 Hz oscillatory flow velocity field in the transversal horizontal $(x, y)$ plane of an array of a line of 7 hairs with a spacing of 500 μm and in the vertical $(x, z)$ plane. Right: PIV measurements of the flow between hairs in a line. **b** Left: $V_x$ component of the 60 Hz oscillatory flow velocity field in the transversal horizontal $(x, y)$ plane of an array of $7 \times 7$ hairs with a spacing of 500 μm and in the vertical $(x, z)$ plane of the first line. Right: corresponding PIV measurement. The PIV Measurement is performed in the axial orientation of the first line of hairs. The far field velocity $V_\infty$ is oriented along $x$ (**c**) Left: same as (**b**) left, but the vertical $(x, z)$ plane is represented for the second line. Right: PIV measurement of the velocity along the second line of hairs in the array

and PIV measurements are satisfying as both model and PIV measurements show a large, and comparable, decrease of velocity between the hairs. This decrease is highly pronounced near the substrate. Figure 12.9b, c show the comparison between model and PIV for the flow in the $(x, z)$ plane of the first and second line of the $7 \times 7$

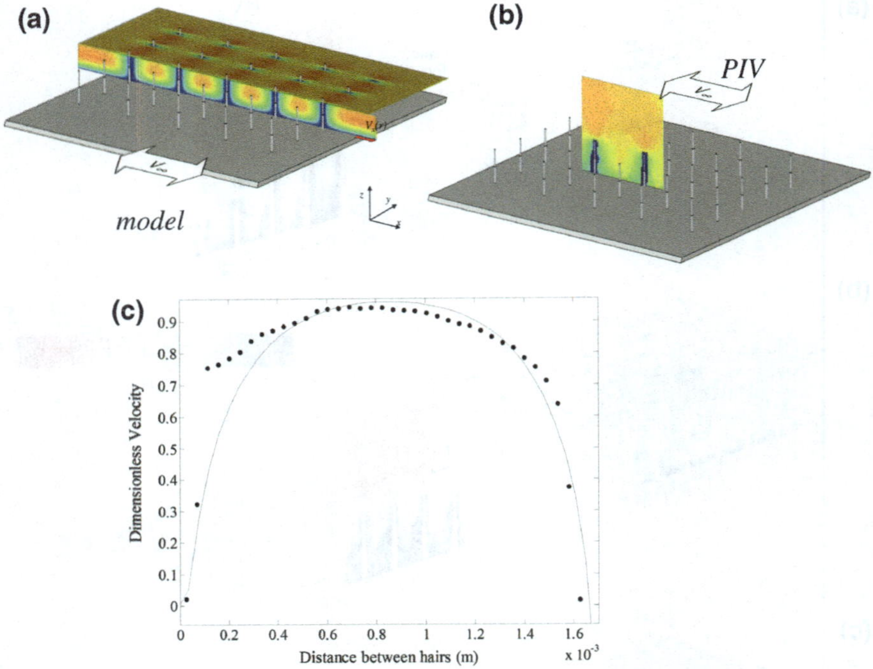


**Fig. 12.10** **a** Calculated $V_x$ component of a 60 Hz oscillatory flow along the transversal horizontal $(x, y)$ and vertical $(x, z)$ planes of an array of $5 \times 5$ hairs with $s = 1700\,\mu\text{m}$. **b** Measured $V_x$ component in the vertical $(x, z)$ plane of the same array. The PIV measurement is performed in the axial orientation of the first line of hairs. The far field velocity $V_\infty$ is oriented along $x$. **c** Comparison of $V_x$ along the $x$-oriented profiles as seen in (**a**) and (**b**). Solid line: model. Dots: PIV measurements

hairs array. Here again, it seems that the model accurately predicts the expected velocity. It appears that the velocity experienced by hairs in the second line of the array (Fig. 12.9c) is higher than experienced by the first line of hairs (Fig. 12.9b).

*Measurements at* $s = 1700\,\mu\text{m}$ *for lines and arrays* ($s/d \approx 34$). We have modeled and measured the flow velocity in an array of $5 \times 5$ hairs in a 60 Hz oscillatory flow with a spacing $s = 1700\,\mu\text{m}$ and the results are shown in Fig. 12.10. We have extracted the $V_x$ velocity along the $x$-axis between two hairs to compare the prediction of the model and the PIV measurements (Fig. 12.10c).

## 12.3.2 Drag-Torque Experiments on Three Hairs in Line

### 12.3.2.1 Design and Fabrication

A dedicated microfabricated chip is designed for characterizing the viscosity-mediated coupling effects. Basically three hair-sensors were placed in rows on the chip surface and inter-hair distances between the hairs were varied in each of these rows. Furthermore, each such row is placed at ≈500 μm (i.e. s/d ≈ 10) away from the next row. The effect of viscous coupling between SU-8 hairs arranged in perpendicular configuration has been modeled by [13]. While there exist some coupling effect it was reported that at distances ($s/d > 5$), the coupling is less than 0.1 at 160 Hz. Hence, this large spatial gap between the rows is expected to yield minimal mutual impact of neighboring-rows. The design of the hair-sensors is slightly modified from their original dimensions [4]. Each hair-sensor comprises of a ≈0.7 mm long polymer hair centrally placed on a suspended membrane which allows the hair to rotate under air flows. In selected rows on the chip, one of the perturbing hairs is fixed to the surface (i.e. without membranes) to facilitate us to analyse the influence of fixed- and free-moving hairs on viscosity-mediated coupling.

A schematic representation of the hair-sensors is shown in Fig. 12.11 (top-right). A brief overview of the fabrication of the hair-sensors [23] is presented here. First, a protective layer of silicon-rich nitride is deposited using LPCVD, followed by the deposition of a ≈800 nm thick poly-silicon layer. This poly-silicon layer acts as a sacrificial layer for the final release of the structures. A ≈1 μm thick silicon-

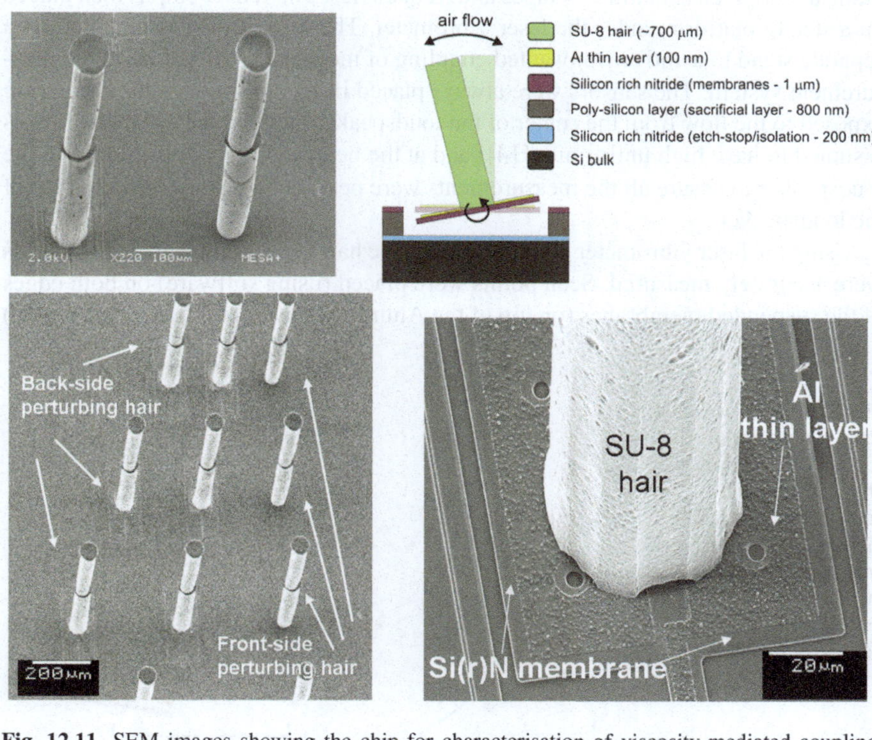


**Fig. 12.11** SEM images showing the chip for characterisation of viscosity-mediated coupling. Top-left: free-moving reference hair-sensor placed near a fixed perturbing hair of same length, Top-right: schematic of a hair-sensor [23]. Bottom-left: hair-sensors are arranged at various inter-hair distances. Bottom-right: close-up of the membrane and hair base

rich nitride layer is deposited and patterned to form the sensor membranes and springs. The membrane is covered by a thin layer of Aluminum (which functionally serves as the top electrode of the sensor capacitors, but in the current studies, only serves as a reflecting surface during optical characterization). Long hairs are made by spinning and patterning two subsequent layers (each of ≈350 μm thick) of negative-tone polymer resist (SU-8) using standard lithography techniques. Finally the sensor membrane is released by selective-etching of the sacrificial poly-silicon layer using a $XeF_2$ etcher. Figure 12.11 also shows SEM images from different portions of a successfully realised chip.

#### 12.3.2.2 Experimental Setup

The hair-sensors are exposed to the near-field of a small loudspeaker (diameter, 4 cm) and their mechanical responses are optically characterised by a laser vibrometer (Polytec, MSA400) [25]. Figure 12.12 shows a schematic representation of the measurement setup. In order to calibrate the generated flow field, the loudspeaker is first placed under the laser vibrometer to characterise the movement of its membrane under given actuating voltages and frequencies. The sensor chip is then placed on a steady platform under the laser vibrometer. The loudspeaker is mounted on a separate stand to avoid any unwanted coupling of mechanical vibrations to the measurement system. The sensors were always placed in such a way that the hairs were exposed to the flow from the center of the loudspeaker membrane (where the flow is assumed to have high uniformity [14]) and at the nearest possible position from the loudspeaker to ensure all the measurements were performed in the near-field [26] of the loudspeaker.

Using the laser vibrometer the dynamics of the hair-sensor under given flow fields were accurately measured. Scan points were placed (using software) on both edges of the suspended membranes (on top of the Aluminum layers, for better reflectivity)

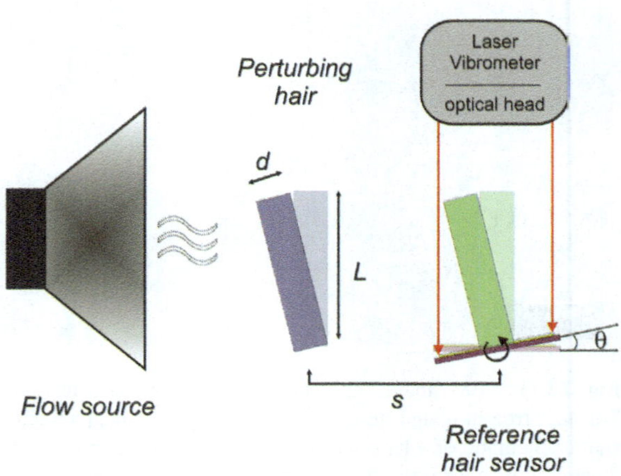


**Fig. 12.12** Experimental setup comprising of a flow source (loudspeaker), the chip with artificial hair-sensors and the laser vibrometer. The rotational angle of the sensor membrane is measured in presence and absence of a flow perturbing hair

and maximum displacements were measured. From these displacements and the distance between the scan points, the rotational angles of membranes were calculated. In order to characterize the viscosity-mediated coupling between the hair-sensors measurements were performed on a reference hair-sensor, with the presence of one or two hair-sensors nearby and repeated in their absence (after manual removal of the perturbing hair(s)). For both measurements the corresponding rotational angle of the reference sensor was determined and hence, the coupling coefficient was obtained. However, the manual removal of the perturbing hair is tedious and care was taken to ensure that both measurements took place in near-identical conditions. The hair was removed using a micro-probe and the chip was precisely placed back into the system aided by the scan points of the laser vibrometer software.

#### 12.3.2.3 Results

The hair-sensor in the middle is our reference and optical measurements were made on it for the following three different cases: (i) when two perturbing hair-sensors were present, (ii) when one of the hair-sensors (at the back-side) was removed and (iii) when both perturbing hairs were removed. The measurements were repeated for three normalised distances ($s/d$ ≈2.1, 2.9 and 8.6). Figure 12.13 shows the normalised frequency response of the reference hair-sensor ($\theta_{2p}$, $\theta_p$ and $\theta_r$), for the three different cases of flow perturbation at three different normalised distances. The error margins in the measurement values account for the possible variations in the flow source and variations in positioning of the sensors (manually done with the aid of the laser vibrometer software, as described before). All the measured responses are compared with the model. The response of the reference hair was observed to be distinct for each case of normalised distance and shows clear influence of the normalised distance between the hair-sensors.

Figure 12.14 shows the coefficients of viscous coupling ($\kappa_1$, $\kappa_2$ and $\kappa_3$) calculated from the above measurement results, for the three above-said normalised distances. The figure shows model predictions as well for ease of comparison.

Figure 12.15 shows the coefficient of coupling for fixed and free-moving front-side perturbing hair measured at 200 Hz. A back-side perturbing hair-sensor was present in all the measurements and only the front-side perturbing hair-sensor (fixed or free) was removed. The measured coefficients of viscous coupling ($\kappa_2$, in this case) at different normalised distances ($s/d$) are shown.

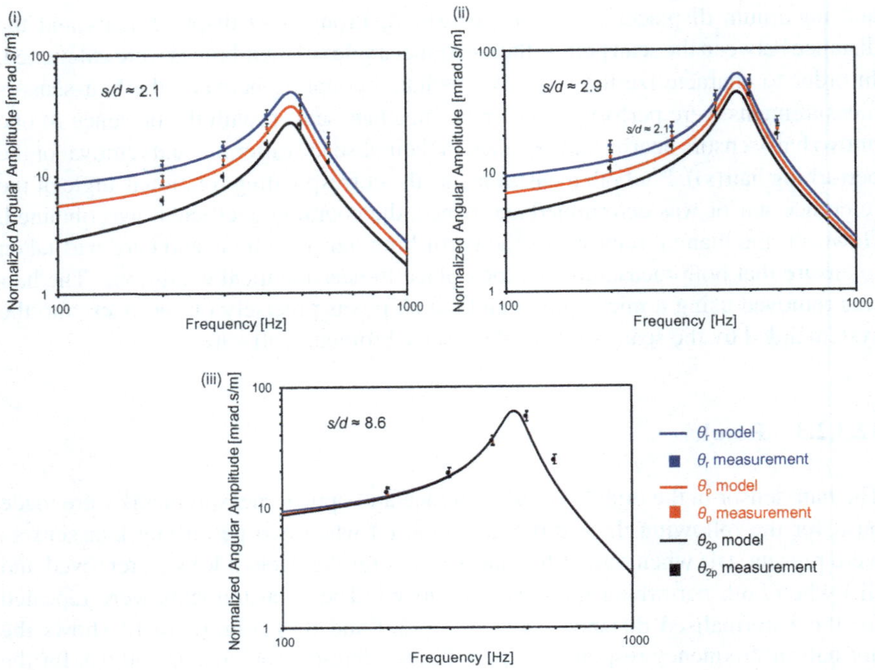


**Fig. 12.13** Normalised angular responses of the reference hair-sensor: (i) when both perturbing hair-sensors were present ($\theta_{2p}$), (ii) when the back-side perturbing hair was removed ($\theta_p$) and (iii) when both perturbing hair-sensors were removed ($\theta_r$). The measurements were made for three normalised distances ($s/d \approx 2.1$, 2.9 and 8.6)

## 12.4 Discussion

### *12.4.1 Flow Fields*

Given the overall satisfying agreement between our measurements and model we us the latter to investigate some of the effects of fluid mediated coupling. As an example of the predictions of our model, Fig. 12.16 illustrates the comparison of the flow velocity profile between an axial alignment of 3 hairs ($s = 500\,\mu$m) and a array of $3 \times 3$ hairs ($s = 500\,\mu$m) in a 60 Hz oscillatory flow. The curves represent the velocity as a function of the axial position between hair 1 and hair 2, for the case of 3 hairs, and hair 2 and hair 5, for the array of 9 hairs.

The reduction of the axial velocity in the axial alignment is significant with velocity reductions at mid-points between two hairs of over 60% relative to the far field velocity. Contrarily, the axial velocity amplitude reduction in the array configuration halfway two hairs is limited to 40% of the far field velocity. It is a counter intuitive aspect of adding hairs, where one would expect that adding more hairs will reduce the velocity inside an array, whereas from this simulation it seems that for

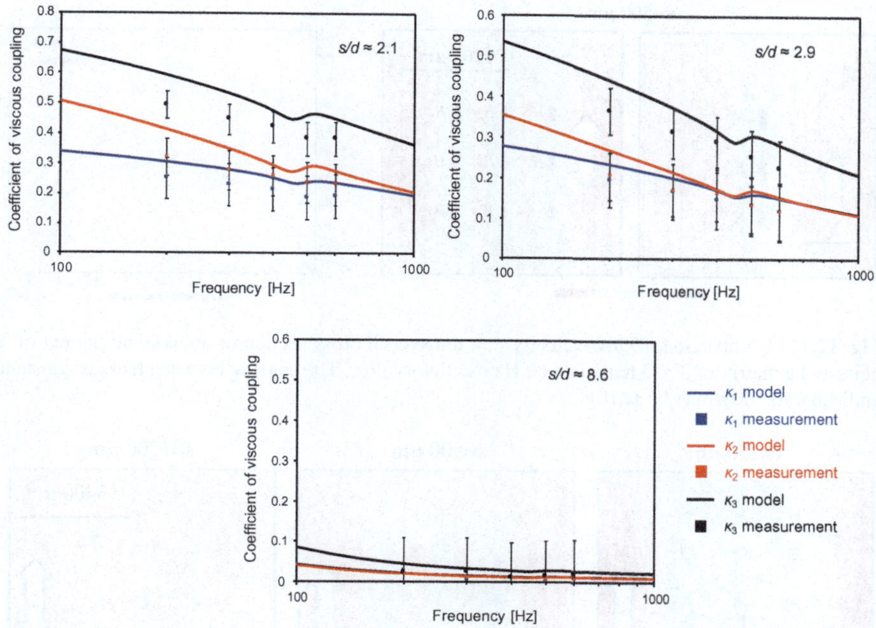


**Fig. 12.14** Coefficients of viscous coupling ($\kappa_1$, $\kappa_2$ and $\kappa_3$) calculated from the measurements performed at three normalised distances. Significant coupling is observed at smaller distances between the hair-sensors and the effect reduces greatly at higher distances

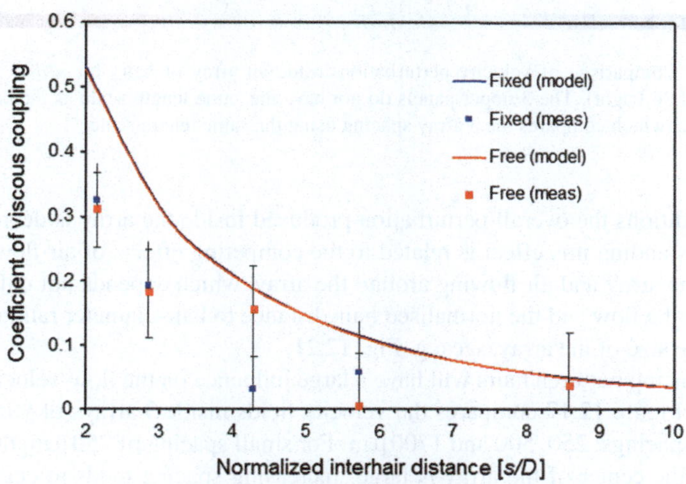


**Fig. 12.15** Comparison of viscous coupling coefficient ($\kappa_2$) at 200 Hz, for a free-moving or fixed front-side perturbing hair on the reference hair-sensor (while in presence of free back-side hair, for both cases)

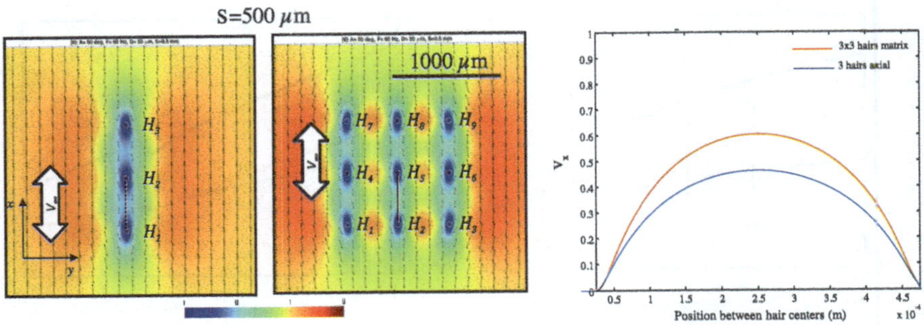


**Fig. 12.16** Comparison of flow velocity in a transversal cross section of an axial alignment of 3 hairs and a matrix of $3 \times 3$ hairs in a 60 Hz oscillatory flow. The spacing between hairs is constant and equal to 500 μm ($s/d = 10$)

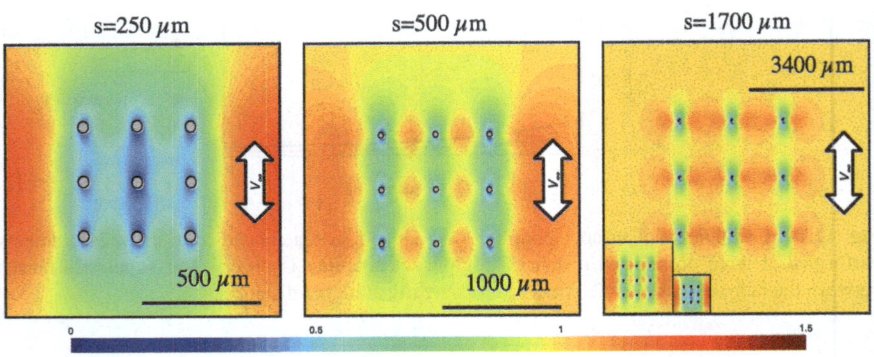


**Fig. 12.17** Comparison of velocity perturbation inside an array of hairs for $s/d = 5$ (left) 10 (center) and 34 (right). The 3 upper panels do not have the same length scale as indicated in the bottom panel which compares the 3 array spacing using the same length scale

some conditions the overall perturbation produced inside the array is decreased. In our understanding this effect is related to the competing effects of air-flow passing through the array and air flowing around the array, which depends not only on frequency of the flow and the normalised hair-distance to hair-diameter ratio ($s/d$) but also on the size of the array, see e.g. Fig. 12.21.

The spacing between hairs will have a large influence on the flow-velocity inside the array. Figure 12.17 compares the velocity fields inside 3 arrays of 9 hairs with different spacings: 250, 500 and 1700 μm. For small spacing of 250 μm, the perturbation in the center of the array is large. Increasing spacing tends to decrease the perturbation. For spacing of 1700 μm the perturbation is weak and approaches the perturbation of isolated single hairs.

In Fig. 12.18, we present a panel of 20 velocity fields simulated for 5 different arrays with spacing ranging from 250 to 2000 μm ($s/d = 5 - 40$) and at 5 different frequencies ranging from 30 to 480 Hz. At low frequency and for dense arrays, hairs experience only a small proportion of the far field velocities due to the large perturba-

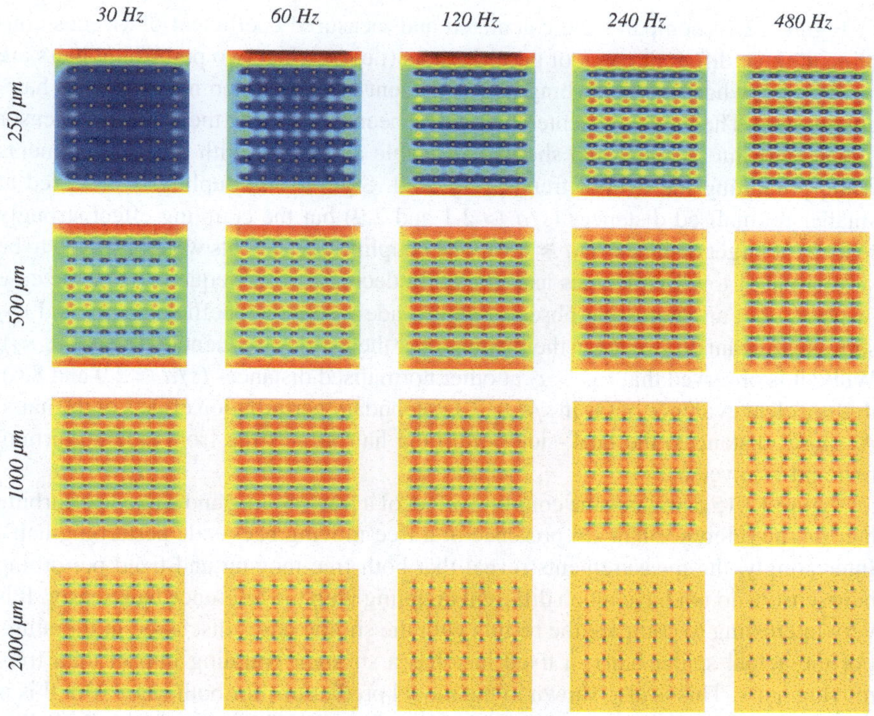


**Fig. 12.18** Predicted velocity fields as simulated for 4 different 7 × 7 arrays of 50 μm diameter hairs with spacing ranging from 250 to 2000 μm at 5 different frequencies ranging from 30 to 480 Hz

tions. Even peripheral hairs are affected by the formation of a global boundary layer over the whole array of hairs which act as a highly coupled system. As expected the increase of hair-spacing leads to a reduction of the perturbation by viscous coupling and leads to increased exposure of the hairs to the flow. This gives insight for the design of flow cameras and it surely highlights the trade-off between spatial and temporal accuracy. A small spaced array will be able to monitor small scale and fast events, whereas large spaced arrays should be capable of following slower events but with a lower spatial resolution.

### 12.4.2 Drag-Torque Experiments

As for the fabricated hairs for the drag-torque experiments it should be remarked that they were larger in diameter than designed. This has caused some of he relative distances $(s/d)$ to be smaller than designed, which is only beneficially for the experiments discussed here.

Figure 12.14 compares the calculated and measured coefficient of viscous coupling at three different cases of perturbations (i.e. (i) when two perturbing hairs are present, (ii) when one perturbing hair is present and (iii) when no perturbing hairs are present). The model presented before has been used taking the actual hair-sensor dimensions into account and shows reasonable agreement with the measurements while predicting the overall trends very well. Significant coupling is observed at smaller normalised distances ($s/d \approx 2.1$ and 2.9) but the coupling effect strongly reduces at larger distance ($s/d \approx 8.6$). The coupling coefficients were observed to be substantial at low frequencies and gradually decrease with frequency. For the case of $s/d \approx 2.1$, an interesting observation is made where the coefficients ($\kappa_1$ and $\kappa_2$) show significant difference in their values over the various frequencies (i.e. $\kappa_1 < \kappa_2$). While it is observed that $\kappa_1 = \kappa_2$ at other normalised distances ($s/d \approx 2.9$ and 8.6), this is a clear evidence of (as in $s/d \approx 2.1$) secondary perturbation effects of the hairs. At closer distances, the back-side perturbing hair affects the front-side perturbing hair and vice versa.

Figure 12.15 compares the coupling effect of a free-moving and a fixed perturbing hair on the reference hair (in presence of a free-moving back-side perturbing hair). Interestingly, the measurements reveal that both free-moving and fixed perturbing hair-sensors do not have much different coupling with the reference hair-sensor. It is very interesting to compare the results with the similar cases discussed in literature. For the actual spider hairs, a fixed hair has a stronger coupling impact than free-moving hairs. This is also shown in the model predictions for both cases. But this is contrary to the measurement observations on the actual filiform hairs by Bathellier et al. [9], where a fixed hair contributes to significant coupling and free-moving hairs contribute very little. But as a stark contrast, our results show more or less similar coupling effects for both cases (see Fig. 12.15). Not surprisingly, we think this has to do with the large stiffness of our sensor springs, compared to those of actual trichobothria [27]. Due to the large torsional stiffness, the motion of the free-moving hairs under the flow conditions, does not make a large difference relative to fixed hairs. This is also evident from the model predictions shown for both cases, which take the actual torsional stiffness values into account.

In the introduction to this work, it was mentioned that Lewin et al. [13] developed an FEM model for viscous coupling. The model was designed to incorporate our artificial sensor's geometrical parameters, i.e. hair length of 1000 and 50 μm diameter. Figure 12.19 shows the results of our measurements compared with both Bathellier's [9] and Lewin's [13] models. Only the coefficient of viscous coupling ($\kappa_1$) between two hair-sensors (one reference hair and one perturbing hair, as described in Appendix 1) is considered in the comparison, as both models were established for two hairs. While the measurements show reasonable agreement with the models, it has to be noted that some of the differences between the model values and the measurements may be attributed to the fact that our measurements were done at 200 Hz as opposed to Lewin's published model results which were evaluated at 160 Hz. Hence, Lewin's results are expected to show a higher estimate than Bathellier's model. Nevertheless, comparing the models with the measured coupling coefficients it seems both models are over-estimating the viscous coupling effects as found for our artificial sensors.

The obtained results give a solid critical remark to previous array designs. As per our design of functional artificial sensor arrays, the hairs ($\approx$700 $\mu$m long) were separated by 300 $\mu$m (i.e. for a hair diameter of 70 $\mu$m, the normalised distance, $s/d \approx 4.2$) and the results clearly show the coupling effect is high ($\kappa > 0.1$) for such a distance. It can be assumed that these sensor arrays do suffer from significant viscous coupling effects. Hence, the most important insight that can be derived from the presented results will be the crucial design rule for minimum distance between the hairs of sensor arrays.

### 12.4.3 *Impact of Viscous Coupling on Spatio-Temporal Flow Sensing*

In the previous sections we have argued that our modelling, both of the air-flow distribution in arrays of MEMS hairs, as well as for the rotation of MEMS hairs, has an encouraging predictive quality; the studies on viscous coupling presented in this chapter have reproduced previous analyses [9, 10, 13, 21] with respect to frequency dependence and the influence of the normalised distance on the fluid mediated coupling between flow-sensitive hairs. However, these previous studies have not touched on the influence of the direction of the flow relative to the orientation of the hairs, nor on the influence of the size of the arrays. Here we take the liberty to use our hair-array model to investigate some of the effects that may occur in arrays of fixed hairs. Especially since observations on insects are extremely difficult, models

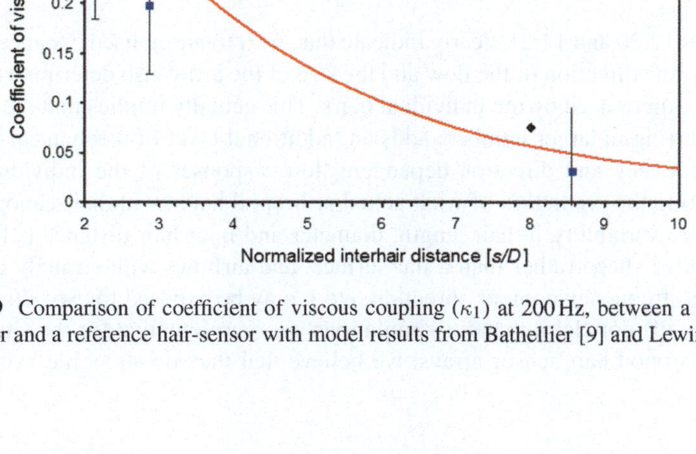


**Fig. 12.19** Comparison of coefficient of viscous coupling ($\kappa_1$) at 200 Hz, between a perturbing hair-sensor and a reference hair-sensor with model results from Bathellier [9] and Lewin [13]

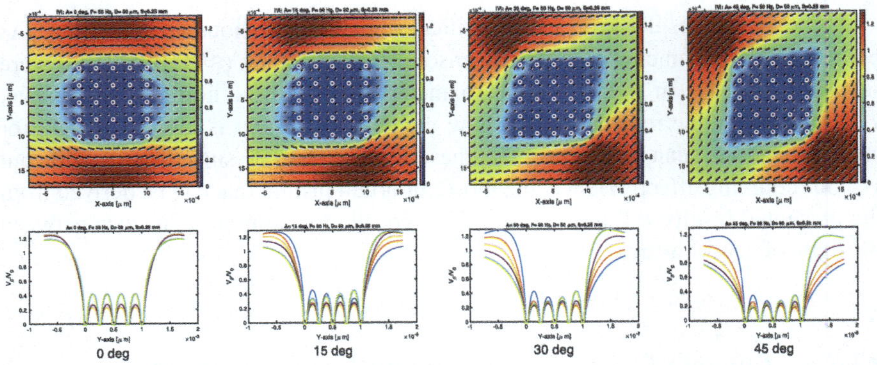


**Fig. 12.20** Normalised flow-velocity for an array of 5 × 5 hairs with $s = 250\,\mu\text{m}$, $d = 50\,\mu\text{m}$, $f = 50\,\text{Hz}$ and $\psi$ is 0, 15, 30 and 45° (from left to right). Top row: normalised velocities. Bottom row: normalised velocity profiles along the 5 columns of the array

may help to coarsely explore the various effects. The analysis and model we have introduced seem helpful in this respect.

Figure 12.20 shows an array of 5 × 5 hairs with harmonic airflow at 50 Hz at 4 different angles. The color-scale indicates the magnitude of the amplitude of the flow, whereas the arrows indicate its direction. Due to viscous coupling air-flow finds it way partly circumventing the arrays resulting in reduced velocities inside the arrays. Hairs in a row or column are asymmetrically exposed to the air-flow, as also clearly indicated by the normalised velocity profiles (parallel component) in the lower part of Figure 12.20.

In Fig. 12.21 we show three arrays of increasing size, having 5 × 5, 9 × 9 and 17 × 17 hairs. From these simulations we see that the effect of displacing the flow along the outside of the arrays increases with array-size, the 17 × 17 arrays showing the largest normalised flow-velocities at the upper-left and lower-right corners of the array, with values of up to 1.77. The lower graph of Fig. 12.21 shows the normalised velocity-profile for the center column of the three arrays. Center values seem rather comparable, but velocities seem to drop somewhat getting more to the outside of the larger arrays.

Figures 12.20 and 12.21 clearly indicate that, next to normalised distance $s/d$ and frequency, the direction of the flow and the size of the array also determine the flow-values as experienced by the individual hairs. This actually implies that the effect of hairs occurring in larger numbers adds an 'additional layer of mechanical filtering' to the frequency and direction dependent flow-responses of the individual hairs, complicating the prediction of what actual arthropod hairs, with hair-canopies with rather more variability in hair length, diameter and inter-hair distance [21], seated on a rounded shape rather than a flat surface, and airflows with strongly changing properties (frequency content, direction, etc.), may be exposed to. So, although we can (and will) not claim that our simulations are representative for the situation of actual arthropod hair-sensor arrays, we believe that they do show the existence of

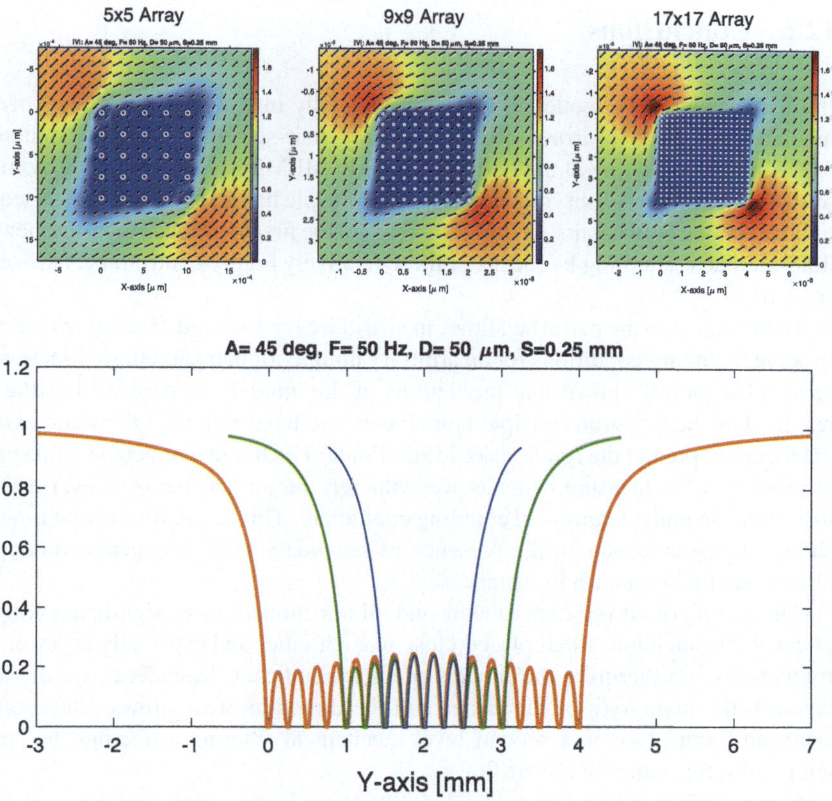


**Fig. 12.21** Normalised flow-velocity for arrays of 5 × 5, 9 × 9 and 17 × 17 hairs with $s = 250\,\mu\text{m}$, $d = 50\,\mu\text{m}$, $F = 50\,\text{Hz}$ and $\psi$ is 45°. Top row: normalised velocities. Bottom row: normalised velocity profiles along the centre column of the arrays

additional filtering effects that occur due to viscous coupling in hair-sensors existing in high density canopies.

The most important insight that can be derived from the presented results is the crucial role played by the minimum distance between the hairs, i.e. the density of the hairs. Recent work also clarified the need to have a variety of hair lengths to optimally measure a transient air flow in its full complexity [28]. This implies the need to understand viscous coupling in detail among a canopy of hairs of different sizes, for which we know much less.

## 12.5 Conclusions

In this work we have modelled and experimentally investigated the effects of fluid mediated coupling in arrays of MEMS based hair-sensors. We have done so by experimentally looking at the perturbed flow in MEMS based hair-arrays by means of PIV as well as by observing hair rotations of triple hair structures with subsequent removal of one or two hairs. In doing so we were the first to produce measurements on fluid-mediated coupling by means of these relatively highly controllable hair-sensor systems

With respect to the perturbed flows in arrays we have extended existing theory [9] to solve boundary conditions for an arbitrary number of hairs and flow-direction and have subsequently shown that predictions of this model compare well to the PIV results. For the situation of triple hair sensors we have extended the same existing 2D theory to predict the rotation angle amplitude. The trends predicted by this model are found in the measurements as well, though the model seems to over-estimate the experimentally determined coupling coefficient. This work, for the first time, has shown strong evidence on the presence of secondary viscosity-mediated coupling effects, at smaller inter-hair distances.

The results of all our experiments and all our models show significant coupling effects between hairs, when placed close to each other and especially at lower flow frequencies. Furthermore the models predict significant dependence on the angle between the main axes of the arrays and the direction of the flows. These effects establish themselves as a second level mechanical filtering, affecting the spatio-temporal information in the air-flows.

Our experimental and theoretical results provide important insights and possible guidelines towards optimal sensor array designs while simultaneously possibly shedding some light on biophysical effects in closely-packed hair-sensor arrays.

**Acknowledgements** We like to acknowledge the financial contribution of the European Union and the Netherlands Organisation for Scientific Research (NWO), section Applied and Engineering Sciences, to the research presented in this work. We are grateful for the very stimulating collaboration we have had with our colleagues in the Cilia and Cicada EU projects and Greg Lewin for running his FEM code with our MEMS sensor data. We thank Y. Brechetand and Yuri Estrin for their continued interest in this work. Finally we are indebted to B. Bathellier, F. Barth and J. Humphrey† and their colleagues for their pivotal work on the subject of viscous coupling in arthropods as well as for the many discussions we had over the years.

## Appendix 1: Theoretical Analysis of Viscosity-Mediated Coupling

The following is a reiteration of the model derived by Bathellier et al. [9] and rewritten to our case of interest. For the purpose of simplification, all the hairs considered here

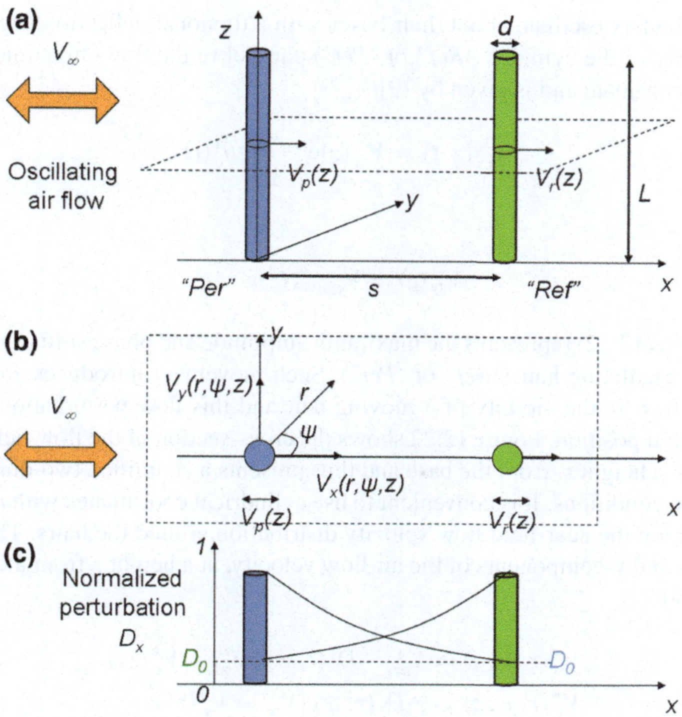


**Fig. 12.22** Simple model to derive the dominant parameters of viscous coupling between a reference hair and a flow-perturbing one [9]

are assumed to be of identical cylindrical geometric shape (however the model of Bathellier et al. is also applicable for hairs of unequal lengths).

Figure 12.22a shows the schematic of the model consisting of two hairs-*Per*' and '*Ref*' (standing for '*Perturbing*' and '*Reference*' hairs) of diameter $d$, placed along the $x$-direction and separated by a distance $s$. Oscillatory air flow is assumed only along $x$-direction for simplicity and hence, the cylinders move in the direction of the flow. The far-field harmonic air flow velocity for the $x$-direction is given as[4]:

$$\vec{V}_{\infty}^{*}(t) = V_{\infty} e^{i\omega t} \vec{e}_x \tag{12.20}$$

---

[4]Note: In the following analysis, quantities with a * superscript represent the time-dependent formulations whereas quantities without * are (complex) amplitudes.

The cylinders oscillate about their bases with a torsional deflection angle, $\theta_c^*(t)$. The velocity of the cylinder ('*Ref*' or '*Per*') parallel to the flow-direction depends on the $z$-coordinate and is given by [9]:

$$V_{cy}^*(z,t) = V_{cy}(z)e^{i\omega t} = z\dot{\theta}_c^*(t) \tag{12.21}$$

where,

$$V_{cy}(z) = V_{cy,\max}(z)e^{i\phi} \tag{12.22}$$

Equation (12.22) represents the maximum amplitude and phase shift of the velocity of the oscillating hair ('*Ref*' or '*Per*'). Such movement introduces a distortion in the air flow in the vicinity of a moving hair and this flow perturbation depends on the spatial position. Figure 12.22 shows the cross-section of the flow field around the hairs at a height $z$, from the base and thus presents a simplified two-dimensional view of the conditions. It is convenient to use cylindrical coordinates with $r$, $\psi$ and $z$ for describing the near-field flow velocity distribution around the hairs. The instantaneous $x$- and $y$-components of the air flow velocity, at a height $z$ from the base are given as [9]:

$$V_x^*(r,\psi,z) = V_\infty^* - D_x(r,\psi)\left(V_\infty^* - V_x^*(z)\right) \tag{12.23}$$

$$V_y^*(r,\psi,z) = -D_y(r,\psi)\left(V_\infty^* - V_x^*(z)\right) \tag{12.24}$$

where $D_x$ and $D_y$ represent the $x$- and $y$-components of the perturbations introduced in the flow-field due to the hair.

The no-slip conditions for this scenario are such that $V_x^*(d/2,\psi,z) = V_{cx}^*(z)$ and $V_y^*(d/2,\psi) = 0$, implying $D_x = 1$ and $D_y = 0$ at the hair surface. The near-field flow distribution around the cylinders is derived by the theory of Stokes' applied to solve the simplified Navier-Stoke's equations [9]. From (12.23) and (12.24), the $x-$ and $y-$components of the perturbation $D$ can be derived in terms of two dimensionless expressions: $\eta = r/d$ and $\lambda = d/2\sqrt{i\omega\rho/\mu}$ and the modified Bessel functions of the second order $K_0$, $K_1$ and $K_2$. They are derived as:

$$D_x = \frac{V_\infty - V_x(z)}{V_\infty - V_{cx}(z)}$$
$$= -\Re\left\{\left(\frac{K_2(\lambda)}{K_0(\lambda)}\frac{1}{4\eta^2} + \frac{2K_1'(2\lambda\eta)}{K_0(\lambda)}\right)\sin^2\psi + \left(\frac{K_2(\lambda)}{K_0(\lambda)}\frac{1}{4\eta^2} - \frac{K_1(2\lambda\eta)}{\lambda\eta K_0(\lambda)}\right)\cos^2\psi\right\} \tag{12.25}$$

$$D_y = \frac{V_y(z)}{V_{cy}(z) - V_\infty}$$
$$= -\Re\left\{\left(\frac{K_2(\lambda)}{K_0(\lambda)}\left(\frac{1}{4\eta^2} - \frac{K_2(2\lambda\eta)}{K_2(\lambda)}\right)\right)\sin(2\psi)\right\} \tag{12.26}$$

In the above expressions, $D_y$ is not significant in determining $x$- or $y$-components of the near-field distribution as it is 0 for $\psi = 0$ or integer multiples of 90°. For our case, where hairs are in a parallel configuration along the direction of flow (hence, $\psi = 0$ and $\eta_0 = s/d$ i.e. the position of the reference hair from the perturbing hair or vice versa), the normalised perturbation ($D_x$) is simplified and given as [9]:

$$D_1 = D_x(s, 0) = D_x(s, \pi) = \Re\left\{\frac{1}{4\eta^2}\frac{K_2(\lambda)}{K_0(\lambda)} - \frac{1}{\lambda\eta}\frac{K_1(2\lambda\eta)}{K_0(\lambda)}\right\} \quad (12.27)$$

Figure 12.22c shows the schematic representation of $D_x$ from both of the hairs.

The flow at any vector position $\vec{r}$ (Fig. 12.22) can be derived by summing up the unperturbed far-field flow velocity and velocity perturbations caused by both the hairs. At any position $\vec{r}$ the flow perturbations along the $x$-direction by the hairs will be denoted as $D_x(\vec{r})$ and $D_x(\vec{r} - \vec{s})$. The flow at any position $r$ along the $x$-direction is given as [9]:

$$V_x(z) = V_\infty - A_p\left\{V_\infty - V_p(z)\right\} D_x(r) - A_r\left\{V_\infty - V_r(z)\right\} D_x(r - s) \quad (12.28)$$

where $V_p$ and $V_r$ are the velocities of the hairs defined by (12.22) and $A_p$ and $A_r$ are two unknown complex terms. The values of these terms can be determined by applying the no-slip condition at the surface of the hairs (i.e. $V_x = V_p$ or $V_r$), which yields two equations. The equations are solved and the values of the complex terms are given as [9]:

$$A_p\left(V_\infty - V_p(z)\right) = \frac{V_\infty}{1 + D_1} + \frac{V_r(z)D_1}{1 - D_1^2} - \frac{V_p(z)}{1 - D_1^2} \quad (12.29)$$

$$A_r\left(V_\infty - V_r(z)\right) = \frac{V_\infty}{1 + D_1} + \frac{V_p(z)D_1}{1 - D_1^2} - \frac{V_r(z)}{1 - D_1^2} \quad (12.30)$$

where $D_1$ [defined in (12.27)] represents the perturbation due to a hair at the position of another.

The hairs respond to the driving flows by angular deflection $\theta$ and the conservation of angular momentum equation is given as [5]:

$$J\frac{\partial^2\theta}{\partial t^2} + R\frac{\partial\theta}{\partial t} + S\theta = T_d \quad (12.31)$$

where $J$, $R$ and $S$ are the intrinsic mechanical parameters of the hair representing the moment of inertia, torsional damping constant and torsional stiffness respectively. The flow-induced drag torque $T_d$ exerted on the hair can be calculated by integrating the drag-force along the length of the hair. In order to calculate the viscosity-mediated coupling, we need to solve (12.31) for two cases: (i) when the reference hair is isolated and doesn't have any contribution from a perturbing hair and (ii) when the reference hair has a perturbing hair placed next to it, in the direction of the flow. The drag-torque acting on an isolated reference hair is given as:

$$T_{\mathrm{d,R}} = \alpha \int_0^L (V_\infty - V_\mathrm{r}(z))z dz \tag{12.32}$$

where $\alpha = 4\pi\mu G - i\pi^2\mu G/g$, $G = g/(g^2 + \pi^2/16)$ and $g = \ln(\lambda)$ [9]. By solving (12.28) using (12.29) and filling in the contributing parameters, the complex rotational angle of the reference hair (isolated, without any flow perturbations) can be written as [9]:

$$\theta_\mathrm{R}(\omega) = \frac{3V_\infty R_\mu}{2LJ} \frac{1}{(\omega_0^2 - \omega^2) + i\left(\frac{\omega}{J}\right)(R + R_\mu)} \tag{12.33}$$

where $\omega_0 = \sqrt{S/J}$ is the resonance frequency of the hair system, $\omega$ is the driving oscillatory flow frequency and $R_\mu$ represents the damping coefficient due to the surrounding medium, which is air [9]. Now, if a perturbing hair is placed in a parallel configuration, i.e. placed along the direction of the flow ahead of the reference hair (Fig. 12.22), the drag-torque acting on the reference hair ('*ref*') in the presence of perturbing hair ('*per*') is given by [9]:

$$\begin{aligned} T_{\mathrm{d,1P}} &= \alpha \int_0^L A_\mathrm{r}(V_\infty - V_\mathrm{r}(z))z dz \\ &= \alpha \int_0^L \left\{ \frac{V_\infty}{1 + D_1} + \frac{V_\mathrm{p}(z)D_1}{1 - D_1^2} - \frac{V_\mathrm{r}(z)}{1 - D_1^2} \right\} z dz \end{aligned} \tag{12.34}$$

where $V_\mathrm{r}(z)$ and $V_\mathrm{p}(z)$ denote the velocity of the hairs at a height z from their base. Using (12.31) in (12.28), the complex rotational angle of the perturbed reference hair (in response to the flow including the perturbations) is given as [9]:

$$\theta_{1\mathrm{P}}(\omega) = \frac{3V_\infty R_\mu}{2LJ} \frac{\varepsilon_1 + \varepsilon_3 \frac{2}{3}\gamma_p}{(\omega_0^2 - \omega^2) + i\left(\frac{\omega}{J}\right)(R + \varepsilon_2 R_\mu)} \tag{12.35}$$

where $\gamma_\mathrm{p}$ is a complex quantity of which the magnitude represents the tip velocity of the perturbing hair, normalised by the far-field velocity $V_\infty$. For the case where the perturbing hair is immobile or fixed, this means $\gamma_\mathrm{p} = 0$ and (12.35) is simplified [5]. In (12.35), the term $\varepsilon_1$ stands for the perturbed flow at the position of the reference (Note that when the distance between the hairs increases, $D_1$ becomes progressively small and $\varepsilon_1$ goes to 1, representing the unperturbed flow). The term $\varepsilon_2$ represents the additional damping of the reference hair due to the presence of the perturbing hair. The term $\varepsilon_3$ represents the dynamic perturbation due to the movement of air medium by the movement of the perturbing hair. These three terms, $\varepsilon_1$, $\varepsilon_2$ and $\varepsilon_3$ represent the viscous coupling between the hairs and depend on the lengths of both

**Table 12.2** Viscous coupling terms for the perturbing and reference hairs. Here $D_1$ is the normalised perturbation coefficient of a perturbing hair at the position of the reference hair [9]

| $\varepsilon_1$ | $\varepsilon_2$ | $\varepsilon_3$ |
|---|---|---|
| $1 - \frac{D_1}{1+D_1}$ | $1 + \frac{D_1^2}{1-D_1^2}$ | $\frac{D_1}{1-D_1^2}$ |

hairs [9]. For the case of equal hair length (both reference and perturbing hairs), the values of these constants are given in Table 12.2.

By making use of the above equations, the response of the reference hair in the presence of a perturbing hair and in its absence can be derived and a coefficient for the viscosity-mediated coupling can be defined as:

$$\kappa_1 = \frac{\theta_R - \theta_{1P}}{\theta_R} \tag{12.36}$$

By measuring the response of the reference, in presence and absence of a perturbing hair, the effect of viscosity-mediated coupling can be experimentally investigated.

## References

1. R.K. Jaganatharaja, Cricket Inspired Flow-Sensor Arrays, Ph.D. thesis, University of Twente. ISBN 978-90-365-3215-0. https://doi.org/10.3990/1.9789036532150 (2011)
2. G. Von der Emde, Warrant, E. (eds.) See the latest compilation in sensory ecology, the most vibrant field, in *The Ecology of Animal Senses: Matched Filters for Economical Sensing* (Springer, Berlin, 2015) ISBN 978-3-319-25490-6. https://doi.org/10.1007/978-3-319-25492-0
3. D. Floreano, R. Pericet-Camara, S. Viollet, F. Ruffier, A. Brückner, R. Leitel, M.K. Dobrzynski, Miniature curved artificial compound eyes. Proc. Natl. Acad. Sci. **110**(23), 9267–9272 (2013)
4. H. Droogendijk, J. Casas, T. Steinmann, G.J.M. Krijnen, Performance assessment of bio-inspired systems: flow sensing MEMS hairs. Bioinspiration Biomimetics **10**(1), 016001 (2015). https://doi.org/10.1088/1748-3190/10/1/016001
5. J.A.C. Humphrey, F.G. Barth, Medium flow-sensing hairs: biomechanics and models, in *Advances in Insect Physiology–Insect Mechanics and Control*, vol. 34 ed. by J. Casas, S.J. Simpson (Elsevier, Amsterdam, 2007), pp. 1–80
6. T. Shimozawa, T. Kumagai, Y. Baba, Structural and functional scaling of the cercal wind-receptor hairs in cricket. J. Comp. Physiol. A **183**, 171–186 (1998)
7. O. Dangles, D. Pierre, C. Magal, F. Vannier, J. Casas, Ontogeny of air-motion sensing in cricket. J. Exp. Biol. **209**, 4363–4370 (2006)
8. J. Casas, T. Steinmann, G. Krijnen, Why do insects have such a high density of flowsensing hairs? Insights from the hydromechanics of biomimetic MEMS sensors. J. R. Soc. Interface **7**(51), 1487–1495 (2010)
9. (a) B. Bathellier, F.G. Barth, J.T. Albert, J.A.C. Humphrey, Viscosity-mediated motion coupling between pairs of trichobothria on the leg of the spider Cupiennius salei. J. Comp. Physiol. A **191**, 733–746 (2005). (b) Erratum: J. Comp. Physiol. A **196**, 89 (2010)
10. B. Cummins, T. Gedeon, I. Klapper, R. Cortez, Interaction between arthropod filiform hairs in a fluid environment. J. Theo. Biol. **247**, 266–280 (2007)

11. J.J. Heys, T. Gedeon, B.C. Knott, Y. Kim, Modeling arthropod filiform hair motion using the penalty immersed boundary method. J. Biomechanics **41**, 977–984 (2008)
12. P.S. Alagirisamy, G. Jeronimidis, V. Le Moàl, An investigation of viscous-mediated coupling of crickets cercal hair sensors using a scaled up model, in *Proceedings of SPIE*, vol. 7401 San diego, USA (2009)
13. G.C. Lewin, J. Hallam, A computational fluid dynamics model of viscous coupling of hairs. J. Comp. Physiol. A **196**(6), 385–395 (2010)
14. J. Palka, R.B. Levine, M. Schubiger, The cercus-to-giant inter-neuron system of crickets. I. Some attributes of the sensory cells. J. Comp. Physiol. **119**, 267–283 (1977)
15. M.A. Landolfa, J.P. Miller, Stimulus-response properties of cricket cercal filiform receptors. J. Comp. Physiol. A **177**, 749–757 (1995)
16. F.E. Theunissen, J.P. Miller, Representation of sensory information in the cricket cercal sensory system. II. Information theoretic calculation of system accuracy and optimal tuning curve widths of four primary interneurons. J. Neurophysiol. **66**, 1690–1703 (1991)
17. M.A. Landolfa, G.A. Jacobs, Direction sensitivity of the filiform hair population of the cricket cercal system. J. Comp. Physiol. A **177**, 759–766 (1995)
18. J.A.C. Humphrey, R. Devarakonda, I. Iglesias, F.G. Barth, Dynamics of arthropod filiform hairs. I. Mathematical modeling of the hair and air motions. Philos. Trans.: Biol. Sci. **340**, 423–444 (1993)
19. G.J.M. Krijnen, J. Floris, M.A. Dijkstra, T.S.J. Lammerink, R.J. Wiegerink, Biomimetic micromechanical adaptive flow-sensor arrays, (Invited) in *Proceedings of "SPIE Europe Microtechnologies for the New Millennium 2007"*, 2–4 May 2007, Maspalomas, Gran Canaria, Spain. 65920F, Proceedings of SPIE, Bioengineered and Bioinspired Systems 6592. SPIE. ISBN 978-0-8194-6726-3 (2007)
20. G.J.M. Krijnen, M. Dijkstra, J.J. Van Baar, S.S. Shankar, W.J. Kuipers, R.J.H. De Boer, D. Altpeter, T.S.J. Lammerink, R. Wiegerink, MEMS based hair flow-sensors as model systems for acoustic perception studies, Nanotechnology 17(4), 28 February 2006, S84–S89 (2006)
21. J.J. Heys, P.K. Rajaraman, T. Gedeon, J.P. Miller, A model of filiform hair distribution on the cricket cercus. PLoS ONE **7**(10), e46588 (2012). https://doi.org/10.1371/journal.pone.0046588
22. H.T. Schlichting *Boundary Layer Theory*, International Edition (McGraw-Hill) ISBN 13: 9780070553347
23. C.M. Bruinink, R.K. Jaganatharaja, M.J. de Boer, E. Berenschot, M.L. Kolster, T.S.J. Lammerink, R.J. Wiegerink, G.J.M. Krijnen, Advancement in technology and design of biomimetic flow-Sensor arrays, in *Proceedings of the IEEE International Conference on Micro Electro Mechanical Systems (MEMS)*, Sorrento, Italy, 2009, pp 152–155
24. T. Steinmann, J. Casas, G. Krijnen, O. Dangles, Air-flow sensitive hairs: Boundary layers in oscillatory flows around arthropod appendages, J. Exp. Biol. **209**, 4398-4408. https://doi.org/10.1242/jeb.02506 (2006)
25. http://www.polytec.com/eu/
26. H-E. de Bree, V.B. Svetovoy, R. Raangs, R. Visser, The very near field, theory, simulations and measurements, in *Proceedings of 11th International Congress on Sound and Vibration*, St. Petersburg (2004)
27. F.G. Barth, U. Wastl, J.A.C. Humphrey, R. Devarakonda, Dynamics of arthropod filiform hairs. II. Mechanical parameters of spider trichobothria (Cupiennes salei Keys.). Philos. Trans.: Biol. Sci. **340**, 445–461 (1993)
28. T. Steinmann, J. Casas, The morphological heterogeneity of cricket flow-sensing hairs conveys the complex flow signature of predator attacks, J. R. Soc. Interface, **14**(131), 1 June 2017, 20170324 (2017)

# Chapter 13
# Architectured Materials in Building Energy Efficiency

**Bernard Yrieix**

**Abstract** The term "architecture" included in the "architectured material" approach receives a particular echo in the building sector. It inevitably evokes the technical and artistic discipline of adapting the building to its use. In a similar way the architecturation of materials approach in response to complex and antagonistic expectations for energy efficiency in building is promising although it is less developed in this sector than in others. However the engines for its development are indeed more present than ever: reduce the total emissions of greenhouse gases, demands for multifunctionality, and requests for inaccessible properties by a single bulk material. If mechanics has long been the central concern, the end of the 20th century marked a turning point towards energy and that is the reason why this chapter deals to this aspect and especially with the building envelope. Many properties come to complicate the game of simple thermic and mix the cards: mechanical (compression, tensile, creep), hydric (water sealing and absorption), hygric (water sorption and permeation), acoustics, aesthetics, cost, optical (visible and infrared), embodied energy and environmental impact, etc. Examples are given in this chapter. This approach started very late, only sixty-seventy years ago with the foaming of polymers and with the manufacturing of mineral wool. In parallel other material where designed at a scale closer to the product scale like hollow bricks, lightweight concretes. It is only very recently that a final improvement is done on traditional insulation materials polymeric foams and glass wool by achieving a very efficient infrared opacification. It is also in the same time that the breakthrough of the super insulation comes based on a specific architecture of the material involving more matter but dealing with the nanometer scale in order to confine the molecules of the gas in pores smaller than their mean free path. Two families of material/products are exemplified: the vacuum insulation panels and the super insulation at atmospheric pressure. For each, several optimisation problems are ongoing or on the table and the architecture of the material and product is the key for success in reaching antagonist attempts. The strongly moving context around the renewables energies is also bringing new playground for architectured materials.

B. Yrieix (✉)
Materials and Mechanics of Components, EDF Lab Les Renardières, EDF R&D,
77818 Moret-Sur-Loing, France
e-mail: bernard.yrieix@edf.fr



Y. Estrin et al. (eds.), *Architectured Materials in Nature and Engineering*, Springer Series in Materials Science 282,
https://doi.org/10.1007/978-3-030-11942-3_13

One first theme is the heat storage where the difficulty for reaching the wanted power could be addressed by this approach. The very wide and new second theme is the field of active walls and specific collectors where the aimed properties ask for filling some gaps in the materials spaces. Finally the approach of architectured materials is slowly irrigating the developments for energy efficiency in building with the knowledge diffusion and thanks to the education students are more and more aware of the possibilities of this approach especially to solve difficulties and then to innovate.

## 13.1 Introduction

The term "architecture" included in the "architectured material" approach receives a particular echo in the building sector. It inevitably evokes the technic and artistic discipline of adapting the building to its use. The general function of a building is to house people, their property and their activities from climate, pests and predators, or natural and human risks. This results in functional and technical specifications which have conflicting expectations and whose satisfaction appealed to both the architecture (of the building) and special materials.

For the Architectured Materials community which adapts the material to its use the antagonist expectations evoke the famous empty spaces of Ashby maps where the perfect material remains to be invented or designed. Research in the field is the key to push the edges of the possible. From this point of view research in the building sector suffers somewhat from the comparison with that of other areas. If it is excellent in narrow areas including some materials, it is mainly industrial type made by industrials and champions of techno-economic process optimization. The result is often that the overall aspects of the expectations and interactions between systems stay beyond their control. This is why the architecturation of materials approach in response to complex and antagonistic expectations is less developed in the building sector than in others such as transport. There is therefore more a slow appropriation of the methods and tools developed elsewhere.

However the engines for its development are indeed present. The first is the very current need for energy efficiency where the building holds a decisive role in the fight against global warming as responsible for 20% of total emissions of greenhouse gases [1]. The second is a very strong need for multifunctionality, very old as we shall see Sect. 2 but also very modern (Sects. 3–5) or futuristic Sect. 6. The third motor is, as elsewhere, the inaccessible properties by a single bulk material, in particular motivated by the evolution of the set of functions requested through the ages. If mechanics has long been the central concern, the end of the 20th century marked a turning point towards energy and that is the reason I will limit this chapter to this aspect of the energy in the building. Specifically the building sector will be that of the envelope, decisive in terms of efficiency, but also that of building equipment's. The building envelope is made of opaque and transparent walls where we distinguish the walls, floors, roofs and openings, which for thermal purposes it is added thermal bridges.

Many properties come to complicate the game of simple thermic and mix the cards justifying that "filling the gaps in materials spaces" remains more relevant than ever! To the main properties I deliberately fixed to be thermal insulation and heat storage, it is necessary to list the following secondary characteristics: mechanical (compression, tensile, creep), hydric (water sealing and absorption), hygric (water sorption and permeation), acoustics, aesthetics, cost, optical (visible and infrared), embodied energy and environmental impact; the list is not exhaustive. Examples are given in the following paragraphs.

## 13.2 Historical Elements and Generalities

### *13.2.1 Traditional Housing*

Let's start with some historical examples of the building envelope. The old traditional habitat has many examples, some of which are listed in Table 13.1. These examples show that the optimization is not as thorough as some explain. The thermal management is ultimately somewhat based on the use of local materials and of course the only ones available. However, the optimization was carried to a high level at the system level and at the architecture of buildings with a generalized concern for bio-climatic design and components durability. Some current developments clearly turn those backs. An example of this that I love is that of the traditional roof in tropical Asia that provides excellent comfort and uses resources wisely. Its structure (trusses and purlins) consists of fine wood whose lifespan exceeds 50 years while the less durable cinnamon wood is used for the cover (rafters and battens supporting palms) whose set is replaced every ten years. Today palms are replaced by steel sheets with two disadvantages: (1) a sustainability inconsistency because wood rafters remained the same lasts much shorter than the sheet metal and (2) a blatant discomfort due to heat radiation from the sheet metal.

In conclusion, the general tendency to revisit history with current knowledge and to assign current approaches to past designers does not stand up to close examination of the facts. Optima achieved are often locally very relevant especially in the compromise between resources and bio-climatic adaptation for older or more traditional habitats. As for the bio-mimicry in materials, it is appropriate not to copy but to inspire.

### *13.2.2 The Post 1973 Period*

The year 1973 saw the first oil crisis hit the Western world recalling that energy is a precious commodity that can become expensive. The forty years that followed saw a succession of energy crises and crystallize environmental awareness around the

Table 13.1 Building envelope of a few traditional habitats

| Climate | Location | Materials used | Useful material or structure properties | Unwanted material properties |
|---|---|---|---|---|
| Dry tropical mediterranean | Near and middle east Sub-Sahara Africa | Raw earth, clay or dung reinforced with wood or straw<br>Wood and bamboo | Low solar absorptivity<br>High IR emissivity<br>High water adsorption & desorption | |
| Wet tropical | South and south-east Asia | Exotic wood or palm<br>Palms for roofing<br>Bamboo | High ventilation rate (including floor)<br>Water and bio stamina | |
| Various | Troglodytes | Rocks | Air tightness<br>High heat capacity (sensible heat)<br>Stable source and well of heat | |
| Cold | Northern countries | Wood<br>Peat | Low conductivity<br>Air tightness<br>Water vapour buffer | |
| | | Raw stones | Durability | High conductivity<br>Air tightness |
| Cold | Northern countries: igloo | Snow | Low conductivity<br>Air tightness<br>Phase change material | High solar reflectance<br>High IR absorbance and emissivity |

management of resources and of global warming. The search for energy efficiency, especially for buildings, was initiated in Europe in 1974 and has continued to grow since at different speed depending on the country. It was first served by the emergence or spread of thermal insulation materials. Figure 13.1 and Table 13.2 set a few dates and shows that modern insulation is recent with the appearance of foams after 1945.

Somehow there was the implementation of a particular material architecture, only able to achieve the very low thermal conductivity as shown in the two maps of materials properties Fig. 13.1. Only a few natural materials have approaching thermal conductivities: they are themselves very porous, fibrous or cellular.

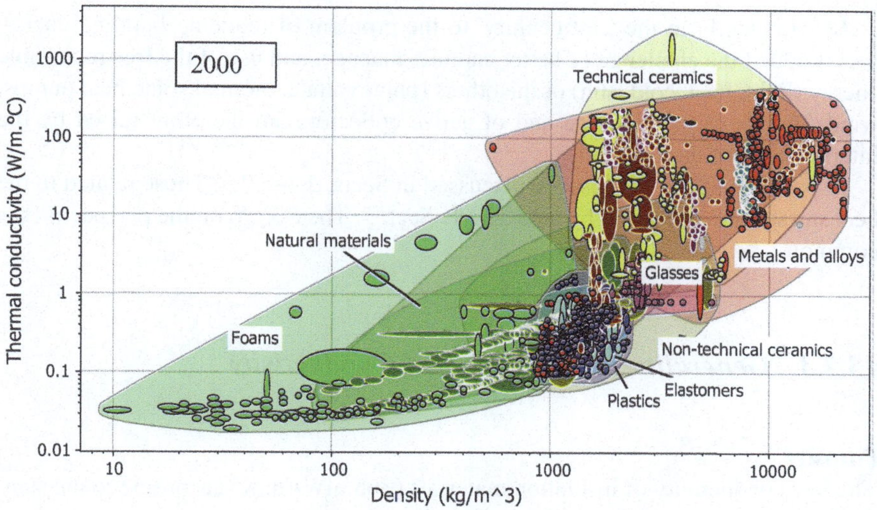


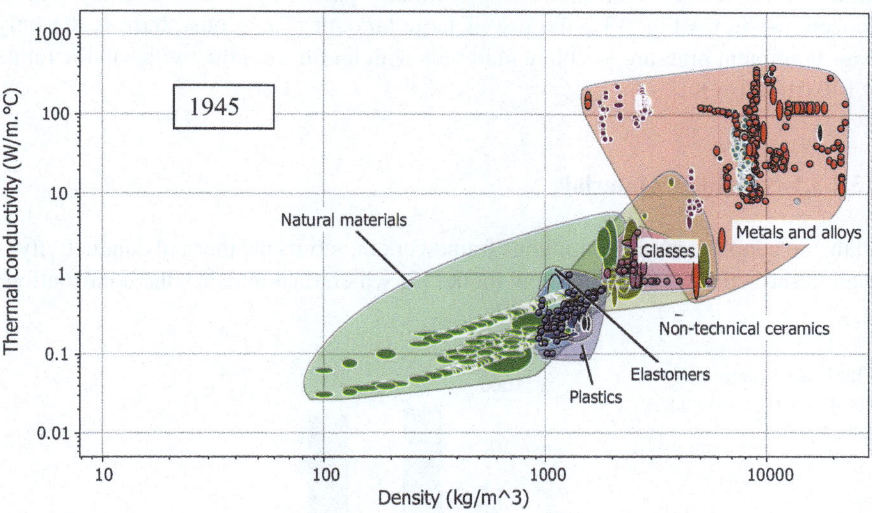


**Fig. 13.1** The appearance of foams date after 1945

**Table 13.2** Modern insulation is very new

| Raw material | | Insulate | | |
|---|---|---|---|---|
| Mineral wool | 1890 | – Industrial wool<br>– In building | 1938<br>1960 | Owens<br>Corning |
| Polystyrene | 1930 | Expanded | 1951 | BASF |
| Polyurethane | 1953 | | | |
| | | Vacuum insulation panel | 1970 | |
| Silica aerogel | 1931 | – Xerogel<br>– Blanket | 1990<br>2000 | Nanopore<br>Aspen |

More recently in the 21st century, to the problem of reducing building energy loss is added the challenge of better capture, manage, and use of the free renewable energy (light, heat, cold, sun) or the others (photovoltaic, thermal solar, heat pumps, wind). Issues of heat storage and of active collectors are therefore raised by the intermittence of these energies.

Aspects related to the loss are discussed in Sects. 2.3–4.2.3. Those related to the heat storage are illustrated by an example Sect. 5. The Sect. 6 on the perspectives is common.

## *13.2.3 Generalities About Thermal Conductivity*

**Porosity**

The low conductivity of insulation materials (<65 mW/(m K) according to the standards) is the result of the mixture of solid and gaseous phases intimately mixed. Indeed, as shown Fig. 13.2 the use of large amount of gaseous phase is the only way at ambient pressure to obtain materials with thermal conductivities in the range 0.1–0.01 W/(m K).

### 13.2.3.1 Cellular Materials

In the no nanostructured continuous framework materials the thermal conductivity is well described by the parallel flow model [2] which distinguishes the contributions

**Fig. 13.2** Conductivity of the solids, liquids and gas

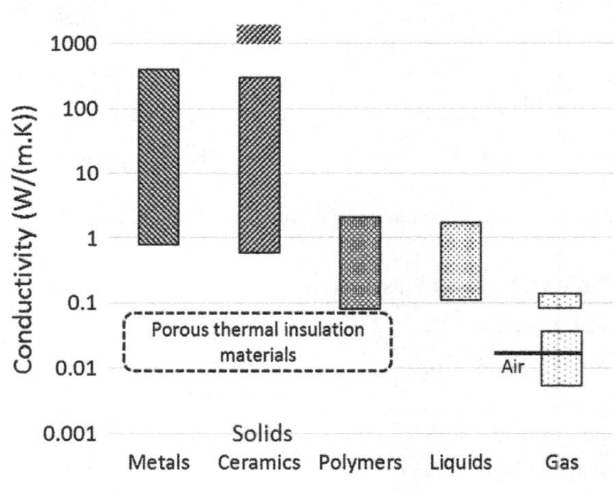

of the internal infrared radiation, the solid phase conduction and the gas phase conduction[1]:

$$\lambda = \lambda_r + \lambda^*_{cs} + \lambda^*_{cg} \tag{13.1}$$

with:

$\lambda_r$ the radiativity given by Rosseland approximation:

$$\lambda_r = \frac{16}{3}\frac{n^2\sigma \overline{T}^3}{K} \tag{13.2}$$

n refractive index (≈1 for low density materials),
K extinction coefficient ($m^{-1}$),
σ Stefan-Boltzmann constant (5.67 $10^{-8}$ W $m^{-2}$ $K^{-4}$)
$T$ is an average temperature within the insulant; $\overline{T}^3 = (T_1 + T_2) \cdot (T_1^2 + T_2^2)/4$. $T_1$ and $T_2$ are the temperatures of the surfaces

$\lambda^*_{cs}$ the solid conduction[2]

$$\lambda^*_{cs} = (1-\theta)^\alpha.\lambda_{cs} \tag{13.3}$$

$\theta$ the porosity

$$\theta = 1 - \rho/\rho_s \tag{13.4}$$

$\rho$ and $\rho_s$ the densities of the material and the skeleton

$\lambda_{cs}$ the conductivity of the solid material.

α ≈ 1 for foams and α ≈ 1.5 … 2 for materials such as aerogels or fumed silica.
$\lambda^*_{cg}$ the gaseous conduction

$$\lambda^*_{cg} = \theta.\lambda_{cg0} \tag{13.5}$$

$\lambda_{cg0}$ the conductivity of unconfined gas

Finally as the extinction coefficient is proportional to the density $K = \sigma_{ext}.\rho$ ($\sigma_{ext}$ is the mass extinction coefficient in $m^2$/kg), the conductivity of highly porous materials with continuous skeleton is of the form: $\lambda = C + B.\rho + \frac{A}{\rho}$ wherein $A$, $B$, and $C$ are constants, $C$ corresponding to the conductivity of the stationary air.

[1]Considering that the pore size of these materials prevents natural convection which is the main aim of their solid phase.

[2]For a single-phase solid. For a multiphase solid, a simple mixture of law suits if one remains in high porosities; if not one can use a homogenization rule adapted to the configuration.

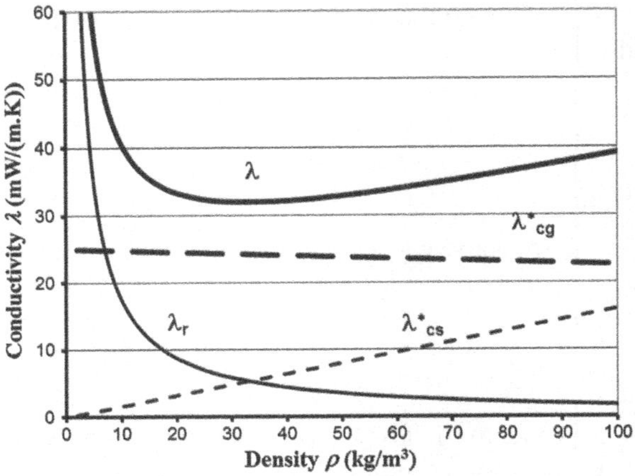


**Fig. 13.3** Contributions to the conductivity at 10 °C of an expanded polystyrene as a function of its density

This first level of microstructure or of the architecture of cellular medium shows that an optimum separates the region of low densities dominated by the thermal radiation from those of higher densities dominated by solid conduction; An example about the expanded polystyrene is given Fig. 13.3.

**Morphology**

The influence of the morphology is also of importance because it could influence both the solid conduction and the radiativity. The simplest way to consider the first is to take into account the tortuosity of the solid phase in the parallel model:

$$\lambda_{cg}^* + \lambda_{cs}^* = \theta.\lambda_{cg0} + (1-\theta)\frac{\lambda_{cs}}{Z} \tag{13.6}$$

with $Z$ the tortuosity as the ratio of the effective distance into the solid phase by the chord distance between two representative points.

The effect of anisotropy could be consider with $Z$ depending on the direction. Of course finite elements calculations are a more precise way to evaluate the conduction contribution but they suffer of the same limitation than homogenization models: the lack of coupling between conduction and radiation.

For the radiativity, the influence of the morphology of the polymer foam has been particularly studied by Glicksman [3] which shows the importance of having the material in the pillars optically thick, rather than in the windows of cells transparent to infrared radiation, but with the limitation of keeping closed the cells of foamed materials with heavy insulating gas. Extensive work was conducted by Coquard and Baillis on the modelling of heat transfer in low density medium [4] by the resolution of the radiation-conduction coupling which leads to the predictive quantification of the conductivity of polymer foam, fibrous mat and later on of nanostructured materials.

#### 13.2.3.2 Fibrous and Granular Materials

Another type of architecture is that of fibrous or granular materials which adds to previous behaviours the influence of contact resistance between fibers or grains.

The solid conduction then takes the form of a sum of conduction along the fibers and conduction at the contacts. The ratio $R_f/R_c$ (Fig. 13.4) of the thermal resistance of the fiber between two contacts and the contact resistance is the marker of the type of behaviour from weak contacts to perfect contacts where we find again the case of continuous skeleton.

Thus the topology of the material becomes essential: fiber section $S_f$, length between contacts $l_c$ or number of contacts per unit length, contact section $S_c$, contact pressure, and of course fiber orientation with respect to heat flow. All these variables are used by the manufacturers of fibrous insulation; in the literature one found empirical models for some families of materials such as glass wool. The same principle applies to granular materials where one considers parallel flows (Fig. 13.5) [5].

The commercial products are like the polymer foams very porous: respectively from 0.93 to 0.98 and from 0.96 to 0.99 for mineral and organic wools.

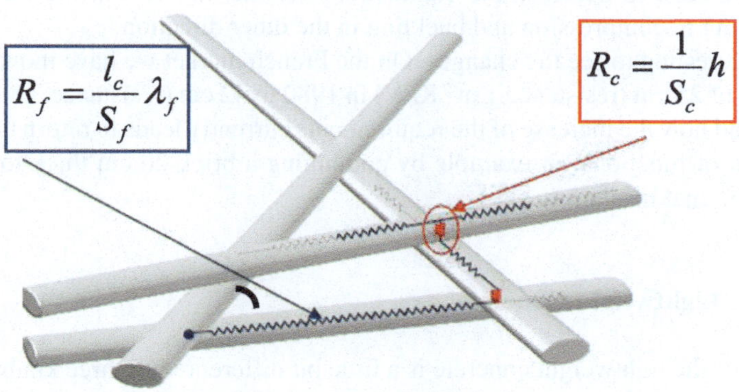


Fig. 13.4 Principle of solid conduction contribution to the conductivity of fibrous media (Courtesy 3SR)

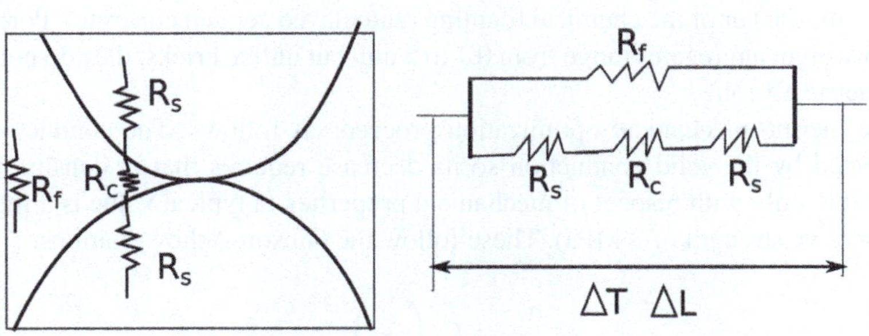


Fig. 13.5 Principle of parallel flux in granular media

## 13.3 Traditional Insulation

### 13.3.1 The Thermal/Mechanical Compromise

Optimizations made over the years bear alternately on materials, products and systems. They target conductivity while the mechanical strength is a constraint to satisfy.

For the building envelope the iconic materials of these approaches are the insulating load-bearing materials used in distributed insulation systems (i.e. in the wall thickness, as opposed to insulation systems from outside or from inside). One find mainly said honeycomb brick "monolithic", light concretes and the frame structures.

#### 13.3.1.1 Bricks

The solid brick earthenware is 7000 years old and always used. The current product has two pores scales: micrometric pores evenly distributed to decrease the conductivity of the ceramic material (and conferring resistance to frost) and centimetric anisotropic cells to decrease the conductivity in one direction while maximizing load-bearing in compression and buckling in the other direction.

It is interesting to see the changes. On the French market we have moved from a thickness of 25 cm (resistance 1 $m^2$ K/W) in 1980 to 42 cm (resistance 3.8 $m^2$ K/W) in 2012 and now the increase of the requirements currently leads to return to a partial separation of functions for example by combining a brick 20 cm thick to a 12 cm thick traditional insulation.

#### 13.3.1.2 Lightweight Concrete

The case of the lightweight concrete is a little bit different with three kinds of products: lightweight aggregates, foamed, and autoclaved aerated. Both are based on the strategy of the cellularisation of the structure by the use either of very lightweight aggregate (e.g. expanded polystyrene beads), or of the physical foaming (foam concrete or mortar) or of the chemical foaming (autoclaved aerated concrete). Pore sizes or lightweight aggregates range from 0.1 to 5 mm but unlike bricks, they do not show centimetric alveoli.

The thermo-mechanical optimization proceeds as follows. The conductivity is dominated by the solid conduction so its decrease requires that in density, which is possible only with respect of mechanical properties (a typical value is a required compressive strength of 3 MPa). These follow the Gibson-Ashby relations:

$$\frac{\sigma_y}{\sigma_{ys}} = C_\sigma \left(\frac{\rho}{\rho_S}\right)^{1.5} \tag{13.7}$$

$$\frac{E}{E_S} = C_E \left( \frac{\rho}{\rho_S} \right)^2 \tag{13.8}$$

where $\sigma_y$, $\sigma_{ys}$, $E$, $E_S$, are respectively the plastic collapse stress of the foam, the yield stress of the solid material, the elastic modulus of the foam, and the elastic modulus of the solid material of which the foam is made.

$C_\sigma$ and $C_E$ the constants dependent on the microstructure; for example $C_\sigma$ decreases with the size heterogeneity of cells or lightweight aggregates.

There are therefore three main levers to optimize: the density to select the pair of mechanical/thermal properties, the yield strength of the matrix, and the heterogeneity of cells to change compromise.

Again, the increased thermal requirements of the building envelope did not allow these materials to meet major market except for autoclaved cellular concrete that have reached conductivities of 90 mW/(m K) for supporting structures and 45 mW/(m K) for purely insulating supplements.

## *13.3.2 Traditional Insulation on the Asymptote*

The efficient traditional insulation materials tend to have relatively low density which places them in the domain where the radiativity dominates the solid conduction. Therefore any reduction in density, for example for cost reasons, results in a significant increase in conductivity. This is the case of the two heavyweights markets: glass wool and expanded polystyrene.

The recent development (after 2000) of these materials can be expressed by the question "how to approach the limit constituted of the conductivity of still air without changing the gas as in the polyurethane?". The answer is in (1): we must reduce both conduction and solid radiativity. But these properties being antagonistic to each other and incidentally with the mechanical properties, it is necessary to resort to new architectures dedicated to cellular materials and fibrous materials.

### 13.3.2.1 Opacification of Expanded Polystyrene

The conductivity of the reference expanded polystyrene EPS is 38 mW/(m K). Contributions to this conductivity of still air, of the solid skeleton and of the radiativity are respectively of 25, 5 and 8 mW/(m K). Decrease the latter almost without adding solid conduction was the challenge. It was to disperse on the walls of cells micrometric filler called infrared opacifier (Fig. 13.6) which considerably increase the extinction coefficient of the material by a combination of absorption and scattering of the radiation. Fillers must comply with specific dimensions in proportion to the emission spectrum of the grey body at these temperatures, a maximum of extinction is obtained at a diameter of 3 μm. The nature of fillers can be varied but in order

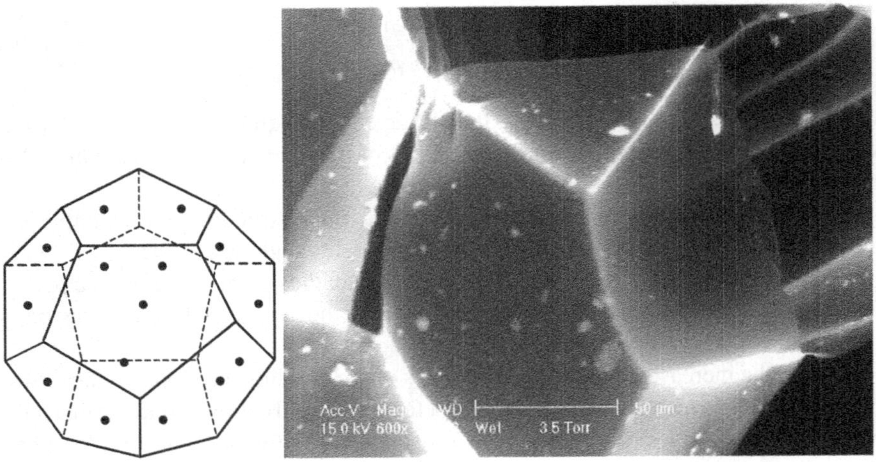

Fig. 13.6 Schematic microstructure and SEM observation of an opacified EPS

to facilitate the manufacturing, graphite and carbon black dominate and explain the black to grey colour taken by the EPS panels.

At the end a synergistic effect is achieved because the sharp decline of the IR semi-transparency of the material allows decreasing in return its density and therefore the solid contribution. Finally opacified EPS are marketed with a certified conductivity of 29 and 30 mW/(m K) for a density of only 25 and 18 kg/m$^3$. We can consider that the asymptote is reached.

#### 13.3.2.2 Improvement of Fibrous

The case of fibrous is close but there are differences. The available area to disperse fillers without contact between them is restricted, the only way to play really is through the diameters of the fibers themselves. In fact the decrease in diameter is accompanied by a diffusion interaction of the radiation more effective and therefore an increase in the extinction coefficient of the material which results in a downward shift of the curves conductivity as a function of density with a clearly marked effect at low densities (Fig. 13.7).

This reduction in the fiber diameter is accompanied by a decrease in the mechanical characteristics of these nonwoven mats which require in turn changing the architecture of the material. So depending on the manufacturing process we will look at a compromise between thermal properties [Density, "Thermal" fiber concentration (small diameter), Fiber orientation, Fiber diameter (low for IR opacification)] and the mechanical properties [Density, "Mechanical" fiber concentration (large diameter), Fiber orientation (Mechanical bonding), Locking structure (thermal bounding or sizing)].

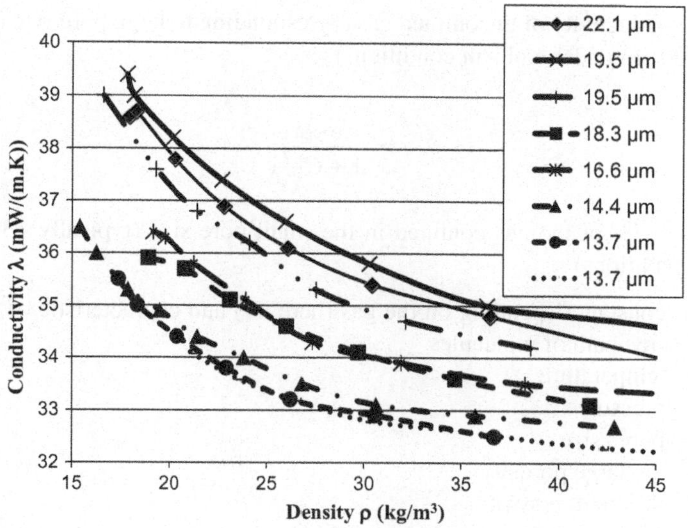


**Fig. 13.7** Evolution of the characteristic curve of conductivity of PET non-woven with the fiber diameter. Conductivity of comparable nonwovens according to equivalent diameter of the fibers

The same approach applies to mineral fibers able more easily to smaller diameter than organic. As for EPS their progress place them also on the asymptote with products marketed with a certified conductivity of 30 mW/(m K).

## 13.4 The Super-Insulation

Do even less conductive than the best conventional insulation materials [go slightly below the conductivity of 25 mW/(m K)] requires replacing the air by a heavier gas and therefore less conductive that should be prevented of diffusion outwardly of the product. Polyurethane, polyisocyanurate and phenolic foams reach and such products are marketed with a certified conductivity which include ageing of 23–25 mW/(m K). New foaming gas will soon make it possible to reach 18–20 mW/(m K).

Do even better means to further reduce the largely predominant gas conduction contribution. To do this two ways exist and are physically equivalent: reduce pressure or confine the gas in pores of a size less than or equal to the mean free path of the molecules. Thus the gas conduction in (1) becomes:

$$\lambda_{cg}^{*} = \theta_{\text{large pores}} \cdot \lambda_{cg0} + \theta_{\text{small pores}} \cdot \lambda_{cg} \tag{13.9}$$

with:

$\lambda_{cg0}$ the conductivity of unconfined gas corresponding to large pore size (typically > 300 nm under ambient conditions)

$$\lambda_{cg} = \frac{\lambda_{cg0}}{1 + C\frac{T}{\phi . P}}, \tag{13.10}$$

the conductivity of the gas confined in the small pore size (typically <300 nm at ambient conditions)

$C$ constant depending on the gas (here air) and characteristic of the mean free path of molecules
$T$ temperature
$P$ pressure
$\phi$ pores size
$\theta_{large\ pores}$ the large porosity
$\theta_{small\ pores}$ the small porosity.

It follows from (10), called Knudsen relationship, that there is equivalence between the pressure and the pore size, the decrease in either leading to decrease the conduction of the gas. In other words, the decrease in conduction with the gas pressure decrease is more pronounced as the pore size is small as shown in Fig. 13.8.

The nanostructuring of pores while maintaining high total porosities request nanostructuring of solid skeleton. Besides the process problems involved, here are three consequences. The first is that the radiativity is changed because the solid characteristic size is too small to efficiently absorb infrared radiation whose mode of interaction with the solid is then Raleigh diffusion [6]. It is therefore necessary in

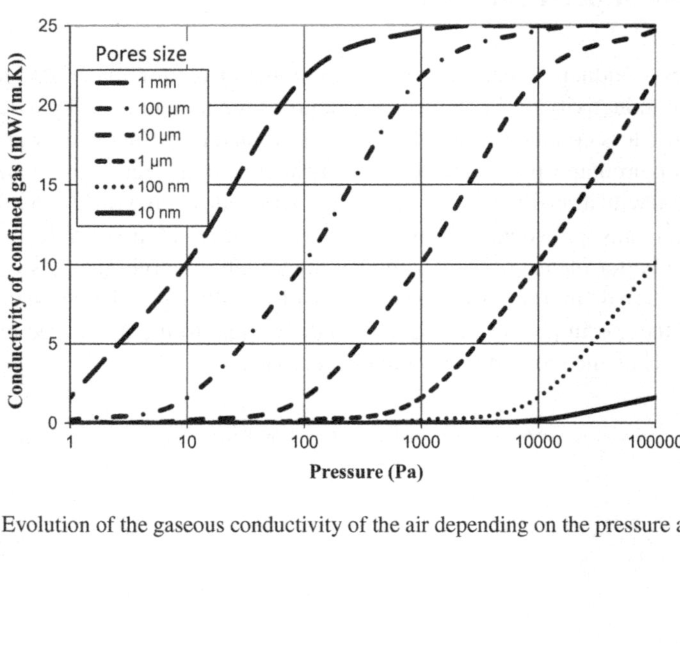


**Fig. 13.8** Evolution of the gaseous conductivity of the air depending on the pressure and the pore size

most cases to opacify these materials the same manner as in Sect. 3.2.1. The second is that this dimension can be of the same order of magnitude as the mean free path of the phonons and thus lead to a sharp decrease in the solid conduction, to take into account in (1). Finally the third one is that the amount of matter required to divide the volume in a huge number of pores cannot be small; as a consequence the density of these materials are 5–10 times higher than that of traditional insulation materials (i.e. 50–150 kg/m$^3$) and so the total porosity is much lower (0.89–0.97).

The textural characteristics of the materials are therefore of prime importance for satisfying simultaneously super-insulation and other properties. This will require particular architectures described in the following paragraphs.

The word super-insulation refers to a material whose conductivity is lower than that of still air at ambient conditions (25 mW/(m K)). There are two families: the vacuum insulation panels (VIPs) and super-insulation at atmospheric pressure (SIAP).

VIP consists of a porous core kept under vacuum by a sealed envelope tight to atmospheric gases. Core materials can be varied; the two main ones being the glass fibers in a particular form and the silica powder, mainly of fumed type. For building applications silica is preferred because it gives a longer life time thanks to a predominant size of pores close to 200 nm which shift to a higher pressure the increase of the conductivity generated by the pressure rise due to the inevitable permeation of atmospheric gases through the barrier envelope [7].

SIAP are aerogels based materials, the only structure able to produce very porous materials (>80%) with predominant pore sizes of <20–70 nm. The dominant material is hydrophobic silica aerogel dried supercritically.

## 13.4.1 Vacuum Insulation Panels

VIP technology is quite mature (TRL 8–9) and corresponds to the lowest conductivities at room temperature on large objects: the center of panel thermal conductivities of new state VIP (in mW/(m K)) are between 2 (specific glass fibers), 4–5 (silica) and 8 (core with micrometric pores). The thermal, mechanical and durability behaviour depends interdependently of their two components: the core material and the barrier envelope [8]. The optimization of the assembly is particularly delicate and is currently the subject of numerous studies like the international program Annex 65 of the IEA-ECB [9].

### 13.4.1.1 Core Material

Besides the low conductivity, physicochemical required properties are mechanical and hygric, and use properties are durability and cost performance.

From the mechanical point of view, it must of course withstand 1 bar pressure, including without creep during service but some applications such as floor requires higher resistance. The main lever to adapt a given type of core is its density; for silica it is made by powder compaction. Furthermore the manufacturing stage requires a certain strength or flexural deformation which obliges to reinforce the silica by long fibers which can conventionally be glass or PET.

The choice of hygric properties is complex and numerous options (more or less hydrophilic or hydrophobic) are possible through the material itself, for example pyrogenic and precipitated silicas, or by the use of getters and/or desiccant for glass fibers. The sustainability of VIP is affected in complex ways still poorly predictable over the long term (i.e. 40 years) especially because the own ageing of components such as the core has wrongly been neglected [10, 11].

The contradictions between properties are numerous: increased density improves mechanical properties and durability, but diminishes conductivity and increases cost; the cheapest silica (precipitated) is mechanically and thermally less efficient but is more hydrophilic which can be sought or not depending on the expected lifetime; strengthening by compaction competes with that by the fibers; precipitated silicas do not age in the same manner as fumed; and finally one of the keys for the mechanic appears to be (1) the surface chemistry and (2) the morphology conferred by the process to indivisible aggregates ($\approx$150 nm) of individual silica particles ($\leq$20 nm) [12].

#### 13.4.1.2 Barrier Envelope

Barrier levels required for very long lifetimes expected for the building (target 50 years) are really very high and limits of current carrying capabilities. The barrier laminate consists of different layers linked on the multiple functions that are required.

The first of course is the barrier against atmospheric gases including water vapour which permeate 100 times faster than oxygen and nitrogen due to the small size of its molecule. Today we use aluminium layers either laminated (6–15 $\mu$m thick) or metalized (40–120 nm thick). The former are more airtight than the latter but induce a thermal bridge intrinsic to the VIP much more damaging to the effective thermal conductivity than the seconds [7]. The alternation (layers) and juxtaposition (defects) of highly permeable and other totally impervious areas associated with the vacuum configuration explain that the solubility diffusion model [13, 14] used in the packaging is not sufficient to reflect the actual behaviour of barrier laminate because of two gaps. Indeed, on the one hand the hypothesis of Henry's Law[3] proves inaccurate in some configurations of the pair polymer/climatic loading (temperature and humidity), and on the other hand the gases of humid air mixture do not behave like perfect gas, without any chemical or physical identified origin [15]. If the first difference is without much effect on the overall behaviour, it is not even for the second which led to large decreases in diffusion coefficients in mixtures compared to pure gases used in conventional tests of permeation, diffusion and solubility. Works are underway to develop a model on a physical basis that is satisfactory and which is

[3]Proportionality between pressure and concentration.

believed to be suitable for the accurate prediction of the long-term performance of VIPs [16] and in return for the optimisation of the whole VIP architecture.

The second function is the strength of the complex regarding the perforation; it is provided by polymer films such as PET or PA. The third is the sealability that is given by a polyolefin inner layer film such as PE or PP. The polymer films also have a role with respect to permeation by their selectivity of water, oxygen and nitrogen; that's why recently some laminates incorporating EVOH appeared to lower oxygen permeation. The assembly of these various films is done by laminating them together and often by using PU adhesive.

To these main functions it should be added cost, processability, thermal conductivity in the plane, resistance to damage during assembly of the PIV, and finally the degradation of polymer and metal layers under the influence of temperature and moisture in use. This last long ignored aspect is now considered [17] and should enable (i) take into account the own aging of the barrier laminate and (ii) to design laminates more resistant to harsh conditions and therefore broaden the possible applications for example to heat water tank, external insulation and severe hot and humid climates.

#### 13.4.1.3 Conclusion for VIP

In this complex space of properties, the overall architecture and the specific architectures of the core and of the envelope, including macroscopic dimensions, play key roles. The modelling of various behaviours seems to be the way to reach reasoned optimizations guided by the intended application and specifications. This work is being strongly influenced by the approach "multifunctional architected materials".

### *13.4.2 Super-Isolation at Atmospheric Pressure*

The basic materials have conductivities between 10 and 15 mW/(m K), while the insulating materials which they are made of have conductivities between 13 and 20 mW/(m K). These materials are less advanced than the VIP and are either based on silica aerogel (TRL = 5–8) or on organic aerogels (TRL = 3–4). The difficulty in this family of materials is that of compromise between mechanical strength and conductivity.

The reference material is silica aerogel whose properties are well known and whose workings for the purpose of thermal applications are well controlled, except the cost point of view. To simplify, we can say that it has excellent thermal performance but suffer from poor mechanical properties and a high cost (about 3000 €/m$^3$). Current researches interest in these two drawbacks that we will develop in the following paragraphs, but without discussing the chemistry of the gel.

Making aerogels involves very heavy limitations on the forms in which the material can be obtained. The manufacturing of these generic steps are:

- Gelation of a precursor in a solvent by pH change;
- Ripening;
- Possible hydrophobization;
- Drying (going from the gel to the aerogel).

This last step is the critical point of manufacturing because of the very high capillary stresses exerted on the solid skeleton due to the very high porosity and small pores size. These stresses are the origin of mechanical collapse of withdrawal solid network and possibly by re-condensation of chemical bonds. There are three main ways of drying: supercritical (reference channel with respect to ultimate properties but at high cost), subcritical at different levels of temperature and pressure, and finally lyophilization.

Intermediate steps of neutralization, washing and solvent exchange are mostly needed.

Because of these manufacturing constraints, there are three forms of super-insulation products that can be classified in order of increasing difficulty of obtaining:

- granules or powder
  - hollow body filler (hollow walls, day lighting panels);
  - incorporate into other matrices (e.g. facade insulation renderings);
  - preparation of binded panel;
- blankets and other impregnated (fibrous mats and cellular matrices);
- monoliths.

The difficulty gradient is partly due to the drying phase which exert very high capillary stresses and partly due to the increasing dimensions involved in the gelling phase, neutralization, washing, hydrophobization, etc. where diffusion in the porous medium is the limiting step. This hierarchy explains that monoliths are rare and/or of small size. The easiest preparation which is obtaining the granular form of the material involves other processing steps to achieve the final product. The whole process is therefore an essential component.

#### 13.4.2.1 Improve the Mechanical Properties

Silica aerogels having the lowest thermal conductivity at ambient pressure has a density of about 80–200 kg/m$^3$. These fairly rigid aerogels are very brittle [18] with harmful consequences both for use: cracking products and dusting with or without attrition, which, even if it is harmless, is extremely unpleasant and damaging to the thermal performance.

The insulation of opaque walls of building demand for large panels (at least 60 cm $\times$ 120 cm) and between 20 and 60 mm thick [assuming a conductivity of 15 mW/(m K)]. Their handling requires a flexural breaking strength higher than

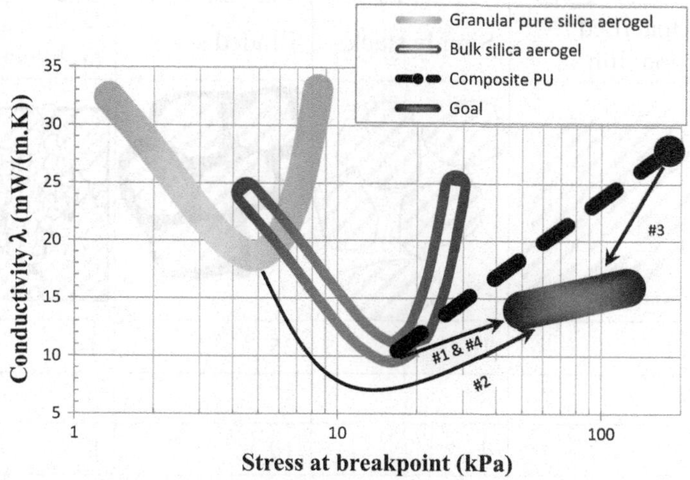


**Fig. 13.9** Approaches to develop SIAP products

50 kPa which is a low value that mechanical reinforcement techniques such as fibers or binders achieve. But in this case the thermal conductivity is greatly impacted. Again the antagonism between the mechanical and the thermal is the challenge which can be illustrated by Fig. 13.9.

We find that the silica monolith lacks the strength and that the simple mixing of grains of silica aerogel with a resin makes it possible to easily obtain this strength but does not retain the sufficiently low conductivity [19]. The explanation is given by Fig. 13.10: the conductivity of a single stack of grains whose monolithic conductivity of 11 mW/(m K) is 16 mW/(m K), due to the increase of the gas contribution in intergranular pores (9). The addition of an organic resin provides cohesion but significantly increases the solid conduction and lead to an overall conductivity of 25 mW/(m K) [see (1) and (3)].

Four lanes (numbered in Fig. 13.9) are proposed to overcome this antagonism and ultimately close the gap conductivity properties/resistance initially empty:

#1 Gelation in situ in a fibrous mat or in an open-cell foam (the product market leader Spaceloft® ASPEN uses this solution).

#2 A specific architecture of the material symbolized in column 5 of Fig. 13.10. Besides incorporating upstream of the choice of materials the manufacturing constraints, the environmental objectives and the target of cost reduction, this solution is based on a dual principle. First maximize compactness for limiting the conduction contribution from the gas phase, this is obtained by extreme control of the stack of components selected over scales from few millimetres to tenth of nanometers; this approach resemble that led to ultra-high property concretes. Secondly use a binder system adapted to the nature of the aerogel on the viewpoints of surface properties (super-hydrophobicity) and of mechanical properties (brittle), and whose morphology after filmification minimizes

| | Opacified monolith | Simple stack | Binded stack | Multiscale architecture |
|---|---|---|---|---|
| |  | 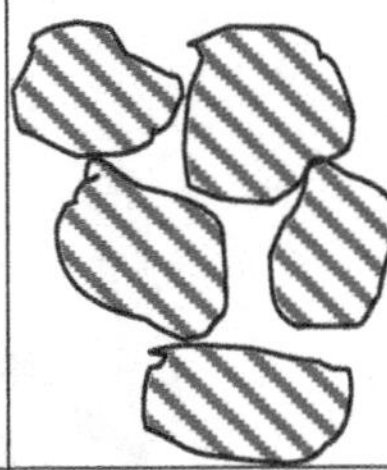 | 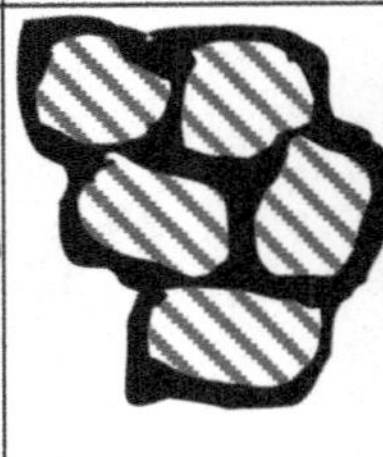 | 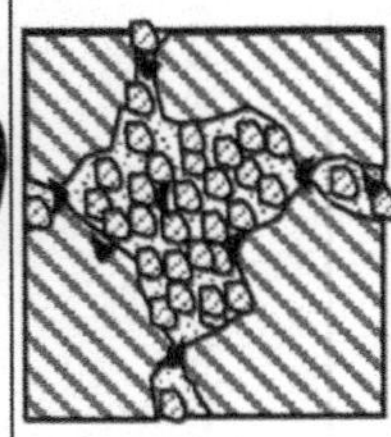 |
| $\lambda$ | 11 | 16 | 25 | 14 |
| $\lambda_{cg}$ | 3 | 8 | 3 | 3 |
| $\lambda_{cs}$ | 4 | 4 | 19 | 8 |
| $\lambda_{r}$ | 4 | 4 | 3 | 3 |

**Fig. 13.10** Influence of mechanical reinforcement of a granular bed by impregnating resin and architected solution [Conductivity in mW/(m K)]

the contribution of solid conduction. This material was patented [20] showing also that the materials architecturation is an opportunity to acquire intellectual property.

#3 The design of organic aerogel that, unlike silica, has very good mechanical properties but may struggle to reach the low conductivities targeted and also to be eco-friendly.

#4 The development of hybrid solutions inorganic—organic intra or intergranular where thermal function is provided by the silica while the mechanical function is provided by the organic part.

#### 13.4.2.2 Reduce the Cost

Today, although their price is falling, SIAP remain expensive as shown Fig. 13.11 which shows the different families of insulation in the space of thickness and cost optimizations. The ultimate goal is to bring representative position of aerogels to the general trend line connecting mineral wools with VIP.

The cost of the performance for 1 $m^2$ of insulation with a thermal resistance equal to 1 $m^2$ K/W (functional unit) being the product of the value per unit volume by conductivity, care must be taken that the lower cost by volume obtained by the control of raw materials and of the process is not accompanied by a drift in conductivity that would nullify gains.

Although compelling, the cost lowering achieved through the path#2 above sees its amplitude limited by the cost of the precursor (TEOS), which represents the vast

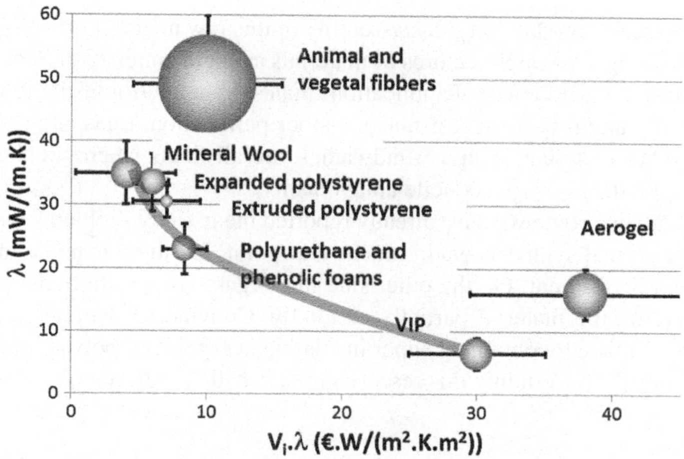


**Fig. 13.11** Position of families of insulating materials in the space Conductivity (λ) depending on the cost of performance per 1 $m^2$ ($V_i.\lambda$) which respectively correspond to optimizations of the thickness and of the cost ($V_i$ is the volume value of the insulation €/$m^3$; the size of the bubbles is proportional to the dispersion of data, data from 2017)

majority of the final overall cost. So, in addition to solving the problem of mechanical strength, the organic route (#3) holds the potential to reduce costs. Indeed, the field of polymers demonstrated its ability to provide multiple structural as functional solutions at much lower costs than inorganic materials.

The organic aerogel both reference and historical is the resorcinol-formaldehyde [21] which date in 1989. Despite the seniority no commercial product exists and only two are at an industrial pilot stage: PU supercritical aerogel Slentite® from BASF for building applications and a polyimide aerogel for space applications [22].

This absence can be explained by three difficulties. The first is to obtain sufficiently finely nanostructured aerogels. Without being able to predict if certain chemical systems work fine and others not. Unlike silica aerogels obtaining the mechanical properties is easy but the thermal conductivity is often too high (too large pores or heterogeneous size, variable radiativity, solid skeleton too coarse and too low porosity).

The second is the toxicity of some reagents [23] or environmental criticism of the synthesis medium [22]. The third is durability including hygric behaviour.

#### 13.4.2.3 Improve the Environmental Performance

The most recent processes of the silica route are well mastered with very little waste and a full solvent recycling. The biggest environmental impact is embodied energy [9] from the precursor synthesis which is difficult to avoid at the moment but which motivates studies of alternative precursors. One trend is manufacturing grains or powders with a greater productivity; the assembly of this raw material into products will require revisiting some architectures of materials as for channel #2 of Sect. 4.2.1 but also to satisfy a wider variety of applications than simple interior insulation. Specific properties of durability, frost resistance, vapour permeation, transparency, association with other materials such as sandwiches, are all fields where architecturation will be among the tools to reconcile antagonisms.

For the organic route we have already reported the toxicity problems with certain reagents or critical synthesis environment viewpoint, but there is no common rule, each system is different. On the other side this organic route offers the possibility of bio-sourcing the material partially or totally. Convincing demonstrations have recently been made by obtaining super insulating aerogel type polysaccharides [24, 25] or tannins [26]. Certainly this research field is called to develop.

## 13.5 Heat Storage

There are three main motivations to store heat for the building:

- Erasing the heating or the air conditioning during electricity consumption peaks to smooth the load curve;
- Improving the efficiency of heat pumps by favouring the operation when temperatures are more beneficial and for longer cycles;
- Solar thermal capture at different time scales: day, week, and season.

In addition to heat loss management, to store heat involves two phenomena: the storage itself in a "tank" and the exchange of energy to load or unload it. The first is a bulk property and the second a transport property. This is what gives all its interest to the architecturation approach to meet both the issue of energy and that of power.

### *13.5.1 Energy*

To illustrate the approach we will consider the simple case of a material plate in which you wish to store and destock energy conveyed by the surrounding environment, and that is to be optimized in volume or mass. Technologically it corresponds to the case of a storage rack.

The energy storable $Q$ is the sum of the sensible heat and the latent heat:

$$Q = \rho.V.\left(c_p.\Delta T + \Delta H.yield\right) \tag{13.11}$$

where sensible heat is the sum of those accumulated before and after any change of state or phase with enthalpy $\Delta H$. *Yield* is the proportion of the state change mobilized by the real thermal loading; it therefore depends on the phase change temperature and on the temperature changes actually imposed by the system.

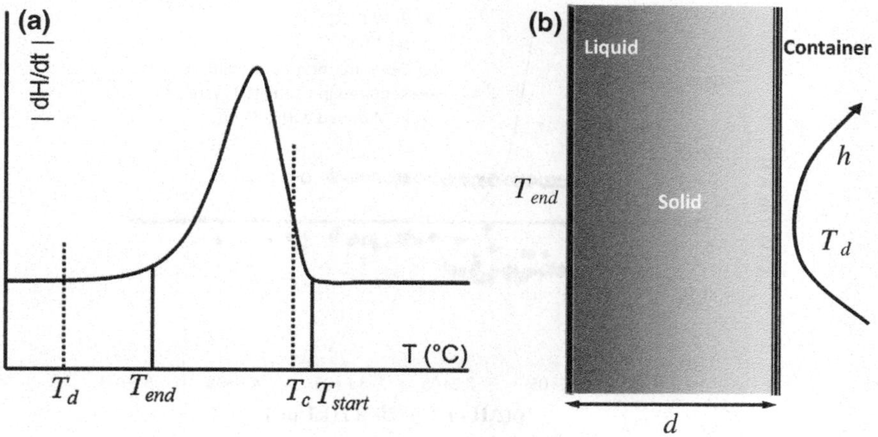


**Fig. 13.12** Setting **a** temperatures of the (11) and (12), and **b** the ultimate state of the discharge for the (13)

To optimize compactness, the material performance index derived for the energy is given by the following equation with the definitions of Fig. 13.12a):

$$PI_{(Energy-Volulme)} = \frac{1}{\rho[Cp.(Tc - Td) + \Delta H.Yield]} \quad (13.12)$$

Plotting the map of conductivity versus storable energy by a change in temperature of 20 K (Fig. 13.13) shows that one can generally form three groups: very good conductive metals, "other" poor conductive material, and phase change materials (PCMs) mostly liquid—solid. They are poor conductors but capable of storing high energy densities whose values are without comparison with all other materials. It is therefore apparent that the PCMs are essential to achieve high energy density stored at low temperature difference.

### 13.5.2 Power

Always to optimize compactness, the material performance index derived for the power discharge is given by the following equation with the definitions of Fig. 13.12b:

$$PI_{(Power-Volume)} = \frac{1}{T_d - T_{end-crystallisation}}\left(\frac{1}{h} + \frac{d}{\lambda_s}\right) \quad (13.13)$$

If one makes the calculation with a ventilated system with performing fins and relatively high air velocity [coefficient $8 < h < 50$ W/(m$^2$ K)] and plate thickness 2 cm, we find it needs a conductivity of the MCP greater than or equal to 1 W/(m K).

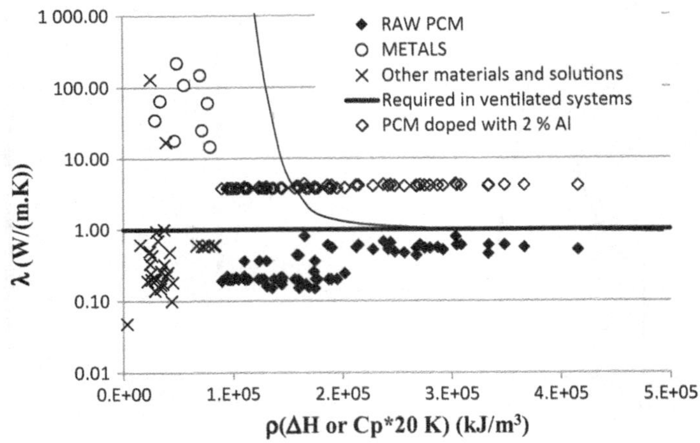


**Fig. 13.13** Materials properties of conductivity and energy storage for a temperature difference of 20 K

Such materials do not exist, we must develop. Two architectures may lend to: at macroscopic scale for example metallic cells or honeycomb, or to a more microscopic level for example by a metallic wool or foam. This new PCM with a doped conductivity (see Fig. 13.13) can reach the requested power levels.

It is important to note that for energy as for power, performance indices are not purely material ($\rho$, $cp$, $\Delta H$, $\lambda s$, $T_{end\text{-}crystallisation}$) but also system ($Tc$, $Td$, $d$, $h$). This means that any change in the material causes the system and vice versa. The other thing that appears is the balance that must be between the thermal resistance of the solid PCM and the contact resistance of the external exchange coefficient $h$.

It is worth noting that if the variables at system level are continuous, it is not the same for the material whose phase change temperatures and enthalpies are discrete values. Moreover, the actual behaviour of PCM is in practice far from the conceptual vision that ensures the seductive appearance because it is important to integrate the breadth of the transformation range, the hysteresis between the charge and discharge, and the speed dependence of all. The architecture of the material but also that of the container, the external exchanger, or that of its control for a given purpose plays a major role in achieving objectives. A good illustration of design methods and sensitivity analysis of the adequacy PCM—System—Application is given by Kuznik [27].

More generally, such a development can therefore only be done for an application well-defined by its specifications and it will be utopian to imagine using an existing material without adapting to the system. This difficulty and the blurring of many specifications explain the failures of the PCMs which outnumber the successes.

## 13.6 Prospects

There are many visions of the long term (2050) trends for buildings but if we only consider the main and most shared streams we can list [28]:

- Towns are bigger than ever and living space is a precious asset;
- Building and energy are always strongly linked;
- Energy is rare, so the local available energies are used and the efficiency is the basement;
- Some convergences appear between the bio sources and both the building and the energy.

On one hand the materials answer to efficiency plus space saving is the superinsulation where VIPs and SIAPs have a bright future already illustrated in the previous sections. On the other hand heat or energy storage as well as other functionalities is the way for a better use of thermal renewable. Some of such thermal components are listed here after with the challenging properties but without materials solutions, just to tease the mind!

### *13.6.1 Thermo-Chemical Storage*

The thermal storage at the season scale is complicated if considering the PCM because of the heat losses. Without such losses, the thermo chemical storage could be convenient especially if it uses a common reagent. That is the case of water vapour available in the air which can react with salts in a reversible hydration. The first tests was not very convincing because of insufficient energy density, of low power and of irreversibility. The material behaviour showed extra colonization of hydrophilic sites, closing of doors between pores and an inappropriate water vapour accessibility. Obviously a hierarchical controlled porosity could be a smart response where the design depends on the application.

### *13.6.2 Ideal Insulation*

The ideal insulation is one which could be huge or zero depending on the needs: huge to avoid any losses, and zero to catch free thermal renewable with all in between scenarios. Such concept of variable and switchable thermal resistance exists and uses both low radiativity and the pressure and the internal dimensions to manage the gaseous and radiative conduction while avoiding solid conduction. In practice the real lock is on the materials and the whole structure which have to deal with thermal properties and mechanical or dielectric properties and where the process ability is another challenging difficulty. Another concept is more on a single particular material, not super-insulant but exhibiting a significant variability of conductivity. The design of ideal but realistic material is a nice application of the multifunctional architectured material approach.

### 13.6.3 Adaptive Solar and Sky Collector

Close to the "cool roof" products, depending on the needs different concepts exist about visible + infrared capture or reflectance. Looking to the sky or to the ground, give radiation sources at different "temperatures". Most of these concepts involves surface with selective spectra properties for example to avoid the diffuse solar radiation while absorbing infrared radiation, or the opposite. In addition, because of dealing with heat transfer, some insulation is needed with non-isotropic behaviour, and because of the integration to building walls or roofs, mechanical, hydric, ageing, UV resistance, anti-fouling, etc.… properties are required.

The history of this kind of component is just beginning but their outlook is attractive because they are based on two new axes which make things possible. Firstly new buildings are near zero energy, so their needs are lower than ever and secondly recent studies demonstrate that it is realistic to cover the heat and cold needs with local renewable energies [29]. Such complex specifications are obviously a nice playground for architectured materials because of the multi-functionality.

### 13.6.4 Storing Walls and Ventilating Walls

These two concepts are quite old but without convincing application except in a certain way the well-known curtain walls. One of the challenges would be to imagine and design a component working as a switchable energy vector through the wall or a part of the wall able to get inside or avoid entrance of free outside energy.

### 13.6.5 Conclusion for Prospects

To conclude on these perspectives, most of these thermal components are at the intersection of getting free inputs, thermal storage and thermal insulation. The very strong coupling between the materials and the systems are both the difficulty and the opportunity to fulfil the application requirements.

## 13.7 Conclusion

In the field of energy efficiency in building the Architectured Materials approach started very late, only sixty-seventy years ago with the foaming of polymers and with the manufacturing of mineral wool. The ultra-porous architecture where then optimized from the thermal point of view dealing with the different contributions to their conductivity: radiative, solid conduction and gaseous conduction. Of course some other properties like mechanic, hygric and cost where also taken into account. In parallel other material where design at a scale closer to the product scale like hollow bricks, lightweight concretes.

It is only very recently, less than twenty years that a final improvement is done on traditional insulation materials polymeric foams and glass wool by achieving a very efficient infrared opacification. The conductivity of the best products on the market is now on the asymptote constituted by the conductivity of their main phase: the still air.

It is also in the same time that the breakthrough of the super insulation comes. It is based on a specific architecture of the material involving more matter but dealing with the nanometer scale in order to confine the molecules of the gas in smaller pores than their mean free path. In such case the gaseous contribution could be zero and then the whole conductivity far from the one of still air. Two families of material/products share the market: the vacuum insulation panels with very low conductivity (2–10 mW/(m K)) and the super insulation at atmospheric pressure which have a conductivity between 14 and 18 mW/(m K) but no severe implementation constraints. For each several optimisation problems are ongoing or on the table and where the architecture of the material and product is the key for success in reaching antagonist attempts.

The strongly moving context around the renewables energies is also bringing new playground for architectured materials. One first theme is the heat storage where the difficulty for reaching the wanted power could be address by this approach. The second theme is very wide and new; it is the field of active walls and specific collectors where the aimed properties ask for filling some gaps in the materials spaces.

Finally the approach of architectured materials is less developed in the building sector than in others such as transport perhaps because building is a very traditionalist industrial sector with very different sizes of company. But as we can see the knowledge diffusion is at work and thanks to the education students are more and more aware of the possibilities of this approach especially to solve Chinese puzzles and then to innovate.

## References

1. IPCC/ GIEC, 3d working group, 2014
2. M.G. Kaganer, *Thermal Insulation in Cryogenic Engineering* (Israel program for scientific translations, Jerusalem, 1969)

3. L.R. Glicksman, Heat transfer and ageing of cellular foam. Cell. Polym. **10**, 276–293 (1991)
4. R. Coquard, D. Baillis, Modelling of heat transfer in low density EPS foams. J. Heat Transf. **128**, 538–549 (2006)
5. D. Kunii, J.M. Smith, Heat transfer characteristics of porous rocks. A.I.CH.E.J. **3**, 71–78 (1960)
6. R. Coquard, D. Baillis, V. Grigorova, F. Enguehard, D. Quenard, P. Levitz, Modelling of the conductive heat transfer through nano-structured porous silica materials. J. Non-Cryst. Solids **363**, 103–115 (2013)
7. H. Simmler, S. Brunner, U. Heinemann, H. Schwab, K. Kumaran, P. Mukhopadhyaya, D. Quénard, H. Sallée, K. Noller, E. Kükükpinar-Niarchos, C. Stramm, M. Tenpierik, H. Cauberg, M. Erb, Vacuum insulation panels. study on VIP-components and panels for service life prediction of VIP in building applications Subtask A. *Technical report, IEA/ECBCS Annex 39 HiPTI-project High Performance Thermal Insulation for Buildings and Building Systems*, September 2005
8. B. Yrieix, B. Morel, E. Pons, VIP service life assessment: interactions between barrier laminates and core material, and significance of silica core ageing. Energy Build. **85**, 617 (2014)
9. IEA-EBC Annex 65. Long Term Performance of Super-Insulating Materials in Building Components and Systems, 2013–2017
10. S. Brunner, K.G. Wakili, Hints for an additional aging factor regarding the thermal performance of vacuum insulation panels with pyrogenic silica core. *Vaccum* **100**, 4–6, (2013) https://doi.org/10.1016/j.vacuum.2013.07.033
11. E. Pons, B. Yrieix, S. Brunner, Evaluation of VIPs after mild artificial aging during 10 years: Focus on the core behavior. Energy Build. **162**, 198–207 (2018)
12. B. Benane, Mécanique des lits de silices granulaires pour l'optimisation des cœurs de panneaux isolants sous vide. Ph.D. thesis, INSA de Lyon, France (Jan. 26. 2018)
13. M.H. Klopffer, B. Flaconnèche, Transport properties of gases in polymers: bibliographic review. Oil & Gas Sci. Technol.—Rev. d'IFP **56**, 223–244 (2001)
14. H.C. Langowski, Permeation of gases and condensable substances through monolayer and multilayer structures, in *Plastic packaging: interactions with food and pharmaceuticals*, 2nd edn. (Weinheim, Germany Wiley-VCH Verlag GmbH & Co. KGaA: 2008), pp. 297–347
15. M. Bouquerel, Influence des transferts hygro-aérauliques sur les transferts thermiques dans les super-isolants nanostructurés sous vide. Ph.D. thesis, INSA de Lyon, France (Dec. 13. 2012)
16. A. Batard, E. Planes, T. Duforestel, L. Flandin, B. Yrieix (2018) Water vapour permeation through high barrier materials: numerical simulation and comparison with experiments. J. Material Science (2018)
17. F. Dubelley, Mécanismes de dégradation des enveloppes barrières pour application panneaux isolants sous vide. Ph.D. thesis, Savoie Mont Blanc Univ. France, (April 6. 2016)
18. J.C.H. Wong, H. Kaymak, S. Brunner, M. Koebel, Mechanical properties of monolithic silica aerogels made from polyethoxydisiloxanes. Microporous Mesoporous Mater. **183**, 23–29 (2014)
19. D. Ge, L. Yang, Y. Li, J. Zhao, Hydrophobic and thermal insulation properties of silica aerogel/epoxy composite. J. Non-Cryst. Solids **355**, 2610–2615 (2009)
20. B. Yrieix, B. Morel, G. Foray, A. Bogner, Matériau super-isolant à pression atmosphérique à base d'aérogel. *Patent FR 2975691 WO2012168617*
21. R. Pekala, Organic aerogels from the polycondensation of resorcinol with formaldehyde. J. Mater. Sci. **24**, 3221–3227 (1989)
22. M.A.B. Meador, E.J. Malow, R. Silva, S. Wright, D. Quade, S.L. Vivod, H. Guo, J. Guo, M. Cakmak, Mechanically strong, flexible polyimide aerogels cross-linked with aromatic triamine. ACS Appl. Mater. Interfaces. **4**, 536 (2012)
23. G. Biesmans, D. Randall, E. Francais, M. Perrut, Polyurethane-based organic aerogels thermal performance. J. Non-Crystalline Solids **225**, 36 (1998)
24. C. Rudaz, R. Courson, L. Bonnet, S. Calas-Etienne, H. Sallee, T. Budtova, Aeropectin: fully biomass-based mechanically strong and thermal super-insulating aerogel. Biomacromol **15**, 2188 (2014)

25. R. Sescousse, R. Gavillon, T. Budtova, Aerocellulose from cellulose-ionic liquid solutions: preparation, properties and comparison with cellulose-NaOH and cellulose-NMMO routes. Carbohydr. Polym. **83**, 1766 (2011)
26. G. Amaral-Labat, A. Szczurek, V. Fierro, A. Pizzi, A. Celzard, Systematic studies of tannin—formaldehyde aerogels: preparation and properties. Sci. Technol. Adv. Mater. **14**, 015001 (2013)
27. F. Kuznik, J.P. Arzamendia Lopez, D. Baillis, K. Johannes, Design of a PCM to air heat exchanger using dimensionless analysis: application to electricity peak shaving in buildings. Energy Build. **106**, 65–731 (2015)
28. B. Yrieix, Future trends in materials for thermal insulation and building envelope. Paper presented, in *Mapping the Future of Materials Science, SF2M Conference*, Sevres, France (September 7–9, 2015)
29. L. Chesné, T. Duforestel, J.J. Roux, G. Rusaouën, Energy saving and environmental resources potentials: toward new methods of building design. Build. Environ. **58**, 199–207 (2012)

# Chapter 14
# Topological Interlocking Blocks for Architecture: From Flat to Curved Morphologies

**Giuseppe Fallacara, Maurizio Barberio and Micaela Colella**

**Abstract** The paper concerns the theme of topological interlocking blocks for architecture and the relationship between flat stereotomic assemblies and curved morphologies. After a brief history of the subject, theoretical foundations and speculative research are presented. The research includes several built full-scale prototypes and architectural elements. The last part of the chapter describes the didactic experiences concerning the theme, during the third year Architectural Design Studio held by the authors, at Politecnico di Bari, Italy.

## 14.1 Introduction

The interest around Stereotomy started to reawaken in the early 1990s, within the researches about the History of Construction. This new favourable cultural climate has allowed rediscovering the discipline, both for its historical value and the unexpressed design possibilities. The diffusion of parametric modelling and digital fabrication tools creates the ideal conditions to design and build new stereotomic prototypes, which are typically characterized by considerable architectural and geometric complexity. In this context, the chapter aims to outline the potential of topological interlocking blocks for architecture through combining stereotomic principles, engineering, material science and architectural aesthetic sensibility.

This paper is the result of the combined work of the three authors. Paragraphs 1 and 2 was written by Maurizio Barberio; paragraphs 3 and 4 by Giuseppe Fallacara; paragraph 5 by Micaela Colella. The authors have revised all the paragraphs, and the paper structure has been conceived together.

G. Fallacara (✉) · M. Barberio · M. Colella
Dipartimento di Scienze dell'Ingegneria e dell'Architettura, Politecnico di Bari, Bari, Italy
e-mail: Giuseppe.fallacara@poliba.it



Y. Estrin et al. (eds.), *Architectured Materials in Nature and Engineering*, Springer Series in Materials Science 282,
https://doi.org/10.1007/978-3-030-11942-3_14

## 14.2 Topological Interlocking Blocks for Architecture: A Brief History

The recent history of architectural applications of "topological interlocking blocks" follows two parallel roads, however destined to meet after a few years: the first road is the architectural one, starting from the study of the architectures of the past and ending at the contemporary design experiments; while the other concerns the disciplines of material science and engineering.

In the field of architecture, this type of structures, derived from the 18th-century patents of Joseph Abeille and Sebastien Truchet, are rediscovered thanks to the studies of several historians of architecture and construction. Jean-Marie Pérouse de Montclos in the book "*L'Architecture à la française, du milieu du xv$^{e}$ à la fin du xviii$^{e}$ siècle*" of 1982 [1], and later Enrique Rabasa Díaz with the article "*La bóveda plana de Abeille en Lugo*" [2] during the Congress on the History of Construction of 1998, rediscover the theme of "*voute plate*". In 2002, the Department of Architecture of the Politecnico di Bari, under the guidance of Prof. Claudio D'Amato, developed a research project focused on studying the flat vault of Joseph Abeille from a historical, architectural design and engineering point of view [3, 4]. The idea of the Department is to start from the study of history to obtain useful principles for contemporary design. An important contribution in this direction was given by the work of Giuseppe Fallacara, who in 2006 started to spread his research on the design of complex vaulted systems obtained through simple topological deformation operations of the Abeille's flat vault [5] (Fig. 14.1). Using this principle Fallacara produced a series of contributions in the form of scientific publications [6–9] and experimental prototypes, some of which will be presented in the following paragraphs.

Also in 2006 Joël Sakarovitch published the results of an important workshop held at the *Grands Ateliers of the Isle d'Abeau* (GAIA), inaugurated in 2002, during the Second International Congress on Construction History [10]. Among the topics of the workshop, there were the fabrication, the construction and the loading tests of a series of Abeille's flat vaults made of local limestone. In 2008, an important collaboration between Italian and French architects, engineers and stonecutters started [11], which led to work on the *Sphera* project, a double curved sphere portion made of topological interlocking blocks (Fig. 14.2). Other engineering researches will be launched later at the GSA Laboratory at École nationale supérieure d'architecture de Paris-Malaquais under the guidance of Maurizio Brocato [12].

Alongside these events, in 2001, an Australian research group [13] "published a paper introducing a concept of topological interlocking as a materials design principle" [14]. Totally unaware of previous patents of Abeille and Truchet [14], the group continues to work on the theme, providing other very important scientific contents [15–17] that will soon meet the interest of other members of the scientific community. In 2012, the paths of architecture and materials science met in Oliver Tessmann's "Topological Interlocking Assemblies", presented during the 30th eCAADe conference. Tessmann adopted the research of the Australian research group within the parametric design and digital fabrication fields [18–20].

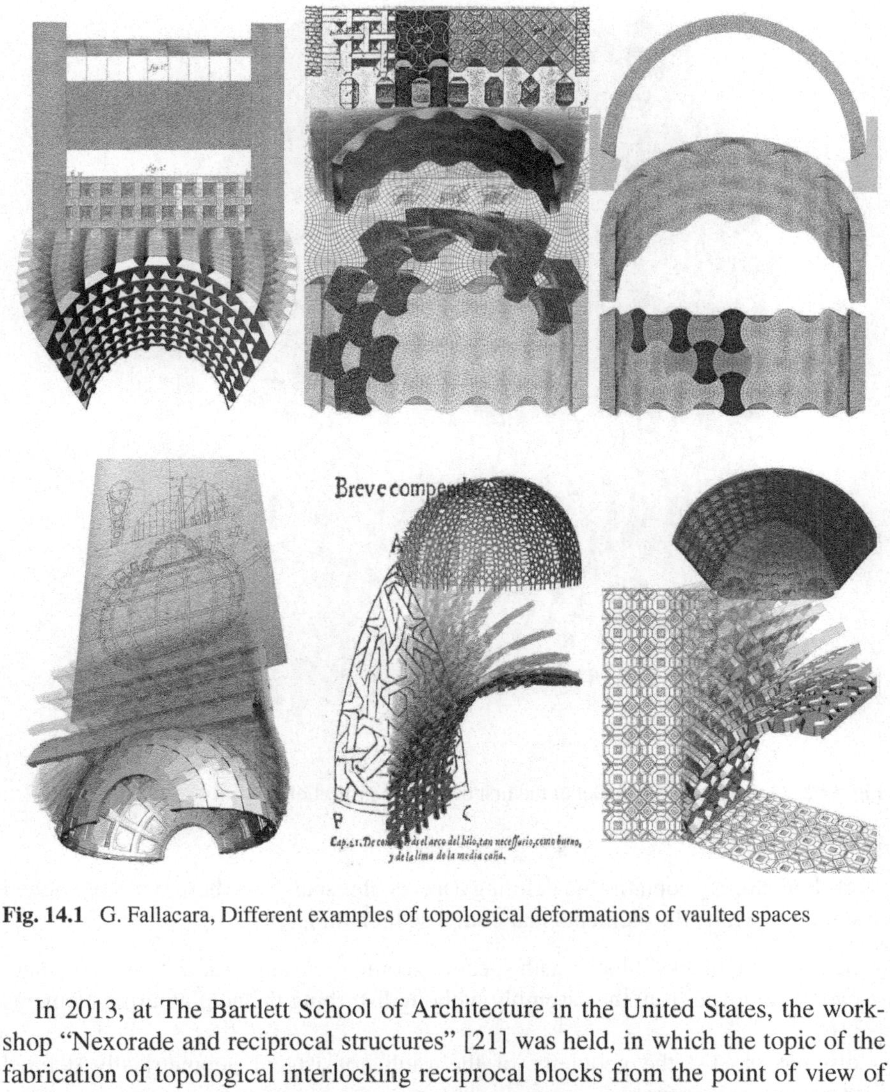

**Fig. 14.1** G. Fallacara, Different examples of topological deformations of vaulted spaces

In 2013, at The Bartlett School of Architecture in the United States, the workshop "Nexorade and reciprocal structures" [21] was held, in which the topic of the fabrication of topological interlocking reciprocal blocks from the point of view of robotic manufacture was dealt with by using polystyrene blocks cut out by means of a robotic arm.

In 2014 the exhibition "Stereotomic Design. Evolution: from the hand to the robot" [22] edited by Giuseppe Fallacara within Marmomacc 2014, sealed the mature merge between history, architecture, engineering and materials science, with a section of the exhibition devoted to "biomechanical stereotomy" (Fig. 14.3). The two parallel paths traced by Prof. Estrin's group in Australia and the nascent New Fundamentals Research Group, led by Prof. Fallacara in Italy, finally converged. In 2015, Fallacara was invited by Prof. Estrin's group to hold together the workshop "New stereotomic vaulted structures with interlocking, discrete and reciprocal blocks. Theory and practice" at the Monash Art Design & Architecture (MADA) (Fig. 14.4). The name of the

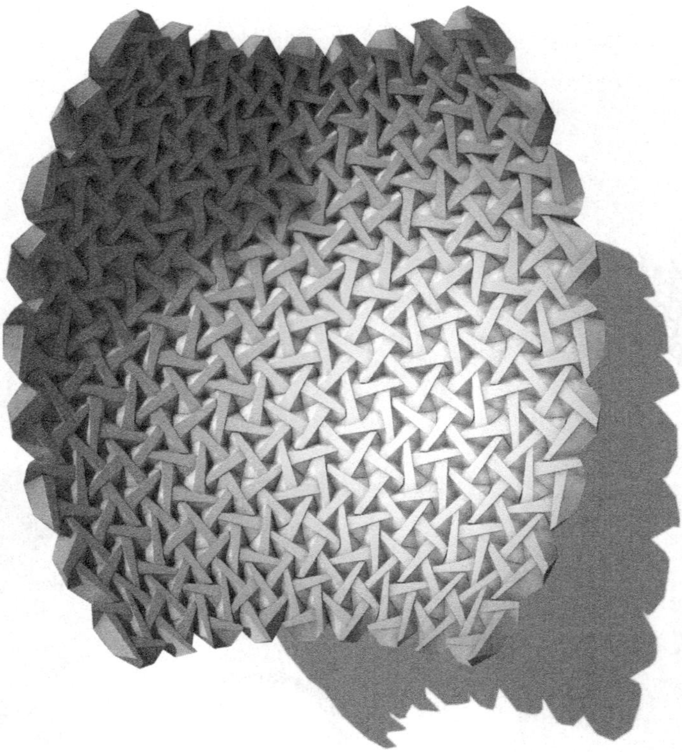

**Fig. 14.2** G. Fallacara, 3D model of the first developed project of *Sphera*

workshop already contains the defining aspects of research on these types of vaulted systems, in which the elements that compose them must be:

- interlocking blocks: blocks with specific geometry arranged such that no element can be removed from the assembly without disturbing the neighbouring elements [16];
- discrete blocks: the geometry of the vault consists of geometrically defined stereotomic elements;
- reciprocal blocks: three-dimensional mutually supporting sloping blocks placed in a closed circuit; the inner end of each block rests on and is supported by the adjacent block.

From all these experiences, the most recent research in the fields of Computer Aided Architectural Design and Digital Fabrication derives, with different contributions in different parts of the world:

- in Israel, concerning the potential of using the concept of topological interlocking as a structural and organizational mechanism for architecture in general, and for building façades in particular [23, 24];

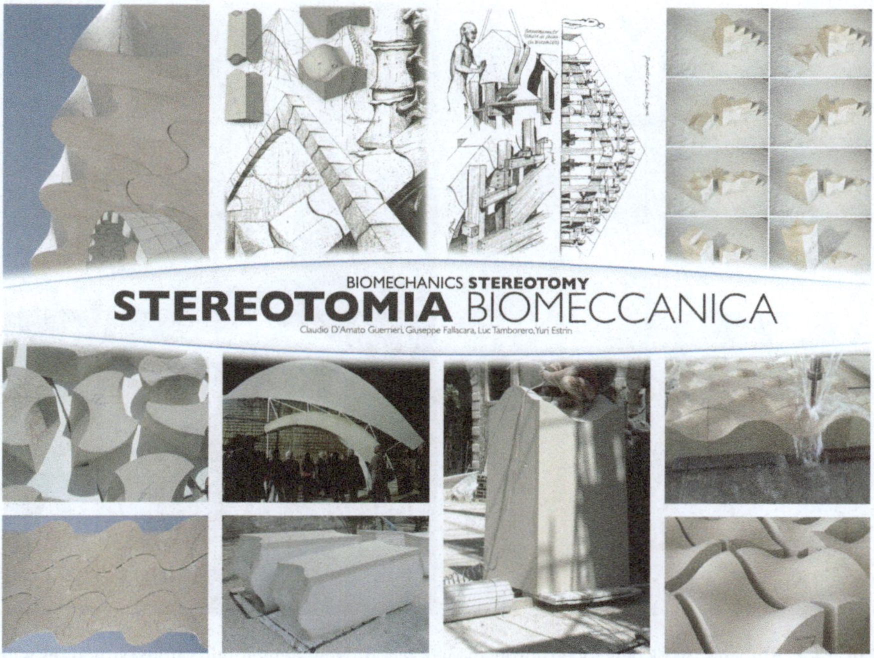


**Fig. 14.3** A part of a poster exhibited at Marmomacc Verona in 2014, dedicated to "Biomechanical Stereotomy" (poster designed by M. Barberio and Micaela Colella; exhibition curated by G. Fallacara and V. Minenna)

- between Malta and Finland, concerning the geometric versatility of Abeille vault for a stereotomic topological interlocking assembly [25];
- in Australia, with a doctoral thesis entitled "Utilisation of topologically-interlocking osteomorphic blocks for multi-purpose civil construction" [26], and about stereotomy of wave jointed blocks [27, 28];
- between Palestine and France, with the "Stone Matters" prototype, designed by AAU Anastas and the Paris-Malaquais GSA Laboratory and built in Palestine in 2017.

## 14.3 Speculative Design and Topological Stereotomy: From the "Flat Vault" to the "Curved Space"

In this paragraph, the theoretical thinking behind some projects, realized between 2006 and 2015, using topological interlocking blocks derived from Abeille's and Truchet's patents, is described. The speculation aims to solve some of the most important theoretical issues related to the realization of vaulted spaces:

**Fig. 14.4** Students' work developed during the workshop "New stereotomic vaulted structures with interlocking, discrete and reciprocal blocks. Theory and practice" (Monash University, 2015)

- Improving mutual anchorage, and therefore the friction, between the ashlars of a vaulted system;
- Reducing wall thickness;
- Designing the mounting of the vault both as ornamental and structural;
- Varying the shape of the mounted system in a smooth and continuous way;
- Rationalizing the number of standard ashlars for the construction of any vaulted space;
- Automatically obtaining all the geometric data both for the manual and the electronic cutting (CAD/CAM/CNC) of the individual blocks;
- Automatically providing all the numerical data aimed at the structural analysis.

A vaulted space can be defined as a portion of space delimited by flat surfaces of tread and perimetration, and by curved covering surfaces, which constitute the true region for the characterization of the perceived area. The curved line, with any mathematical complexity, can be designed and built in two basic ways: with a tool that can draw the curved (i.e.: compass, algorithm, etc.), or by bending a straight line. The latter mode is totally attuned to the topological view. The method is starting from a very simple consideration, viz. the observation that most of the vaulted systems can be imagined as a discontinuous structure or masonry wall that has been folded and/or deformed to obtain its final geometrical conformation, with due simplifications and specifications. For example, a limestone wall can be theoretically bent to become a barrel vault and the latter can be bent to become an annular one. Proceeding as well, by vertically translating its extremes, the annular vault can become a helical barrel vault (Fig. 14.5).

This method uses the technique of bending, thanks to software providing three-dimensional modelling and the parametric-variational one, thus being based on the topological geometry and its functions of deformation. As already stated, according to the topological concept of folding, a flat sheet of paper and one rolled-up to form a cylindrical surface are equivalent from a topological point of view (Fig. 14.6).

By using some tools for spatial transformation and volumetric deformation of a shape, a correspondence between a flat surface and a spatial one is realized. The

**Fig. 14.5** Topological deformations applied to a simple stone wall by: using this technique it is possible to easily deform a wall thus obtaining a barrel vault

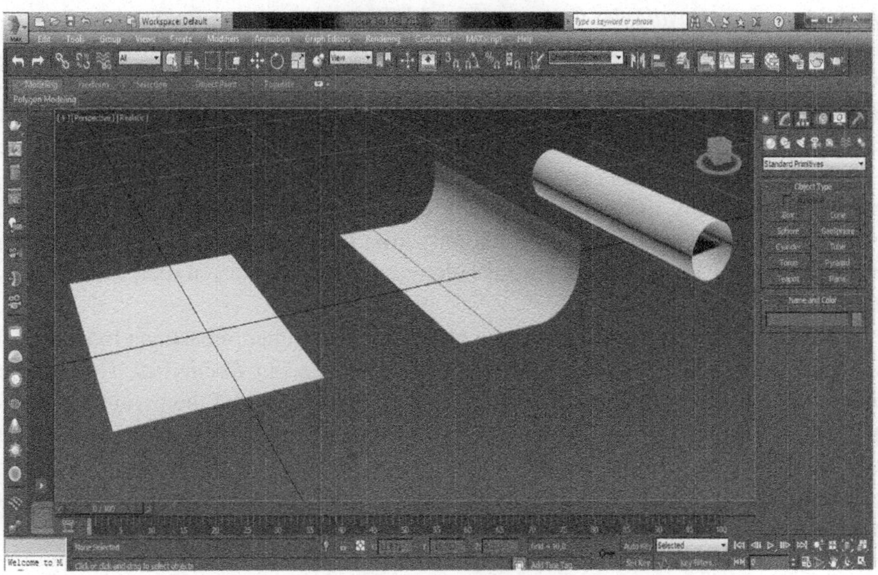


**Fig. 14.6** The topological concept according to which a flat sheet of paper can be rolled-up to form a cylindrical surface

modelling process is thus considered as a flexible sculpture of the three-dimensional digital data. Every simple starting shape, such as a parallelepiped, or a sphere, or a set of basic solids, can be modified by means of parametric topological deformations to achieve the degree of complexity of the desired final object. In so doing, we achieve the final object in an indirect way, and not directly through the canonical modelling that, in the case of complex objects, would involve major difficulties for the three-dimensional modelling. Such deformations are made possible thanks to the subdivision of the surface by tessellation, according to a predetermined polygonal pattern: each intersection between the vertices is behaving like an articulated joint. As a result, the more segments there are in a surface, the more deformable it is. It should be noted that, contrarily to the physical operation of bending a material, a simple or double infographics fold does not produce tears or overlapping of matter; it only produces a transformation of the geometry, which becomes the object of evaluation and study.

Having intuitively understood the deformation processes that have led to the final configuration of a stereotomic system, we can start modelling a simple and appropriate solid by moving backwards. Subsequently, we can alter or deform it in various ways, also by assigning more modifiers to the same object. This method has been used for the design of numerous vaulted spaces of the new generation and for the production of expositive prototypes presented during cultural events since 2005. In these cases, the objects of the topological manipulation have been referring to two well-known French patents dating back to 1699 and describing a "flat vault". The French engineer Joseph Abeille (1669–1752) ideated the first license considering this construction. The license was published in 1699 in *Machines et inventions approuvées par l'Académie Royale des Sciences* (Fig. 14.7).

The standard ashlar, repeated in a rectangular matrix to realize this flat vault, was a polyhedron with the two axial sections being isosceles trapezoidal-shaped, and oriented towards opposite directions. As previously mentioned, Abeille's planar ashlars are parts of tetrahedra, mutually joining according to the same logic of platonic solids. The static behavior of such solution is the same as that of a bidirectional plane, working identically in both directions. Each of the ashlars is supported in such a way that the vault becomes functional only after completion. Such static condition allows us to consider such structure as belonging to the category of reciprocal frames (Fig. 14.8).

In 1704 the Carmelite Jean Truchet, also known as Father Sebastien (1657–1729) in *"Mémoire concernant les voutes plates"* published in *Recueil de l'Académie Royale des Sciences* described a patent to solve the problem of the pyramidal holes present in the solution of Abeille. Truchet's ashlar possesses a squared lower base and an upper one resulting in a mixed-line (composed by four identical circular arcs, arranged in pairs on parallel sides of the square to form concave and convex curves and jointed by means of ruled surfaces).

RECUEIL DES MACHINES. 159

1699. No. 50.

# VOUTE PLATE,

INVENTÉE

## PAR M. ABEILLE.

CETTE Voute eſt de niveau, tant à ſon parement de douelle, qu'à celui de l'extrados; les claveaux qui la compoſent ſont tous ſemblables, & n'ont que ſix faces ou panneaux, ainſi qu'un aube; ils forment des quarrés parfaits comme ABCD dans toute l'étendue du parement de douelle, & des rectangles EFGH au parement de l'extrados; les quarrés à la douelle ſont d'alignement en tous ſens, & les rectangles à l'extrados font avec de petits carreaux entremêlés, un compartiment régulier, de ſorte que cette voute forme tout enſemble, & un plafond ABIKL, pour l'étage inférieur, & un pavé EFMNO, pour l'étage ſupérieur. Fig. I. II.

Les quatre panneaux de joints de chaque claveau ſont en coupe; il y en a deux qui ſont inclinés en talus PP, deux qui ſont en ſaillie depuis les côtés du quarré de douelle QQ.

Le quarré du parement de douelle des claveaux étant déterminé à une certaine grandeur, l'épaiſſeur de ces claveaux aura les trois quarts de la longueur du côté de ce quarré, & la coupe des panneaux des joints ſera d'un tiers de cette épaiſſeur, ſoit aux panneaux en talus, ſoit aux panneaux en ſaillie; ce qui donnera des angles égaux pris les uns depuis le parement du quarré de douelle, & les autres depuis le parement d'extrados alternativement,

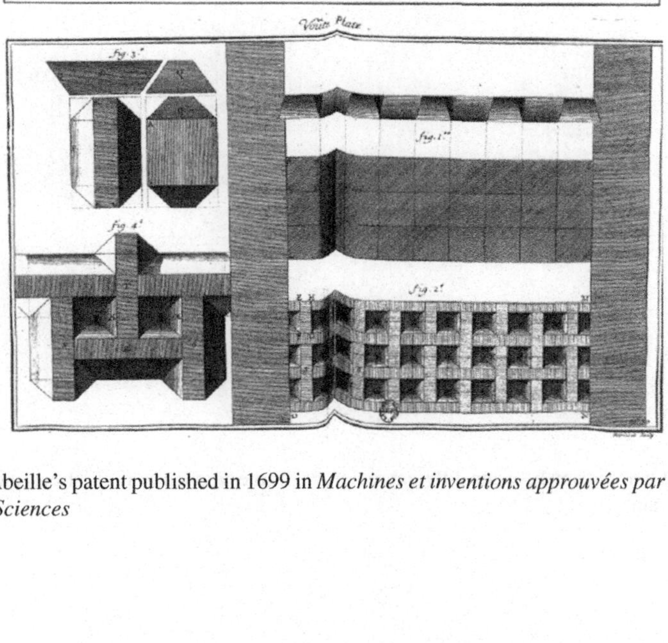


**Fig. 14.7** Abeille's patent published in 1699 in *Machines et inventions approuvées par l'Académie Royale des Sciences*

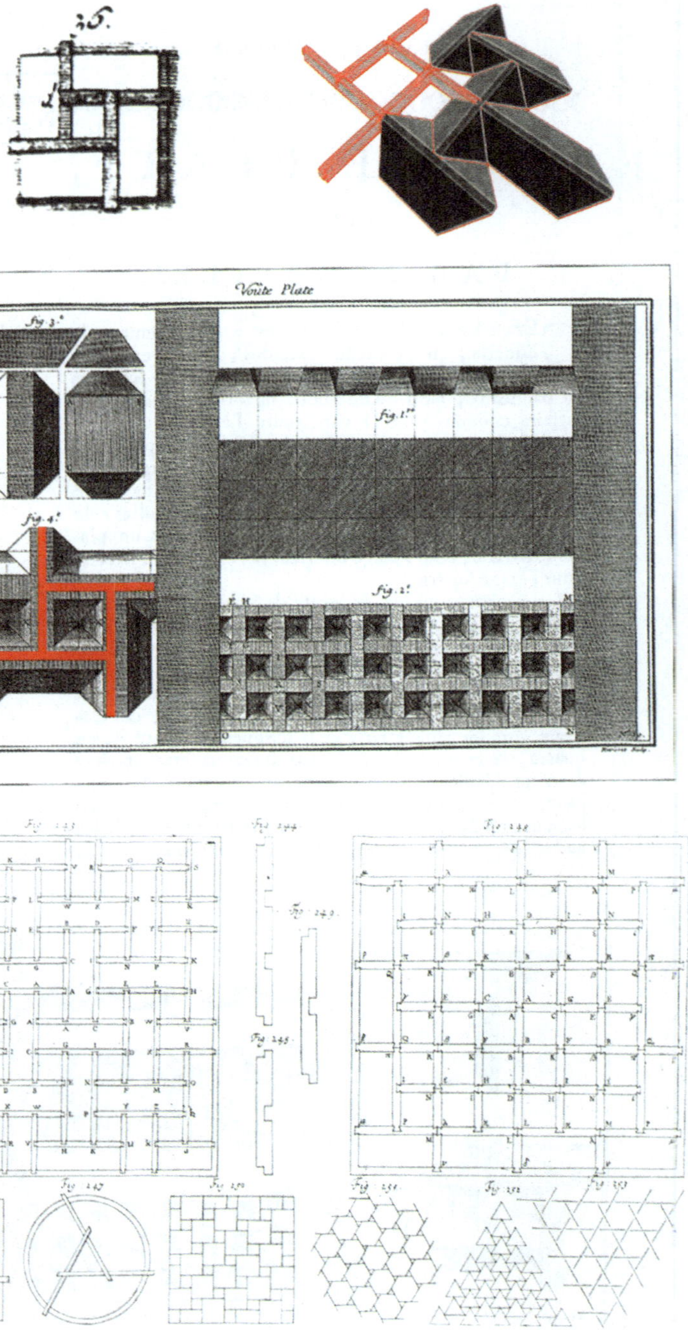


**Fig. 14.8** A comparison between different reciprocal frame wooden configurations and a flat reciprocal stone vault

## 14.4 Built Prototypes and Structures

As said, the experimentation was conducted by submitting the flat vaults by Abeille and Truchet to a deformation by folding. This technique allowed us to obtain a new conformation of the individual stone blocks, by transforming the plane into a half-cylinder or a spherical shape. The projects conducted included:

- *Portale Abeille*, Corderie dell'Arsenale, X Biennale Internazionale di Architettura Venezia, 2006;
- *Alexandros Obelisk*, Corderie dell'Arsenale, X Biennale Internazionale di Architettura Venezia, 2006;
- *Ponte Truchet*, Summer School, Universidad San Pablo CEU Madrid, 2007;
- *Sphera*, Stage Sphera, Grands Atelier Isle d'Abeau Lyon, 2008;
- *The Bin Jassim Dome*, Doha, 2012;
- *Wave Block Wall*, 2014;
- *Osteomorphic Catenary Arch*, 2015.

The *Portale Abeille* (Abeille's Gate) is a barrel vault that, due to its complex stonework, transfers pressure in two directions: along the curve of the arch and perpendicular to the arch (Fig. 14.9). This variation to the traditional static behaviour of the barrel vault was made possible by using specific patented stonework techniques, designed for the construction of a flat stone roof. The stereotomic system known as *voûte plate* is one of the most interesting technical-stylistic responses offered by stone masonry to the world of construction, a solution that has, however, been used very rarely. The flat vault defies common sense in its conjunction of two geometrically contradictory terms: 'vault', which by definition is a three-dimensional surface and 'flat' which by its very nature is two-dimensional. The historical problem raised in stereotomic design is how to construct a ceiling using a thrusting vault that is lowered

**Fig. 14.9** C. D'Amato and G. Fallacara, *Portale Abeille*, Biennale di Venezia, 2006

to a maximum level (zero rise). The static principle at the core of this type of vault presupposes that the loading path above the structure is both vertically and horizontally conducted, using a careful weaving together of wedge shaped ashlars, creating a successfully compressed vault. The ashlars are reiterated as a series to form a flat vault in a polyhedron form whose two axial sections, in the shape of an isosceles triangle, thrust in opposite directions. As described in Sect. 14.3, the static function of this solution is that of a flat, bidirectional plane that works in two directions in exactly the same way: each individual ashlar is maintained by, and maintains, the others in such a way that the vault works only when the construction of the whole is completed. Two surfaces are visible when the vault is completed: the square, homogenous and continuous web of the extrados, and the intrados, characterized by an interesting pattern of plaiting with pyramid-shaped features. The experimentation into the creation of *Portale Abeille* was based on the use of topological parametric-variational software which allowed us to bend and deform the patented vault, transforming the flat vault into a semi-cylindrical form as a new ashlar conformation. In order to transform the flat vault into a barrel vault, we identified a formal geometric variability range using the definition of modification parameters. Deciding on a single curve for the vault meant using two serial ashlar types for the creation of the barrel vault. An increase in deformation and curvature results in an ensuing rise of the number of ashlar types in the new vaulted system.

The *Alexandros Obelisk* (Fig. 14.10) is an attempt to resolve the problem of constructing a large and tall object in a small space: the lobby to the Arsenale building, constructed on the occasion of the Architecture Biennale, Venice. Moving away from notions typically connected with the construction of monoliths, like an obelisk, it was decided to divide the work height-wise, into three large stone blocks. Structural research into joints was able to resolve the necessity of movement in a slim obelisk

**Fig. 14.10** C. D'Amato and G. Fallacara, *Alexandros Obelisk*, Biennale di Venezia, 2006

structure. By studying historical obelisks in stone, we were able to identify edges and corners, where most of the force is concentrated, as the critical weak points in the structure, causing fracture and breaking. The joints designed for this project resemble the anatomical conformation of the knee joint, allowing movement of the blocks without the risk of damaging the stone. Like cartilage, neoprene was used in the joints, imitating the same spatial configurations of the contact surfaces of the large blocks. A steel cable was threaded through the vertical axis of the obelisk to prestress the whole structure. The tension of the steel cable, when added to the total weight of the blocks cut from Leccese limestone, was the same as if the whole structure had been constructed in granite. Another element of interest from the point of view of statics, is the suction effect created between the joints. The perfect joining of the blocks causes a pneumatic suction effect that, functioning as a 'glue', theoretically obviates the need for cables to reinforce the structure. This prototype represents a constructive solution of great interest in the construction of large pillars or stone columns.

*Ponte Truchet* (Truchet's Bridge) is an oblique barrel vault structure developed according to the solution for a flat vault patented by Truchet as the Abeille vault (Fig. 14.11). The prototype was made according to the same modelling and topological variation techniques used in the design of the Portale Abeille, the difference being that the Truchet ashlars are far more complex due to the surface curvature. The logic behind the joints used in the Alexandros obelisk was also adopted here to build Ponte Truchet with an oblique projection. It is well known that the barrel vault has the problem of 'empty thrust', a problem that led stone masons in the XIX and XX centuries to study the best way to resolve such a static issue: stones blocks into orthogonal and helicoidal shapes. These solutions are refined geometric-constructive solutions to an issue which, in that period, was resolved generally with steel or reinforced concrete. In order to pass from the flat vault of Truchet to an oblique barrel vault, we worked with orthogonal lines between the blocks, the flat

**Fig. 14.11** Ponte Truchet, Summer School, Universidad San Pablo CEU Madrid, 2007

system was deformed several times to obtain a number of modular ashlars making 50% of the blocks in the entire thrust system. The blocks were cut using CAD/CAM and numerical control machines, because masons using manual cutting techniques could not reproduce the geometry. Regarding the final construction, as the keystone could not be inserted vertically, the structure was divided into two parts onto which the stone blocks were positioned, then one half of the structure was pulled towards the other, closing the vault. The hermetic closure of the two parts of the system, fitting together perfectly, effectively resolved static problems.

The interest in the possible formal modifications of Abeille's ingenious patented vault led us to examine the relationship between the surface of the vault, the type of stonework employed and the minimum number of modular ashlars necessary for the complete tiling of the structure. The desire to use the fewest possible geometric elements to 'fill' the space, without creating problems of continuity, responds to the necessity to simplify the phases of construction and mounting in any vaulted structure. The "*Sphera*" project was an experiment to create a prototype of a spherical vault using the same ashlars as those patented by Abeille while at the same time applying the simplicity required by contemporary construction specifications (Fig. 14.12). The tiling of a spherical surface with a single geometrical shape is, as is well known, impossible. Thus, we decided to use the joints between ashlars as

**Fig. 14.12** Sphera, Stage Sphera, Grands Atelier Isle d'Abeau Lyon, 2008

irregular compensatory elements, capable of covering the spherical surface: identical ashlars, different joints. It is due to the particular geometrical conformation of Abeille's ashlars that the irregularity of the joints is effectively masked.

The first example of such an innovative family of domes was born in August 2012 in Qatar, as part of a hammam project designed by architect Joseph Caspari. The occasion came into being thanks to the support of Européenne de Marbre, the firm that first proposed the model through the publication of the manual "Plaited stereotomy" and with the creation of a scale model made using rapid prototype printing. The *Bin Jassim Dome* (Fig. 14.13) concerns different sectors, involving various fields such as tessellation, platonic bodies, stereotomy and nexorades. It took many years of research and the cooperation of many researchers to complete the project: the idea of such a vault stemmed from the three-dimensional models created using topological modelling by Giuseppe Fallacara (Fig. 14.14). The final model of the vault was engineered by Maurizio Brocato at the geometric-structural architecture laboratories of the Ecole d'Architecture Paris-Malaquais using a number of softwares, such as Mathematica and catem3m. The second constraint is represented by the keystones' catalogue: eight different pieces whose positioning is marked by a colour code printed on polystyrene. The position of the pieces has also to be taken into account since it is quite difficult to identify the direction of the heads due to their almost identical symmetry. Once the two above mentioned constraints were mastered, the work was completed within 22 working days. Compared to a full dome, the Bin Jassim dome presents some remarkable advantages, its weight being nearly divided in two on the keystone sides: 6.75 tons instead of 11.8 tons. The surface of the seal is also much less bulky and is fragmented, which helps the installation and control of fittings. The final work allowed us to keep a 5 mm wide visible seal opening and the stones were damaged very little during the laying, despite their sharp edges and acute angles. On completion, the work project followed the canons of practicability, easiness, celerity and quality provided for by the contractor.

*Wave Block Wall* is an application of the patent by Yuri Estrin and his research team (A. Dyskin, E. Pasternak, Han Chuan Khor, G. Simon, A. Molotnikov and L. Goldin). This prototype is based on the principle of topological interlocking and analyzes the properties of the mortarless structures whose design is based on this principle (Fig. 14.15). Our work focuses on structures built of osteomorphic blocks—the blocks possessing specially engineered concavo-convex contact surfaces allowing assembling various 2D and 3D structures (Fig. 14.16). These structures are easy to build and can be made demountable. They are flexible, resistant to macroscopic fractures and tolerant to missing blocks. The blocks are kept in place without keys or connectors that are the weakest elements of the conventional interlocking structures. The overall structural integrity of these structures depends on the force imposed by peripheral constraint.

The *Osteomorphic Catenary Arch* is an arch whose curve is reminiscent of a long chain, held by both ends and left hanging (Fig. 14.17). It looks like a parabola, but has a different geometry. The catenary arch is also called balanced arch because its shape allows a uniform load redistribution. Unlike other types of arcs, it does not need buttresses or other supporting elements and abutment. The catenary curve constitutes

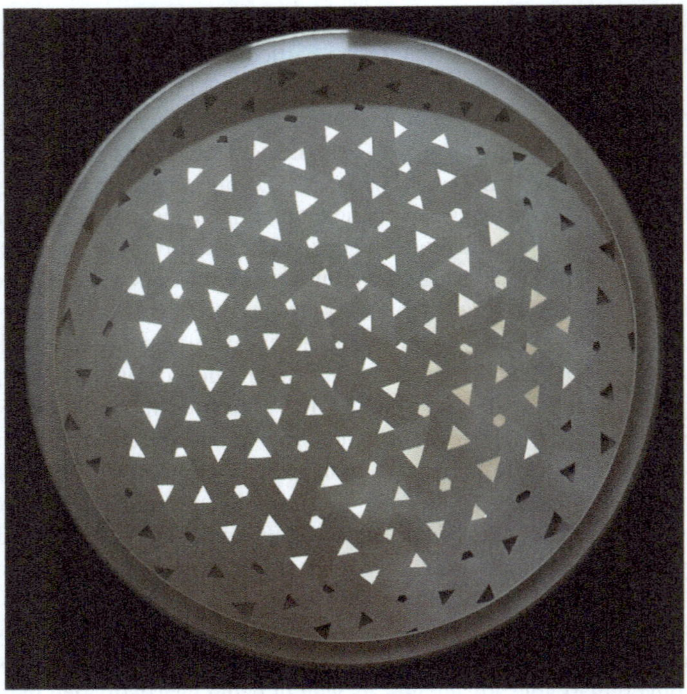

**Fig. 14.13** Intrados of The Bin Jassim Dome, Doha, 2012

**Fig. 14.14** Extrados of the Bin Jassim Dome, Doha, 2012

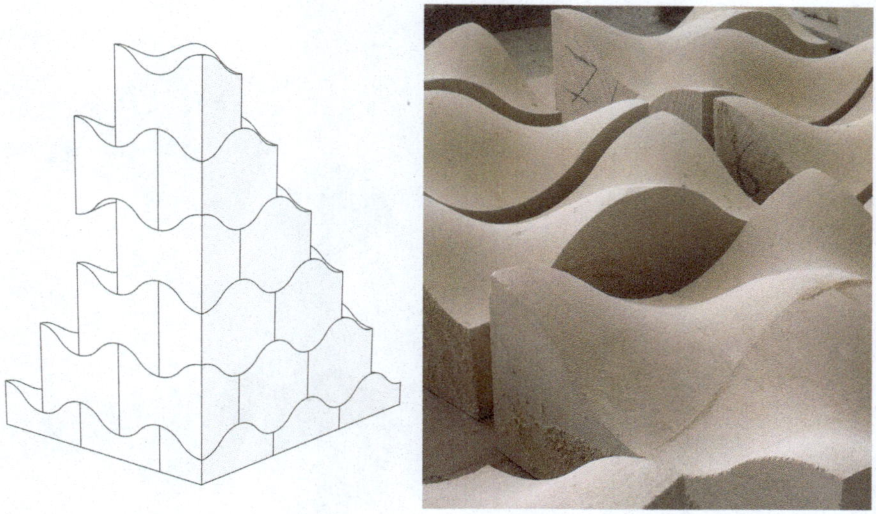

**Fig. 14.15** Wave Block Wall (stone prototype developed by Pimar), 2014

the base to match the osteomorphic stone blocks, similarly to the blocks having wave-shaped form composing the wall system described above. In this case, all the stone blocks undergo a topological deformation so that they are geometrically similar but have dimensional and angular variations. The realization of the individual blocks, by great geometric complexity, is only possible using computerized numerical control machines. Each individual block was individually numbered in order to be easily identified for the next assembly step. Then, the blocks were placed on the ground and assembled to take the form of the catenary arch. The construction system also applies to barrel-vaulted systems with a catenary section in a concept of mechanical mounting.

## 14.5 Didactic Experiences

All the research conducted converges naturally in the university education: in the third year architectural design studio of the DICAR Department at Politecnico di Bari [29–31], and in other training workshops. The goal is to give students adequate preparation for dealing with contemporary architecture challenges, through experimental teaching methods that aim to bring the university design courses in the architect's work practice. To pursue this goal, it is important to help students "think tridimensionally", and for this reason, studying and designing vaulted spaces has a fundamental training value.

This type of exercises proved to be important because the students had to take into account more aspects simultaneously:

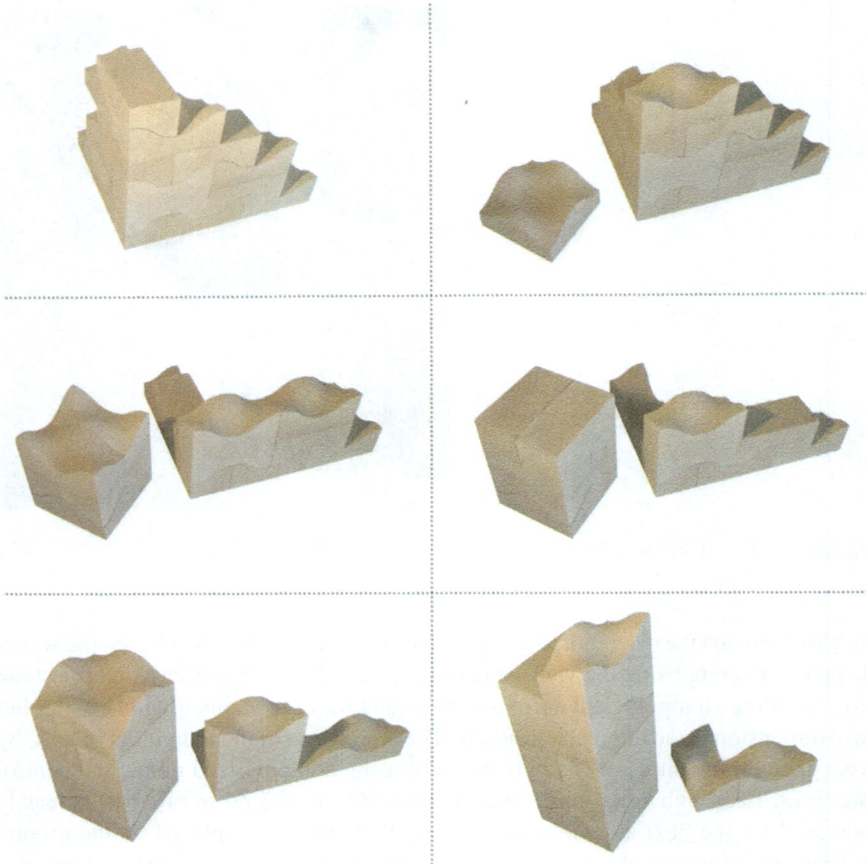

**Fig. 14.16** Different configurations possible using the Wave Block (stone prototype developed by Pimar), 2014

- prefigurative invariant, that is a subdivision capacity in appropriate sections of a vaulted system;
- technical/geometric invariant, that is the capacity of geometric, precise definition of an architectural system and of ashlars and its realization (projective technique and cutting technique);
- static invariant, that is the capacity of providing static balance of the architectural system of drystone joint (graphic and mechanical static of rigid structures).

Thus, while modelling, the students focused on the subject of digital stereotomy through the study of topological deformations, using 3DsMax® and Grasshopper® software in order to obtain a complex folded space (Fig. 14.18). From the start, the students were encouraged to experiment with both modelling techniques. The two programs foresee two different methods of work. For example, the first one foresees

**Fig. 14.17** Osteomorphic Catenary Arch (prototype developed by Generelli SA), 2015

the study and modelling of flat, interlocked or interconnected stereotomic assemblies, subsequently deformed. Although topological deformations have a very strong educational and training value, and the use of digital instruments such as 3DsMax® has great potential, thanks to the rapidity of execution and an extreme intuitiveness of use, the outcomes often have a non-negligible gap from real construction constraints. It has been reported that a significant number of students, excited by the freedom of deformation, forget the architectural outcome. Omitting the prefiguration of space phase, and employing an excessive number of parametric deformations, students produce morphologies that are sometimes closer to sculpture field than architecture. To prove this, the chosen shapes often result in poor static and inadequate thicknesses. In this training process, the guidance from the faculty becomes indispensable, because it helps students to understand problems and to formulate suitable solutions.

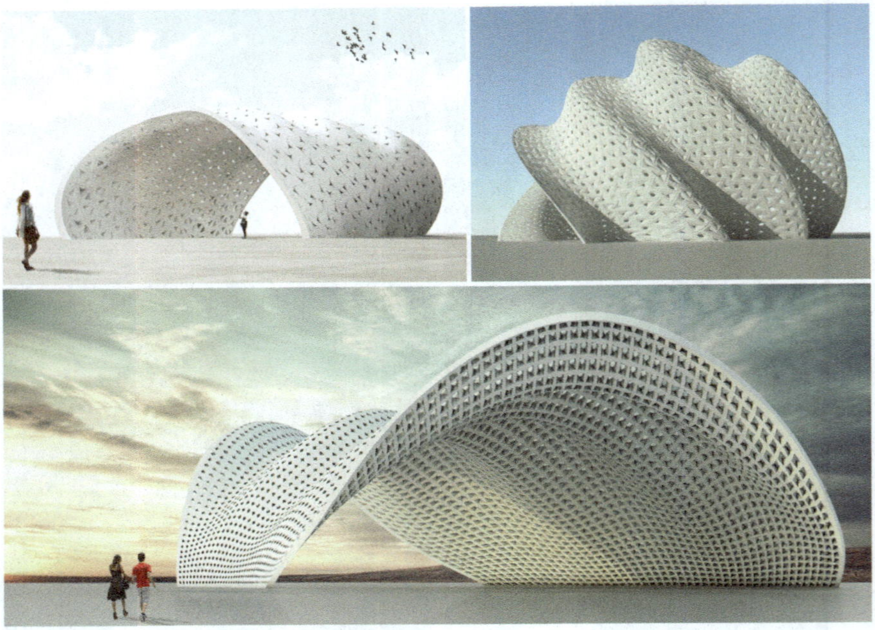

**Fig. 14.18** Some examples of complex architectural shapes made of topological interlocking blocks, developed by students during the third year Design Studio led by Prof. Giuseppe Fallacara, with Maurizio Barberio and Micaela Colella at Politecnico di Bari

When the surfaces to deform become very complex (free-form surfaces, spherical, etc.), their use is severely limited by the fact that some blocks will turn out to be too deformed, making some of these unfit to be prototyped with the current tools for architectural fabrication. For this reason, the course provides a series of introductory lectures to Grasshopper®, a plug-in for Rhinoceros®. Both the deformers of 3DsMax® and Grasshopper® work parametrically but, unlike the former, Grasshopper® allows writing of custom scripts in a simplified way, thanks to a more intuitive graphic interface. The lessons provided some simple tools that, if properly applied, could have great potential. Among these instruments, the attention has been focused on the 'Box Morph' and 'Surface Morph' components, generally used to populate any surface with 3D or 2D patterns. In this way, the students need to think simultaneously about the general form of architectural space and the components in which the structure is divided, concentrating their attention on the main purpose of the exercise, that is, to think and design an architecture. However, in this kind of projects there are always two crucial design problems to address: the static and geometric optimization of shape. Also in this case, the course offers students some intuitive tools that they could quickly manage, whose results (although not scientifically exhaustive) should stimulate them to further investigations. For example, for the static optimization it was decided to use the add-on 'Kangaroo Physics' [32], widely known in the computational design field for the simulation tools of physical phenomena, including the digital form-finding, essential for the optimization of the shape of a roof in order to have only compressive stresses.

Only after carrying out, in the first part of the course, of a large number of exercises concerning vaulted spaces, students are more ready to face the design of a more complex and functional building. After this training period, they will approach the design with more consciousness, considering all the different aspects that need to be taken into account to design a building that is actually usable and feasible.

## 14.6 Conclusions

As outlined in this paper, vaulted spaces made of topological interlocking blocks are relevant for architecture for at least two reasons:

- for their intrinsic value as vaulted systems, both from structural and aesthetic point of view;
- for their extrinsic value in the academic education field, in which architecture students are deliberately "forced" to consider static, constructive and formal aspects simultaneously.

For this reasons, it can be concluded that not only do morphologies made from topological interlocking blocks represent a relevant subject for the architecture field, but they are also suitable for interdisciplinary research in collaboration with engineers and material scientists.

## References

1. J.M. Pérouse de Montclos, *L'Architecture à la française, du milieu du XV^e à la fin du XVIII^e siècle* (Picard, Paris, 1982)
2. E. Rabasa Díaz, *La bóveda plana de Abeilla en Lugo* (1998)
3. E.D. Nichilo, Learning from stone traditional vaulted systems for the contemporary project of architecture. The experimental construction site at the ponton de oliva, in *Proceedings of First International Congress Construction History.* (Instituto Juan de Herrera, Madrid, 2003), pp. 743–754)
4. G.R. Uva, Learning from traditional vaulted systems for the contemporary design. an updated reuse of flat vaults: Analysis of structural performance and recent safety requirements, in *Proceedings of First International Congress Construction History.* (Instituto Juan de Herrera, Madrid, 2003), pp. 2015–2021
5. G. Fallacara, Digital stereotomy and topological transformations: reasoning about shape building, in *Proceedings of Second International Congress Construction History,* (Queen's College Cambridge, Cambridge, 2006), pp. 1075–1092
6. G. Fallacara, *Verso una progettazione stereotomica. Nozioni di stereotomia, stereotomia digitale e trasformazioni topologiche: ragionamenti intorno alla costruzione della forma* (Aracne Editrice, Roma, 2007)
7. R. Etlin, G. Fallacara, L. Tamborero, *Plaited Stereotomy-Stone Vaults for the Modern World* (Aracne, Roma, 2008)

8. G. Fallacara, Toward a stereotomic design: experimental constructions and didactic experiences, in *Proceedings of the Third International Congress on Construction History* (2009), p. 553
9. G. Fallacara, *Stereotomy-Stone Architecture and New Research* (Presses des Ponts et Chaussées, Paris, 2012)
10. J. Sakarovitch, Construction history and experimentation, in *Proceedings of the Second International Congress on Construction History* (2006), pp. 2777–2792
11. F. Fleury, Evaluation of the perpendicular flat vault inventor's intuitions through. Large scale instrumented testing, in *Proceedings of the Third International Congress on Construction History, Cottbus* (2009)
12. M. Brocato, L. Mondardini, Geometric methods and computational mechanics for the design of stone domes based on Abeille's bond. Advances in Architectural Geometry **2010**, 149–162 (2010)
13. A.V. Dyskin, Y. Estrin, A.J. Kanel-Belov, E. Pasternak, A new concept in design of materials and structures: assemblies of interlocked tetrahedron-shaped elements. Scripta Mater. **44**(12), 2689–2694 (2001)
14. Y. Estrin, Topological interlocking and osteomorphic blocks, in *Stereotomic Design*, ed. by G. Fallacara, V. Minenna (Gioffreda Editore, Maglie (LE), 2014), pp. 11–13
15. A.V. Dyskin, Y. Estrin, A.J. Kanel-Belov, E. Pasternak, Topological interlocking of platonic solids: a way to new materials and structures. Philos. Mag. Lett. **83**(3), 197–203 (2003)
16. A.V. Dyskin, Y. Estrin, E. Pasternak, H.C. Khor, A.J. Kanel-Belov, The principle of topological interlocking in extraterrestrial construction. Acta Astronaut. **57**(1), 10–21 (2005)
17. Y. Estrin, A.V. Dyskin, E. Pasternak, Topological interlocking as a material design concept. Mater. Sci. Eng. C **31**(6), 1189–1194 (2011)
18. O. Tessmann, Topological interlocking assemblies, in *Physical Digitality—Proceedings of the 30th International Conference on Education and Research in Computer Aided Architectural Design in Europe*. (Prague, Czech Republic, 2012), pp. 12–14
19. O. Tessmann, M. Becker, Extremely heavy and incredibly light: performative assemblies in dynamic environments, in *18th International Conference on Computer-Aided Architectural Design Research in Asia: Open Systems, CAADRIA 2013; Singapore; Singapore; 15 May 2013 through 18 May 2013*, pp. 469–478. The Association for Computer-Aided Architectural Design Research in Asia (CAADRIA), Hong Kong, and Center for Advanced Studies in Architecture (CASA), Department of Architecture, School of Design and Environment, National University of Singapore, Singapore (2013)
20. O. Tessmann, Interlocking manifold kinematically constrained multi-material systems, in *Advances in Architectural Geometry 2012* (Springer Vienna, 2013), pp. 269–278
21. T. Schwartz, L. Mondardini, Integrated design and robotized prototyping of Abeille's vaults. in *Robotic Fabrication in Architecture, Art and Design 2014* (Springer International Publishing, 2014), pp. 199–209
22. M. Colella, Stereotomic design exhibition. evolution: from the hand to the robot, in *Stereotomic Design*, ed. by G. Fallcara, V. Minenna (Gioffreda Editore, Maglie (LE), 2014), pp. 38–52
23. M. Weizmann, O. Amir, Y.J. Grobman, Topological interlocking in architectural design (2015)
24. M. Weizmann, O. Amir, Y.J. Grobman, Topological interlocking in buildings: a case for the design and construction of floors. Autom. Constr. **72**, 18–25 (2016)
25. I. Miadragovič Vella, T. Kotnik, A. Herneoja, T. Österlund, P. Markkanen, Geometric versatility of Abeille Vault. *eCAADe 2016* (2016)
26. H.T.D. Yong, *Utilisation of Topologically-Interlocking Osteomorphic Blocks for Multi-Purpose Civil Construction*. Ph.D. Thesis, The University of Western Australia (2011)
27. S. Weir, D. Moult, S. Fernando, Stereotomy of wave jointed blocks, in *Robotic Fabrication in Architecture, Art and Design 2016* (Springer International Publishing, 2016), pp. 284–293
28. S. Fernando, D. Reinhardt, S. Weir, Waterjet and wire-cutting workflows in stereotomic practice, in *Protocols, Flows and Glitches, Proceedings of the 22nd International Conference of the Association for Computer-Aided Architectural Design Research in Asia (CAADRIA) 2017*, ed. by P. Janssen, P. Loh, A. Raonic, M.A. Schnabel (2017), pp. 787–798

29. G. Fallacara, M. Barberio, M. Colella, *Mater[i]a: Conoscenza e Progetto. Nuovo polo museale multifunzionale per Matera 2019* (La Stamperia Edizioni, Matera, 2016)
30. M. Barberio, M. Colella, G. Fallacara, Stereotomy, sustainable construction and didactics. case study: a new museum for matera, European Capital of culture 2019, in *3rd International Balkans Conference on Challenges of Civil Engineering, 3-BCCCE, 19–21 May 2016, Epoka University, Tirana, Albania*. ALB (2016)
31. G. Fallacara, M. Barberio, M. Colella, *Con_corso di progettazione. Learning by designing.* (Aracne Editrice, Roma, 2017)
32. D. Piker, Kangaroo: form finding with computational physics. Archit. Des. **83**(2), 136–137 (2013)

# Index





Y. Estrin et al. (eds.), *Architectured Materials in Nature and Engineering*, Springer Series in Materials Science 282,
https://doi.org/10.1007/978-3-030-11942-3

The manufacturer's authorised representative in the EU is Springer Nature Customer Service Centre GmbH, Europaplatz 3, 69115 Heidelberg, Germany. If you have any concerns regarding our products, please contact ProductSafety@springernature.com

Printed and bound by CPI Group (UK) Ltd, Croydon, CR0 4YY

22/09/2026

02221050-0003